Electrical Measurement Techniques

Lars Bengtsson

Electrical Measurement Techniques

For the Physics Laboratory

Springer

Lars Bengtsson
Department of Physics
University of Gothenburg
Göteborg, Sweden

ISBN 978-981-99-8186-1 ISBN 978-981-99-8187-8 (eBook)
https://doi.org/10.1007/978-981-99-8187-8

This Springer imprint is published by the registered company Springer Nature Singapore Pte Ltd.
The registered company address is: 152 Beach Road, #21-01/04 Gateway East, Singapore 189721, Singapore

Paper in this product is recyclable.

Preface

This book is about electrical measurement techniques with a focus on measurement systems in a physics lab. It is also on an 'advanced' level, meaning that it assumes the reader has math and physics skills corresponding to a bachelor's degree in science or engineering.

When I started as a Ph.D. student in experimental atomic physics many years ago, I was very well prepared 'physics-wise', but when I started working in the atomic laser lab, I soon realized that there were so many things I needed to know that were not included in the physics program's curriculum. I knew all about Newton mechanics, thermodynamics, atomic physics, and wave equations (or at least I thought I did), but I knew nothing about instruments' bandwidth, transmission cables, sensors, microchannel plates, vacuum gauges, piezo crystals, probes, filters, spectral analyzers, signal processing, analog-to-digital converters, time-to-digital converters, uncertainty budgets, lock-in amplifiers, and PID controllers. Every day there was something new to learn, and it was quite overwhelming and sometimes a little frustrating. I went to graduate school to learn more about physics but spent most of my time learning about electrical engineering stuff.

I had to figure out all these things by myself and it took precious time away from the things I really wanted to study, and I remember wishing that there was a book that summarized it all, like 'Electrical engineering for physicists'. Well, now there is. This book summarizes what a Ph.D. student in experimental physics needs to know from the electrical engineering curriculum to work in a physics lab.

The book contains many examples and problems. The problems are *solved*; from experience, I know that this is appreciated by the readers. It contains both 'practical' aspects of the equipment in a physics lab (like bandwidth, probes, transmission cables, controllers, etc.) as well as signal processing theory (like transform theory, filtering, convolution, correlation, and curve fitting), but the intended focus is always on the *understanding*. According to Bloom's taxonomy triangle, a student's first encounter with a subject is characterized by *remembering*, i.e., root learning and mechanical solving of standard problems. This is what characterizes bachelor classes. Most of the mathematics in this book is not new to you; if you have a bachelor's degree in physics, you have seen the math before, but for most students, that implies a cognitive understanding on Bloom's *remember* level.

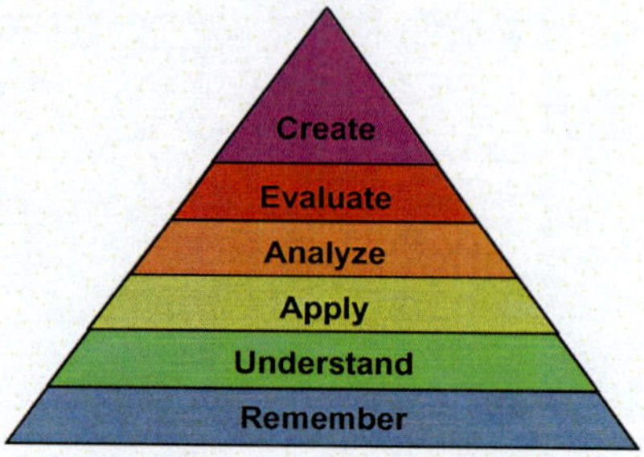

Bloom's taxonomy (revised)

In this book, you will see the same math again, but since you have already processed this math on the *remember* level, you are now ready to take it to the next level(s), the *understanding* and *application* levels. That is the intention of the theoretical parts of this book; to take the math you already know to a higher cognitive level. That means that exercises are not focused on 'mechanical procedures' but are designed to promote a deeper understanding.

Göteborg, Sweden
October 2023

Lars Bengtsson

Contents

Acronyms

AC	Alternating Current
ACF	Auto-Correlation Function
ADC	Analog-to-Digital Converter
BAG	Bayard–Alpert Gauge
BIPM	Bureau International de Poids et Mesures
BNC	Bayonet Neill-Concelman
CAN	Controller Area Network
CEM	Channel Electron Multiplier
CJC	Cold Junction Compensation
CMRR	Common Mode Rejection Ratio
CMOS	Complementary Metal Oxide Semiconductor
CLT	Central Limit Theorem
CW	Continuous Wave
DAC	Digital-to-Analog Converter
DAQ	Data Acquisition
DC	Direct Current
dB	decibel
DCV	Direct Current Voltage meter
DFT	Discrete Fourier Transform
DLS	Dynamic Light Scattering
DMM	Digital Multimeter
ECG	Electro Cardio Gram
EM	Electro Magnetic
EMC	Electromagnetic Compatibility
emf	electromotive force
ENOB	Equivalent Number of Bits
FIR	Finite Impulse Response
FFT	Fast Fourier Transform
GUM	Guide to the expression of Uncertainty in Measurement
HV	High Vacuum
IA	Instrumentation Amplifier

iid	independent and identically distributed
IIR	Infinite Impulse Response
LIA	Lock-In Amplifier
LC	Level Crossing
lm	lumen
LTI	Linear and Time-Invariant
LV	Low Vacuum
LVDT	Linear Variable Differential Transformer
lx	lux
MCP	Microchannel Plate
MEMS	Micro Electro Mechanical Systems
NIM	Nuclear Instrumentation Module
NIST	National Institute of Science and Technology
op amp	Operational amplifier
OS	Overshoot ratio
OSR	Oversampling Rate
pcb	Printed circuit board
PCS	Photon Correlation Spectroscopy
PIN	P-doped, Intrinsic, N-doped
PLL	Phase-locked loop
PMT	Photomultiplier Tube
ppm	Parts per million
PSD	Phase-Sensitive Detector
PSD	Position-Sensitive Detector
PTAT	Proportional To Absolute Temperature
PWM	Pulse Width Modulation
RBW	Resolution Bandwidth
rms	root mean square
RTD	Resistance Temperature Detector
S&H	Sample and Hold
SAR	Successive Approximation Register
$\Sigma\Delta$	Sigma Delta
SNR	Signal-to-Noise Ratio
SoC	Start of Conversion
TDC	Time-to-Digital Converter
TDR	Time Domain Reflectometry
TP	Twisted-Pair
TTL	Transistor–Transistor Logic
UAF	Universal Active Filter
VLSI	Very Large-Scale Integration
VM	Voltage Meter

Chapter 1
Introduction

Abstract This chapter describes some basic concepts like common and normal mode voltages, common mode rejection ratio, and signal-to-noise ratio. The dB unit is defined, differential- and single-ended signals and rise and fall times versus bandwidth are discussed. Finally, the propagation of a signal through a measurement system is considered and each component's influence on the signal is highlighted.

1.1 Electrical Measurement Systems

A successful 'measurement' depends on a long chain of components and their interactions. It typically starts with a 'sensor' that converts some physical quantity (like temperature, acceleration, sound, etc.) to an electrical quantity (like voltage, current, resistance, etc.). After the sensor comes the 'signal conditioning'. The signal conditioning processes the 'raw' sensor signal in hardware (mostly analog). The signal conditioning electronics' job is to convert the sensor signal to a 'standard format', like $0 - +5$ V or 4–20 mA. This usually includes both passive (resistors and capacitors) and active (operational amplifiers) components.

Next, for several reasons, the signal-conditioned signal is 'filtered'. The main reasons are to a) suppress unwanted interferences (see Chap. 2) and b) to prevent 'aliasing' in sampling systems (see Chaps. 7 and 9). After the filter, the signal is 'sampled' by an 'analog-to-digital converter' (see Chap. 11); this is where the signal is 'digitized', i.e., transferred to the system computer. From here on, all signal processing is 'digital' (as in 'computer algorithms'). Digital signal processing includes digital filters (Chap. 10), spectral analysis (Chap. 8), and correlation (Chap. 16). You also need to know some 'post-processing' techniques (non-real time), such as uncertainty analysis and confidence interval estimations (Chap. 14), and curve fitting (Chap. 17).

Throughout the entire 'chain', from the sensor to the computer sampling, the signal is exposed to 'noise', and it is important that you know a) where it comes from, i.e., the most common noise sources (see Chap. 2) and how to protect your measurement signal from the most common noise sources (see Chaps 2, 5, and 6).

L. Bengtsson, *Electrical Measurement Techniques*,
https://doi.org/10.1007/978-981-99-8187-8_1

Before we get into the details of the different measurement chain components, we need to define some basic concepts to make sure we have the right vocabulary.

1.2 Common and Normal Mode

Voltages (and currents too) can be in either 'common' or 'normal' mode. To understand the difference, we should probably use 'potentials' rather than 'voltages'; voltage is a difference in potential. If we measure the voltage across a pair of conductors, the 'normal mode' voltage is the potential difference between the conductors, and the 'common mode' voltage is the potential that is 'common' on both conductors. This is illustrated in Fig. 1.1.

A better name for the 'normal mode' voltage would be 'differential mode', which is indeed used in some contexts, but 'normal mode' seems to be the most used name. In a typical measurement, we measure the normal mode voltage, with a voltage meter, and if the voltage meter (VM) is 'perfect' it will measure only the potential difference, i.e., the normal mode voltage. The voltage meter *subtracts* the potential on one wire from the other.

$$U_{VM} = (u_0 + u_1) - u_0 = u_1 = u_{nm} \tag{1.1}$$

However, subtractions in electronics are never perfect and in a non-perfect voltage meter, there will be a 'cm residual' in U_{VM}:

$$U_{VM} = u_{nm} + F_{cm} \cdot u_{cm} \tag{1.2}$$

F_{cm} is the 'common mode suppression number' and is an important parameter for any voltage meter; the lower the better is the voltage meter (and the more expensive it is). Manufacturers don't specify the F_{cm} number though, they specify the *CMRR*,

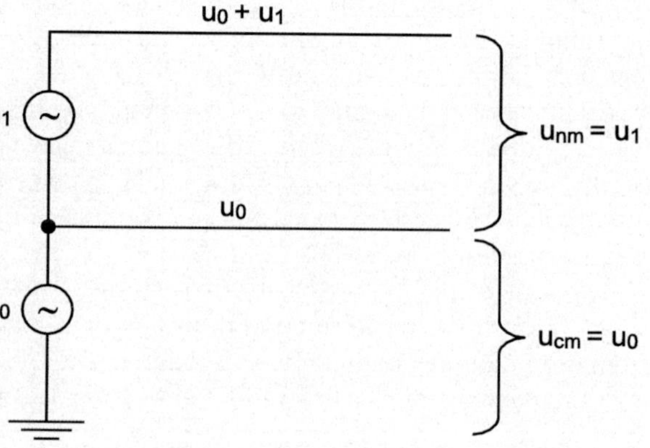

Fig. 1.1 Signal model

the *Common Mode Rejection Ratio* of the voltage meter. CMRR is defined as

$$\text{CMRR} = 20\log\frac{1}{F_{\text{cm}}}\text{dB} \tag{1.3}$$

A 'good' desktop DMM (digital multimeter) has a CMRR of 140 dB (for example, Keysight model 34461A), but this number typically drops rapidly with frequency; the 'AC' CMRR for the 34461A model is 70 dB. A 'good' handheld DMM has a typical CMRR of 120 dB (for example, Fluke 179). Finally, even if the CMRR decreases with frequency, they are almost always designed to suppress the power line frequency (50/60 Hz), because that is where the common mode noise comes from in most cases; for 50/60 Hz, the CMRR is usually as good as the DC suppression.

1.3 Signal-To-Noise Ratio

In the general case, the measurement signal will always be a complex of 'signal' and 'noise', or the 'good' part and the 'bad' part. The noise is what will prevent us from measuring our quantity with perfect (infinite) accuracy, the more noise, the less accuracy. Actually, we can live with a lot of noise if we also have a lot of 'signal'; it is the magnitude of the noise compared to the signal level that is the interesting number. We quantify the 'signal situation' with the 'signal-to-noise' ratio:

$$\text{SNR} = 20\log\frac{\text{signal rms}}{\text{noise rms}}\text{dB} \tag{1.4}$$

In Eq. (1.4), it is usually understood that it is the 'normal mode' noise we mean; if we refer to the common mode noise, we will specify that explicitly.

1.4 Decibel Units

Equation (1.4), the 'dB' unit is really 'dimensionless'; it is a logarithmic measure of a relation between two voltages. However, the dB unit is also sometimes used to express absolute voltages. For example, 1 dBm corresponds to the voltage that develops exactly 1 mW in a resistor R. R is usually, but not necessarily, a 50 Ω resistor. Since $P = U{\cdot}I = U^2/R$, we have that

$$1\text{ mW} = \frac{U^2}{50} \Rightarrow U = \sqrt{0.05} = 0.2236\text{ V} \tag{1.5}$$

Hence, 5 V equals 20·log(5/0.2236) = 27 dBm. Sometimes, you also see the 'dbV' unit. The dBV unit relates the voltage to 1.00 V.

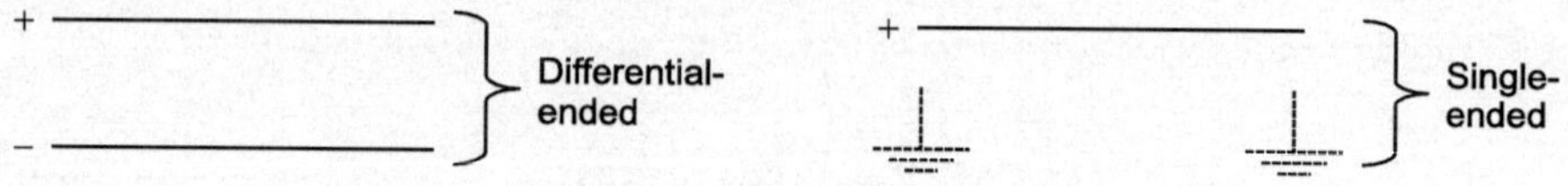

a A differential-ended signal **b** A single-ended signal

Fig. 1.2 **a** A differential-ended signal. **b** A single-ended signal

1.5 Differential-Ended Versus Single-Ended

In electrical measurement laboratories, the terms 'differential-ended' and 'single-ended' signals are used regularly. A 'differential-ended' signal has two wires and none of them are ground. The signal is delivered as a potential difference between two wires, see Fig. 1.2a. A differential-ended signal is also sometimes called 'non-referenced'.

A 'single-ended' signal, on the other hand, is a single wire; it is understood that the signal is the potential on this wire relative ground.

1.6 Signals

1.6.1 Risetime and Falltime

We have already used the term 'signal' repeatedly, but we have not yet properly defined it. A 'signal' could be a lot of things, but in this context, it will be understood to be a variation of voltage in time. An AC voltage if you like, but 'AC signals' are generally interpreted as sinusoidal voltages and our scope of signals is much wider here.

One of the most basic properties of a signal (and one of the most important ones to us), is its *risetime*. A signal's risetime is defined as the time it takes for the signal to go from 10 to 90% of its maximum voltage. Correspondingly, the *falltime* is defined as the time it takes to go from 90 to 10% of the maximum (Fig. 1.3).

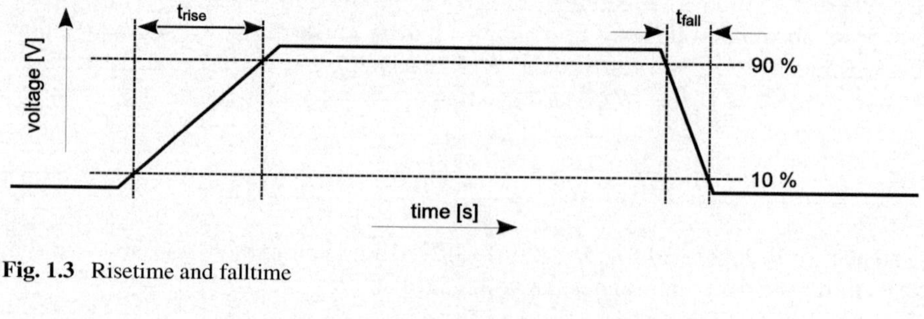

Fig. 1.3 Risetime and falltime

If only the risetime is specified, you may assume that the falltime equals the risetime.

1.6.2 Bandwidth

So why is the risetime of a signal so important? It is important because, from the signal's risetime, we can calculate its *bandwidth*. We will talk *a lot* about *frequencies* in this book (for both signals and systems) and knowing a signal's bandwidth is paramount for how you design your 'measurement chain'. We will define 'bandwidth' properly later (for both signals and systems), but for now, we settle with the following definition: A 'signal' is in the general case 'complex', it consists of several components, where a 'component' is understood to be a sinusoidal signal. A signal is in general a sum of a lot of sines, and the signal's bandwidth is simply the frequency of the sinusoidal with the highest frequency.

There is a simple relationship between a signal's bandwidth and risetime:

$$B = \frac{0.35}{t_{\text{rise}}} \tag{1.6}$$

We don't derive that expression here (but it is just straightforward electricity calculus).

1.7 Systems

You can't talk about 'signals' without also talking about systems. A 'system' is anything that the signal passes through in the measurement chain. 'Systems' are not only amplifiers and filters but also include the transmission lines and the instruments. All systems are typically specified by their bandwidth, and it is important to understand what impact each system has on the signal. Filters and amplifiers are designed to have a specific impact on the signal, but transmission lines and instruments should ideally have no impact on the signal. However, 'no impact' implies infinite bandwidth, and we never have that.

If the system's bandwidth is $< \infty$ (which it always is), it will *slow down* the signal. 'Slow down' as in 'the signal's rise time will increase' for each system it passes. When a signal propagates through a measurement chain, it is slowed down by the chain components, and risetimes are added in *squares*. Figure 1.4 illustrates a signal chain.

First, we get each system's risetime from their bandwidth (use Eq. (1.6)) and then we add the squares:

$$t_{\text{rise,out}}^2 = t_{\text{rise,in}}^2 + t_1^2 + t_2^2 + t_3^2 + t_4^2 \tag{1.7}$$

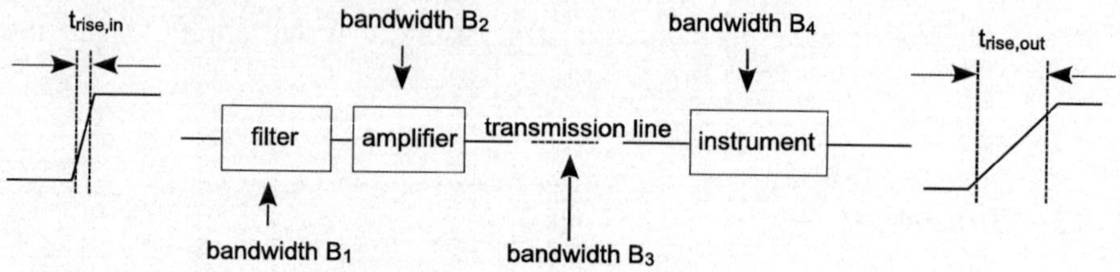

Fig. 1.4 Measurement chain: Risetime propagation

where $t_n = 0.35/B_n$. In the general case,

$$t_{\text{rise,out}} = \sqrt{t_{\text{rise,in}}^2 + \sum_i \left(\frac{0.35}{B_i}\right)^2} \tag{1.8}$$

1.8 Solved Problems

Problem 1.1 The voltage meter in Fig. 1.5 is a 6½ digit DMM with a CMRR of 130 dB. (**a**) What voltage will the voltage meter display? (**b**) How much of this voltage is due to the NM and CM parts, respectively?

Solution (**a**) In Fig. 1.5, we have a common mode voltage of 30.43916 V and a normal mode voltage of 30.47325–30.43916 = 0.03409 V. A CMRR of 130 dB is translated to a CM suppression of $10^{-130/20} = 3.162 \cdot 10^{-7}$.

The voltage meter will measure

$$U_{\text{m}} = 0.0340900 + 3.162 \cdot 10^{-7} \cdot 30.43916 = 0.0340900 + 0.000009625 =$$

$$= 0.034099625 \text{ volts}$$

But the question was: 'What voltage *will the voltage meter display*?'. We have a 6½ digit DMM; the range will be 100 mV, so the display will show **034.0996 mV**. (A '½ digit' means that the first digit (the most significant digit) can only be '0' or '1'. In our example, it must be '0', since we use the 100-mV range).

(**b**) Of the 34.0996 mV on the display, 0.009625/34.09966 = 0.03% is due to the CM residual.

Fig. 1.5 DC voltage measurement

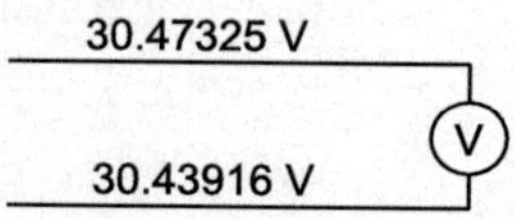

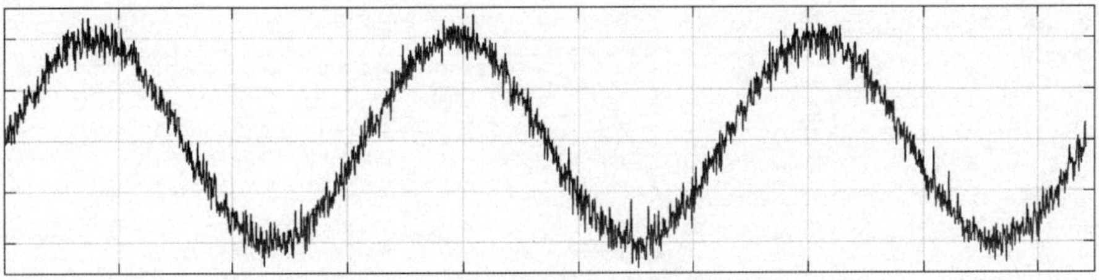

Fig. 1.6 Our standard problem; a sine with noise

Problem 1.2 Figure 1.6 illustrates a problem that we will treat repeatedly in this book: A sinusoidal signal with 'white' noise. In Fig. 1.6, the amplitude of the sine is 1 V and the white noise is 'gaussian' with a zero mean and a variance of 0.01 V^2. What is the signal-to-noise ratio in this signal?

Solution A sine with amplitude A volts has an rms voltage of $A/\sqrt{2}$, and the rms of Gaussian ('normal') noise is the square root of the variance:

$$\mathrm{SNR} = 20\log\frac{1/\sqrt{2}}{\sqrt{0.01}} = 17\ \mathrm{dB}$$

Problem 1.3 Convert the voltages 30 dBm and 15 dBV to voltages [V].

Solution In the dBm case, we assume that a 50 Ω resistor is used as reference:

$$U = 0.2236 \cdot 10^{30/20} = \underline{7.07\ \mathrm{V}}$$

15 dBV corresponds to

$$15 = 20\log\frac{U}{1} \Rightarrow U = 10^{15/20} = \underline{5.62\ \mathrm{V}}$$

Problem 1.4 Figure 1.7 illustrates a 'perfect' square signal (risetime = 0). What would it look like on a 100 MHz oscilloscope?

Solution The scope will 'slow down' the signal with 0.35/0.1 ns = 3.5 ns (Fig. 1.8).

If the input signal has risetime 0 s, we will only see the scope's 'reaction time' on the screen.

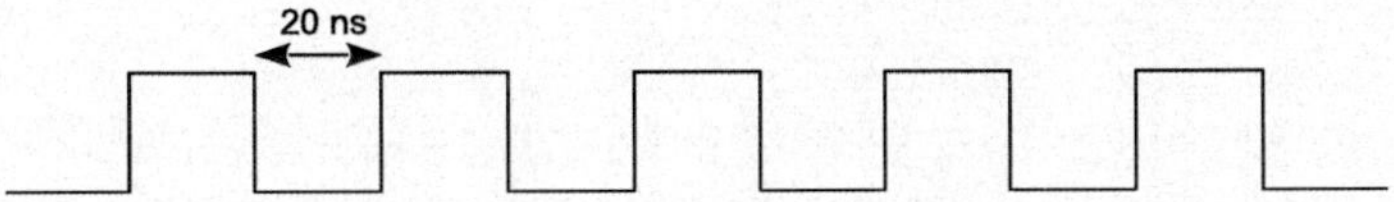

Fig. 1.7 A 'perfect' 25 MHz square wave

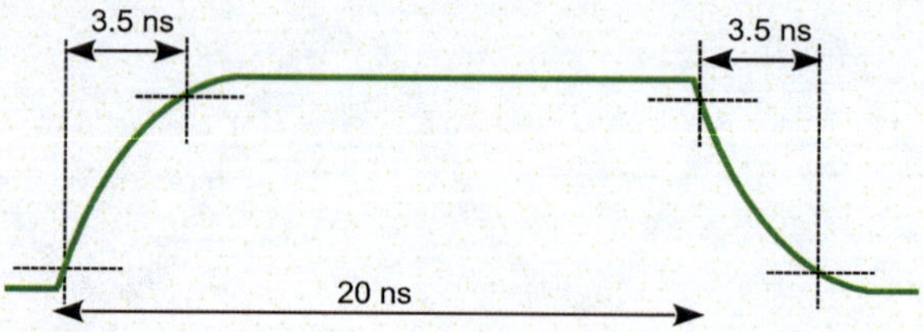

Fig. 1.8 The oscilloscope slows down the signal

Problem 1.5 In an experiment, the signal risetime is expected to be approximately 5 ns. What bandwidth does the oscilloscope need not to have any significant impact on the result?

Solution First, we need to define exactly what we mean by 'significant impact'. The exact definition depends on the circumstances; here we require that the scope's contribution to the total risetime must be less than 10%; $t_{\text{out}} \leq 5.5$ ns:

$$5.5^2 = 5^2 + t_{\text{scope}}^2 \Rightarrow t_{\text{scope}} = 2.3 \text{ ns} = \frac{0.35}{B} \Rightarrow B = 152 \text{ MHz}$$

The oscilloscope needs a bandwidth of at least 150 MHz.

Chapter 2
Noise: Sources and Remedies

Abstract This chapter illustrates how noise can couple to a measurement system in different ways and how noise can be prevented from entering the measurement system by de-coupling techniques. Noise sources are described, different kinds of crosstalk are discussed, and the importance of grounding and shielding is highlighted. This chapter also explains why Faraday cages are used and the advantages of coax cables and twisted-pair cables.

2.1 Introduction

Noise is omnipresent in all measurements. It may or may not be a problem; it may be too small compared to other sources of uncertainty to have any significant impact on the result, or it may have a frequency that is outside the measurement signal's bandwidth. However, in the general case, noise is significant in the signal's bandwidth, and we must 'deal with it'. The *first* action should *not* be to apply signal processing (in hardware or software), the first action is to try to *prevent* the noise from entering the system. To prevent noise from entering the system, we must understand how it got there in the first place. Three conditions must be fulfilled for noise to be a problem in a measurement system: (a) There must be a noise *source*, (b) there must be a *coupling* between the noise source and the system, and (c) the noise frequency must be within the signal's bandwidth.

Obviously, the *first* action should be to try to identify the noise source; if we can identify it, maybe we can eliminate it. For example, the signal wire might just be too close to some unshielded, high-frequency power cable (like the spark plug cable in a car); rearranging the setup may be all we need to do. In other situations, we can identify the source, but we can't do anything about it (like the 50/60 Hz interference from the local power line).

In such cases, where we can't do anything about the source (or may not even be able to identify it), we are left with the only option of trying to *break the coupling* between the source and the system. To do that, we need to understand how noise couples to our system and that can only be done in a handful of ways.

L. Bengtsson, *Electrical Measurement Techniques*,
https://doi.org/10.1007/978-981-99-8187-8_2

However, some noise cannot be 'de-coupled' since it is 'internal', it is generated by the system itself and that noise we will have to learn to live with. There are some things we can do to reduce it, but we will always have some internal noise. We will look at the internal noise first.

2.2 Internal Noise

2.2.1 *Johnson Noise*

Johnson noise is an omnipresent source of noise. It is a natural phenomenon that was first described by John B. Johnson at Bell Labs in 1926. Johnson noise is 'bandlimited white noise' that originates in 'anything with an ohm-resistance'. The rms of the white noise is

$$u_{\text{rms}} = \sqrt{4\pi kRTB} \tag{2.1}$$

where k is Boltzmann's constant ($1.38 \cdot 10^{-23}$ J/V), R is the resistance, T is the temperature (in Kelvin), and B is the bandwidth of the instrument (Fig. 2.1).

From Eq. (2.1), we can see that we can reduce the Johnson noise in three different ways; (a) it depends on temperature so cooling the system will reduce the noise, (b) it depends on the resistance which implies that we shouldn't use excessive impedances unless necessary. We learned in basic electricity that 'good' voltage meters should have a high input impedance. Well, Eq. (2.1) contradicts that; high input impedances increase the white Johnson noise.

And (c), we can limit the instrument's bandwidth. For that reason, high-bandwidth instruments, like oscilloscopes, almost always have a 'bandwidth-limiting' option, see Fig. 2.2. Intuitively, a 'high-bandwidth instrument' sounds good, but keep in mind that the Johnson noise increases with the bandwidth.

We can re-write Eq. (2.1) as

$$u_{\text{rms}} = \sqrt{4\pi kRT} \cdot \sqrt{B} = \alpha\sqrt{B} \Rightarrow \alpha = \frac{u_{\text{rms}}}{\sqrt{B}}\left[\frac{\text{V}}{\sqrt{\text{Hz}}}\right] \tag{2.2}$$

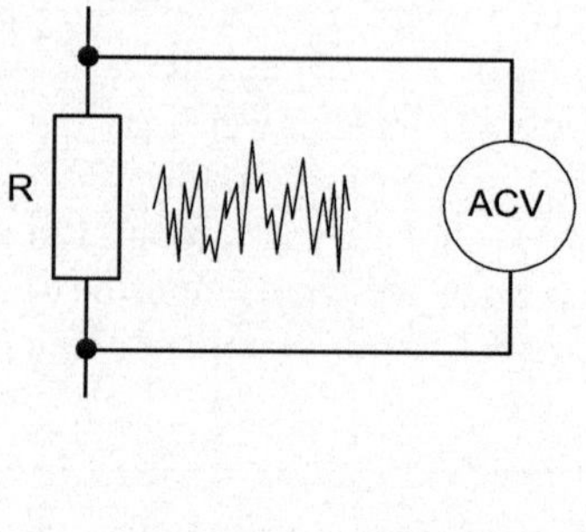

Fig. 2.1 Johnson noise is omnipresent wherever we have a resistance

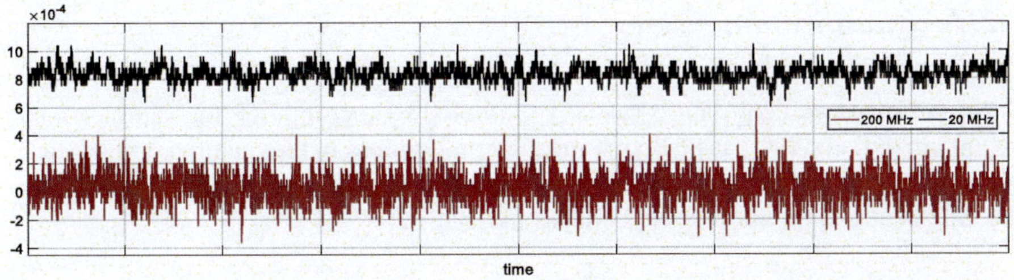

Fig. 2.2 The Johnson noise in an oscilloscope (open input) for 20 MHz bandwidth (top) and 200 MHz bandwidth (bottom). (The offset for the 20 MHz waveform was added in MATLAB)

where α is the *noise factor*. Manufacturers of sensors and amplifiers always specify the noise factor. They can't know the bandwidth of your application, so the noise factor is all they provide. The end user must multiply the noise factor with the square root of the measurement system's bandwidth to estimate the noise in the sensor/amplifier. So, whenever you see a number with the unit V $\mathrm{Hz}^{-1/2}$, you must multiply with the square root of your bandwidth.

2.2.2 Shot Noise

Another kind of noise is the *shot noise* that appears in low-current measurements, nA or less. Shot noise is caused by the fact that current involves the transportation of electrons which have a quantized charge ($1.6 \cdot 10^{-19}$ As) and shot noise is simply random variations of the charge density. Just like the Johnson noise, the shot noise increases with the bandwidth of the instrument.

2.2.3 1/f-Noise

$1/f$-noise (or 'flicker noise') is a strange phenomenon that we really don't know where it comes from. It decreases with frequency (as $1/f$) and has been observed in applications far from electronic systems, such as music, biology, and economics [1]. One of its most characteristic properties is that the noise power per decade is constant: There is as much noise power in the interval 10–100 Hz as in the interval 10–100 MHz. Obviously, this noise will only be a problem when you measure low voltages (sub-μV) at low frequencies. Because it is dominated by low frequencies, it is sometimes called 'pink noise' (as in 'red-shifted').

2.2.4 *Quantization Noise*

Since our measurement data is (usually) sampled by a computer, the samples must be 'quantized' and there is always some information loss in the quantization process. This noise is called 'quantization noise' and depends on the quantizer's resolution. We will treat quantization noise in detail in Chap. 11.

2.3 Coupling By Radiation

2.3.1 *Electric Dipole Antennas*

An electromagnetic field propagating in the z direction, emits an electric and magnetic field in the x and y directions, respectively, both perpendicular to the direction of propagation. For now, we focus on the electric field only, see Fig. 2.3.

In Fig. 2.4, we have placed an electric dipole antenna in the electric field, oriented so that it points in the direction of the E-field variation.

In Fig. 2.4, the E-field direction points upwards, and hence the electrons in the antenna conductor are driven downwards; there will be an accumulation of negative charges at the bottom end of the antenna and a lack of negative charges at the top end. Also, the charge transportation will be registered as a short current transient by the amp meter.

Figure 2.5 illustrates the same dipole antenna a moment later, when the E-field over the antenna points in the other direction. The electrons will be driven to the other end of the antenna and the amp meter will again register a short current transient, now with the opposite sign.

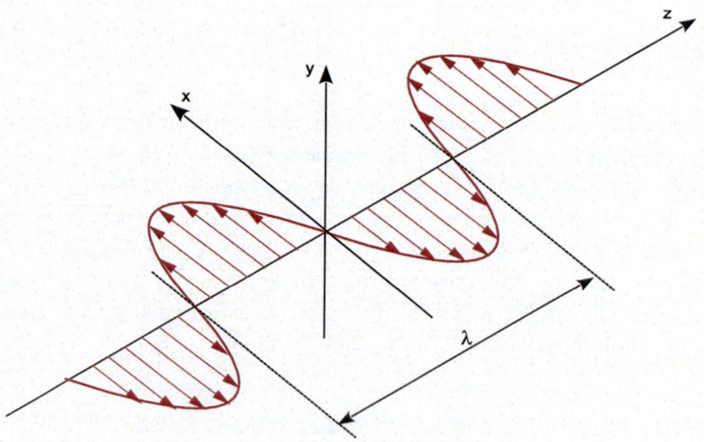

Fig. 2.3 A propagating electromagnetic wave emits an electric field

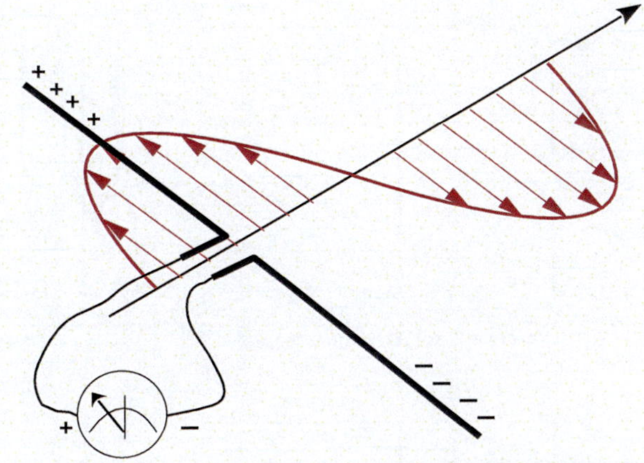

Fig. 2.4 An electric dipole antenna along the E-field direction

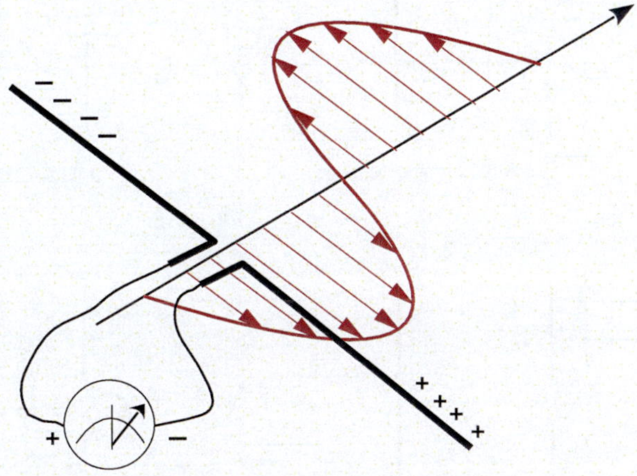

Fig. 2.5 The E-field moves the electrons back and forth in the antenna

We can see that this will repeat as long as the wave exists; the propagating electromagnetic wave will induce an AC current in the dipole antenna with a frequency that depends on the wavelength λ of the wave: $f = c_0/\lambda$. We get a maximum current in the antenna if the length of the antenna is $\lambda/2$.

However, our objective here is not to design antennas, but to understand how noise can couple to our measurement system. From Figs. 2.4 and 2.5, we can see how the presence of an electric field could induce a current in one of our signal wires in the system. Figure 2.6a–g illustrate how an AC current in one wire ('transmitter') emits an electric field and how it is picked up by an adjacent wire ('receiver'). This is an example of *crosstalk by E-radiation*.

The remedy could be quite simple; from Figs. 2.4, 2.5, and 2.6, we can see that for a current to be induced, the receiver 'antenna' (our signal wire) must be aligned with the E field direction. Hence, rearranging the signal transmission wires might help (unless there are multiple and multidirectional sources).

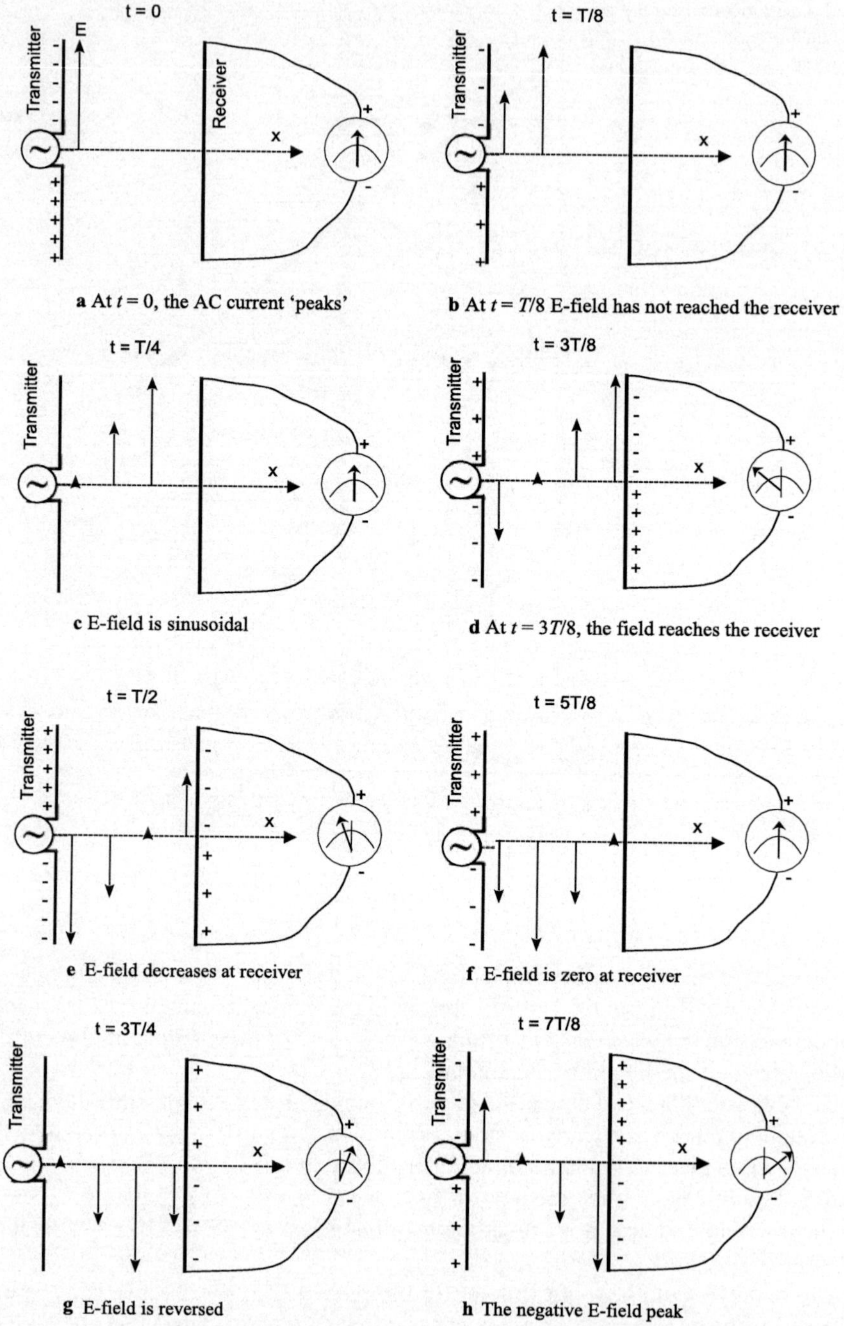

Fig. 2.6 **a** At $t = 0$, the AC current 'peaks'. **b** At $t = T/8$, E-field has not reached the receiver. **c** E-field is sinusoidal. **d** At $t = 3\,T/8$, the field reaches the receiver. **e** E-field decreases at receiver. **f** E-field is zero at receiver. **g** E-field is reversed. **h** The negative E-field peak

Secondly, we should also realize that this may not be a significant problem at all. The wave frequency must also match the wire length ($\lambda/2$) for any significant currents to be induced. For 'centimeter cables' that indicates GHz frequencies which is outside the bandwidth of most measurement systems (see Problem 2.1). If you have transmission cables 'tens of meters' long, the matching frequency could very well be within your system's bandwidth (the longer the cable, the more likely it is that it will pick up a frequency that interferes with your system).

When the frequency/wire length relation is 'right' and rearranging the transmission lines doesn't help (or is not possible), you need to prevent the electric field from interacting with your transmission line in the first place. How do you do that? The answer is a *Faraday cage*.

The Faraday cage was invented by Michael Faraday in 1836 to prove his hypothesis that there is *no electric field inside a closed metal surface*. The reason is that the electric field will redistribute the surface charge to produce an electric field in the opposite direction, canceling the field inside the surface. This is illustrated in Fig. 2.7.

Hence, we can protect our signal wire by an enclosing, conducting shield. The shield doesn't have to be perfect; it can be a grid. The general rule is that the grid holes' diameter should be < one-tenth of the wavelength of the radiation it is intended to block. For example, to block radiation from a 5G cell phone, operating at 39 GHz, a grid size of less than 1 mm is required. (On the other hand, 39 GHz is most likely outside your measurement system's bandwidth.)

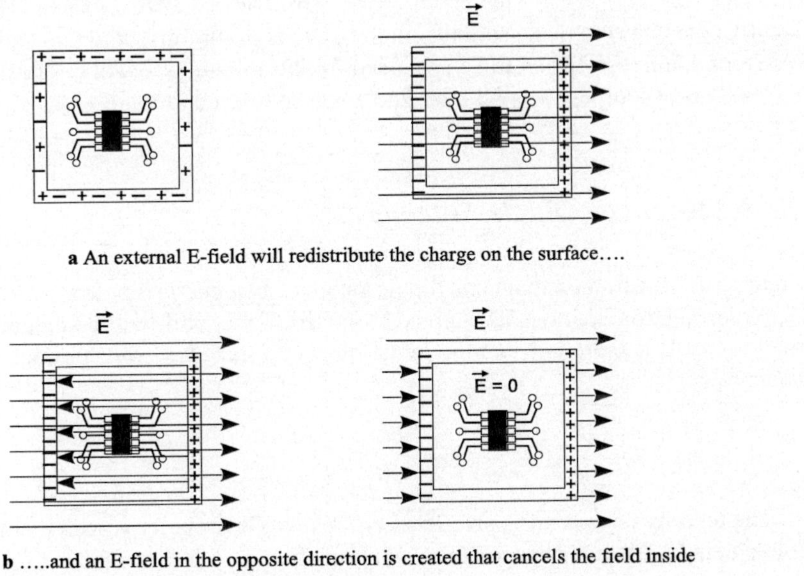

Fig. 2.7 **a** An external E-field will redistribute the charge on the surface. **b** An E-field in the opposite direction is created that cancels the field inside

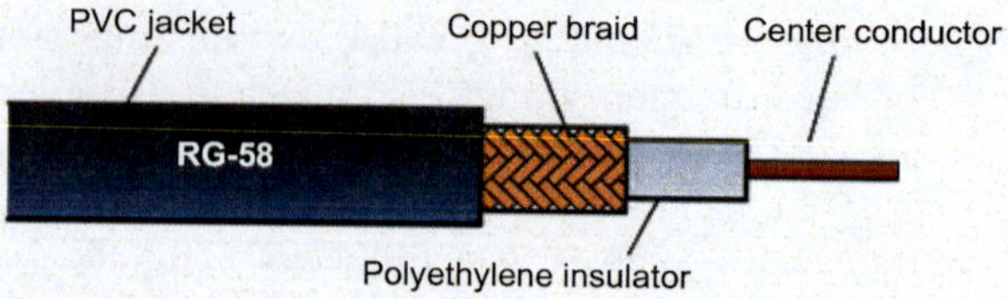

Fig. 2.8 A coax cable is a Faraday cage

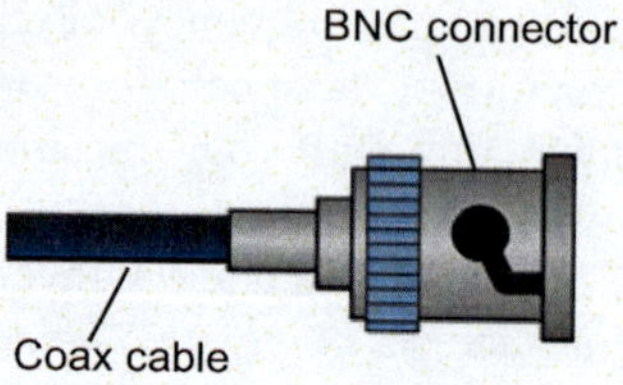

Fig. 2.9 Coax cables have BNC connectors

This is also one of the reasons why we use coax cables in the lab. Coax cables have a surrounding copper braid that forms a Faraday cage for the center conductor. A lot of instruments also have a metal casing to block E-field interferences.

Notice a few details about the Faraday cage. First, it is easily breached; all it takes is that you penetrate it with a conducting wire, which of course could be a problem in an electrical measurement system. The general rule is that the entire system should be in a cage, all the way from the sensor to the sampling computer interface (see Fig. 2.35). That is why coax cables have a BNC connector (Bayonet Neill-Concelman) to make sure that nothing leaks in (or out!) (Fig. 2.8 and 2.9).

Second, it doesn't need to be grounded for the shielding to work. (However, due to other noise sources, we will later find a reason to ground it anyway).

2.3.2 *Magnetic Dipole Antennas*

Figure 2.10 illustrates a magnetic dipole antenna; a *magnetic* dipole is a *circuit loop*.

According to Faraday's law, an electromotive force, emf, will be induced across the loop ends if there is a *change* in the magnetic flux $\Phi = B{\cdot}A$, through the loop area:

$$|\varepsilon| = \frac{d\Phi}{dt} = A \cdot \frac{dB}{dt} \tag{2.3}$$

The polarity of the emf is given by Lenz's law: The polarity is such that it creates a magnetic field that opposes the change of flux. What is important for us, and our electrical measurement system, is that a change in the magnetic field induces a voltage across the ends of an open circuit (and a current in a closed loop) which will add to our measurement signal. Just like an electromagnetic wave carries an electric

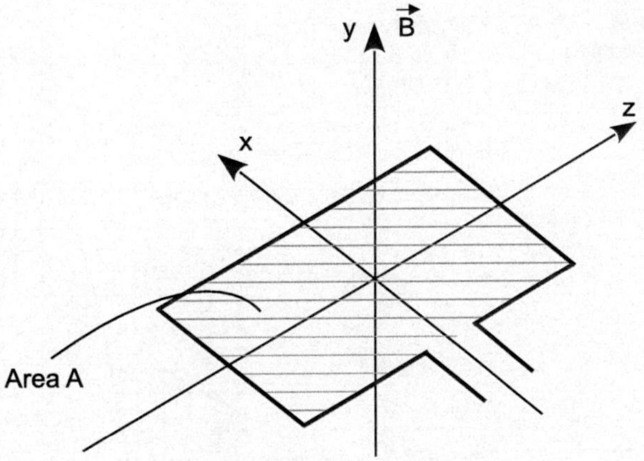

Fig. 2.10 Magnetic dipole antenna

field, it also carries a magnetic field. Figure 2.11 illustrates the magnetic field of the electromagnetic wave in Fig. 2.3.

Figure 2.12a and b illustrate how this field can induce an interference in a circuit. Magnetic fields cannot be blocked by a Faraday cage unless it is made of 'mu-metal' which is a nickel–iron ferromagnetic alloy. Mu-metal shields are expensive and usually not your first option.

Apart from electromagnetic waves in space, there are other sources of magnetic fields. Figure 2.13 illustrates the magnetic field around a current conducting wire.

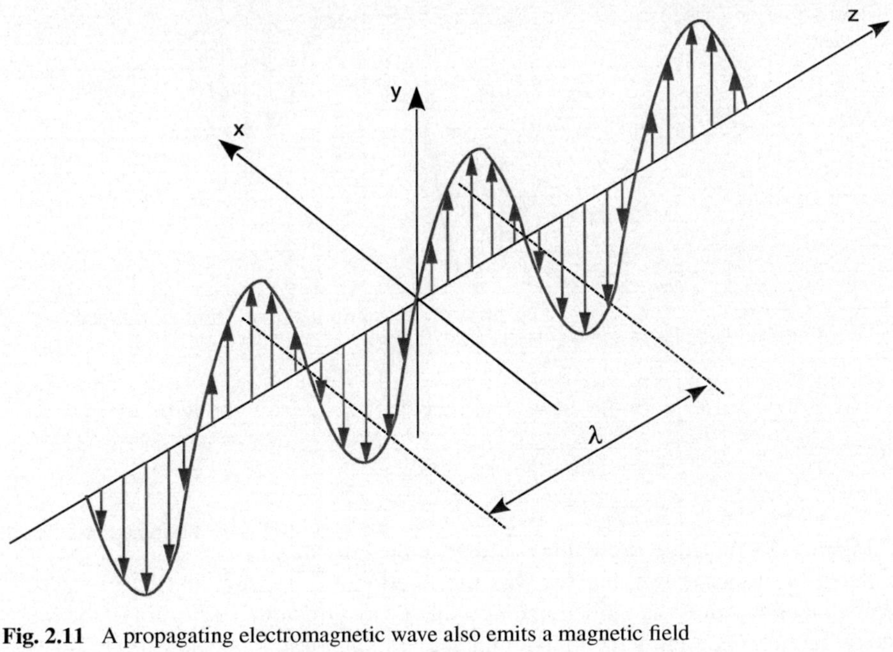

Fig. 2.11 A propagating electromagnetic wave also emits a magnetic field

Fig. 2.12 **a** An opposing current is induced. **b** An opposing current is induced

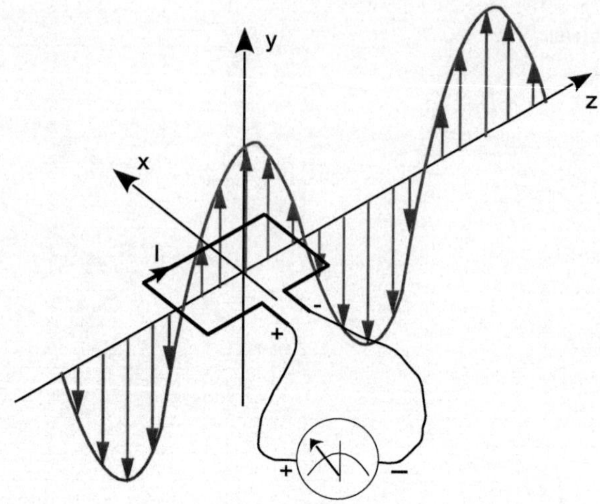

a An opposing current is induced

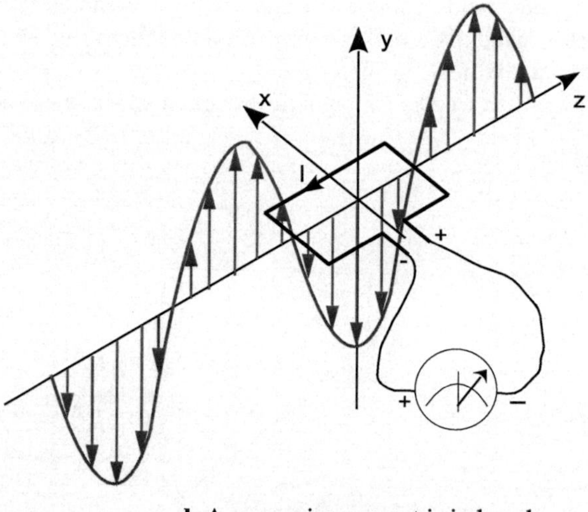

b An opposing current is induced

The B-field varies with the current and the distance x from the wire as

$$B(t) = \frac{\mu_0 I(t)}{2\pi x} \tag{2.4}$$

Figure 2.14 illustrates how this could generate *crosstalk by B-field radiation.*

Also, in this case, we can see that the interference depends on the geometric setup; the circuit must be perpendicular to the B field to induce a current in the loop and therefore, rearranging the loop might help. If that doesn't help, and assuming we

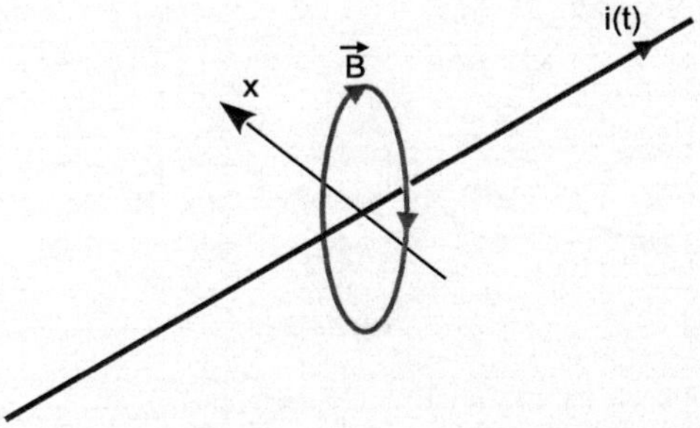

Fig. 2.13 Magnetic field around a current conductor

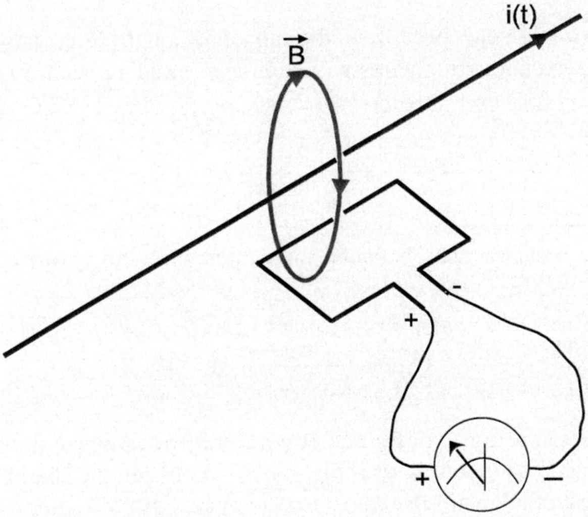

Fig. 2.14 Magnetic field crosstalk

can't afford a mu-metal shield, we need another solution. The solution is in Eq. (2.3): If we can't make $dB/dt = 0$, we can try to make $A = 0$. That is what we do when we use a *twisted pair* cable (TP cable), see Fig. 2.15.

In a TP cable, not only do we make the loop area ≈ 0, if there are any remaining areas they will cancel each other, since the induced current in two adjacent loops will have opposite direction. A TP cable is a very efficient way to protect your signal wire from *B*-field interferences. NB. If you also place the TP cable close to the ground plane, you also cancel potential common mode interferences induced by the magnetic field.

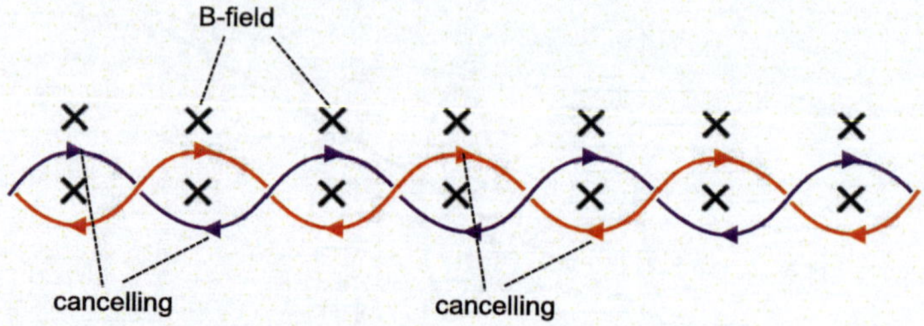

Fig. 2.15 A twisted-pair cable cancels the B-field interferences

2.4 Capacitive Crosstalk

Capacitive crosstalk occurs because of the capacitance between conducting surfaces. First, between two conducting wires with diameter d and a distance D apart, there is a capacitance C_{12} (per unit length) (Fig. 2.16):

$$C_{12} = \frac{\pi\varepsilon_0}{\ln\left(\frac{2D}{d}\right)}\left[\mathrm{F\,m^{-1}}\right] \tag{2.5}$$

There is also a capacitance between the center wire and its metallic shield (see Fig. 2.17):

$$C_{\mathrm{cyl}} = \frac{2\pi\varepsilon_0\varepsilon_r}{\ln(R/r)}\left[\mathrm{F\,m^{-1}}\right] \tag{2.6}$$

In Eq. (2.6), r is the radius of the center wire, R is the radius of the cylinder shield and ε_r is the dielectricity constant for the material between the shield and the center wire (usually polyethylene in the coax case, see Fig. 2.8).

To see how this enables crosstalk, we first place two unshielded wires next to each other, see Fig. 2.18. According to Eq. (2.5), there is a capacitance between the wires and hence an AC current in Wire 1 has a way into Wire 2. This will cause an interfering voltage across the load impedance in Wire 2. Figure 2.19 illustrates the equivalent circuit.

To remedy the capacitive crosstalk, we first apply a Faraday shield, see Fig. 2.20. However, according to Eq. (2.6), there is also a capacitance between the wire and the

Fig. 2.16 Two conductors are a capacitance

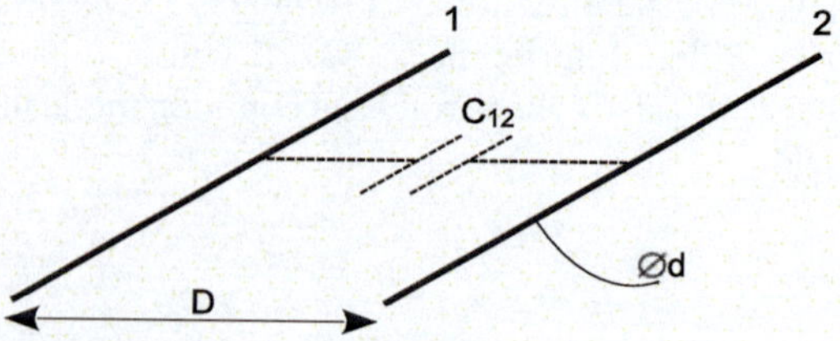

Fig. 2.17 Capacitance between shield and center wire

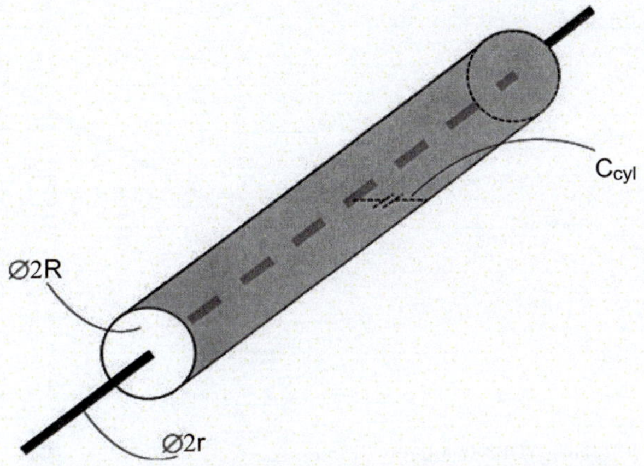

Fig. 2.18 Capacitive crosstalk

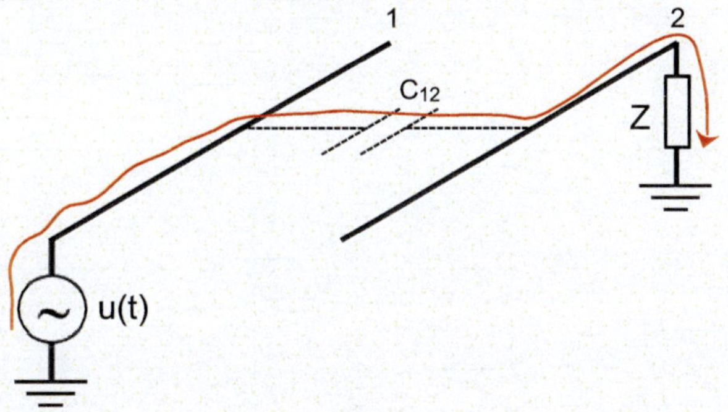

Fig. 2.19 Signal model of capacitive crosstalk

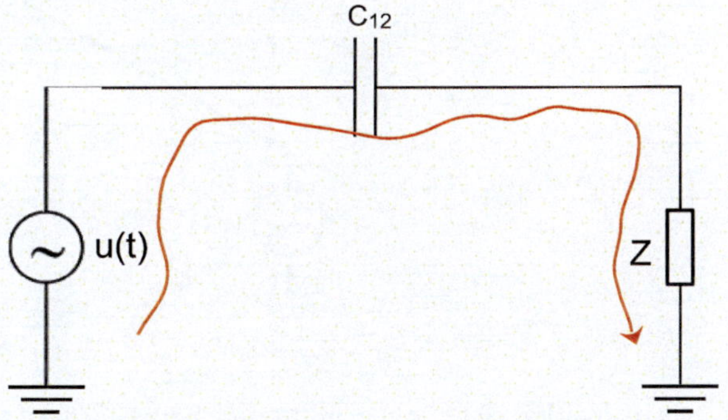

shield, so the only thing we accomplish is an extra capacitor in series, see Fig. 2.21; the current in Wire 1 can still find a way into Wire 2.

So, a Faraday cage *does not* protect your system from capacitive crosstalk (only against E-field crosstalk). The trick that enables the shield to protect your system also against capacitive crosstalk is to *ground the shield*, see Fig. 2.22.

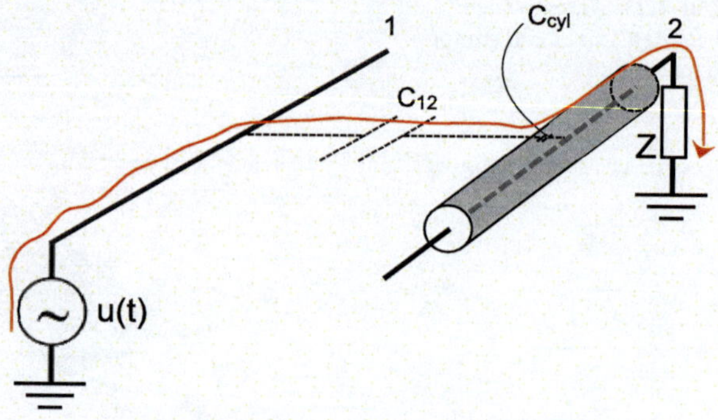

Fig. 2.20 A Faraday cage doesn't help

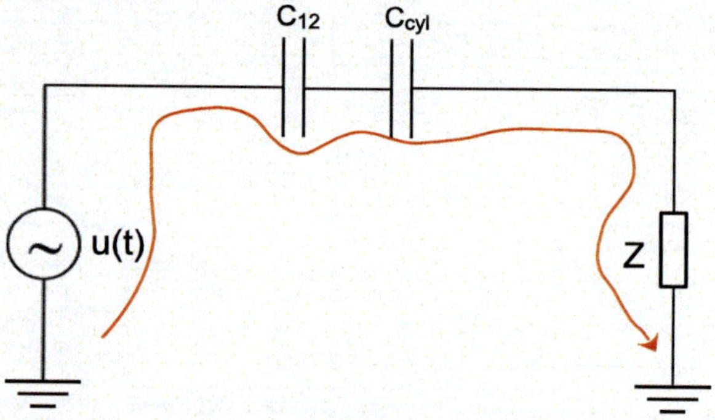

Fig. 2.21 Crosstalk path is still open

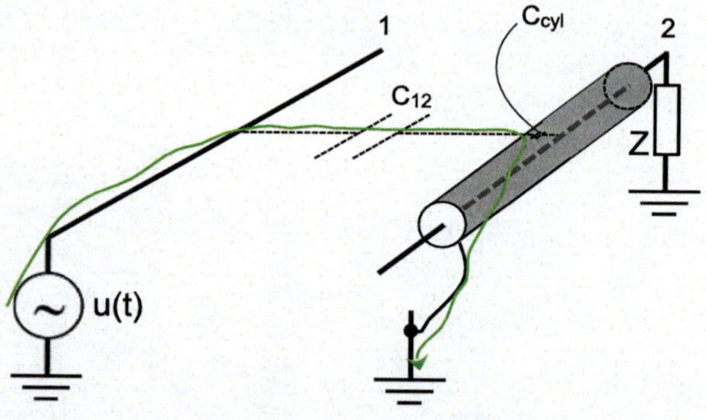

Fig. 2.22 Ground the shield!

The shield surface is the mid-point between the capacitors in Fig. 2.21 and by grounding the shield, currents from Wire 1 trying to 'sneak in' to Wire 2 are effectively short-circuited to ground, see Fig. 2.23.

Finally, please note that capacitive crosstalk is a high-frequency problem; the impedance of a capacitor is $1/j\omega C$, it decreases with frequency and hence high-frequency signals have an easier way into the neighbor wire than low-frequency

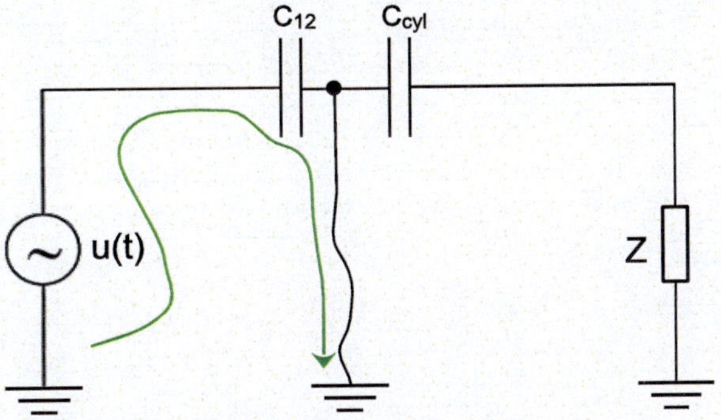

Fig. 2.23 Grounding the shield will short-circuit the crosstalk path

signals. As a matter of fact, this is true for most crosstalks; the problem increases with frequency.

2.5 Inductive Crosstalk

Capacitive crosstalk is based on the existence of a capacitance between two conducting surfaces. *Inductive* crosstalk is based on the fact that every conducting wire has a certain, non-zero, inductance per length unit, see Fig. 2.24. For example, a common RG-58 coax cable has a series inductance of approximately 250 nH/m.

From basic electricity, we also know that two coils close to each other form a *transformer*, see Fig. 2.25, and the mutual inductance M between them is a measure of how much of the voltage over the primary coil that is transferred to the secondary coil:

$$M = k \cdot \sqrt{L_1 \cdot L_2} \tag{2.7}$$

where k is a constant depending on 'geometric and environmental' parameters.

The voltage induced in the secondary coil is

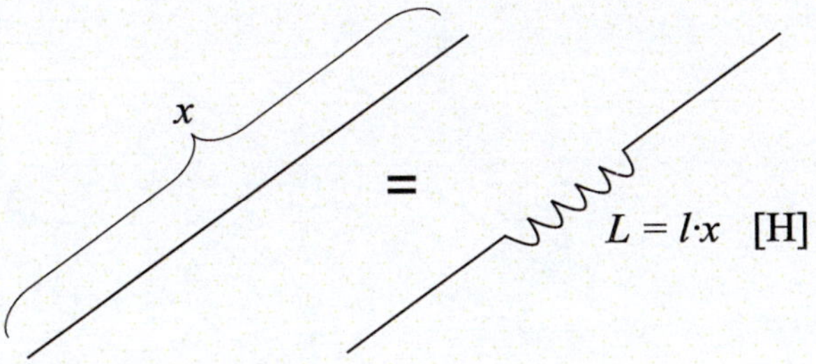

Fig. 2.24 A conductor has some inductance l H/m per unit length

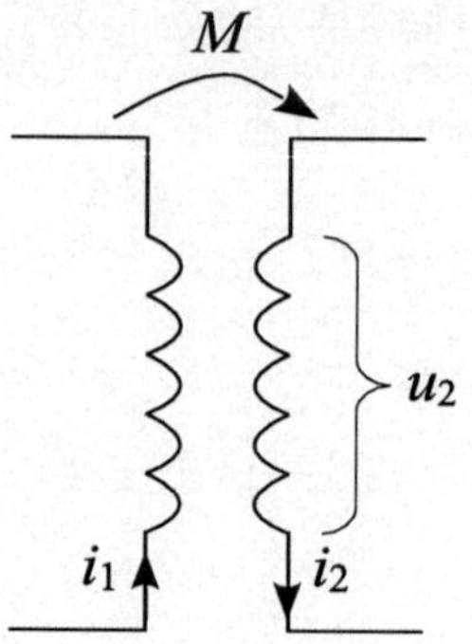

Fig. 2.25 A transformer

$$u_2 = M\frac{di_1}{dt} \tag{2.8}$$

If we assume that the current i_1 in Wire 1 is sinusoidal, $i_1(t) = i_0 \sin \omega t$, the derivative is $\omega i_0 \cos \omega t$ and

$$u_2 = M\omega i_0 \cos \omega t \tag{2.9}$$

Notice in Fig. 2.25 that there is also a current i_2 induced in the secondary coil and its direction is determined by Lenz's law; its direction is such that it counteracts its origin.

Now we have all we need to explain inductive crosstalk. If we place two conductors close to each other, they will form a transformer because of their inherent inductance per unit length, see Fig. 2.26, and Eq. (2.8) tells us that a current (i.e., a *current change*) in Wire 1 will induce a voltage in Wire 2, (which according to Eq. (2.9) increases with frequency; inductive crosstalk is also a high-frequency problem).

The remedy in this case is a little more sophisticated than earlier. First, we place a new conductor *between* Wire 1 and Wire 2. We will call it the 'shield conductor', or just the 'shield', see Fig. 2.27. Just like there is a mutual inductance between Wires 1 and 2, there will be a mutual inductance between Wire 1 and the shield, and between Wire 2 and the shield.

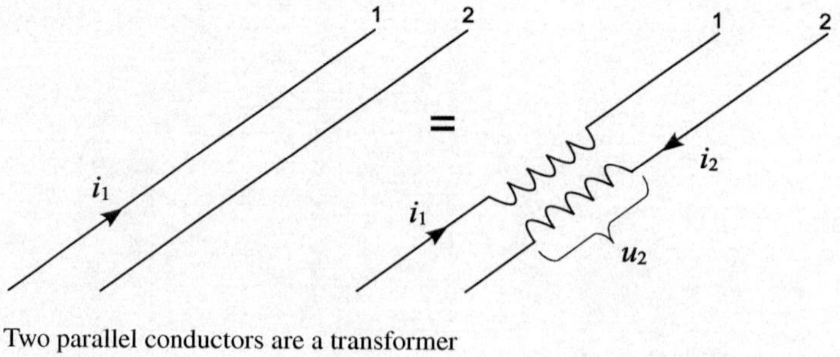

Fig. 2.26 Two parallel conductors are a transformer

Fig. 2.27 There are mutual inductances between each pair of wires

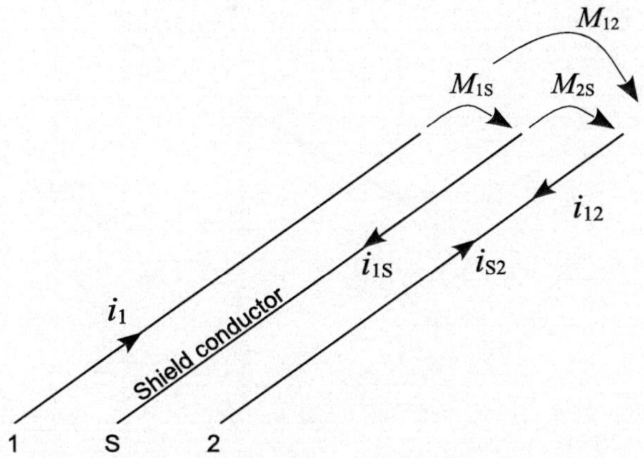

Just like the current in Wire 1 induces a current i_{12} in Wire 2, it will also induce a current i_{1S} in the shield conductor. But because there is a mutual inductance between the shield and Wire 2, the current i_{1S} will induce a current i_{S2} in Wire 2, and because i_{1S} is in the opposite direction of i_1, the induced current from the shield will be in the opposite direction of i_{12}, see Fig. 2.27.

So, the shield induces a current in Wire 2 that has the same frequency as the current induced by wire 1, but it has a phase shift of 180°, which means that it will interfere 'destructively' with i_{12}. i_{S2} will cancel i_{12} completely if $M_{12} \cdot i_1 = M_{S2} \cdot i_{1S}$. $i_1 > i_{1S}$, but $M_{S2} > M_{12}$, so there is a good chance that they will cancel. If we do it right.

How do we do it 'right'? As a matter of fact, how do we do it at all? First, it doesn't seem very practical to just place an extra 'dummy' wire next to our signal wire. Second, it probably wouldn't work anyway. An absolute condition for this trick to work is that the shield wire can conduct a current; it needs to be a closed loop.

We can achieve that without adding an extra 'dummy' cable. From Fig. 2.22, we learned that we need a grounded shield anyway to protect our system from E-field radiation and capacitive crosstalk. Well, there is our 'dummy' shield already! All we must do is to ground it in *both ends* (make it a closed loop) to also protect us against inductive crosstalk, see Fig. 2.28.

However, grounding the shield at both ends might introduce new noise and we will investigate that in the next section.

2.6 Common Impedances

A current must always have a return path; what goes out must come back. Current will always find a way back and it will choose path(s) according to Kirchhoff's current law. A current carrying wire must always have a return path to 'close the loop'. When

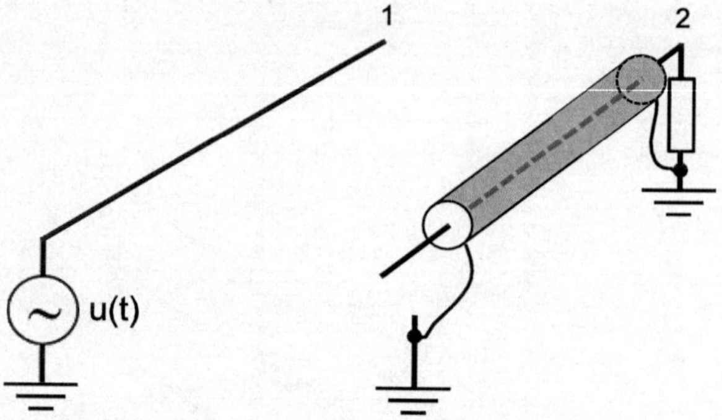

Fig. 2.28 If we ground the shield at both ends, we are protected against E-field, capacitive, and inductive crosstalk

you use a coax cable, the shield (the copper braid surrounding the center wire) is the return path for the current, see Fig. 2.29.

In the general case, we use a 'common ground' as the return path for the current, see Fig. 2.30.

Using common ground as the return path in a measurement system is in general not a good idea. Two circumstances, which are quite common, can make the ground itself a source of crosstalk and noise. First, suppose that the ground path conductor is not 'perfect', i.e., it has a resistance > 0 ohms (which it almost always has). Second, since it is a *common* ground, other signals *also* use it for current return (i_x), see Fig. 2.31 (where R_{wire} is the resistance in the wire).

If we apply Kirchhoff's voltage law in Fig. 2.31, we get

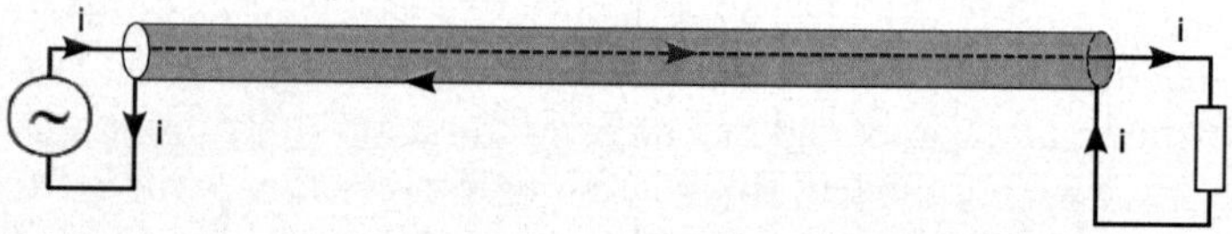

Fig. 2.29 What goes out must come back; the shield is also the return path

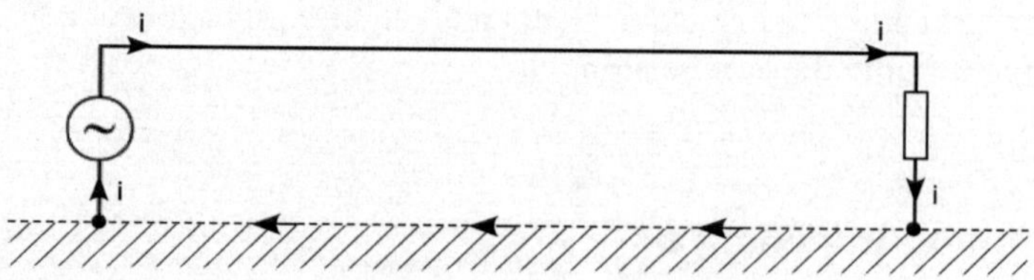

Fig. 2.30 'Ground' is a common return path

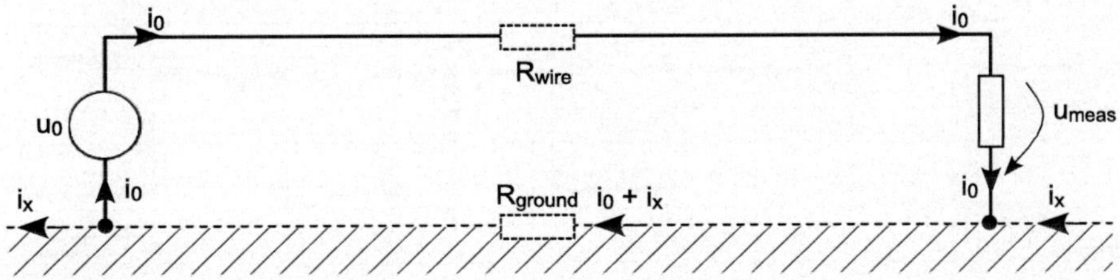

Fig. 2.31 Common ground as return path

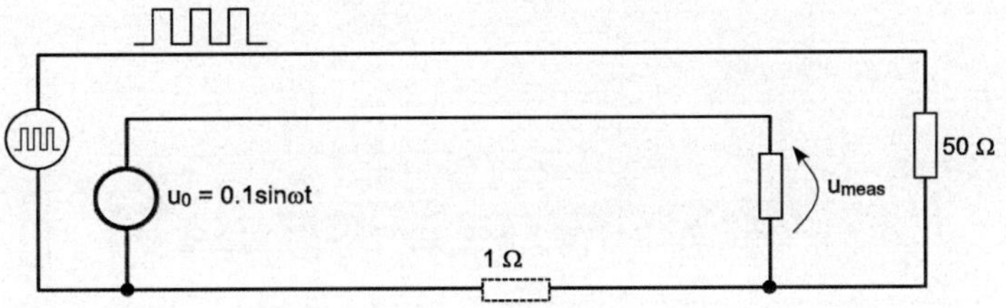

Fig. 2.32 Multi-signal system with common return path

$$u_0 - i_0 R_{\text{wire}} - u_{\text{meas}} - (i_0 + i_x) R_{\text{ground}} = 0$$

$$u_{\text{meas}} = \underbrace{u_0}_{\substack{\text{'True'}\\ \text{signal}\\ \text{value}}} \underbrace{- i_0 \left(R_{\text{wire}} + R_{\text{ground}}\right)}_{\text{cable loss}} \underbrace{- i_x R_{\text{ground}}}_{\text{crosstalk}} \tag{2.10}$$

In Eq. (2.10), $i_0\left(R_{\text{wire}} + R_{\text{ground}}\right)$ is the 'cable loss' in the circuit. We always have some of that. This is not 'noise'. After all, it is caused by the signal itself, and it is predictable (we can compensate for it). The problem in Eq. (2.10) is the term $i_x R_{\text{ground}}$ which is caused by an external current that has nothing to do with u_0 or our system. $i_x R_{\text{ground}}$ represents 'crosstalk by common impedance'.[1]

This explains why the shield against inductive crosstalk in Fig. 2.28 is a potential problem; both ends of the return path are grounded which is an invitation to other currents using the common ground to enter our system.

Example 2.1 In a multi-signal system, a signal wire carrying a small sinusoidal signal shares return wire with a fast TTL clock signal, see Fig. 2.32. The TTL signal wire is '50-Ω terminated' to reduce pulse reflections (see Chapter 5), which means that the current in the clock wire (during the 5-V pulses) is 5/50 = 100 mA. Make a prediction of the clock signal's impact on the measurement of the sine if the resistance of the return wire is 1 Ω.

[1] Sometimes called 'common ground crosstalk'.

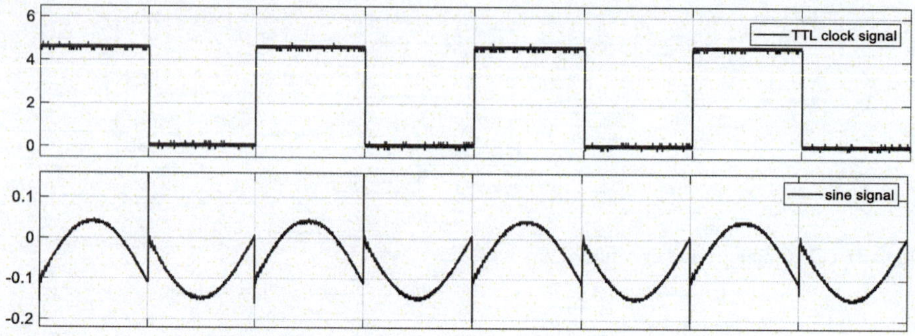

Fig. 2.33 A 'real case' example

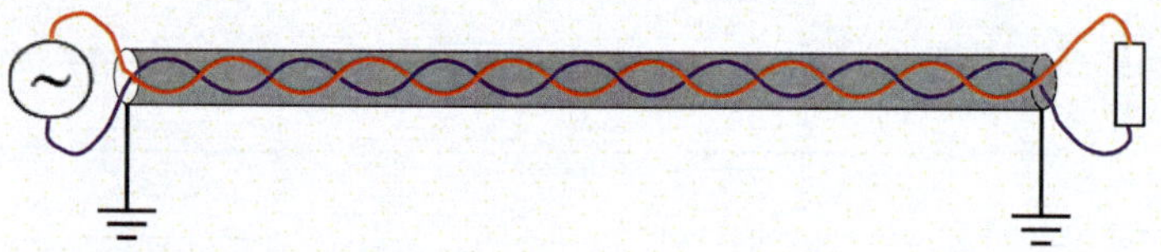

Fig. 2.34 A shielded TP cable

Solution The $i_x R_{ground}$ voltage in Eq. (2.10) is 100 mA $\times$ 1 Ω = 100 mV. Hence, according to Eq. (2.10), we will measure a signal that is 100 mV lower than expected during the positive duty cycles of the clock signal. Figure 2.33 illustrates a real example recorded with an oscilloscope, using three 4.5 m wires with a cross-sectional area of 0.08 mm^2.

When it comes to the use of shields, like the ones we have been using in Figs. 2.22 and 2.28, to protect our system against capacitive and inductive crosstalk, there are two cases that need to be treated differently; whether the shield is 'just a shield' or if it also carries the return current. This difference is paramount because it determines how you can use it.

If the shield is also the return path, as in a coax cable, the shield should *only* be grounded at one end. Never ground a coax cable at both ends. If you have problems with inductive crosstalk and need to ground the shield at both ends, you can't use a coax cable: you must use a 'shielded pair-cable', where the shield is *not* the return path. That also has another advantage; you can twist the signal pair wires to also get B-field protection.[2]

A shielded TP cable, grounded at both ends, protects your system against 'everything', but coax cables have higher bandwidth and support longer cable lengths (Fig. 2.34).

[2] You don't 'twist' the cables yourself; you buy a 'shielded twisted-pair' cable.

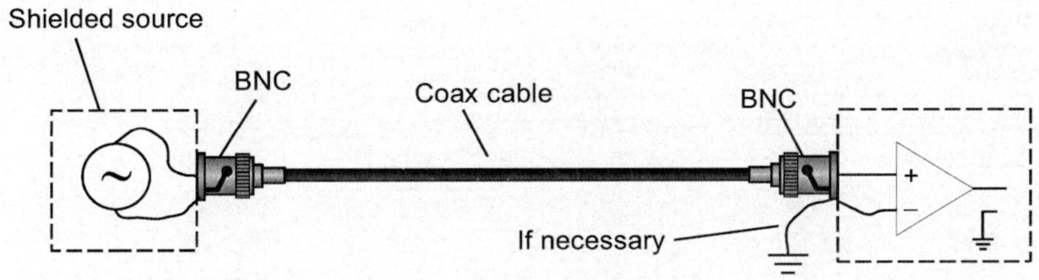

Fig. 2.35 Keep the signal in a Faraday cage all the way

2.7 Summary and Recommendations

When you plan the setup of an electrical measurement system, you need to keep external crosstalk interference in mind. Some of them you need to consider from the outset, because they are omnipresent in almost all environments, while with others, you just wait and see if they show up.

Of all the potential crosstalk sources we have presented in this chapter, capacitive crosstalk and common impedance crosstalk are the most common problems. (Protecting your system against capacitive crosstalk takes care of E-field crosstalk at the same time.) You should take deliberate actions to prevent capacitive and common impedance crosstalk when you plan a measurement setup.

What remains is B-field and inductive crosstalk. In my experience, they are not a major problem in most physics labs; don't make any special plans to protect your system against B-field and inductive crosstalk, just keep them in mind if you still have interference problems after taking precautions against capacitive and common impedance crosstalk.

That means that we don't start with a shielded twisted-pair cable, you start with a coax cable and one signal ground point only (if necessary). If you have two different ground points, you open up for common impedance crosstalk.

Figure 2.35 illustrates what should be your first option ('plan A').

The signal should preferably be transported as a 'non-referenced' differential signal to the receiving DAQ (Data AcQuisition) system's differential amplifier input.

If the source is inherently 'referenced' (ground related), you might have to 'de-reference' it, either using an opto coupler (if the signal is digital) or an isolation transformer (if the signal is analog), see Fig. 2.36.

Only when the coax system in Fig. 2.35 fails, you consider a shielded TP cable. If the shielded TP cable system also fails (or if its bandwidth is too small or if it can't offer long enough cables), an alternative solution could be a fiber optics solution. Fiber optic transmission cables are immune to all the above crosstalk interferences. They also have extreme bandwidth and allow long cable lengths but are expensive and somewhat more complicated to handle. Fiber optics are not perfect though, they have their own issues (like dispersion, for example).

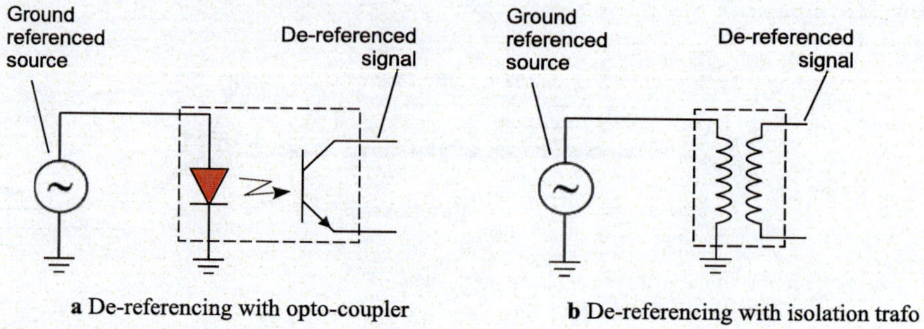

Fig. 2.36 **a** De-referencing with opto-coupler. **b** De-referencing with isolation trafo

However, whatever you do, there will almost always be some unwanted signal component(s) in your measurement signal that you want to get rid of and when you have set up your system according to the recommendations above and still have some noise (you always have the 'internal' noise), then, but only then, you will have to start the 'signal processing', which is what most of this book is about.

2.8 Solved Problems

Problem 2.1 Assuming a common trace length on a pcb (printed circuit board) is 5 cm (two inches). For what EM frequencies would such a board be particularly vulnerable to electric field interferences?

Solution It is particularly vulnerable if the pcb trace length equals $\lambda/2$, i.e., for EM waves with a wavelength of 10 cm. That corresponds to a frequency of

$$f = \frac{3 \cdot 10^8}{0.1} = \underline{3\ \text{GHz}}$$

Problem 2.2 A desktop DMM with an input impedance of 10 MΩ has an open input (no input signal). When set to DCV range and 'statistics mode' it displays a standard deviation of 16 μV. Estimate the instrument's bandwidth in the DCV range.

Solution Assuming room temperature (300 K), we solve for B in Eq. (2.1):

$$B = \frac{u_{\text{rms}}^2}{4\pi k RT} = \frac{\left(16 \cdot 10^{-6}\right)^2}{4\pi \cdot 1.38 \cdot 10^{-23} \cdot 10 \cdot 10^6 \cdot 300} = \underline{500\ \text{Hz}}$$

Problem 2.3 A typical kitchen microwave oven operates at 2.45 GHz. What grid size would you recommend for a protective metal mesh in a microwave oven?

Solution An electromagnetic wave with a frequency of 2.45 GHz has a wavelength of $3 \cdot 10^8 / 2.45 \cdot 10^9 = 12$ cm. One-tenth of 12 cm is 1.2 mm. (Compare that to the size of the grid in the front door of your microwave oven.)

Reference

1. Kiely, R. 2017. Understanding and eliminating 1/f noise. In *Analog Dialog*, p. 4.

Chapter 3
Sensors

Abstract This chapter first describes thermocouples starting from the famous experiments by Seebeck and Thomson. The basic concepts of thermocouples are described such as hot and cold junctions, the Seebeck coefficient, and thermocouple 'types' are explained. This chapter also explains what the 'cold junction compensation' is and the law of intermediate temperatures is illustrated. Resistance temperature detectors (such as Pt-100) are described, and the necessity of accurate resistance measurements is explained (the '4-wire method'). The measurement of extremely high and extremely low temperatures is covered at the end of Sect. 3.2. Section 3.3 introduces the versatile strain gauge principle and its many applications. It is also explained why strain gauges are (almost) always connected to a Wheatstone bridge and how it can be used to measure a wide range of physical quantities (like force, pressure, liquid level, torque, etc.). Piezoelectric crystals and Hall sensors are explained and light sensors (photodiodes, position-sensitive detectors, and photomultipliers). In the particle detector section, channeltrons and microchannel plates are explained and in the final Sect. 3.9, the most common vacuum gauges are presented.

3.1 Introduction

In this context, a 'measurement' refers to the measurement of a *physical* quantity, like temperature, pressure, acceleration, sound intensity, or light intensity. Most of these physical quantities can be measured using mechanical gauges (for example, we can measure temperature using a mercury thermometer) but in a physics lab, the destination for data is almost always a computer (of some kind) and that requires that we have access to the measurand in electrical form. The device that transforms a variation in a physical quantity into a variation in an electrical quantity is called the *sensor*. The words 'gauge' and 'transducer' are also common in this context, but we will mostly use the word 'sensor' here.

The preferable electrical quantity is (almost) always volts [V], because, first, we know how to measure voltage very accurately and, second, the 'analog-to-digital' elements that 'sample' the signal need voltage as the input quantity (see Chap. 11).

L. Bengtsson, *Electrical Measurement Techniques*,
https://doi.org/10.1007/978-981-99-8187-8_3

However, a lot of sensors do not produce voltage as the primary output (they may produce a current or a change in resistance or capacitance), and in those cases we need some 'supporting' electronics to generate a voltage output. This supporting electronics is referred to as the 'signal conditioning' electronics.

There are *a lot* of physical quantities and, for most sensors, there are more than one sensor technique available, so learning about sensors appears to be quite a challenge. However, if you count all sensors in all physics labs, you will find that there is one kind of sensors that dominates completely: Temperature sensors. No matter what the 'physics' is about, temperature must almost always be measured somewhere. So, if you only have time to learn about one sensor technique, you should start with temperature sensors. (Then you understand maybe as much as 30% of all sensors in a lab.) Hence, temperature sensors are what we will start with.

If you have time to learn one more sensor technology, I recommend you learn about the 'strain gauge principle' since this versatile sensor technique is the basis for a lot of sensors for different physical quantities.

3.2 Temperature Sensors

3.2.1 Thermocouples

Thermocouples are one of the most common sensors in a physics lab and this is probably the first sensor you should learn about as a physicist. In 1821, the German physicist Thomas Johann Seebeck discovered that if you make a closed circuit of two different conductors and keep the two junctions at different temperatures, there will be a current in the circuit (Fig. 3.1).

In most textbooks, the Seebeck effect is used to explain thermocouples. In this book, I will instead use the *Thomson* effect (since I think it makes the understanding a little less mysterious). The Thomson effect was discovered in 1854 by the British physicist William Thomson (Lord Kelvin). He did similar experiments on a *single* wire and found that when current was flowing in a conductor, one end got warm, and one end got cold. He also found that this process was reversible; if the two ends of a conductor are held at different temperatures, a current will be induced in the conductor. If we don't have a closed loop, the regrouping of charge will induce

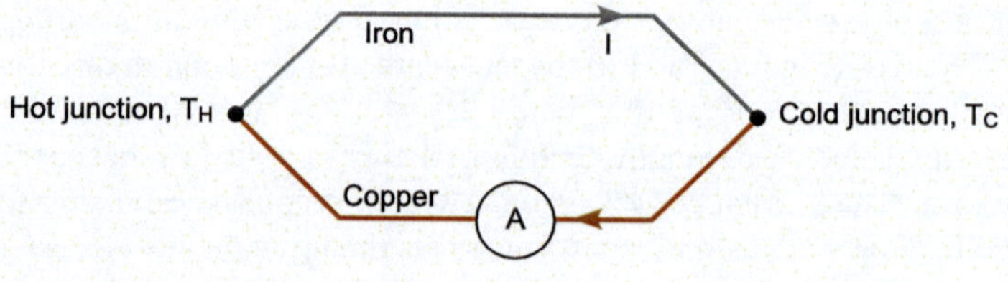

Fig. 3.1 The Seebeck effect

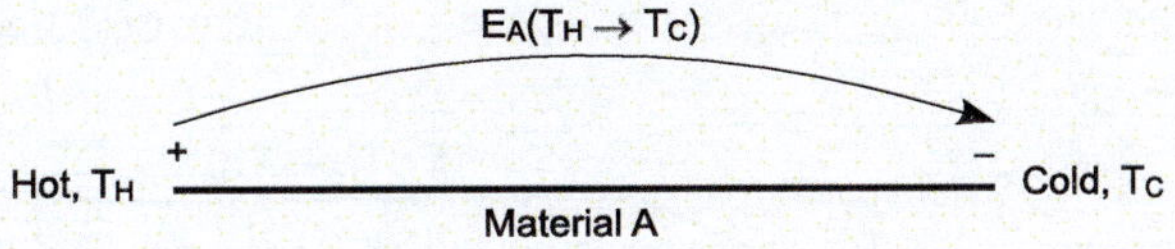

Fig. 3.2 The Thomson effect

a voltage across the ends, an *emf*. We will denote this emf $E_A(T_H \rightarrow T_C)$ ('from temperature T_H to temperature T_C along material A) (Fig. 3.2).

Next, we need to figure out how to use the Thomson effect to measure temperature. Figure 3.3 illustrates a (naïve) experiment. A voltage meter is used to measure the emf across a single wire. We connect the voltage meter to the end points using a wire of (unknown?) material C. The Thomson effect is as valid for the wires of material C as it is for the wire of material A. The voltage meter will measure the voltage

$$U_m = E_C(T_0 \rightarrow T_H) + E_A(T_H \rightarrow T_C) + E_C(T_C \rightarrow T_0) \quad (3.1)$$

(T_0 is the temperature of the voltage meter.) From expression (3.1), we can see that we would have to consider the contribution from the voltage meter wires, and that would make this solution impractical (to say the least); we would always have to make sure we have the right wires to the voltage meter. For that reason, we use a pair of wires of dissimilar materials (a *thermocouple*) as illustrated in Fig. 3.4. This straightforward design eliminates the voltage measurement's dependence on the C wires.

We can see that by writing out the expression for the voltage U_m:

$$U_m = E_C(T_0 \rightarrow T_C) + E_A(T_C \rightarrow T_H) + E_B(T_H \rightarrow T_C) + E_C(T_C \rightarrow T_0) \quad (3.2)$$

And since $E_C(T_0 \rightarrow T_C)$ of course $= -E_C(T_C \rightarrow T_0)$ the E_C terms will cancel and we can write Eq. (3.2) as

$$U_m = E_A(T_C \rightarrow T_H) + E_B(T_H \rightarrow T_C) = E_B(T_H \rightarrow T_C) - E_A(T_H \rightarrow T_C) \quad (3.3)$$

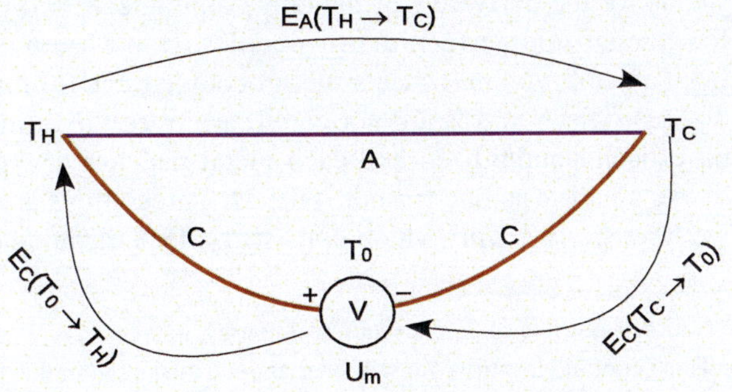

Fig. 3.3 A simple Thompson effect experiment

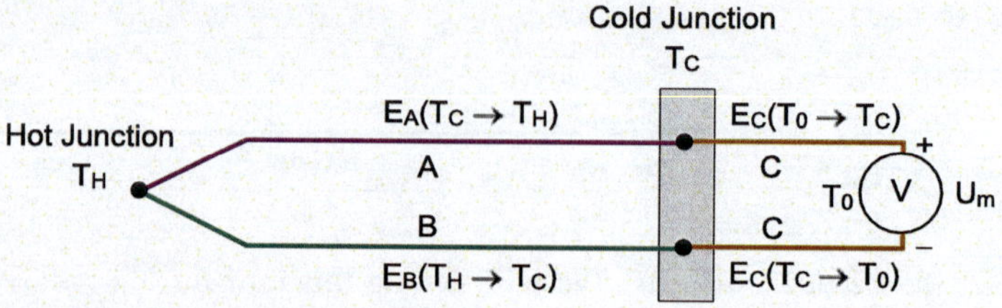

Fig. 3.4 A thermocouple

Table 3.1 Thermocouple data

Type	Metal pair	Seeb. coef. [μV/°C]	Range [°C]	Accuracy [°C]
E	Chromel/Const	59	–270 … +870	±1.7
J	Fe/Const	50	–200 … +760	±2.2
K	Chromel/Alumel	39	–270 … +1260	±2.2
S	Pt–Rh/Pt	5	–50 … +1600	±1.5
T	Cu/Const	39	–270 … +370	±1.0

With the design in Fig. 3.4, the voltage we measure is independent of the C wires and the temperature of the voltage meter, and that is what we need for the Thomson effect to be a useful sensor technique.

There are some things we need to know before we use thermocouples. First, we notice from Eq. (3.3) that the voltage U_m (the *thermo emf*) represents a temperature *difference*; we must *know* the temperature T_C at the cold junction. The cold junction temperature is measured separately by another sensor (see next section). Second, the thermo emf is exceedingly small, typically a few or tens of μV/°C (see Table 3.1). Third, the thermo emf is *not* a linear function of the temperature. In fact, NIST[1] recommends that they are best described by a ninth-order polynomial. Figure 3.5 illustrates the emf of a 'type K' thermocouple in the range –10 to +40 °C (' (**'o'**)') and a linear approximation; notice the deviation at higher temperatures. (We will explain the 'type' letter later.)

Even though thermocouples are not linear, they are often characterized by a sensor coefficient called the *Seebeck coefficient*, which has the unit μV/°C. This number stands for the derivative of the emf graph at $\Delta T = 0$ °C. NB. This number is only for comparison between thermocouples. Don't try to use it to derive a temperature from an emf; you *must* use a thermocouple table for that!

Materials A and B are not paired arbitrarily in a thermocouple. The metal pairs have been standardized and each metal pair has a 'type' letter. It is also common to use an alloy as one (or both) metal. Three alloys are particularly common. First, we have *Constantan*[2] which consists of 45% nickel and 55% copper. Then there is

[1] National Institute of Science and Technology, nist.gov.

[2] This is not the last time we will hear about Constantan in this book; it is also the most common material in strain gauges, see Sect. 3.3.

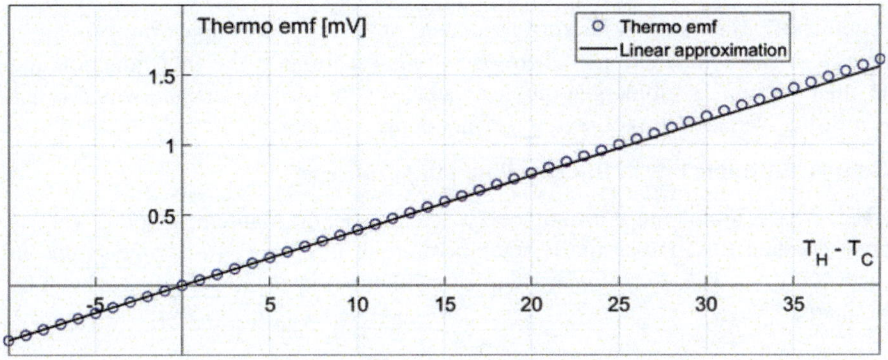

Fig. 3.5 Temperature dependence of a type K thermocouple

Alumel which is 95% nickel and 5% aluminum and *Chromel* which has 90% nickel and 10% chromium.

Figure 3.6 compares the five most common thermocouples' emf graphs in the range −10 °C to +50 °C and Table 3.1 summarizes their parameters.

A common question in the physics lab is: 'What thermocouple type should I use?'. Well, first, it must of course cover your temperature range. The type S thermocouple can measure the highest temperatures and type K, or T, are usually used for exceptionally low temperatures. The type E and type J thermocouples can be used in oxidizing atmospheres (type E also in inert atmospheres). Type K is the most common of all thermocouples; it is inexpensive and accurate, and it can also be used in nuclear applications because of its radiation 'hardness'. Type T thermocouples have the smallest range but are the most accurate and have excellent reliability/repeatability. If you don't know or don't care, you start with a type K thermocouple. In fact, a lot of DMMs have thermocouple inputs and that is almost always for a type K thermocouple.

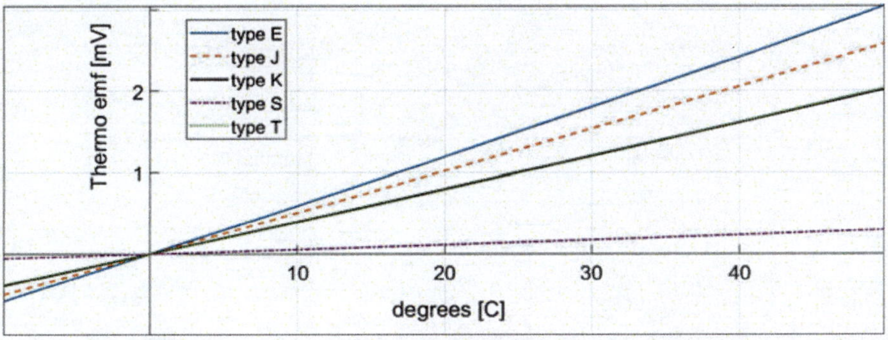

Fig. 3.6 Comparing the most common thermocouples between −10 °C and +50 °C (type K and type T overlap in this range)

Figure 3.7 illustrates an interesting (and common) thermocouple arrangement. It consists of two AB junctions, where one of the junctions is the 'hot' junction and the other junction is submerged into ice water (= 0 °C). To understand why this is a clever trick, we need the following thermocouple law.

The thermocouple law of intermediate temperatures:

In Fig. 3.8a, a temperature difference $T_H - T_{IM}$ generates an emf $E_A(T_H \rightarrow T_{IM})$ across conductor A, and similarly, a temperature difference $T_{IM} - T_C$ generates an emf $E_A(T_{IM} \rightarrow T_C)$. In this case, T_{IM} is an 'intermediate' temperature and can be eliminated:

$$E_A(T_H \rightarrow T_{IM}) + E_A(T_{IM} \rightarrow T_C) = E_A(T_H \rightarrow T_C) \tag{3.4}$$

This is illustrated in Fig. 3.8b.

Back to Fig. 3.7. The voltage meter will measure the thermo emf

$$U_m = E_A(T_C \rightarrow T_H) + E_B(T_H \rightarrow 0\,°C) + E_A(0\,°C \rightarrow T_C) \tag{3.5}$$

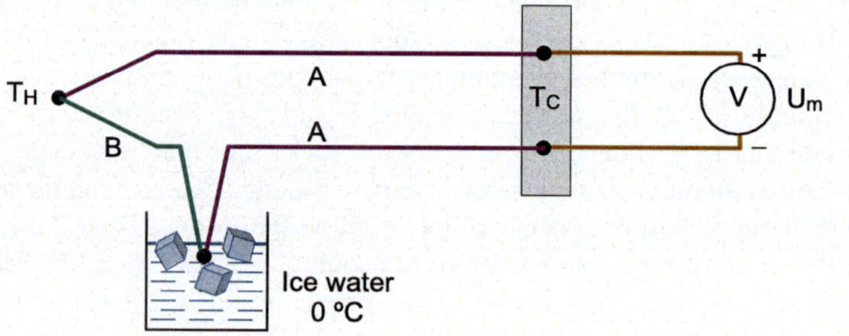

Fig. 3.7 The ice water trick

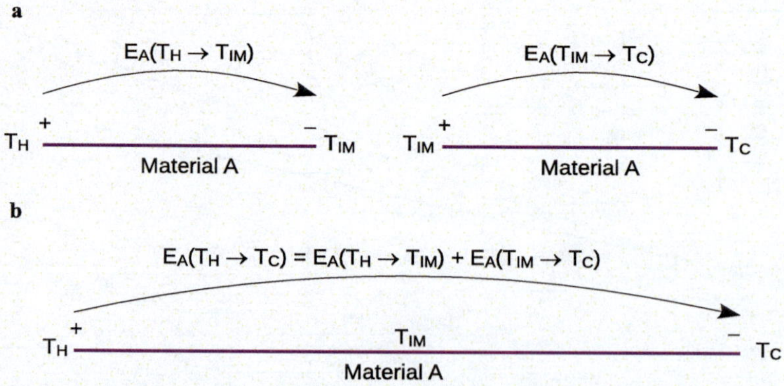

Fig. 3.8 **a** 'Intermediate' temperature. **b** The law of intermediate temperatures

If we use the law of intermediate temperatures on the emfs across the A conductors, T_C will be the 'intermediate temperature', and we get

$$E_A(0\,°C \to T_C) + E_A(T_C \to T_H) = E_A(0\,°C \to T_H) = -E_A(T_H \to 0\,°C)$$

And hence, we can write Eq. (3.5) as

$$U_m = E_B(T_H \to 0\,°C) - E_A(T_H \to 0\,°C) \tag{3.6}$$

In expression (3.6), U_m is independent of the cold junction temperature; we have turned the temperature measurement into an *absolute* measurement. In general, thermocouples only measure a temperature *difference*, but that doesn't mean that you can just add the cold junction temperature to the temperature you get from the thermo emf. The proper way to do it is to convert the cold junction temperature to voltage, add that voltage to the thermo emf, and then convert the summed voltage to a temperature (using a table). This is called 'cold junction compensation' (see Problem 3.5).

3.2.2 *Metal Temperature Sensors*

Temperature sensors based on pure metals are called *resistance temperature detectors* or just 'RTDs'. The resistance of all metals has a positive temperature dependence[3]; the resistance increases when the temperature increases. So, we could use any metal as a temperature sensor, but only three are really used, platinum, copper, and nickel. In fact, in industrial applications, only (almost) platinum sensors are used, so that is what we will be focusing on here. It is highly unlikely that you will ever see anything else in your physics lab.

A platinum temperature sensor is denoted 'Pt-100' or 'Pt-1000'. 'Pt' is of course for 'Platinum' and the number, 100 and 1000, respectively, is the sensor's resistance at 0 °C. Unlike thermocouples, the temperature dependence of metals is very linear. The resistance's dependence on the temperature is given by Eq. (3.7):

$$R = R_0(1 + \gamma T) \tag{3.7}$$

where T is the temperature in °C, R_0 is the resistance at 0 °C, and γ is the sensor coefficient, and for platinum, $\gamma = 3.85 \cdot 10^{-3}\ °C^{-1}$. Hence, for a Pt-100 RTD, we can write Eq. (3.7) as

$$R = 100\big(1 + 3.85 \cdot 10^{-3} T\big) = 100 + 0.385T \tag{3.8}$$

[3] Germanium and silicon have negative temperature coefficients, but they are not 'metals', they are 'metalloids', and there are also non-metals with negative temperature coefficients (like carbon).

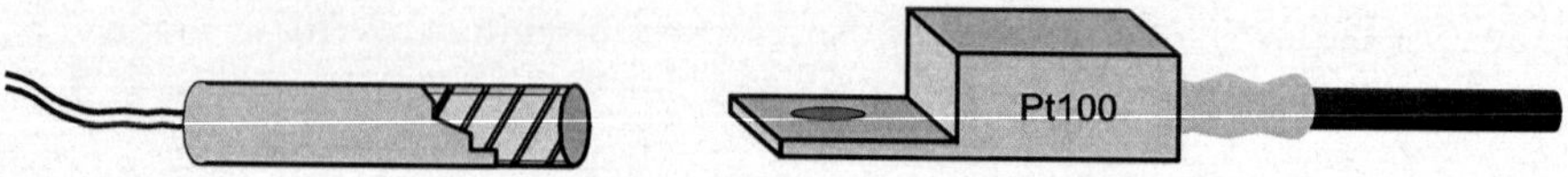

Fig. 3.9 RTDs come in different shapes

From Eq. (3.8), we can see that the sensitivity is 0.385 Ω/°C (and hence ten times higher for a Pt-1000 RTD). Platinum RTDs are mainly used in the temperature range −50 °C to +500 °C. They can operate outside this range, but outside this range, thermocouples typically perform better.

A Pt-100 RTD is usually made of thin platinum wires that are wrapped around some heat-resisting material and then encapsulated in a protective housing, see Fig. 3.9. They have become very popular in industrial applications in ranges below 600 °C because of their excellent accuracy and long-term stability.

RTDs are also often used to measure the cold junction temperature in thermocouple applications.

RTDs are more accurate and reliable than thermocouples, but they have one inherent disadvantage compared to thermocouples; the output quantity is *resistance*, not voltage, and we simply have better instruments to measure voltage compared to resistance. Measuring resistance can be a little precarious and needs to be done carefully. Since this is an especially important aspect of RTDs, we investigate the details in the next section.

3.2.3 Measuring Resistance

Figure 3.10 illustrates the simplest way to measure resistance using a common, portable DMM. Handheld DMMs usually only have two connectors and, when you set the unit selector knob to 'Resistance', it will generate a probing current *and* measure the voltage drop across the external resistance *using the same two connectors*, see Fig. 3.10.

From Fig. 3.10, it is obvious that the resistance measured will also include the resistance of the wires. That may or may not be a problem, it depends on the size of the resistance of the wires and the sensor and on the required accuracy. The following example will illustrate this.

Example 3.1 In a temperature measurement, a Pt-100 RTD is placed five meters from the DMM, and a TP cable is used where the copper wires' cross-sectional area is 0.25 mm^2. What will the error in the temperature measurement be due to the contribution from the wires?

Solution The resistivity of copper is $1.77 \cdot 10^{-8}$ Ωm. The resistance of one wire is

$$R_{\text{wire}} = \rho \frac{L}{A} = 1.77 \cdot 10^{-8} \frac{5}{0.25 \cdot 10^{-6}} = 0.354\,\Omega$$

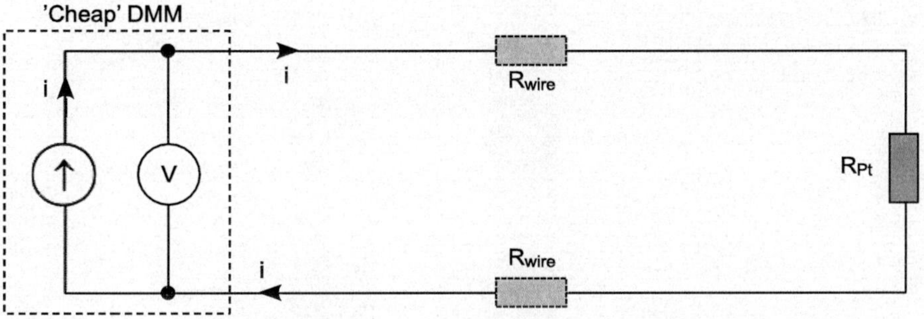

Fig. 3.10 The 2-wire method

And since we have two wires, the total contribution from the wires is 0.708 Ω. Since the sensitivity coefficient of a Pt-100 RTD is 0.385 Ω/°C, 0.708 Ω corresponds to a temperature error of 0.708/0.385 = 1.8 °C. (It would be 18 °C for a Pt-1000!)

The error of 1.8 °C in the previous example, may or may not be a problem, but consider that the distance between the sensor and the DMM was 'only' five meters. In a physics lab, it can be considerably longer. In some labs, the sensor is inside a vacuum chamber in another room. Second, it depends on the application. If you just heat an oven, it might not matter much, but if you try to control a boiler, a few degrees are critical.

To avoid the problem, you must use the *4-wire method*. This is illustrated in Fig. 3.11.

In Fig. 3.11, the current source and the voltage meter have separate wires all the way to the sensor. At a first glance, this may not seem to improve things; we just introduced two more wire resistances. However, if you analyze the two current circuits in Fig. 3.11, you see that there will be no current in the inner circuit, $i_V = 0$ A. The reason is that the inner circuit has a voltage meter in series and voltage meters have *very* high impedance. Hence, there is no voltage drop across the R_{wire} resistances in the inner circuit and the voltage meter will only measure the voltage drop across the RTD. The R_{wire} contributions are effectively eliminated.

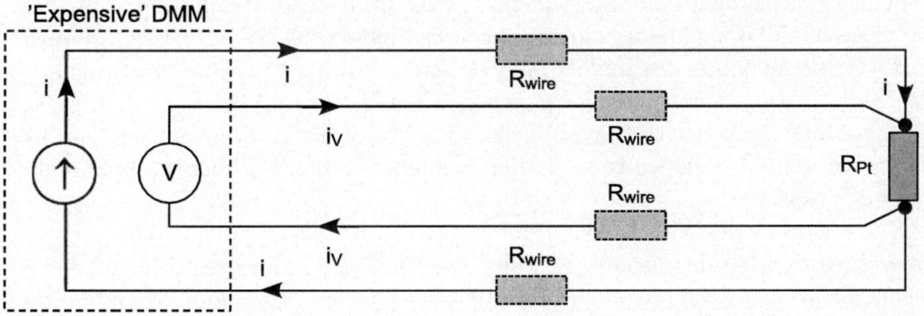

Fig. 3.11 The 4-wire method

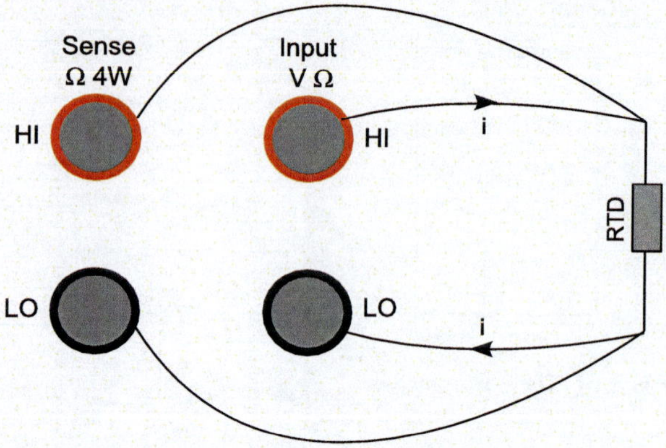

Fig. 3.12 Keysight's/Agilent's DMM interface for 2-/4-wire measurements

If we compare Figs. 3.10 and 3.11, we can also see the disadvantage of the 4-wire method; apart from two extra wires, it also requires an 'expensive' DMM. To do a 4-wire resistance measurement you almost certainly need a desktop DMM, like the popular Agilent/Keysight 344xx model. The front panel resistance interface of this DMM is illustrated in Fig. 3.12. The probe current comes out of the right-hand side pair of connectors and the left-hand side 'Sense' pair measures the voltage. Notice the '4W' label. Most students think this means '4 Watts'. Now you know better.

3.2.4 Bandgap Sensors

Thermocouples and RTDs are extremely popular in physics labs, but they are not the most common temperature sensors when it comes to commercial applications outside the laboratory or in industrial applications. An inherent disadvantage is that they are not semiconductors and cannot be integrated into a silicon wafer, which is a typical demand in commercial products. Also, commercial products seldom require the extreme ranges offered by thermocouples and RTDs.

In commercial applications, the 'bandgap' sensor is very popular. The basic principle behind a bandgap temperature sensor is that the forward voltage of a pn junction (a silicon diode) is very temperature-dependent (approximately −2 mV/°C, compare that with thermocouples in Table 3.1). In principle, you could just bias a silicon diode but that is not recommended. The expression for the forward voltage's dependence on the temperature is overly complicated [1] and it has a disadvantage; just like the thermocouple, it depends on a reference temperature. For that reason, another approach is used.

The forward voltage also depends on the current used. Therefore, two pn junctions are used with different currents, and then the *difference* in forward voltages will be independent of any reference temperature. These devices are sometimes referred to as PTATs, Proportional To Absolute Temperature.

Bandgap sensors can be integrated in silicon and are used in applications up to 200 °C. Other advantages are that they are inexpensive and very sensitive.

3.2.5 *Cryogenic Temperatures*

At cryogenic temperatures (<-153 °C), you need to take extra care to use the right sensor. Some of the sensor technologies described above can be used also at cryogenic temperatures but remember that also the housings must be able to endure the stress implied by the extremely low temperatures. Having said that, silicon diodes can be designed for temperatures down to 1.5 K and some thermocouples can also be used at cryogenic temperatures (type N and T). A special cryogenic RTD has been designed using a platinum/cobalt alloy which can be used down to 1.4 K. In environments that include magnetic fields, ruthenium oxide RTDs are recommended.

3.2.6 *Extremely High Temperatures*

At the other end of the scale, we have extremely high temperatures (> 800 °C), that need to be measured in for example metal processing and plasma physics. There are some thermocouples that can be used up to 1800 °C (type B and type S), but above 1800 °C, *pyrometers* are used.

The word 'pyro' is Greek for 'fire'. Pyrometers are based on two classic laws of physics. All objects warmer than 0 K emit a broad spectrum of infrared radiation. Wien discovered that the wavelength peak of this radiation *decreases* when the temperature *increases,* and Stefan–Boltzmann discovered that the total energy that is emitted from the object (per surface area and unit time) increases rapidly with increasing temperature ($\sim T^4$). Both these laws can be observed in Fig. 3.13.

Since it is easier to measure the increase in emitted energy rather than finding the wavelength peak, it is Stefan–Boltzmann's law that is usually used in pyrometers. Pyrometers are non-contact devices that measure the energy emitted from a black body by focusing the infrared light onto a *thermopile*. A thermopile consists of several thermocouples connected in series, see Fig. 3.14.

The infrared light is absorbed by the material in the hot junction layer, and this will heat the material; the temperature of the hot layer junction will be proportional to the radiation intensity which, according to Stefan–Boltzmann's law, is proportional to the temperature. The thermopile consisting of N thermocouples in series produces a thermo emf that is N times the emf of a single thermocouple. Commercial pyrometers with a resolution of 0.1 °C are available.

One problem that needs to be addressed when you use a pyrometer is the *emissivity* of the object whose temperature you want to measure. The emissivity is a number between 0 and 1 and reflects the object's effectiveness in emitting energy as thermal radiation. Pyrometers are typically calibrated for emissivity = 1 (i.e., a 'black body'),

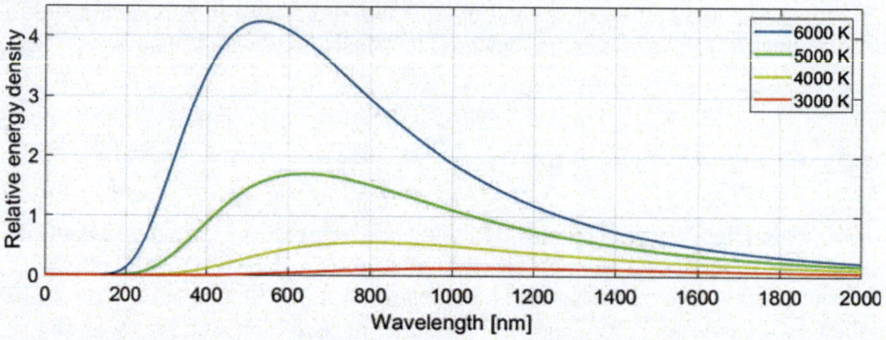

Fig. 3.13 Energy emitted from black body. Notice that the wavelength peak shifts left with increasing temperature and that the total energy emitted (the area under the curves) increases with temperature

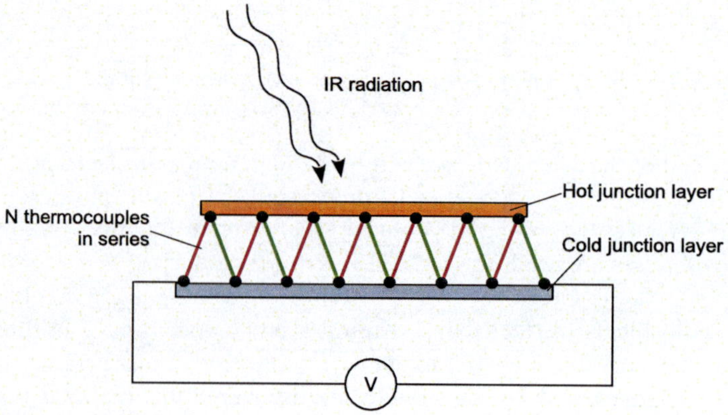

Fig. 3.14 A thermopile

but the emissivity of a shiny metal surface can be as low as 0.1. On the more advanced pyrometers, you can set the emissivity number.

One more thing: Pyrometers do not work through glass, so even if there is a window in your vacuum chamber, you cannot use it to read the temperature of an object inside with a pyrometer. If that is what you need to do, the window must be made of an infrared transparent material such as silicon or sapphire, depending on the temperature range. Potassium bromide has a very wide transparency range in the infrared but is more expensive.

Some labs use 'disappearing-filament' pyrometers where you simply heat a filament until its color matches that of the object (they take advantage of Wien's law rather than Stefan–Boltzmann's law). This could be a good alternative if you need to measure for example the temperature of a filament inside a vacuum chamber that only has plain glass windows.

3.3 The Strain Gauge Principle

3.3.1 Strain Gauges

The resistance of a conductor with cross-sectional area A and length L is

$$R = \rho \frac{L}{A} \tag{3.9}$$

where ρ is the resistivity of the conductor material. If the conductor is subjected to some tension, for example, if we pull both ends, see Fig. 3.15, then the resistance will change.

The resistance will change because the parameters in Eq. (3.9) are affected by the tension; the conductor will be a little longer ($L = L_0 + dL$), the area will decrease ($A = A_0 - dA$) and for some materials even ρ will change ('piezoresistive' materials). The most common material used in sensors ('strain gauges') is Constantan (yes, the same alloy that we use in some thermocouples). Constantan is used because it has a low-temperature coefficient and high 'strain sensitivity', i.e., the resistance changes a lot when it is subjected to strain or tension.

For Constantan, it is mostly a change in the length that is causing the change in resistance. The 'gauge factor' is defined as the quotient between the relative change in resistance and the relative change in length:

$$k = \frac{dR/R}{dL/L} \tag{3.10}$$

(dL/L is, by definition, the 'strain', hence the name 'strain gauge'.) For Constantan, k is approximately 2. Instead of a circular conductor, as indicated in Fig. 3.15, a strain gauge is made from a thin foil that is folded back and forth and then placed between two substrates, see Fig. 3.16.

To use it in an application, it must be glued to the object. The gluing is important; you must use a special glue to make sure the strain gauge is subjected to the same strain as the object. And herein lies the problem. Assuming the object whose strain we are trying to measure, is not made of Constantan (it never is, think 'car chassis') then the object will not be of the same material as the strain gauge and that implies that the object and the strain gauge do not have the same temperature coefficient. If the temperature changes, the object, and the strain gauge will not expand/

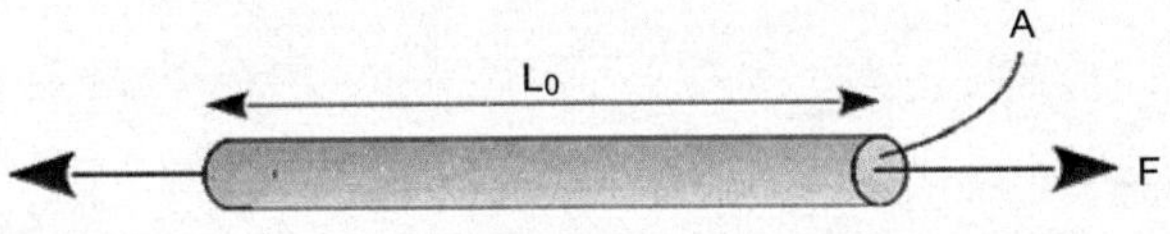

Fig. 3.15 Conductor subjected to tension

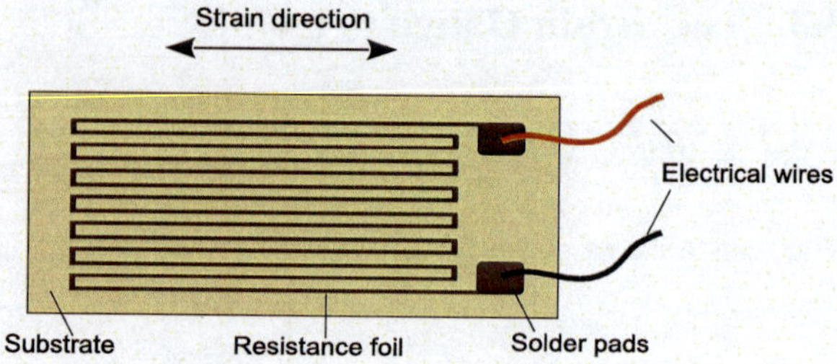

Fig. 3.16 Strain gauge. They come in a huge range of dimensions, but a typical sensor element is 1 × 2 cm

contract equally, and since they are glued hard together, that difference in expansion/contraction will cause a strain in the gauge. This is called a 'false' strain or an 'ostensible' strain.

That means that if we only measure the resistance of the strain gauge, there is no way to tell if a change in resistance is caused by a 'real' strain or a 'false' strain. We must be smarter than that.

The trick is to apply the strain gauges pairwise so that they 'counteract'; when there is a 'real' strain, one is stretched and the other one is compressed. By subtracting the resistance of one from the other, not only do we eliminate 'false' strain we also amplify the signal by a factor of 2. Figure 3.17 illustrates an example.

First, if we assume that the resistance of the gauge 'at rest' is R_0, we can write the resistance as

$$R = R_0 + dR = R_0\left(1 + \frac{dR}{R_0}\right) = R_0(1 + \Delta) \tag{3.11}$$

where $\Delta = dR/R_0$ is the relative change in resistance. In Fig. 3.17, we have glued one strain gauge on the top side of the girder and one on the bottom side. If we subtract them, we get

$$R_m = R_0(1 + \Delta_1) - R_0(1 + \Delta_2) = R_0(\Delta_1 - \Delta_2) \tag{3.12}$$

When the girder is subjected to a force, it will bend downwards and that will *stretch* the gauge on the top side and *compress* the gauge on the bottom side. We will assume here that the gauge on the bottom side is compressed as much as the gauge on the top side is stretched. That means that the gauge on the top side has a positive

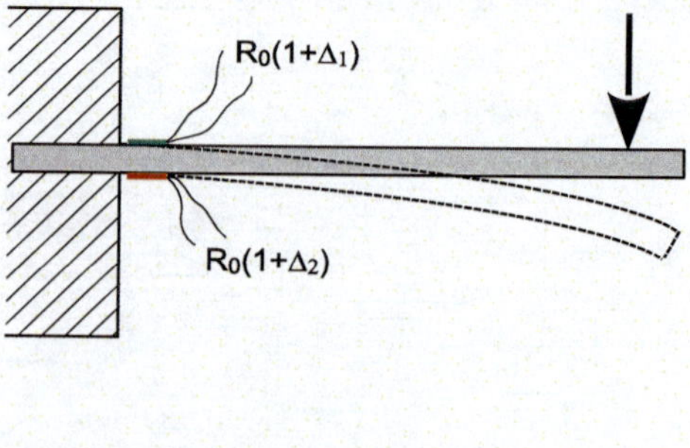

Fig. 3.17 Strain gauges are applied pairwise

Δ and the one on the bottom side has a negative Δ; $\Delta_2 = -\Delta_1$. Hence, for a 'real' strain, Eq. (3.12) becomes

$$R_m = R_0(\Delta_1 - (-\Delta_1)) = 2R_0\Delta_1 \tag{3.13}$$

On the other hand, if we have a 'false' strain due to a temperature change, both gauges will be stretched/compressed in the same direction, i.e., $\Delta_2 = \Delta_1$, and Eq. (3.12) becomes

$$R_m = R_0(\Delta_1 - \Delta_1) = 0 \tag{3.14}$$

By applying the gauges pairwise in a counteracting way and then subtracting their resistances, we get a reading that is independent of variations in temperature.

However, we are still measuring resistance and we would really like to measure voltage. There are two reasons for that. First, we can measure voltage more accurately than resistance, and second, we can amplify voltage.

So, if we would make a wishing list, we would like to have an electric circuit that can produce a *voltage* that is proportional to $\Delta_1 - \Delta_2$, i.e., a circuit that produces

$$u_m = k(\Delta_1 - \Delta_2) \tag{3.15}$$

That would be a *voltage* that only reacts to 'real' strains and would produce 0 V for 'false' strains. Does such a circuit exist? Yes, it does. It may be one of the most common circuits in electrical measurements. It is called the *Wheatstone bridge.*

3.3.2 The Wheatstone Bridge

Figure 3.18 illustrates the Wheatstone bridge where the strain gauges from Fig. 3.17 have been connected in the upper branch of the bridge.

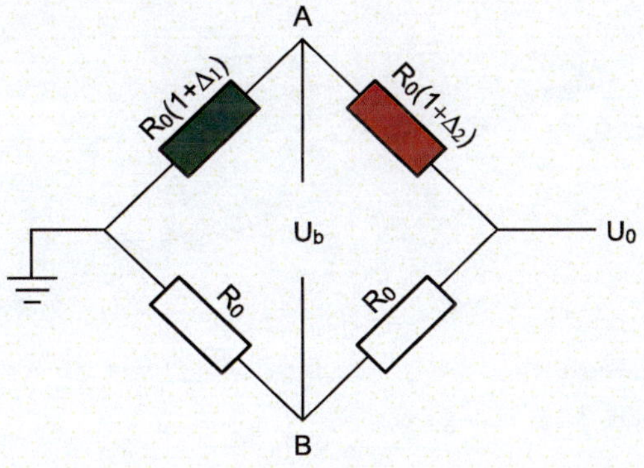

Fig. 3.18 The Wheatstone bridge

The 'bridge voltage' is the potential difference between points A and B:

$$u_b = U_A - U_B = \frac{R_0(1+\Delta_1)}{R_0(1+\Delta_1)+R_0(1+\Delta_2)}U_0 - \frac{R_0}{R_0+R_0}U_0 =$$
$$= \left(\frac{1+\Delta_1}{2+\Delta_1+\Delta_2} - \frac{1}{2}\right)U_0 = \frac{2+2\Delta_1-2-\Delta_1-\Delta_2}{2(2+\Delta_1+\Delta_2)}U_0$$
$$= \frac{U_0}{2}\cdot\frac{\Delta_1-\Delta_2}{2+\Delta_1+\Delta_2}$$

Now we will make an approximation. Δ is the relative change in the resistance and it is *very small.* (Pull a copper wire and try to imagine how much it expands…) Δ is of the order of permille (‰). That means that the denominator above, $2 + \Delta_1 + \Delta_2 \approx 2$, and we can write the bridge voltage above as

$$u_b = \frac{U_0}{4}(\Delta_1 - \Delta_2) \tag{3.16}$$

The bridge voltage is proportional to the difference in relative change in resistance, and hence it does not react to 'false' strains due to changes in temperature.

The bridge in Fig. 3.18 is a 'half-bridge'. In a 'full bridge', we apply four strain gauges to the girder, two on each side, see Fig. 3.19.

Figure 3.20 illustrates how the gauges are connected in the Wheatstone bridge. The bridge voltage in Fig. 3.20 is two times higher than in Eq. (3.16).

At first sight, the beam with strain gauges in Fig. 3.17 might appear to have a limited number of applications; you can measure the strain in a cantilever, but how often do you need to do that? Well, that is just wrong. This 'strain gauge principle' is one of the most versatile and common sensor techniques in electrical measurement systems. It is used to measure a wide range of physical quantities, like acceleration, position, pressure, torque, viscosity, flow, humidity, etc. One of the reasons it has become so popular is the emergence of MEMS technology (MicroElectro Mechanical Systems); the beam with the four strain gauges can be miniaturized to sub-mm scales. Implementing strain gauges in semiconductor material also makes it possible

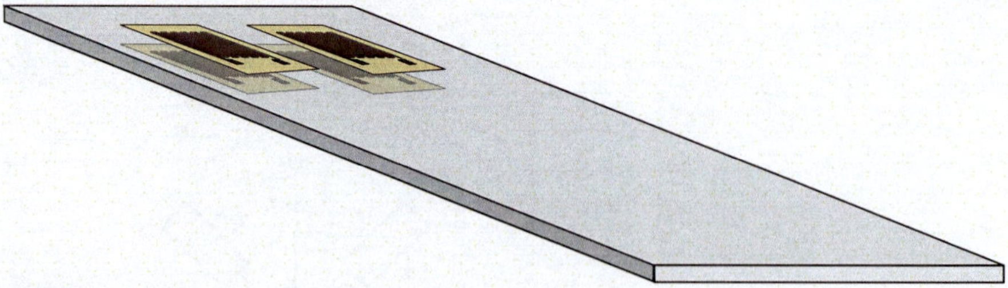

Fig. 3.19 Four gauges

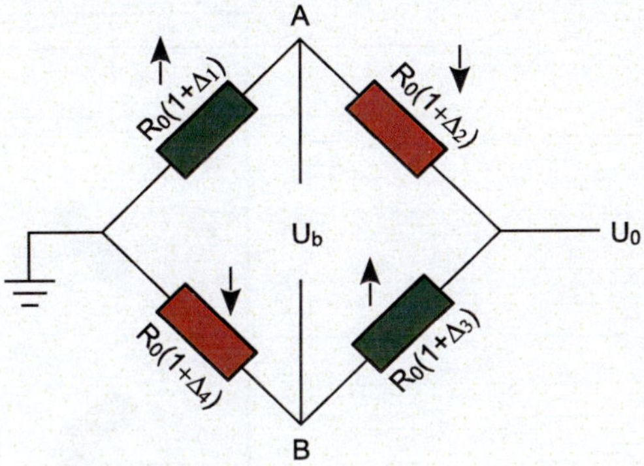

Fig. 3.20 A full bridge

to integrate them on silicon. Below we will present several common applications of the strain gauge principle to give you an idea of the versatility.

3.3.3 Accelerometers

Figure 3.21 illustrates an accelerometer based on the strain gauge principle. A miniature cantilever with four strain gauges is placed in an isolated housing. To increase the sensitivity, a 'seismic mass' is placed on the end of the cantilever and the entire system is filled with some oil to 'damp' the system.

The strain gauges are internally connected in a Wheatstone bridge, see Fig. 3.22. As a matter of fact, this is (typically) how you can identify sensors based on the strain gauge principle; the electrical interface consists of four wires. Two should be connected to 'power' and the other two are the Wheatstone bridge voltage (that should be connected to an instrumentation amplifier, see Chap. 4).

The entire sensor is of the order of 10 mm and Fig. 3.23 illustrates a typical sensor. They are used abundantly in car crash testing and vibration monitoring and

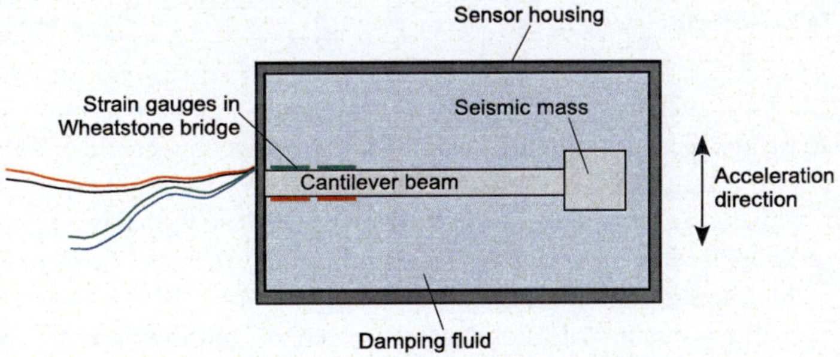

Fig. 3.21 Accelerometer

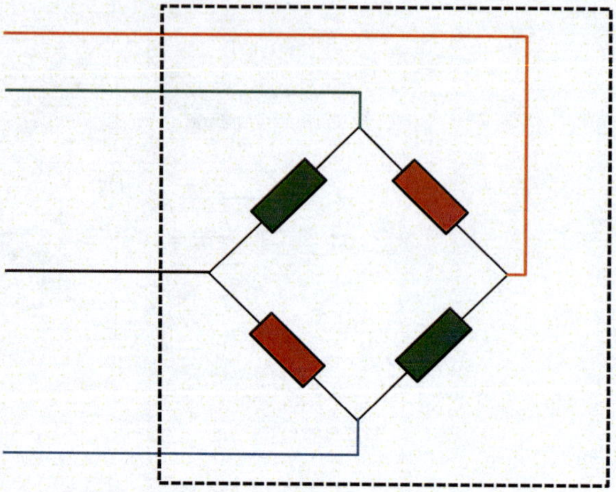

Fig. 3.22 Most likely you have a strain gauge sensor when the electrical interface consists of four wires; connect red/black to power and green/blue to the amplifier (mind the polarity)

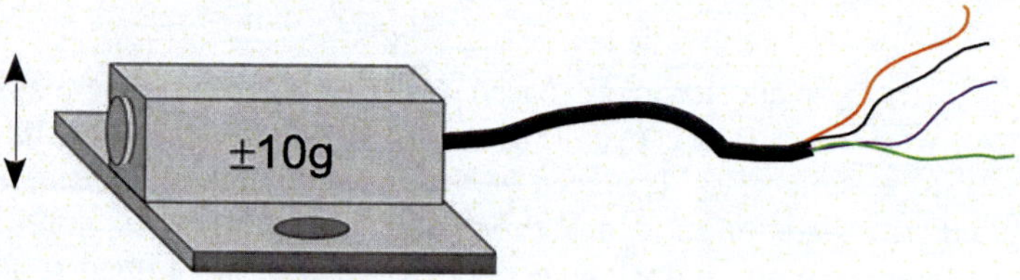

Fig. 3.23 Piezoresistive accelerometer: H × W × L = 5 × 10 × 12 mm

are available in ranges from a few *g* up to thousands of *g* and are also available in 'multi axis' versions (2- or 3-axis).

3.3.4 Pressure Sensors

In Fig. 3.24, piezoresistive strain gauges have been integrated onto a silicon membrane to form a pressure gauge. The strain gauges are connected in a Wheatstone bridge (as in Fig. 3.22); the signal interface is four wires.

The four strain gauges are placed on the silicon membrane so that two are stretched and two are compressed when the membrane is subjected to a force due to air pressure. Figure 3.25 illustrates the membrane in the sensor housing.

Figure 3.25 illustrates an 'absolute' pressure sensor, but they are also available as 'differential' sensors. Differential sensors would also have a pressure inlet on the bottom side of the sensor housing in Fig. 3.25. Figure 3.26 illustrates a differential

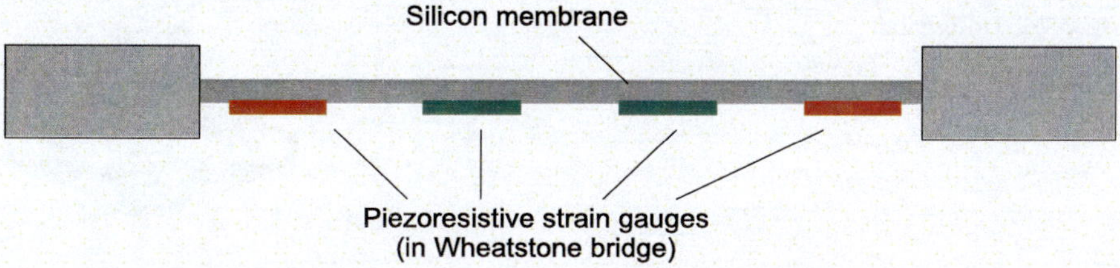

Fig. 3.24 Piezoresistive elements on silicon

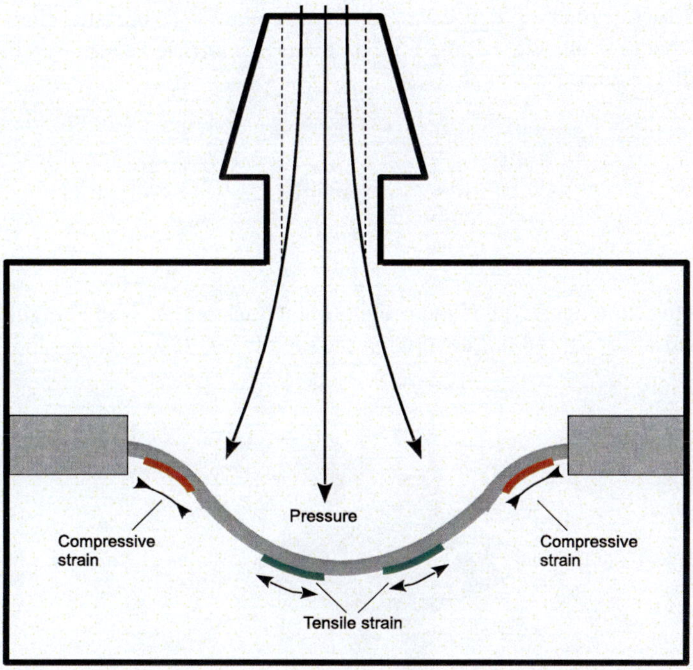

Fig. 3.25 The strain gauges are stretched/compressed due to the air pressure

pressure sensor from NXP Semiconductors. (Notice the four signal interface pins.) These sensors are available in ranges up to about 200 kPa.

3.3.5 Flow Sensors

Once we can measure pressure, we can measure flow. In this context 'flow' means volume flow, [m^3/s], and 'fluid' refers to gas or liquid (but you may assume liquid if it makes understanding easier). Flow measurements are often based on Bernoulli's

Fig. 3.26 Differential pressure gauge based on the strain gauge principle

equation (which is really a 'conservation of energy law'). To measure flow in a fluid pipe, an obstacle is introduced; the pipe diameter is narrowed down, see Fig. 3.27.

Bernoulli's equation states that

$$\underbrace{\frac{p_1}{\rho g}}_{\text{'pressure' energy}} + \underbrace{\frac{v_1^2}{2g}}_{\text{kinetic energy}} + \underbrace{h_1}_{\text{potential energy}} = \frac{p_2}{\rho g} + \frac{v_2^2}{2g} + h_2 \tag{3.17}$$

where ρ is the fluid density, g is the gravitational constant, h_x is the height and p_x is the static pressure that the fluid exerts on the pipe walls. If we assume that $h_1 = h_2$,

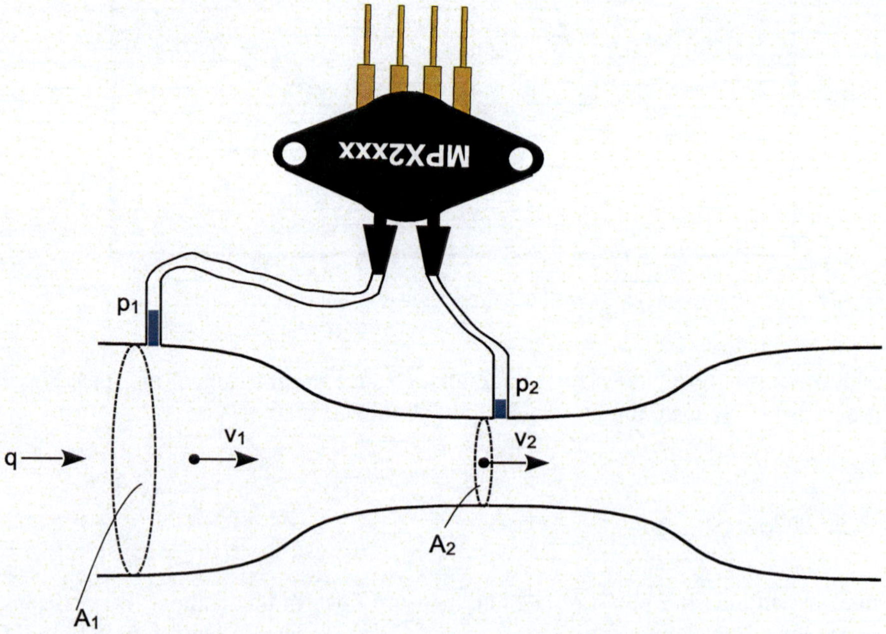

Fig. 3.27 Flow measurement (using a 'Venturi pipe')

we can write Eq. (3.17) as

$$\frac{p_1}{\rho} + \frac{v_1^2}{2} = \frac{p_2}{\rho} + \frac{v_2^2}{2} \Rightarrow \frac{1}{\rho}(p_1 - p_2) = \frac{1}{2}\left(v_2^2 - v_1^2\right) \tag{3.18}$$

In general, the relationship between the flow, q, the pipe area A and the fluid velocity v is

$$q = A \cdot v \, [\mathrm{m}^3/\mathrm{s}] \tag{3.19}$$

Since the flow q must be a constant everywhere in the pipe, we must have that

$$v_x^2 = \frac{1}{A_x^2} q^2 \Rightarrow v_2^2 - v_1^2 = \left(\frac{1}{A_2^2} - \frac{1}{A_1^2}\right) q^2 \tag{3.20}$$

Inserting Eq. (3.20) into Eq. (3.18) and solving for q gives us

$$q = \sqrt{\frac{2}{\rho\left(\frac{1}{A_2^2 - A_1^2}\right)}} \cdot \sqrt{p_1 - p_2} = k\sqrt{\Delta p} \tag{3.21}$$

From Eq. (3.21), we can see that the flow is proportional to the square root of the difference in static pressure in the pipe between the two points with different cross-section areas. Two capillary tubes are inserted at both points and the pressure difference is measured with a differential pressure gauge as illustrated in Fig. 3.27. The tube in Fig. 3.27 with the 'slender waist' is called a 'Venturi pipe'.

3.3.6 *Fluid Level Sensors*

Same thing with fluid level; once you can measure pressure, you can measure fluid level. In Fig. 3.28, a tube is placed in the (empty) vessel, and when it fills with fluid, the air trapped inside the tube is compressed and the pressure gauge will generate a signal proportional to the fluid level.

3.3.7 *Torque Sensors*

Torque is another physical quantity that is readily measured with strain gauges. Figure 3.29 illustrates a shaft, or a spindle, subjected to some torque M. The torque will induce two opposing strains in the shaft. The tensile strain is in the direction of the torque with an angle of 45° versus the shaft direction, see Fig. 3.29. The compressive strain is in the orthogonal direction, at an angle of 90° versus the 'tensile' strain. (Roll

Fig. 3.28 Fluid-level sensor

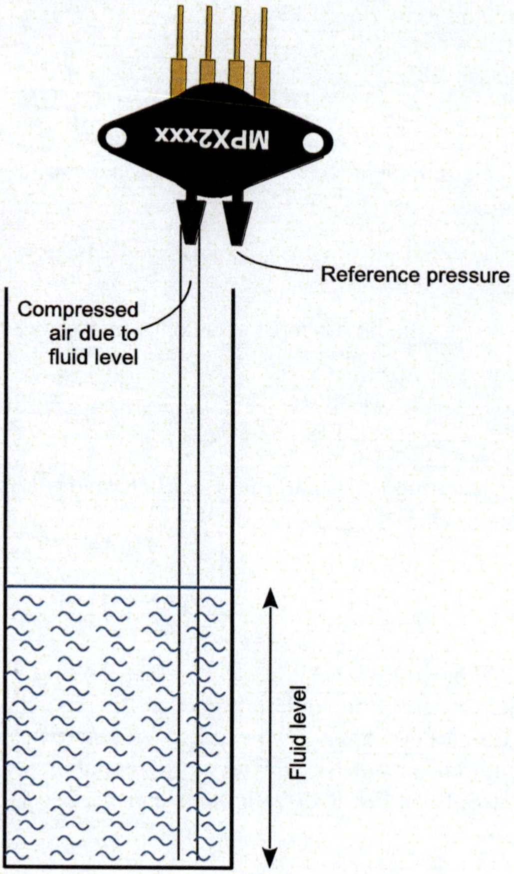

a piece of paper, twist it, and observe the tensions.) Hence, we can get counter-acting strain gauges by applying them pairwise at an angle of 90° as illustrated in Fig. 3.29. For rotating shafts, the signals are transferred to the 'outside' by sliprings.

Fig. 3.29 Measuring torque

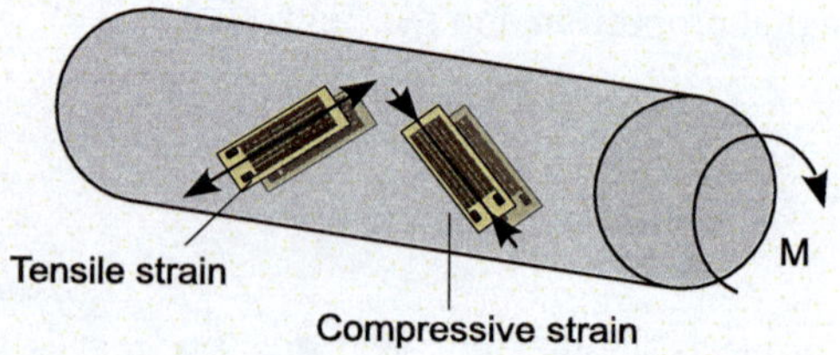

3.3.8 Viscosity Sensors

If you have a degree in physics, you have probably measured viscosity in some undergraduate lab exercise by measuring the time it takes for an object to sink in some fluid. That is fine, but it doesn't provide a 'sensor' for us. The strain gauge does; if you can measure torque, you can measure viscosity. If you rotate a propeller (at a constant speed) in the fluid, the torque exerted on the propeller shaft is proportional to the viscosity.

3.3.9 Load Cell

Figure 3.30 illustrates the load cell principle. When the 'cell' is subjected to a load the vertical strain gauges will be compressed, and the horizontal ones will be extracted, and we can again place the strain gauges in a Wheatstone bridge to get a signal that only reacts to 'real' strains and not to false strains due to a variation in temperature.

Load cells are mostly used for weighting (people, cars, and trucks) but can also be used to measure volume/level in a tank (volume/level is proportional to weight). It is also used to provide force feedback in robotic applications and rocket thrust measurements.

We could give you more examples of applications for the strain gauge principle, but we think we have made the point; the strain gauge principle is a very versatile sensor technique and omnipresent. Instead, we will present some other sensor techniques that are almost as versatile.

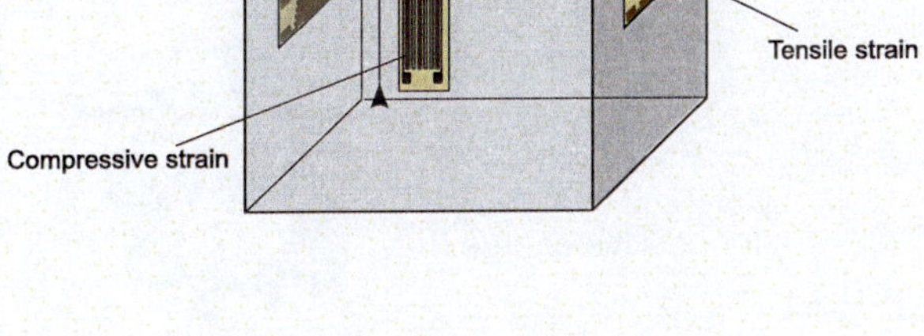

Fig. 3.30 The load cell principle

3.4 Piezoelectric Crystals

A ‘piezoelectric’ crystal has a symmetric crystal lattice of Silicon and Oxygen atoms, see Fig. 3.31. When the crystal is ‘at rest’ the positive Silicon atoms and the negative Oxygen atoms are arranged so that there is no net charge on any surface; the center of charge for positive and negative atoms coincide.

However, if the crystal is deformed by an external force, see Fig. 3.32, the atoms in the lattice are displaced and the centers for positive and negative charges no longer coincide and a net charge can be detected on the surface.

The amount of charge is proportional to the external force deforming the crystal. Just like the strain gauge case, we have a method to detect force, and we saw in the previous sections how that can be used to measure many other quantities. The charge depends very linearly on the force (just like the strain gauge arrangement), but the piezoelectric crystal has one advantage over the strain gauges; it is unaffected by temperature variations.

However, it has some disadvantages too; we need to measure the charge that is generated by the force, and that is not as straightforward as it might seem. To understand the problem, we need a signal model of the crystal; we model it as a current/charge source that produces a charge $Q = kF$ (F is the force on the crystal) and we model the crystal surfaces as a capacitor C_X, see Fig. 3.33.

That means that the voltage across the crystal is $U_X = Q/C_X = kF/C_X$, so by just measuring the voltage across the crystal would give us a number proportional to the force. The problem is that the capacitance C_X is very small and even if it is a non-conducting material, there is still some ‘isolation resistance’ between the surfaces

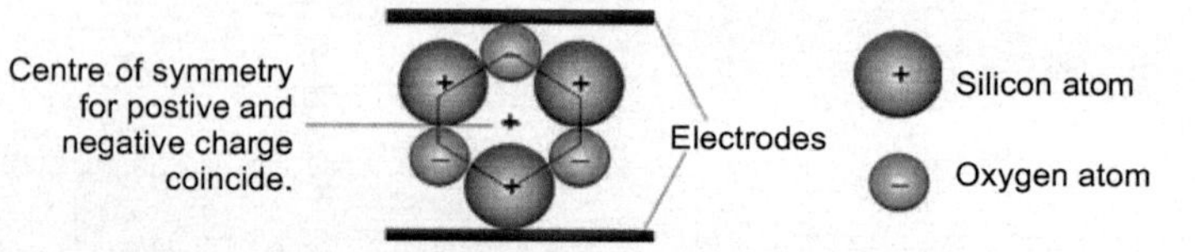

Fig. 3.31 A piezoelectric crystal

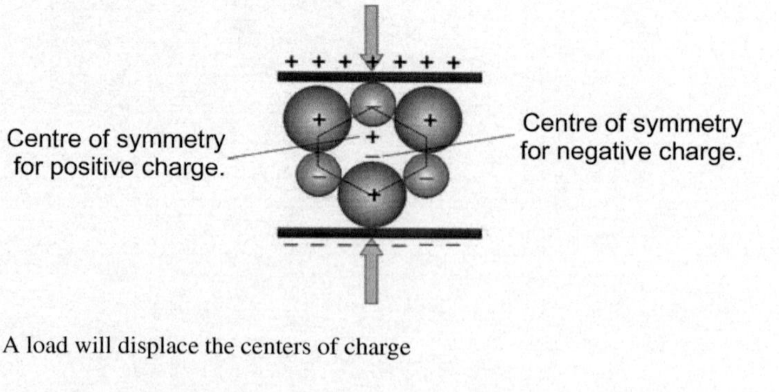

Fig. 3.32 A load will displace the centers of charge

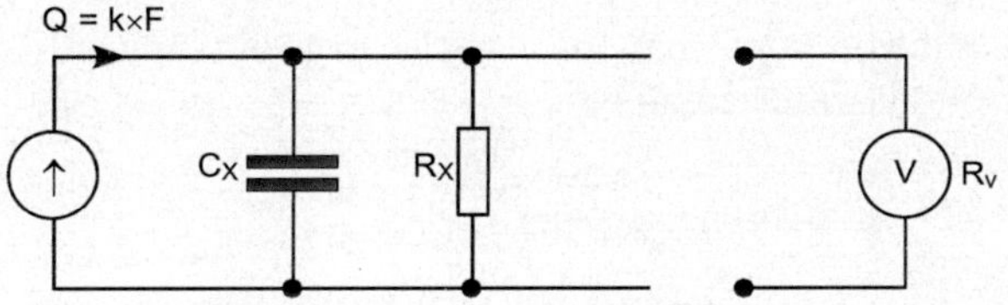

Fig. 3.33 The piezoelectric crystal signal model

(modeled by R_x in Fig. 3.33). Hence, the crystal will discharge through R_x and the voltage will decrease exponentially:

$$U_x(t) = \frac{kF}{C_x} e^{-t/\tau} \tag{3.22}$$

where $\tau = R_x C_x$. The isolation resistance is very large (10^{12} Ω) but C_x is of the order of pF, so $\tau \approx 1$ s; the voltage drops to 36% in just one second. As a matter of fact, the situation is much worse than that. To measure the voltage across the surfaces, we need to connect a voltage meter, see Fig. 3.33, and when we do that, the input impedance of the voltage meter is connected in parallel with R_x, and a typical voltage meter has an input impedance $R_v = 1$ MΩ. The consequence is that instead of 10^{12} Ω, the charge will discharge through R_v and τ in Eq. (3.22) is of the order of 1 μs! The charge is gone long before we have a reasonable chance to record it.

Measuring the charge on a piezoelectric crystal is obviously not that straightforward. The trick is to 'fool' the charge to leave the crystal surface immediately. That is what a 'charge amplifier' does, see Fig. 3.34.

First, the negative op amp input is at 'virtual' ground. That means that the charge Q created on the crystal has three paths to ground: Through C_x, through R_x, or straight forward through no impedance at all. According to Kirchhoff's current law, that means that *all* charge will go straight forward to the negative input of the op amp.

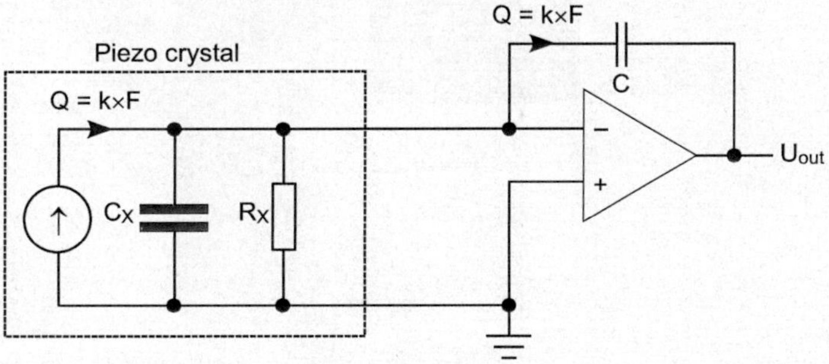

Fig. 3.34 A piezoelectric crystal with charge amplifier

Once it reaches the op amp, it has nowhere else to go but ‘upwards’ to the external capacitor C; the output voltage will be

$$U_{\text{out}} = -\frac{Q}{C} = -\frac{kF}{C} \tag{3.23}$$

It might seem like we just transferred the discharging problem from one capacitor to another, but that is wrong; C is an external capacitor that we can choose arbitrarily. It will have an isolation resistance too, but with the arrangement in Fig. 3.34, we can improve τ by many orders of magnitude.

Still, τ will not be infinite, and piezoelectric crystals are not suitable for measuring static forces, but it is a popular sensor technique in dynamic applications because of its robustness against temperature variations, linear nature, and reliability.

3.5 Hall Sensors

When a charged particle moves in a magnetic field a force will act on it and bend the trajectory according to Fleming’s right-hand rule.

In Fig. 3.35, a current is injected into a flat conductor (a metal strip) that at the same time is subjected to a magnetic field. Due to the magnetic field, the negative electrons will be forced to one side of the conductor and hence there will be a voltage across the conductor’s sides (perpendicular to the current’s direction and to the magnetic field lines). This is called the ‘Hall effect’ after Edvin Hall who discovered this phenomenon in 1879 (while working on his doctoral thesis in physics).

The voltage across the conductor is proportional to both the current and the magnetic field, which indicates that the Hall effect has two major applications: Measuring currents (i.e., charge flow) and measuring magnetic fields. In the first

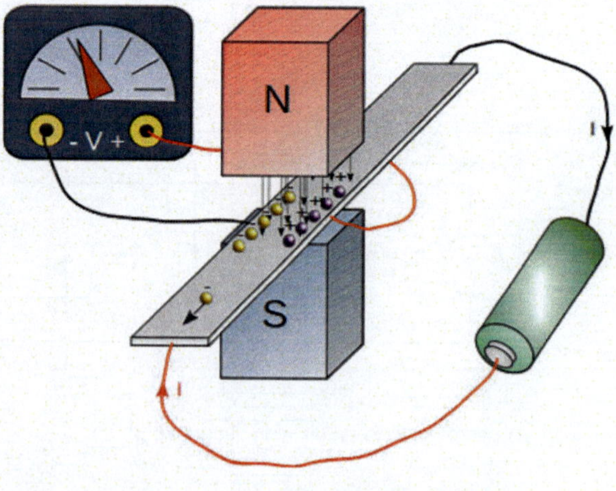

Fig. 3.35 The Hall effect principle

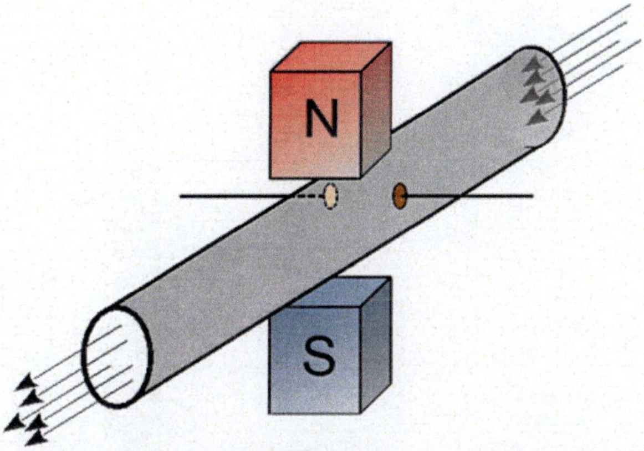

Fig. 3.36 Non-invasive flow meter

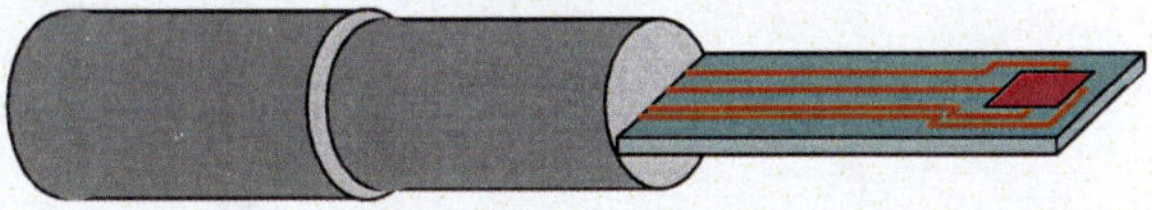

Fig. 3.37 Hall probe

case, we use a constant magnetic field and, in the other case, we use a constant current.

Figure 3.36 illustrates a non-invasive flow meter (for conducting fluids, like water). When the fluid in the flow pipe passes through the (constant) magnetic field, the charges will be forced to the sides of the tube where electrodes pick up the charge and the voltage across the tube will be proportional to the flow q. Notice the advantage compared to the flow meter in Fig. 3.27; the Hall effect flow meter is non-invasive and does not interfere with the fluid (but it only works for conducting fluids with free ions).

Figure 3.37 illustrates a Hall probe used to measure magnetic fields. A small metal plate/film is integrated on a probe stick and a constant current is sent through the plate and the voltage is measured across the plate.

3.6 Position Sensors

There is a plethora of position sensors on the market, and you really need to specify your needs in terms of accuracy, range, sensitivity, and reliability before you start searching for a position sensor. We will only cover one here; the Linear Variable Differential Transformer (LVDT). LVDTs emerged already in the 1930s to meet the need for displacement measurements in the process industry. The basic principle is illustrated in Fig. 3.38. A primary coil is wound on the same bobbin as two secondary coils and the ferromagnetic core is long enough to cover the primary coil and one of the secondary coils at both extremes. The primary coil is excited by an AC signal

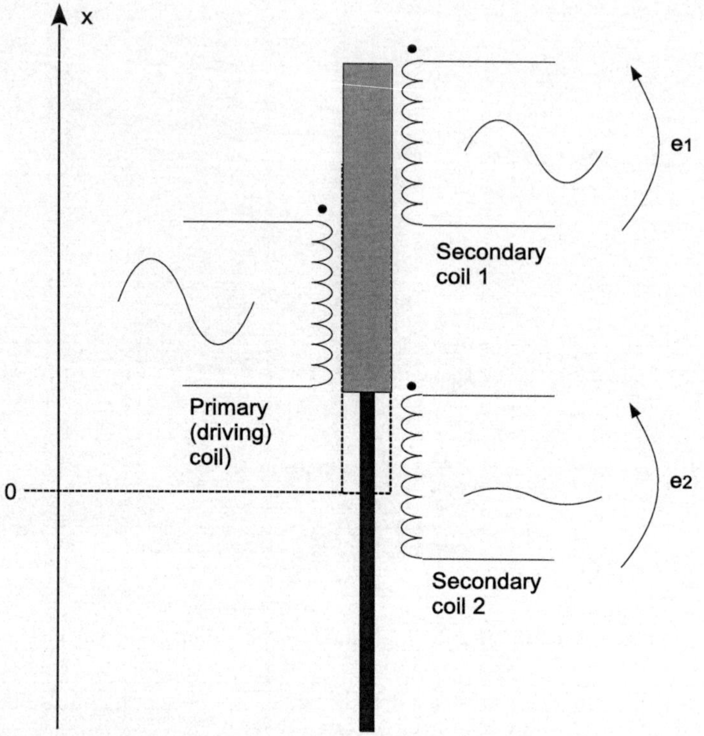

Fig. 3.38 The LVDT principle

and as the core moves from one extreme to the other, the amplitudes of the signals transferred to the secondary coils will vary linearly with the core's distance from the center position. The difference in amplitudes of the two secondary coils' signals is an absolute measure of the core's displacement from the center position.

LVDTs are very linear (better than 1%), robust, and accurate. The resolution is in the low micrometer range. The sensitivity is of the order of millivolts per millimeter ('millivolts' referring to the difference in amplitudes between the secondary coils' signals). LVDTs are not only used for displacement measurements but also as the sensing element in pressure gauges, force measurements, detection of gravitational waves, and calibration of atomic force microscopes. Hydraulic control systems and haptic robot interfaces are other areas of application.

The signal conditioning required is an excitation source for the primary coil and demodulation circuitry for the secondary coils. The amplitudes of the secondary coils' output depend on the core's position x, but will also be influenced by the core material, the excitation frequency, the temperature, and the secondary coils' design parameters (windings, length, diameter, etc.). It has been reported though, that the quotient between the difference and sum of the secondary coils' signals is independent of temperature, excitation current, and excitation frequency [2]:

$$\frac{e_1 - e_2}{e_1 + e_2} = kx \tag{3.24}$$

Equation (3.24) is the key to successful demodulation of the LVDT signal.

3.7 Photo Sensors

3.7.1 Light Units

Before, we get into photosensors, we need to define the units we use to describe light intensity. The power of a light source is in general measured in Watts [W] (= 'radial' flow), but it is common to use 'luminous' flow for light sources in the visible wavelength range. The unit is 'lumen' [lm] (which is an SI unit) and is weighted according to the human eye's response to different wavelengths. For example, a standard 40 W bulb (omnidirectional) emits 400 lm.

Then there is 'lux' [lx] which is the (perceived) light power per unit area, i.e., 1 lx = 1 lm/m^2. This is called 'illumination' or 'luminous intensity'. Direct sunlight corresponds to 50,000–100,000 lx and typical office lighting is about 300–500 lx.

Example 3.2 What is the luminous intensity of a 40 W bulb at 1 m?

Solution We know from above that it emits 400 lm. Assuming that the bulb is 'omnidirectional', at 1 m, the 400 lm is distributed over a sphere with an area of $4\pi r^2 = 4\pi 1^2 = 12.57$ m^2. Hence, the illumination at 1 m is 400/12.57 = 32 lx.

There are many kinds of photosensors, like photoresistors, photodiodes, phototransistors, etc., but here we will limit our presentation to the most common photodetectors used in a physics lab; photodiodes (of different kinds) and photomultipliers.

3.7.2 Photodiodes

If a photon of sufficient energy hits the depletion area between the p- and n-doped area in a diode, an electron–hole pair is created and because of the electric field in the depletion region, the hole will move to the anode, and the electron will move to the cathode. If the diode electrodes are part of a closed circuit, a photocurrent will occur that is (a) in the 'backward' direction and (b) proportional to the light illuminance. This is illustrated in Fig. 3.39.

A photodiode is operated in one of two different modes: The 'photovoltaic' mode or the 'photoconductive' mode. In the photovoltaic mode, the anode and cathode are kept at the same potential (sometimes called the 'zero-biased' mode). The advantage of this mode is that it minimizes the 'dark current', i.e., the self-induced current

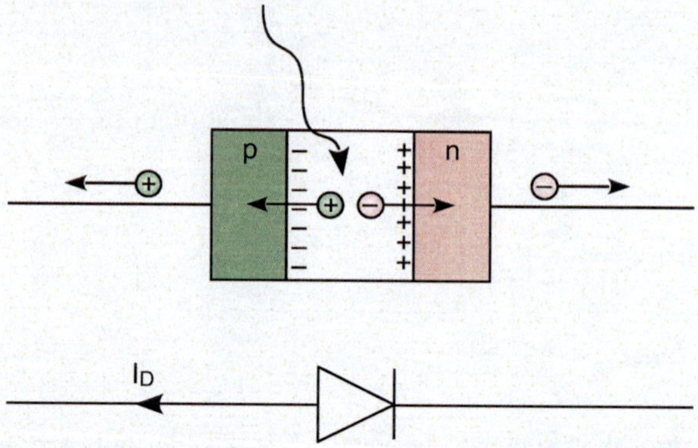

Fig. 3.39 The photodiode principle

due spontaneous creation of electron–hole pairs. Figure 3.40 illustrates an example where a photodiode is operated in the photovoltaic mode.

Figure 3.41 illustrates a photodiode operating in the photoconductive mode where the photodiode is reversed-biased (cathode is at a higher potential than the anode).

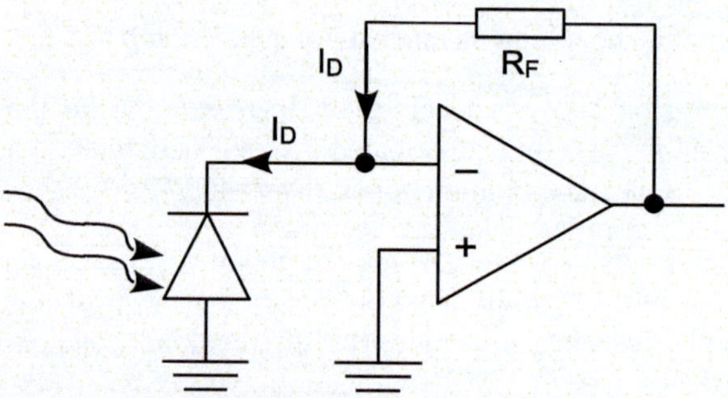

Fig. 3.40 Photovoltaic mode. Notice that the anode and cathode have the same potential. The output range is determined by the feedback resistor

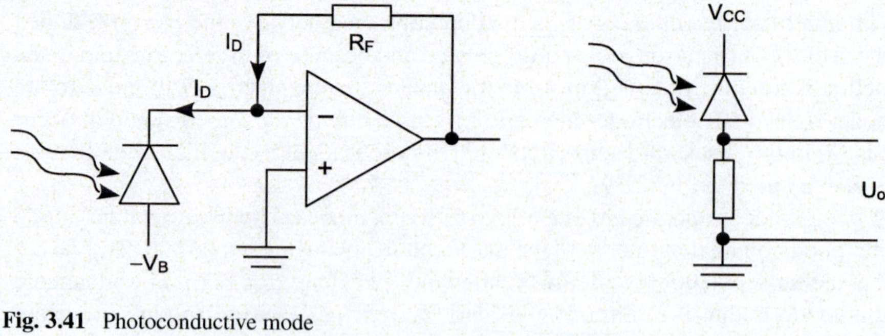

Fig. 3.41 Photoconductive mode

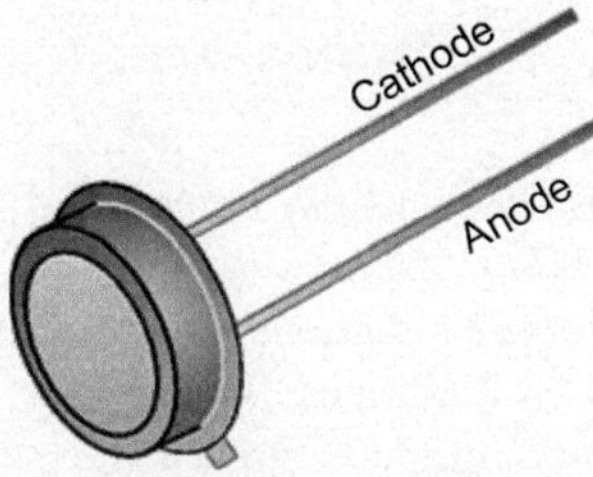

Fig. 3.42 The photodiode BPW21

The advantage of the photoconductive mode is that the reverse-biased voltage widens the depletion area and makes the diode more sensitive (more hole-electron pairs can be created) and it also improves the response time. The reason for the faster response is that when the depleted area is widened, the pn junction capacitance is decreased. The disadvantage of the photoconductive mode is that it increases the dark current.

The sensitivity of a typical photodiode is of the order of 10 nA/lx. Figure 3.42 illustrates the popular BPW21 photodiode that is optimized for green light and has a sensitivity of 9 nA/lx.

A development of the photodiode is the PIN photodiode which consists of three layers, heavily doped p- and n-layers, and a lightly doped intrinsic area. This has two advantages. First, the depletion area where electron–hole pairs can be created is increased, and the wider depletion region decreases the capacitance even more which gives a faster response.

3.7.3 Avalanche Photodiodes

Avalanche photodiodes have yet another layer (four layers) and they are operated at much higher reverse-biased voltages (near the break-down voltage). The idea of an avalanche photodiode is that when an electron–hole pair is created by incident light, the electrons are accelerated to a very high speed (because of the high reverse-biased voltage). When the electrons are accelerated through the fourth layer (a lightly doped p-layer) they will collide with the atoms and because of their high velocity, they will create new electrons (due to impact ionization). This will generate a multiplication effect and the multiplication increases with the reverse-biased voltage. The reverse-biased voltage is of the order of 100–500 V and gain factors of up to 200 can be achieved, which makes them very sensitive.

The disadvantages of avalanche photodiodes are that they are noisy, non-linear and you must handle a high DC voltage in your setup.

3.7.4 Position-Sensitive Detectors

Position-sensitive detectors (PSDs) are photodiodes that produce currents depending on *where* the incident light hits the active surface. It consists of two photodiodes with a common cathode, see Figs. 3.44 and 3.45.

As illustrated in Fig. 3.45, both photodiodes will generate a current when an incident light beam hits the surface, and the *difference* between these two currents is proportional to x (the distance between the incident light beam and the center of the active photodiode area). However, the difference between the currents also depends on the light's intensity. To get a reading that is independent of the light intensity, the manufacturer recommends that you divide by the sum of the currents:

$$\frac{I_{x1} - I_{x2}}{I_{x1} + I_{x2}} = \frac{2x}{L_x} \tag{3.25}$$

In the S3932 model illustrated in Fig. 3.43, the active area is 12 mm long. They are also available as two-dimensional detectors (x and y detectors). The common cathode is usually reverse-biased (5–10 V). Figure 3.46 illustrates the recommended signal conditioning where the currents are converted to voltages.

Once you have the voltages, you can proceed with either analog electronics to generate the sum and difference currents (see solved Problem 3.4), or you can sample them and do it in software. (Since Eq. (3.25) involves division, it is recommended that you do it in software.)

A very common application of PSDs is distance measuring by 'triangulation'. Figure 3.47 illustrates the principle.

Depending on the distance to the obstacle, the reflected light will hit the PSD in a different position relative to the center and the PSD output will be proportional to the distance to the obstacle. For example, this is used by some projectors to measure the distance to the screen (enabling auto-focusing) and by vacuum robots to detect obstacles.

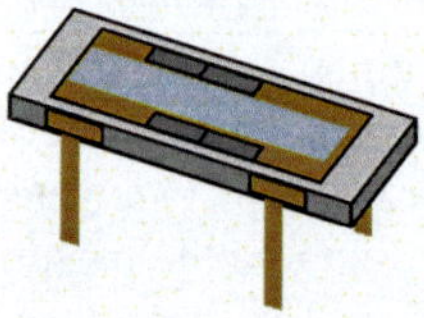

Fig. 3.43 PSD S3932 from Hamamatsu

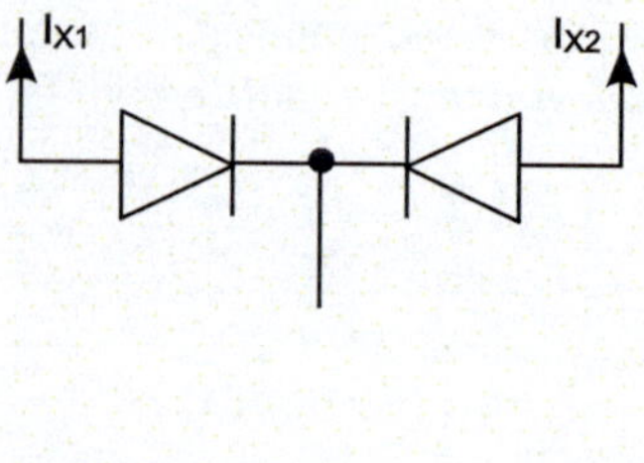

Fig. 3.44 Two photodiodes with common cathode

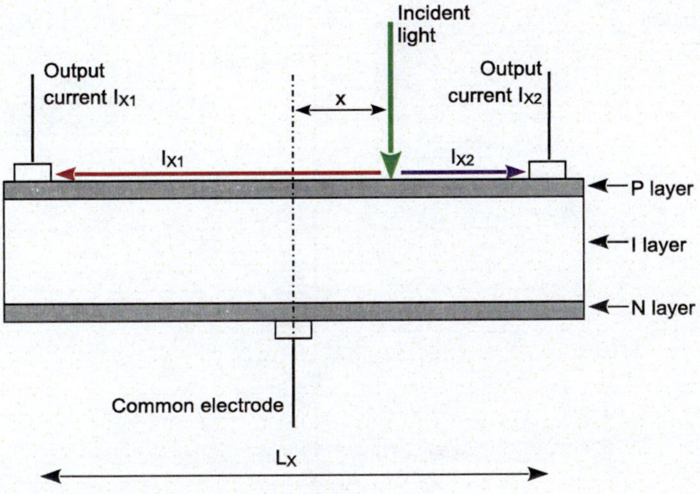

Fig. 3.45 PSD: current difference is proportional to x

Fig. 3.46 Typical signal conditioning for PSD

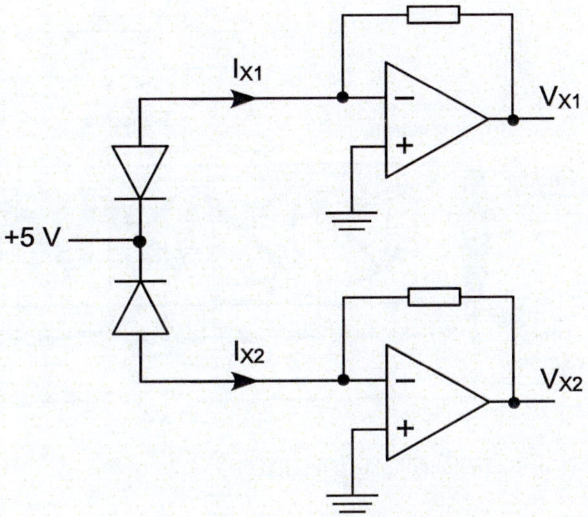

3.7.5 *Photomultipliers*

Figure 3.48 illustrates a photomultiplier tube (PMT). The front end of a PMT consists of a photocathode. When hit by a photon, an electron is emitted (a 'photoelectron'). Inside the tube are several 'dynodes' with successively higher potential: The photocathode is at approximately –1 kV and the potential of the succeeding dynodes is about 100 V higher (each). When the first photoelectron is emitted, it is accelerated

Fig. 3.47 Optical triangulation

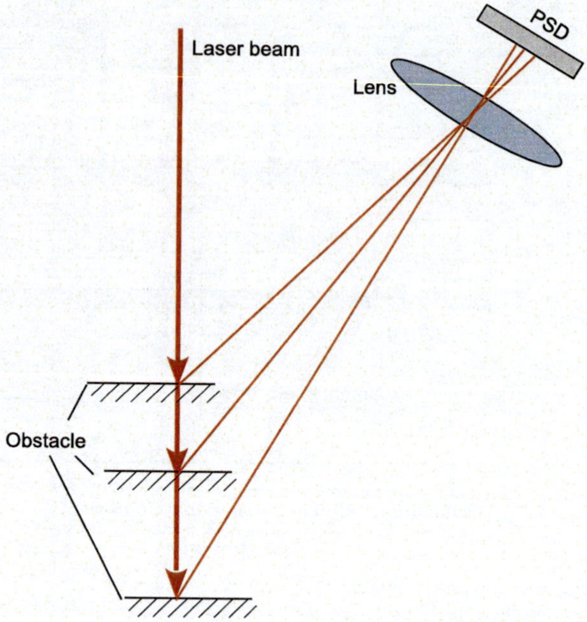

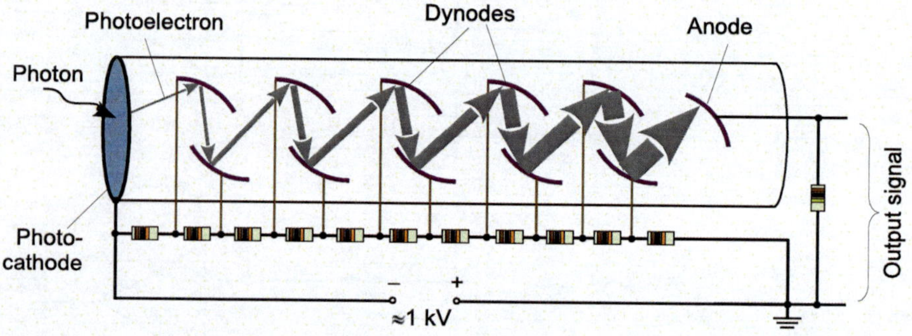

Fig. 3.48 Photomultiplier tube (PMT)

towards the first dynode. The dynodes' surface is made of a material that is specifically designed to emit secondary electrons. Common materials are AgMgO, BeO, or GaAsP. A typical PMT has 12–14 dynodes and the quantum efficiency[4] can be as good as 30% and gains of 10^6 are common, which means that PMTs can be used for photon counting (single photon detection).

[4] The quantum efficiency is the fraction of incident photons that generate a primary electron emission from the photocathode.

PMTs are delicate instruments. First, the tube is evacuated, and the sealing must never be broken. Second, they should *never* be exposed to daylight, even when not powered. They are also very expensive and suffer from high dark currents.

3.8 Particle Detectors

3.8.1 Channel Electron Multipliers

Channel Electron Multipliers, CEMs (or 'channeltrons') are particle detectors that use a similar electron multiplication technique as PMTs. CEMs come in many different sizes and shapes, and Fig. 3.49 illustrates a common standard CEM.

It is primarily used to detect ions (positive or negative) but can also be used to detect electrons or photons. The incoming ion hits a collector cone coated with a high secondary electron emission material and then electrons are accelerated backwards by an electric field and multiplied as they propagate through the channel to the end collector. Unlike the PMT, the CEM does not have discrete dynodes, instead it has one single 'continuous dynode' which creates a continuous electric field inside the channel, forcing the electrons to propagate towards the anode end.

The inside channel is approximately 1 mm in diameter. Notice in Fig. 3.49 the incurvation of the channel. This is necessary to prevent 'ion feedback'. Because of the multiplying effect, the electron density can be very high at the channel output, and this can cause adsorbed gases on the channel wall to desorb and ionize. This results in positive ions in the channel propagating back to the input, producing extra secondary electrons which generate noise in the output signal. The incurvation prevents the ions from gaining enough energy to produce secondary electrons. Without the incurvation, the gain would be limited to $<10^5$. Curved CEMs can have a gain factor of 10^8.

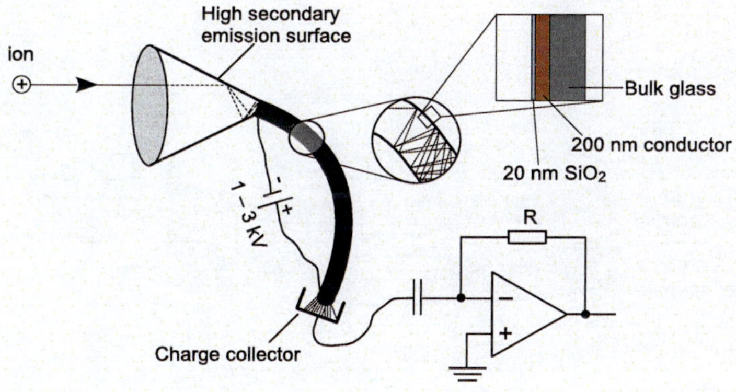

Fig. 3.49 Channel electron multiplier

3.8.2 Microchannel Plates

Microchannel plates (MCPs) are a development of CEMs. As the name indicates, MCPs consist of several (parallel) channels, which provide a spatial resolution (provided that the charge collector is designed for that). Figure 3.50 illustrates a single MCP.

An MCP is typically 0.5–2 mm thick and the diameter can vary from 10 mm up to 200 mm. Each microchannel's diameter is 5–20 μm in diameter (less than a human hair). Just like the CEM channels, the inside of the channels is covered with a material with a high secondary electron emissivity. The top and bottom sides are high-voltage biased to produce an electric field across the continuous-dynode of the order of 10^6 V/m. Still, the gain of a single MCP is only of the order of 10,000. For that reason, two (or more) MCPs are stacked to improve the gain, see Fig. 3.51.

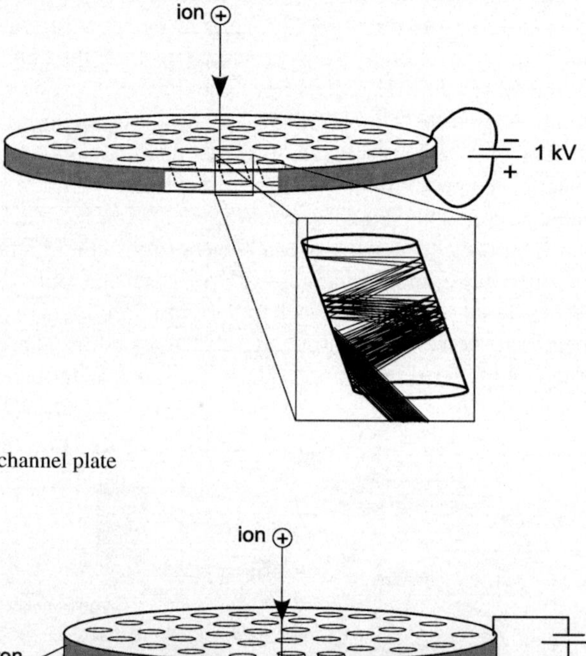

Fig. 3.50 Microchannel plate

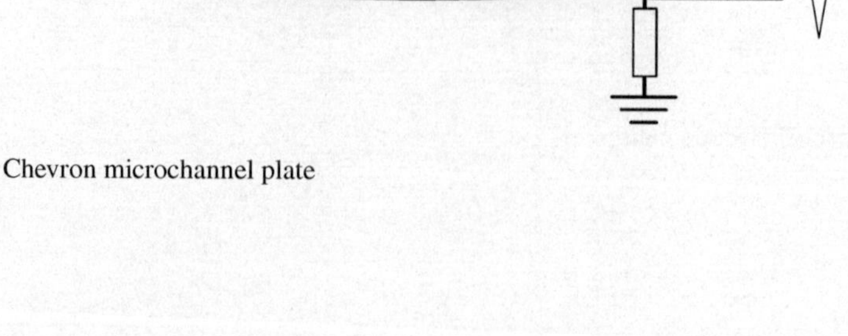

Fig. 3.51 Chevron microchannel plate

Notice in Fig. 3.50 how all the channels are tilted by a small angle against the normal (8–13°). This is to guarantee that an incident particle hits the channel wall and initiates an electron avalanche (instead of just passing straight through). Figure 3.51 illustrates a 'Chevron' MCP.

A Chevron MCP consists of two microchannel plates where the channels' angles are in 'opposite' directions. This has two advantages. First, with two MCPs in series, gains up to 10^7 are possible. Second, the 'opposing' channel angles reduce ion feedback and reduce the noise in the output signal. 'Z' MCPs stack three microchannel plates.

MCPs are used as particle detectors in mass spectrometers and space-based instruments for detection of photons and high-energetic particles, but because of their spatial resolution capacity, they can also be used as intensity amplifiers in night-vision googles.

3.9 Vacuum Gauges

3.9.1 *Introduction*

Vacuum chambers are omnipresent in physics labs, and where there are vacuum chambers there is a need to monitor the pressure, i.e., we need a vacuum gauge. But before we get into the details of vacuum gauges, let's first talk about pressure units. Vacuum/pressure is one of the areas where the use of the SI unit (Pa, Pascal) is not necessarily the most common unit. There is a plethora of vacuum pressure units, and the preference depends on the context. 'Context' can refer to region and/ or application. My personal (prejudiced?) opinion is that Europeans use 'mbar', Americans use 'Torr' and Asians use 'Pa'. Fortunately, 1 mbar is approximately equal to 1 Torr (1 Torr = 1.333 mbar), and the pressure in vacuum chambers are usually only measured in orders of magnitude anyway, so, if the pressure in your vacuum chamber is 10^{-6} Torr or 10^{-6} mbar, doesn't matter, it is the same (for all we usually care). I will use mbar in this presentation.

We distinguish between 'low' vacuum (LV), $\geq 10^{-3}$ mbar and 'high' vacuum (HV) $\leq 10^{-3}$ mbar. This is the only distinction we need when it comes to choosing a vacuum gauge. For LV, you use a 'thermal conductivity gauge' (called a 'Pirani' gauge) and for HV you must use a 'gas ionization' gauge.

3.9.2 *The Pirani Gauge*

There are several ways to implement a thermal conductivity pressure gauge, but the most common one is the 'Pirani' gauge. In a Pirani gauge, a thin filament (usually Platinum) is heated to approximately 50 °C. We know already (from Sect. 3.2.2) that the resistance of a metal filament depends on the temperature. When the filament is heated, some of the heat will be dissipated to the ambient gas (the air) and this will cool the filament (and cooling means decreasing resistance). The heat is carried away by conduction and the gas's ability to carry heat is called 'thermal conductivity'. The thermal conductivity is proportional to the gas density, which of course depends on the pressure. When the gas is evacuated, the thermal conductivity is reduced and the cooling effect on the filament is reduced, and the wire gets warmer and therefore the resistance increases. In conclusion, the filament resistance goes *up* when the pressure *decreases.*

A Pirani gauge consists of two filaments. Both are heated by the same current, but only one is subjected to the vacuum chamber atmosphere. The other one is a reference gauge that compensates for variations in the ambient temperature. The reference gauge is sealed (under a 'reference' vacuum). Figure 3.52 illustrates a Pirani vacuum head.

Hence, to measure the vacuum, we must measure the resistance of the filament (relative the reference filament) and we know already that this is best implemented in a Wheatstone bridge. However, there are several ways to implement the bridge. We could keep the Wheatstone supply voltage constant and measure the resistance or we could keep the current constant. A common implementation is a 'self-balancing' bridge, see Fig. 3.53.

In the bridge in Fig. 3.53, the output signal from an op amp is fed back as the supply voltage to the bridge. Since it has negative feedback, the op amp will do what it takes to keep the inverting (U_-) and the non-inverting (U_+) inputs equal; the op amp will 'balance' the bridge.

When the pressure in the vacuum chamber decreases, R_V (the filament resistance) increases, which means that U_- (inverting op amp input) increases. To compensate for that the current through the filament must be reduced, which means that the supply voltage (the op amp output voltage) must be reduced; the op amp output voltage

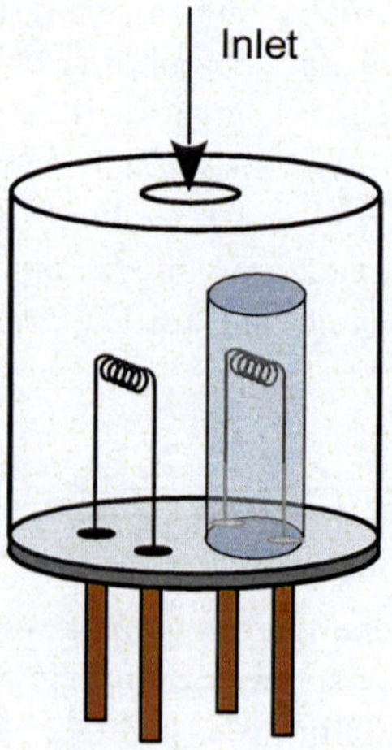

Fig. 3.52 A Pirani gauge head; the reference gauge is sealed

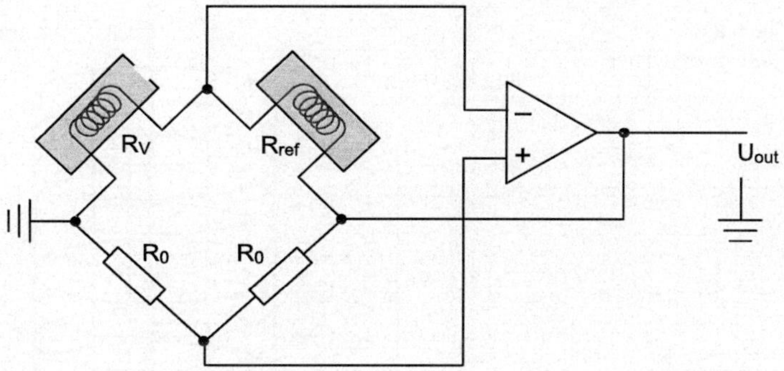

Fig. 3.53 Self-balancing Wheatstone bridge

decreases when the pressure in the vacuum chamber decreases. The op amp's output is proportional to the pressure in the vacuum chamber. (The dependence is not linear and needs to be calibrated.)

This vacuum gauge was invented by Marcello Pirani in 1906 in Germany and is still one of the most common vacuum sensor techniques for 'pre-vacuums' down to 10^{-3} mbar.

3.9.3 Gas Ionization Gauges

There are several ways to design ionization gauges too, but the most common one is the 'nude hot-cathode ionization' gauge designed by Bayard & Alpert in the 1950s. (BAG = Bayard & Alpert gauge.) 'Nude' refers to the fact that it does not have any protective glass shielding. Technically, it is a 'triode' since it has three electrodes. The first electrode is a 'hot' cathode (a helical tungsten wire) that emits electrons. The second electrode is a 'grid' with a positive-biased potential to attract (accelerate) the electrons and the third electrode is a charge collector (the 'anode'). Figure 3.54 illustrates the gauge head.

In Fig. 3.54, the grid is a cylinder, and the anode is just a 'pin'. The cathode filament has a potential of +30 to +50 V and the grid has a potential of +180 to 230 V. When the filament is heated (by a 10-mA current) electrons are emitted and accelerated towards the grid. Most of the electrons pass right through the grid and will interact with the gas atoms inside the grid cylinder. An atom/molecule hit by a high-energy electron will be ionized (at some probability rate) and that will generate a positive ion. This positive ion will be attracted to the 0 V anode pin and generate a current in the anode circuit. The size of this current will be proportional to the gas density in the chamber. The 'conversion factor' is of the order of 100 mA/mbar. At a pressure of 10^{-10} mbar, the current is of the order of 100 pA.

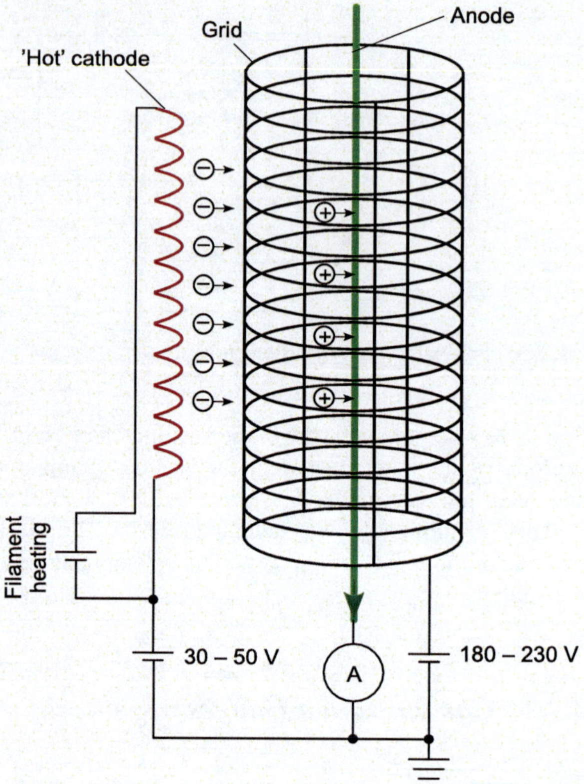

Fig. 3.54 The nude hot-cathode ionization gauge

The range for a typical ionization gauge is 10^{-3} to 10^{-10} mbar and hot-cathode gauges should *never* be exposed to atmospheric pressure (when they are powered) since this can damage them permanently.

3.10 Solved Problems

Problem 3.1 In a thermocouple experiment two conductors A and B were combined as illustrated in Fig. 3.55 and this produced the thermo emf U_{AB} and this experiment was repeated with conductors B and C, see Fig. 3.56, which generated the thermo emf U_{BC}.

Then finally, conductors A and C were combined to form a third thermocouple, see Fig. 3.57. Prove that U_{AC} will equal $U_{\text{AB}} + U_{\text{BC}}$.

Solution The sum of the emfs in the first two experiments is:

$$U_{AB} + U_{BC} = \{E_B(T_H \to T_C) - E_A(T_H \to T_C)\}$$

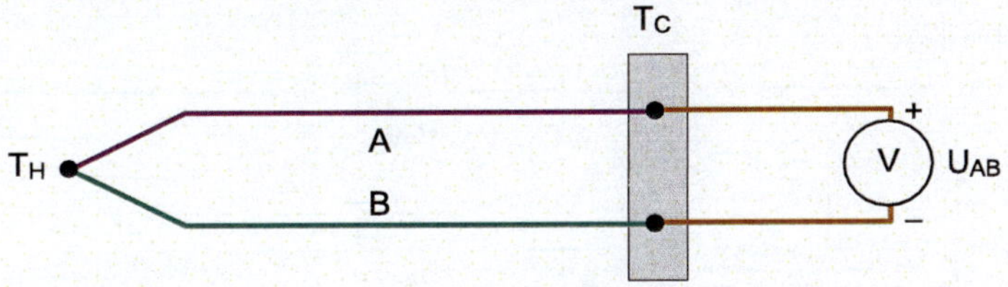

Fig. 3.55 Thermocouple 1

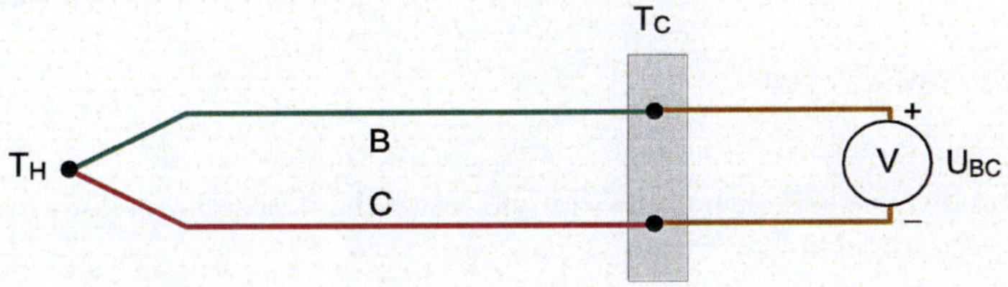

Fig. 3.56 Thermocouple 2

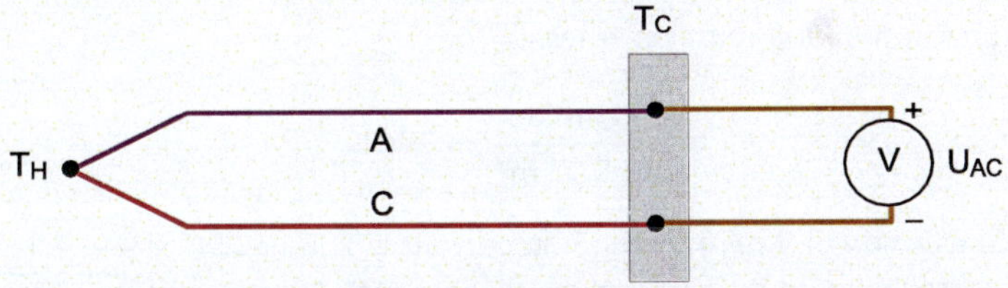

Fig. 3.57 Thermocouple 3

$$+\{E_C(T_H \rightarrow T_C) - E_B(T_H \rightarrow T_C)\} =$$
$$= E_C(T_H \rightarrow T_C) - E_A(T_H \rightarrow T_C) = U_{AC}$$

Problem 3.2 Figure 3.11 illustrates the 4-wire method for resistance measurements. The disadvantage of this method is that it requires an 'expensive' DMM. Prove that you can eliminate the wire resistance using a 'cheap' DMM (see Fig. 3.10) and only three wires.

Solution Figure 3.58 illustrates the solution.

When the switch is in position '1' the DMM will measure $R_1 = 2R_{\mathrm{wire}}$ and when the switch is in position '2' it will measure $R_2 = 2R_{\mathrm{wire}} + R_{\mathrm{Pt}}$. Hence, by subtracting the first measurement from the second, we will get $R_{\mathrm{m}} = R_2 - R_1 = R_{\mathrm{Pt}}$.

Problem 3.3 According to local traffic regulations, your bicycle must have a white headlight and a red taillight. The headlight "must be strong enough to be visible from 300 m". In physical units, that translates to 100 lumens, minimum. Assuming you

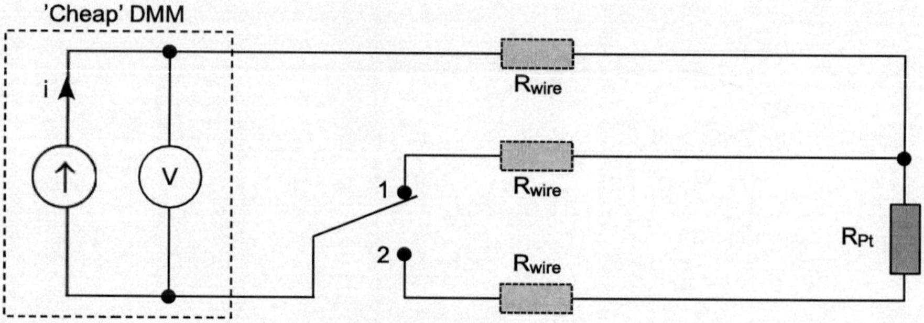

Fig. 3.58 The 3-wire method

have a photodiode BPW21, how would you test if the headlight on your bike meets the local traffic regulations?

Solution I would use the circuit in Fig. 3.40 with a feedback resistor of 1 MΩ. The setup would be in a completely dark room with the headlight as the only source of light and I would place the photodiode 10 cm in front of the headlight. At this distance, the illumination must be at least

$$\frac{100\,\text{lumen}}{4\pi \cdot 0.1^2\,\text{m}^2} = 796\,\text{lx}$$

With a sensitivity of 9 nA/lx, the photocurrent would be $796 \times 9 = 7.16\,\mu\text{A}$ and the output voltage would be $7.16\,\mu\text{A} \times 1\,\text{M}\Omega = 7.16\,\text{V}$. Hence, if the voltage at the op amp output exceeds 7.16 V, my headlight complies with the local regulations.

Problem 3.4 Prove that the circuit in Fig. 3.59 produces both the sum and difference signals for the PSD.

Solution The left part of Fig. 3.59 is just the current-to-voltage conversion. We only need to prove that the right-hand side produces the sum and the difference. The op amp circuit in the upper right corner is the 'difference' circuit (Fig. 3.60).

To see that, first notice that the potential on the op amp's '+' input is $V_{X2}/2$ which then is also the potential at the '−' input. That means that the current I_1 is

$$I_1 = \frac{V_{X1} - \frac{V_{X2}}{2}}{R}$$

This current has nowhere else to go, but to the output. The potential at the op amp's output is

$$U_{\text{diff}} = \frac{V_{X2}}{2} - I_1 R = \frac{V_{X2}}{2} - \left(V_{X1} - \frac{V_{X2}}{2}\right) = V_{X2} - V_{X1} = -(V_{X1} - V_{X2})$$

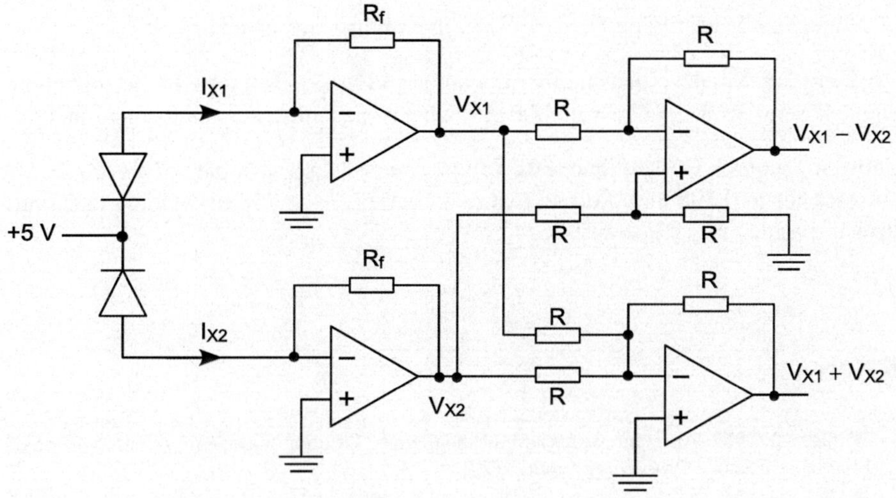

Fig. 3.59 Signal conditioning for PSD

Fig. 3.60 The difference circuit

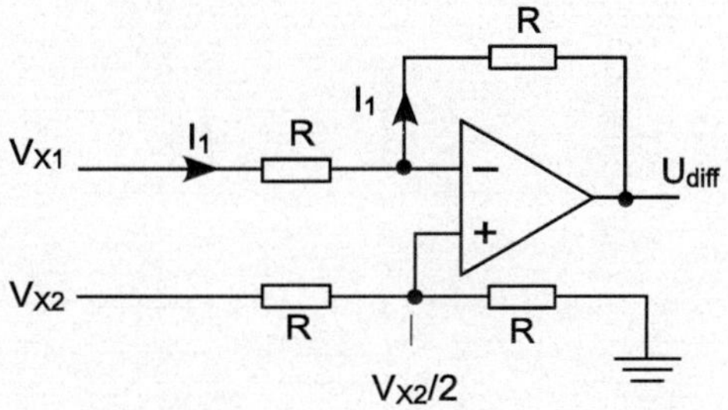

Next, we analyze the lower right-hand side op amp (Fig. 3.61).

In this circuit, the '−' input of the op amp has potential 0 V. That means that the currents I_1 and I_2 are just V_{X1}/R and V_{X2}/R, respectively. These currents must go to the op amp's output and the potential at the output is

$$U_{\text{sum}} = 0 - (I_1 + I_2) \times R = -(V_{X1} + V_{X2})$$

Fig. 3.61 The summing circuit

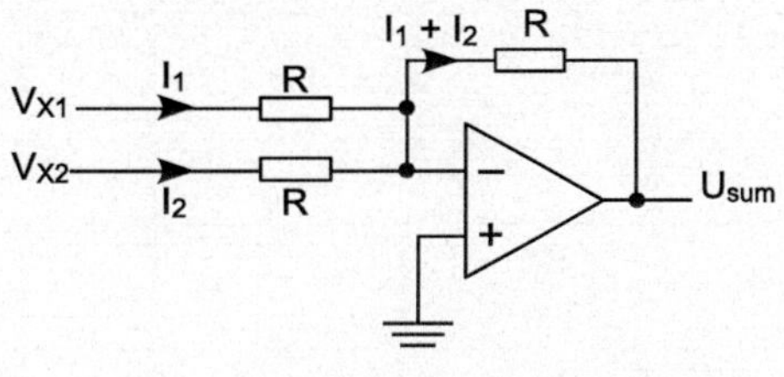

Notice that if we divide U_{diff} by U_{sum}, the minus signs cancel.

Problem 3.5 Suppose that the thermocouple in Fig. 3.4 is a type K thermocouple and that $T_C = 20$ °C. If $U_m = 25.000$ mV, what is the temperature at the hot junction.

Solution A quick Google search for a 'type K thermocouple table', gives that 20 °C corresponds to 0.798 mV. Adding that to 25 mV gives 25.798 mV. Going back into the same table gives $T_H = \underline{621\ °C}$.

References

1. Widlar, R.J. 1967. An exact expression for the thermal variation of the emitter base voltage of bi-polar transistors. *Proceedings of the IEEE* 55 (1): 2.
2. Saxena, S.C, and S.B.L. Seksena. 1989. A self-compensated smart LVDT transducer. *IEEE Transactions on Instrumentation and Measurement* 38(3): 6.

Chapter 4
The Instrumentation Amplifier

Abstract This chapter introduces the instrumentation amplifier. This is perhaps the most important of all amplifiers in electrical measurement systems. The common mode rejection ratio is redefined and its relationship to the signal-to-noise ratio is emphasized. Instrumentation amplifiers are implemented with op amps, and the most common implementation circuits are illustrated.

4.1 Introduction

Different kinds of amplifiers are used in almost all measurement systems and in this chapter, we will introduce one of the most common and versatile amplifiers, the instrumentation amplifier. The instrumentation amplifier is a differential-ended amplifier, i.e., the input voltage is not referenced to ground. We saw in the previous chapter that there are plenty of sensor implementations that rely on such amplifiers (the strain gauge, thermocouples, etc.). From here on, we will refer to the instrumentation amplifier as the 'IA' and its symbol is illustrated in Fig. 4.1.

As you can see, it has the same symbol as the operational amplifier, but unlike the op amp, this amplifier does not need feedback, because the open-loop amplification is 'small'. It amplifies the potential difference between the plus and minus inputs by a 'reasonable' number (10–1000). A huge amplification is not what characterizes the IA.

The IA is all about CMRR (see Sect. 1.2), i.e., the quality of the subtraction. The subtraction of potentials in electronics is never perfect and there will always be a small common mode residual. In Chap. 1, we introduced the signal model that we repeat in Fig. 4.2. The noise is the common mode voltage, and the signal is the normal mode voltage. (See, for example, the Wheatstone bridge signal in Fig. 3.20.) Because of the imperfection in the subtraction, the signal model in Fig. 4.1 is too naïve; instead, we use the model in Fig. 4.3.

F_{NM} represents the amplification of the normal mode voltage (the 'signal') and F_{CM} represents the suppression of the common mode voltage (the 'noise'). The output voltage is

L. Bengtsson, *Electrical Measurement Techniques*,
https://doi.org/10.1007/978-981-99-8187-8_4

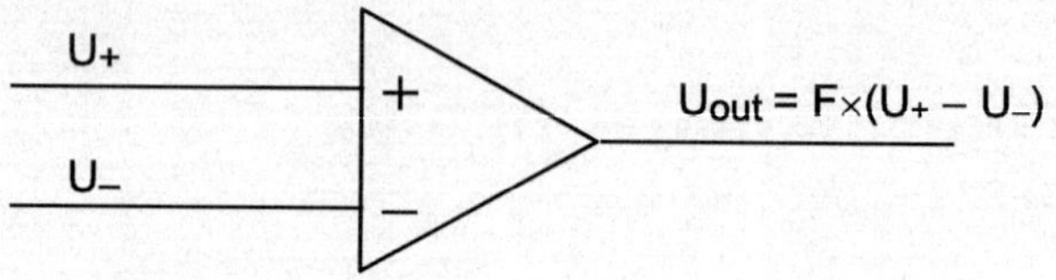

Fig. 4.1 The instrumentation amplifier

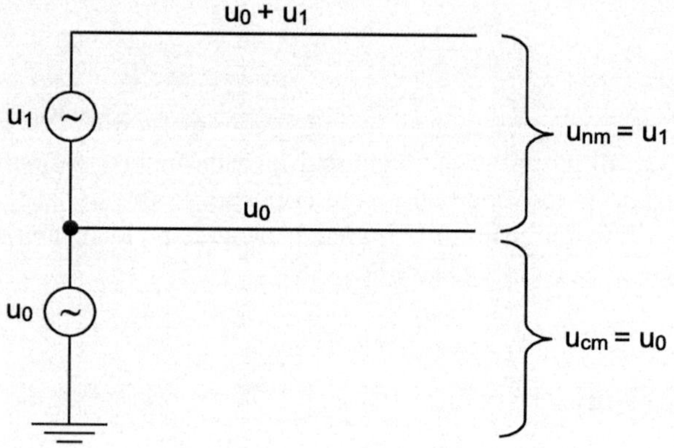

Fig. 4.2 Signal model

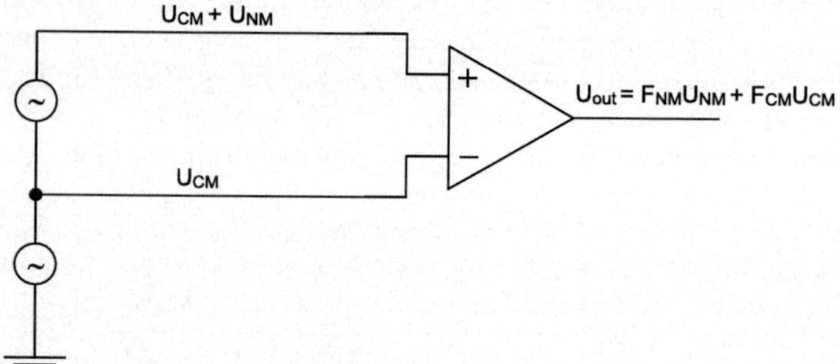

Fig. 4.3 There will always be a small cm residual

$$U_{\text{out}} = F_{\text{NM}} U_{\text{NM}} + F_{\text{CM}} U_{\text{CM}} \tag{4.1}$$

where $F_{\text{CM}} U_{\text{CM}}$ represents what is left of the common mode voltage after the amplifier.

Instrumentation amplifiers are most of all characterized by how well they suppress the common mode voltage compared to how much they amplify the normal mode. We defined the common mode rejection ratio (CMRR) already in Chap. 1 for a DMM. We use the same number to represent the quality of an IA:

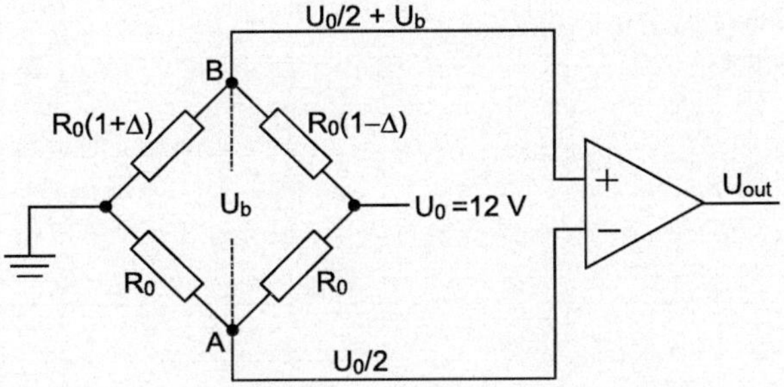

Fig. 4.4 The Wheatstone bridge signal is amplified by an IA

$$\text{CMRR} = 20\log\frac{F_{\text{NM}}}{F_{\text{CM}}}[\text{dB}] \tag{4.2}$$

The following example will illustrate the use of an IA.

Example 4.1 Figure 4.4 illustrates how a Wheatstone bridge produces two voltage potentials: One with the common mode potential and one with the common mode plus the normal mode potential.

Assuming we have a bridge voltage of 1 mV and an IA with CMRR = 105 dB and $F_{\text{NM}} = 100$, what is the output voltage from the IA?

Solution First, we find the F_{CM} from Eq. (4.2): $F_{\text{CM}} = F_{\text{NM}} \times 10^{-\text{CMRR}/20} = 100 \cdot 10^{-105/20} = 5.6 \cdot 10^{-4}$. Hence, normal mode voltages are amplified by 100 and common mode voltages are suppressed by a factor of 0.00056. The common mode voltage is $U_0/2 = 12/2 = 6$ V. The output voltage is

$$\begin{aligned} U_{\text{out}} &= F_{\text{NM}}U_{\text{NM}} + F_{\text{CM}}\text{U}_{\text{CM}} = 100 \cdot 0.001 + 6 \cdot 0.00056 \\ &= 100 + 3.4\,\text{mV} = 103.4\,\text{mV} \end{aligned}$$

4.2 Implementations

4.2.1 Classic IA Circuit

Before we get into the details of IA circuits, we need to understand the classic differential amplifier in Fig. 4.5.

The potential on the op amp's + input is $U_{\text{in}'2}R_2/(R_1 + R_2)$. Since this is also the potential on the op amp's − input, the current i is

$$i = \frac{U_{\text{in}'1} - U_{\text{in}'2}\frac{R_2}{R_1+R_2}}{R_1}$$

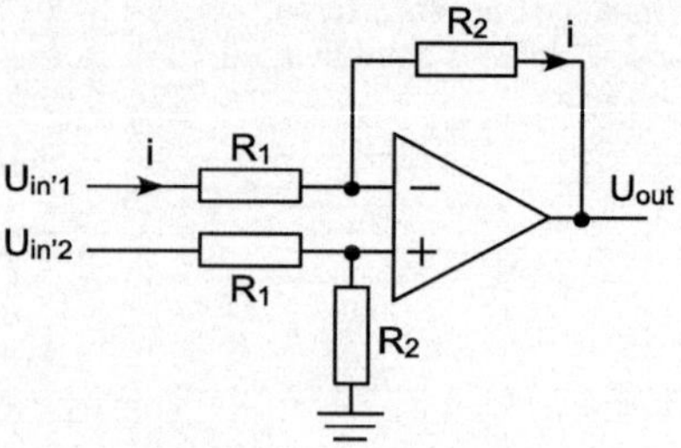

Fig. 4.5 Classic differential amplifier

Hence, the output voltage is

$$U_{\mathrm{out}} = U_{-} - i \cdot R_2 = U_{\mathrm{in}'2}\frac{R_2}{R_1 + R_2} - \frac{R_2}{R_1}\left(U_{\mathrm{in}'1} - U_{\mathrm{in}'2}\frac{R_2}{R_1 + R_2}\right)$$
$$= U_{\mathrm{in}'2}\frac{R_2}{R_1 + R_2}\left(1 + \frac{R_2}{R_1}\right) - \frac{R_2}{R_1}U_{\mathrm{in}'1} = \frac{R_2}{R_1}(U_{\mathrm{in}'2} - U_{\mathrm{in}'1}) \tag{4.3}$$

Hence, we have a differential amplifier where we can choose the amplification arbitrarily with the resistors R_1 and R_2.

This is what we were looking for, but we are not done yet; the circuit in Fig. 4.5 has a disadvantage. Since the + and − inputs of the op amp are 'virtually' short-circuited, the input impedance of the circuit is $R_1 + R_1 = 2R_1$ and we must have a large R_1 resistor to have a large input resistance. But then R_2 must be very large to get an amplification of 10–1000, and it is likely to be impractically large. For that reason, we are looking for another solution. This solution is illustrated in Fig. 4.6.

In Fig. 4.6, we can see that the problem with the input impedance is remedied since the signal inputs are now connected directly to the inputs of op amps. Let's see what the output voltage is (to make sure we still have a differential amplifier).

Assuming $U_{\mathrm{in1}} > U_{\mathrm{in2}}$, the current i_b through the R_b resistor is

$$i_b = \frac{U_{\mathrm{in1}} - U_{\mathrm{in2}}}{R_\mathrm{b}}$$

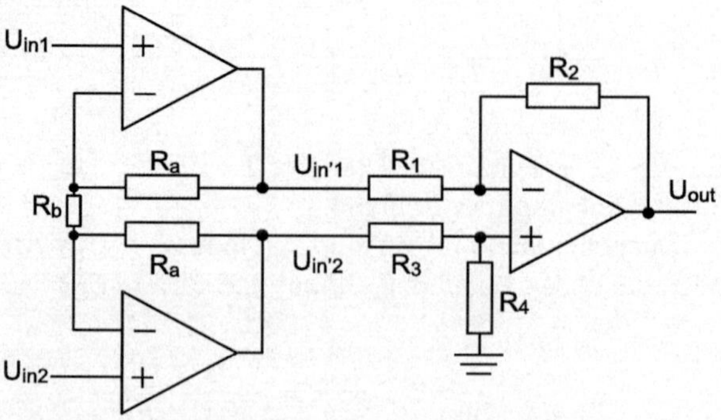

Fig. 4.6 Instrumentation amplifier circuit

Then the voltages $U_{in'2}$ and $U_{in'1}$ are

$$\left.\begin{aligned} U_{in'1} &= U_{in'1} + i_b R_a \\ U_{in'2} &= U_{in'2} - i_b R_a \end{aligned}\right\} U_{in'2} - U_{in'1} = U_{in2} - U_{in1} - 2 i_b R_a$$

$$\Rightarrow U_{in'2} - U_{in'1} = U_{in2} - U_{in1} - \frac{2R_a}{R_b}(U_{in1} - U_{in2})$$

$$= U_{in2}\left(1 + \frac{2R_a}{R_b}\right) - U_{in1}\left(1 + \frac{2R_a}{R_b}\right) = \left(1 + \frac{2R_a}{R_b}\right)(U_{in2} - U_{in1})$$

Inserting this into Eq. (4.3), we get the output voltage:

$$U_{out} = \frac{R_2}{R_1}\left(1 + \frac{2R_a}{R_b}\right)(U_{in2} - U_{in1}) \tag{4.4}$$

The normal mode amplification is $F_{NM} = (1 + 2R_a/R_b)R_2/R_1$ and there are obviously several ways to vary the amplification, but, if we change R_1, R_2, or R_a, it involves *two* resistors. Changing R_b only involves *one* resistor. For that reason, IAs with variable F_{NM} usually allow the user to apply an external R_b resistor.

Example 4.2 Figure 4.7 illustrates an IA-integrated circuit. By varying R_b, what is the range of possible normal mode amplifications?

Solution $R_2/R_1 = 10$, so if $R_b = \infty$, then $F_{NM} = 10$ and if $R_b = 0$, the $F_{NM} = \infty$. The range is $F_{NM} \in [10, \infty]$.

The popular IAs from Texas Instruments (INA128) and Analog Devices (AD622) both have exactly the configuration illustrated in Fig. 4.6. Linear Technology though, has a different implementation that only requires two op amps (LT110x). However, this is at the expense of not offering arbitrary amplifications. The circuit is illustrated in Fig. 4.8.

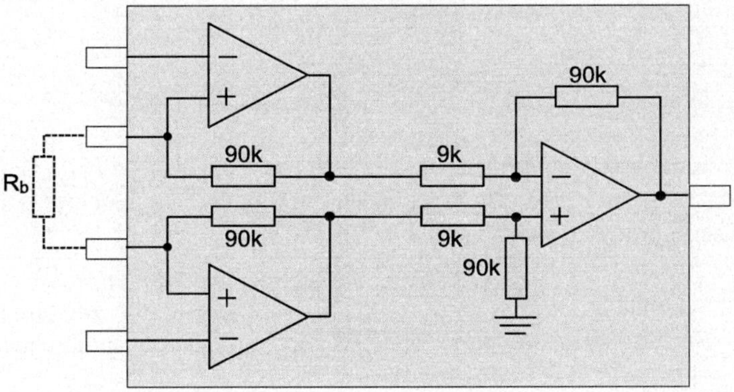

Fig. 4.7 Integrated IA

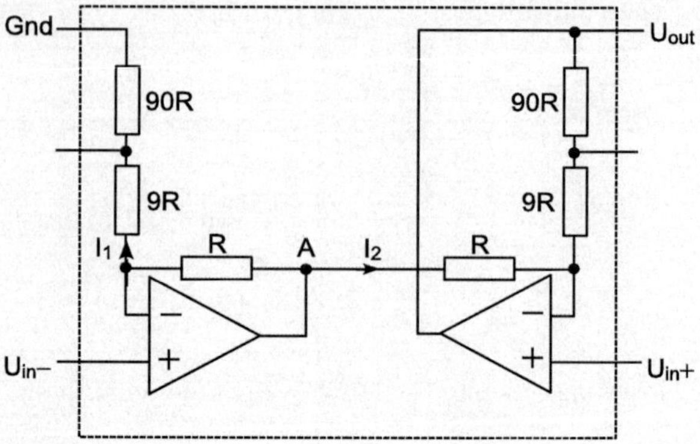

Fig. 4.8 2-op amp IA

As in the previous solution, the inputs are directly connected to the input of op amps, so the input impedance is fine. The current I_1 is $U_{\text{in}-}/99R$. Hence, the potential at point A is

$$U_A = U_{\text{in}-} + I_1 R = U_{\text{in}-}\left(1 + \frac{1}{99}\right) = U_{\text{in}-}\frac{100}{99}$$

The current I_2 is

$$I_2 = \frac{U_A - U_{\text{in}+}}{R} = \frac{U_{\text{in}-}\frac{100}{99} - U_{\text{in}+}}{R}$$

And finally, U_{out} is

$$U_{\text{out}} = U_{\text{in}+} - I_2 99R = U_{\text{in}+} - 100U_{\text{in}-} + 99U_{\text{in}+} = 100(U_{\text{in}+} - U_{\text{in}-})$$

We conclude that the IA in Fig. 4.8 has a normal mode amplification of 100, and it is easy to derive that it would be reduced to 10 if we short-circuit the $90R$ resistors (and F_{CM} would also decrease by a factor of 10).

Example 4.3 In Fig. 4.9, we use an IA to amplify the emf from a thermocouple. The circuit is radiated with 50 Hz EMC interferences from the surrounding power grid which induces a common mode voltage of 1 V in the circuit. The application requires a normal mode amplification of the thermo emf of 50. What CMRR does the IA need if we want the common mode contribution in the output signal to be less the 5%? (The thermo emf is around 0.8 mV.)

Solution The normal mode contribution is 50×0.8 mV = 40 mV. The output signal is 40 mV + $F_{\text{CM}} \cdot 1$ = 40 mV + F_{CM}. The CM part of the output should be less than 5%:

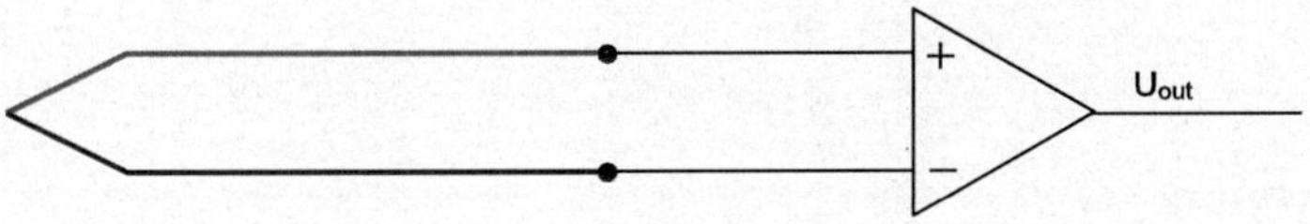

Fig. 4.9 Thermocouple amplifier

$$\frac{F_{CM}U_{CM}}{F_{NM}U_{NM} + F_{CM}U_{CM}} = \frac{F_{CM}}{0.04 + F_{CM}} \le \frac{5}{100}$$
$$\Rightarrow 95F_{CM} = 0.2 \Rightarrow F_{CM} = 2.1 \cdot 10^{-3}$$

$$\text{CMRR} \ge 20\log\frac{50}{2.1 \cdot 10^{-3}} = 88 \text{ dB}$$

We should look for an IA with a CMRR of at least 90 dB.

4.3 CMRR Versus SNR

If we revisit Fig. 4.3, we can see that the 'signal' is the NM part of the voltage before the IA and the 'noise' is the CM part; the signal-to-noise ratio *before* the IA is

$$\text{SNR}_{\text{before}} = \frac{U_{NM}}{U_{CM}} \tag{4.5}$$

In the output signal, the 'signal' is $F_{NM}U_{NM}$, and the noise is $F_{CM}U_{CM}$; the signal-to-noise ratio *after* the IA is

$$\text{SNR}_{\text{after}} = \frac{F_{NM}U_{NM}}{F_{CM}U_{CM}} \tag{4.6}$$

Next, we take the quotient of the signal-to-noise before and after the IA:

$$\frac{\text{SNR}_{\text{after}}}{\text{SNR}_{\text{before}}} = \frac{\frac{F_{NM}U_{NM}}{F_{CM}U_{CM}}}{\frac{U_{NM}}{U_{CM}}} = \frac{F_{NM}}{F_{CM}} = \text{CMRR} \tag{4.7}$$

The quotient of the signal-to-noise before and after the IA is equal to the CMRR. This is how I recommend you think of CMRR. Rather than thinking of CMRR as a quotient between two amplifications, it tells you how much the signal-to-noise ratio is improved (with respect to the *common mode noise*).

Example 4.4 Recalculate Example 4.4 using Eq. (4.7).

Solution The $\text{SNR}_{\text{before}} = 0.8 \text{ mV}/1\text{V} = 0.8 \cdot 10^{-3}$. In the output signal, the CM part should only correspond to 5% of the total signal:

$$\frac{\text{CM}}{\text{CM+NM}} \leq \frac{5}{100} \Rightarrow \frac{\text{NM}}{\text{CM}} \geq \frac{95}{5} = \text{SNR}_{\text{after}}$$

The CMRR we need is

$$\text{CMRR} = 20\log\frac{95/5}{0.8 \cdot 10^{-3}} = \underline{88\ \text{dB}}$$

4.4 Solved Problems

Problem 4.1 An IA has a normal mode amplification of 100 and a CMRR = 80 dB. How much does it attenuate the common mode signals?

Solution $80 = 20\log\frac{100}{F_{\text{CM}}} \Rightarrow F_{\text{CM}} = 100 \cdot 10^{-80/20} = 0.01$

Problem 4.2 If we use the IA in problem 4.1 in an application where $U_{\text{CM}} = 2$ V and $U_{\text{NM}} = 5$ mV, what will the output signal be, and what are the normal and common mode contributions to the output signals?

Solution $U_{\text{out}} = 100 \times 0.005 + 0.01 \times 2 = 0.5 + 0.02 = \underline{0.52\ \text{V}}$. The NM contribution is 0.5 V, and the CM contribution is 0.02 V.

Problem 4.3 An IA is used to amplify the signal from an ECG measurement, see Fig. 4.10. What CMRR is required for the CM contribution at the output to be less than 1%?

Solution The $\text{SNR}_{\text{before}} = 3\ \text{mV}/1.5\ \text{V} = 2 \cdot 10^{-3}$. In the output signal, the CM part should only correspond to 1% of the total signal:

$$\frac{\text{CM}}{\text{CM+NM}} \leq \frac{1}{100} \Rightarrow \frac{\text{NM}}{\text{CM}} \geq \frac{99}{1} = 99 = \text{SNR}_{\text{after}}$$

The CMRR we need is

$$\text{CMRR} = 20\log\frac{99}{2 \cdot 10^{-3}} = \underline{94\ \text{dB}}$$

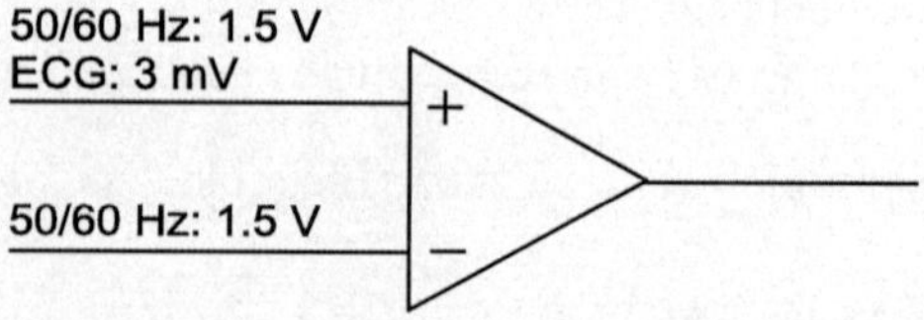

Fig. 4.10 IA in ECG measurement

Chapter 5
Transmission Lines

Abstract This chapter explains why the impedance 50 Ω is so ubiquitous in physics labs. First, the transmission line is introduced and then the extremely important concept of its *characteristic impedance* is defined. Fresnel's law is used to find the reflection coefficient for electric signals in a transmission line (Eq. 5.2) and that will explain why transmission lines need to be *terminated*. One section treats the problem of how to properly split and splice transmission lines. Finally, one of the most common applications of signal reflections in transmission lines is presented, namely, the *Time Domain Reflectometer*.

5.1 Introduction

Once you start working in a physics lab, connecting cables between instruments, lasers, and vacuum chambers, it doesn't take long before you hear people talking about '50 Ω' and that number keeps popping up in manuals and datasheets and it is almost like 50 Ω is a magic number. There is nothing magic about it, but it is paramount that you understand the fuss about 50 Ω. This chapter will remedy that and unravel the '50-Ω secret'.

5.2 The Characteristic Impedance

It all starts with a very simple experiment. In Fig. 5.1, a signal source with internal impedance 100 Ω is connected to a 100-Ω load impedance via a switch. The switch is closed at $t = 0$.

In Fig. 5.1, we can easily conclude that exactly at $t = 0$, the 10 V will immediately be equally distributed across the internal impedance Z_i and the load impedance Z_{load}; at $t = 0$, U_{load} goes immediately from 0 to 5 V.

The experiment in Fig. 5.1 is straightforward, 5 V across the load and 5 V across the internal impedance.

L. Bengtsson, *Electrical Measurement Techniques*,
https://doi.org/10.1007/978-981-99-8187-8_5

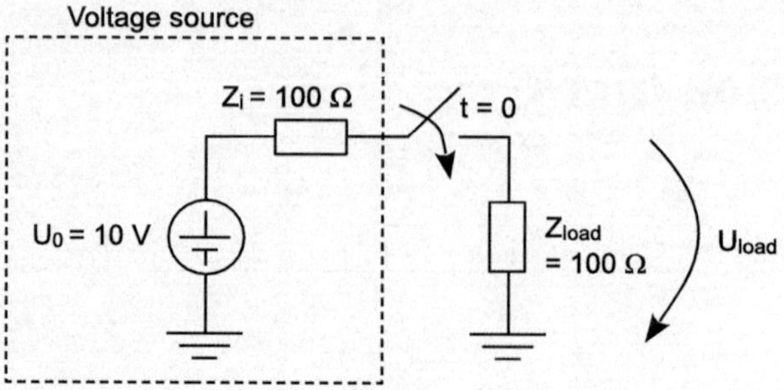

Fig. 5.1 Signal model

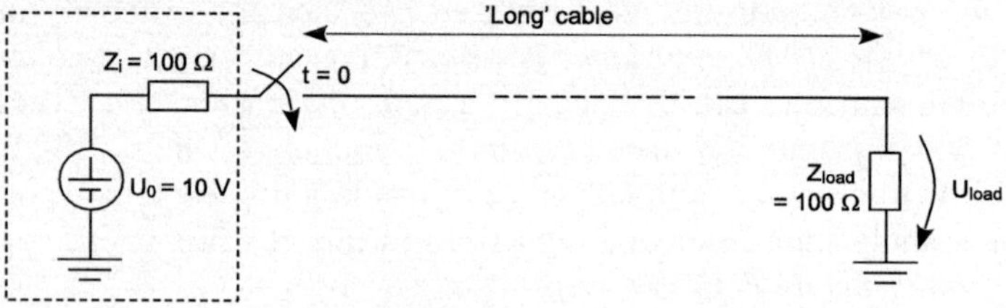

Fig. 5.2 Same experiment but with a small twist

However, as the next example will illustrate, it doesn't take much for your undergraduate electricity skills to 'break down'. In Fig. 5.2, we repeat the exact same experiment, only with one *small* difference; we place the load impedance at some distance. This distance is 'long enough not to be ignored'.

What happens to the voltage across the load impedance now (at $t = 0$)? That's easy: Nothing! Because it takes some time for the voltage to propagate to the load impedance. *Something* will happen *later*, but exactly at $t = 0$, *nothing* happens.

Here is a harder question: How large is the voltage across the *internal* impedance Z_i, at $t = 0$ (the moment immediately after the switch is closed)?

This is the key question here (and the key to the whole 50-Ω fuss). At $t = 0$, the voltage source cannot see the load impedance, so how much should it 'keep' across Z_i and how much should it send down the cable to Z_{load}? Most students answer, '0 V' or '10 V', but that is just wrong. It is more complicated than that. At $t = 0$, the voltage source 'sees' the internal impedance and ….. *the cable*!!! To answer the question, we must understand how the *cable* behaves at $t = 0$ (from the voltage source's point of view).

The answer is that at $t = 0$, the cable is 'perceived' as an *impedance to ground*, see Fig. 5.3. This 'perceived' impedance is the cable's *characteristic impedance* and is denoted Z_0. Here are some facts you should know about Z_0:

- Equation (3.9) tells us the 'ohm resistance' of a conductor of some length and diameter. Most students think that the characteristic impedance has something to do with Eq. (3.9), or even think that Z_0 is the same as Eq. (3.9). This is completely

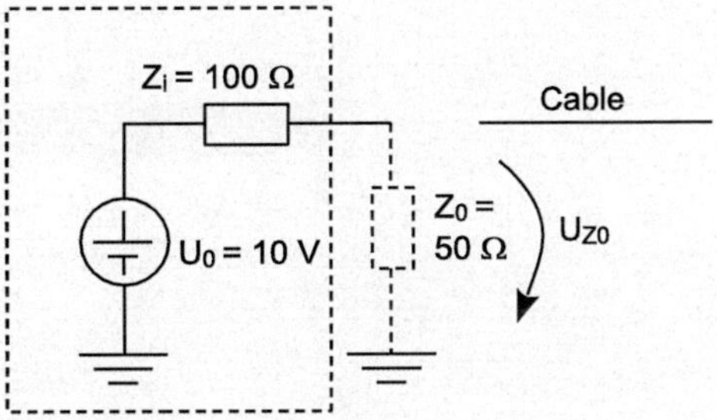

Fig. 5.3 At $t = 0$, the cable is perceived as an impedance to ground

wrong! Z_0 *has nothing to do with the ohm resistance in Eq. (3.9)*!! (Of course it doesn't! Eq. (3.9) depends on the cable length. In Fig. 5.3, the cable length is still unknown, and still, we have a number for Z_0.)

- Z_0 is the 'wave impedance'; how hard it is for the EM field to propagate down the cable.
- Z_0 still depends on manufacturing parameters, but 'standard' cables have 'standard' Z_0 values. The standard cable in a physics lab is the RG58 coaxial cable which has a characteristic impedance of ..... 50 Ω!

This is the origin of the '50-Ω' fuss; since RG58 cables are omnipresent in all labs, a lot of equipment have been customized to work in that 'environment'.

Obviously, from Fig. 5.3, we can see that at time $t = 0$, the voltage U_0 will be distributed across Z_i and Z_0. The voltage across Z_0 is

$$U_{Z0} = \frac{Z_0}{Z_0 + Z_i} U_0 \tag{5.1}$$

U_{Z0} is the voltage that will propagate down the cable towards the load. (U_{Z0} has *nothing* to do with the size of the load Z_{load}.) The rest of U_0 (i.e., $U_0 - U_{Z0}$) stays over Z_i. We will illustrate the consequences of the characteristic impedance by a detailed example.

Example 5.1 Suppose we have the system in Fig. 5.4. If U_0 is a step voltage that goes from 0 to +5 V at $t = 0$, plot the signal levels at the 'near' end (u_n) and the 'far' end (u_f) as a function of time (in the same diagram).

Solution Since we have an RG58 coax cable, $Z_0 = 50\ \Omega$. According to Fig. 5.3, we have the equivalent circuit in Fig. 5.5 at $t = 0$.

Equation (5.1) gives us

$$U_{Z0} = \frac{50}{20 + 50} \cdot 5\,\text{V} = +3.57\,\text{V}$$

This is the voltage that will propagate down the coax cable towards the far end, and since this is the voltage at the *near* end, u_n will go from 0 V to +3.57 V at $t =$

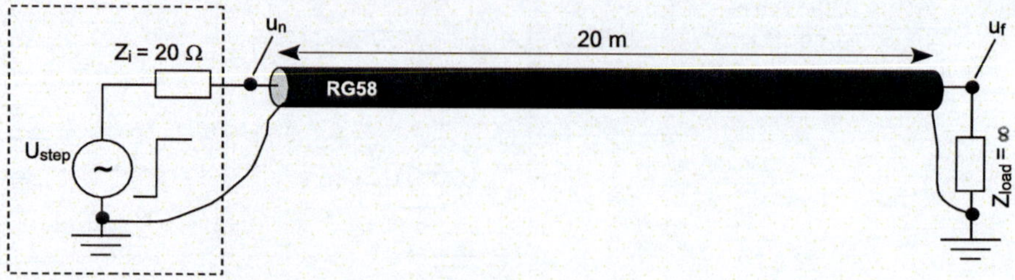

Fig. 5.4 Voltage source and transmission cable. Load impedance = ∞ means the far end is 'open'

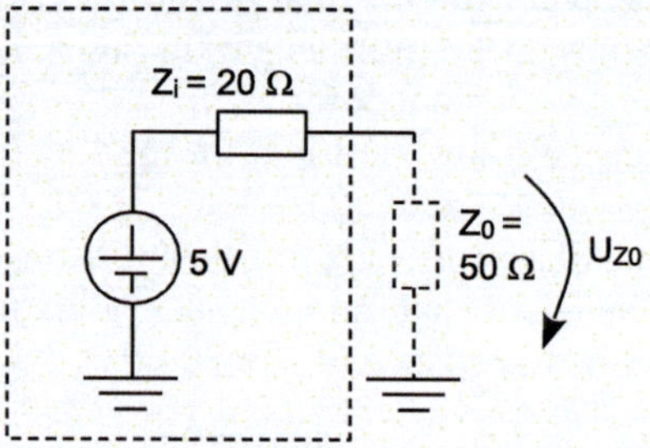

Fig. 5.5 System at $t = 0$

0. Nothing happens at the far end (yet); u_f still $= 0$. Figure 5.6 illustrates the near- and far-end signals a few moments after $t = 0$.

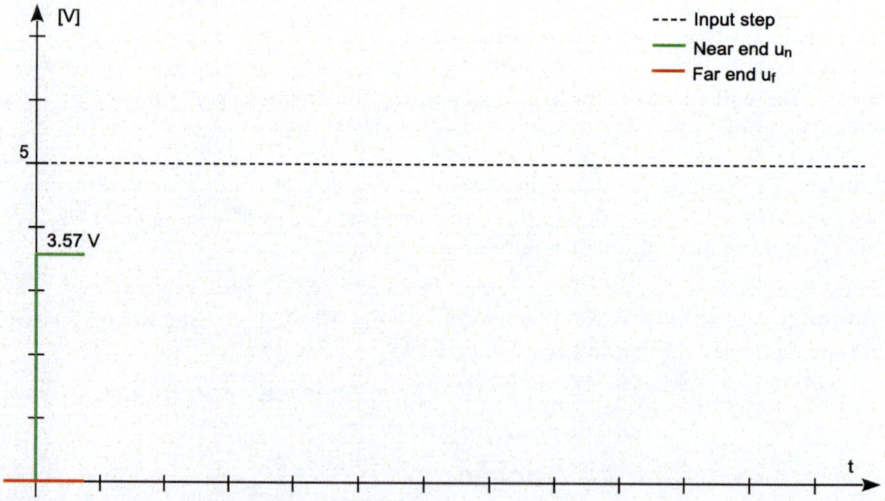

Fig. 5.6 Signal levels just after $t = 0$

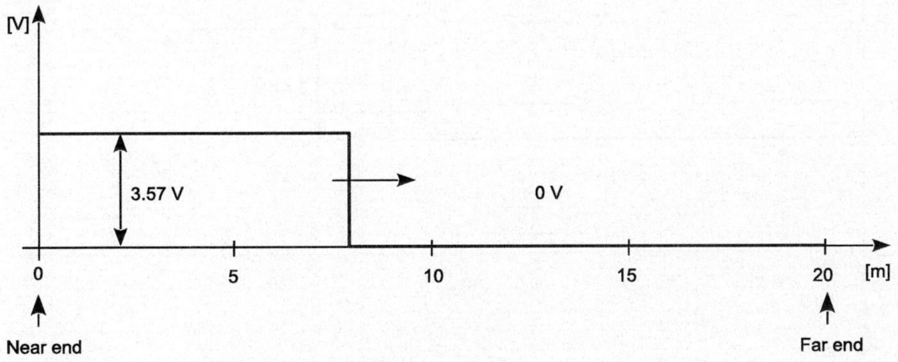

Fig. 5.7 The signal situation on the cable just after $t = 0$

Next, we need to understand what happens in the cable after $t = 0$. U_{Z0} represents the voltage that moves towards the far end; a 'wavefront', 3.57 V high, propagates down the cable, see Fig. 5.7. This 'wavefront' leaves the voltage level +3.57 V behind it on the cable (but the voltage *in front* of the wavefront is still 0 V).

Nothing will happen at the far end until the wavefront in Fig. 5.7 reaches the far end. Before we worry about what happens when the wavefront hits the far end, we must first figure out how long it takes until this happens. How fast does the wavefront in Fig. 5.7 propagate? The wavefront in Fig. 5.7 represents the propagation of the EM field in the cable, and EM fields travel at the speed of light. That is, the speed of light in the cable, which for an RG58 coax cable is $0.66 \times c_0 = 2 \cdot 10^8$ m/s. So, if the cable is 20 m long, it will take $20/2 \cdot 10^8 = 100$ ns. After 100 ns, the wavefront hits the load impedance. To understand what happens at that moment, we will rephrase that: After 100 ns, the *wave* enters a new *medium*. And we know from wave theory that when a wave enters a new medium, there will be *wave reflection*; some wave energy will be absorbed (or transmitted) and some will be reflected. *Fresnel's law* gives us the reflection coefficient as the quotient between the difference and sum of the refractive indices, but we don't know the refractive indices here.

Fortunately, the refractive indices are proportional to the wave impedances, so we can write Fresnel's law as

$$\gamma = \frac{n_2 - n_1}{n_2 + n_1} = \frac{Z_2 - Z_1}{Z_2 + Z_1} \tag{5.2}$$

At the far end, $Z_2 = Z_{load} = \infty$ and $Z_1 = Z_0 = 50\ \Omega$, so the reflection coefficient at the far end is

$$\gamma_f = \frac{\infty - 50}{\infty + 50} = +1$$

That means that *all* of the incoming wave is reflected. (Of course, it is. The far end is *open*. Where else would it go?). So, after 100 ns we will have a wave of +

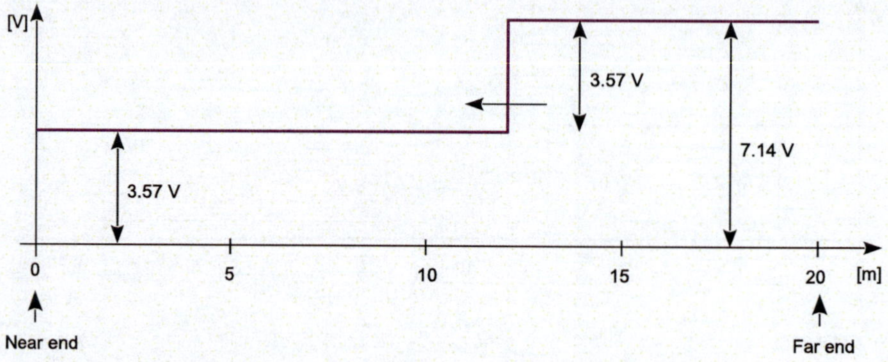

Fig. 5.8 The signal situation on the cable just after 100 ns

3.57 V going *back* to the near end. The '+' sign indicates that there is no phase shift, so the 'back' going waves will interact constructively with the incoming waves, and hence the back/left-going wavefront will leave a voltage level behind of 3.57 + 3.57 = 7.14 V. This is illustrated in Fig. 5.8.

And, since we measure u_f at the far end, $u_f = 7.14$ V after 100 ns. We can now update our timing diagram in Fig. 5.6, see Fig. 5.9. Nothing happens at the near end (yet). The near-end signal is stationary until the new wavefront reaches the near end.

(Notice in Fig. 5.9 that neither signal is nowhere near the 'expected' 5-V line. Yet…).

After 200 ns, the wavefront in Fig. 5.8 reaches the near end, and we will again have wave reflection. The reflection coefficient at the near end is

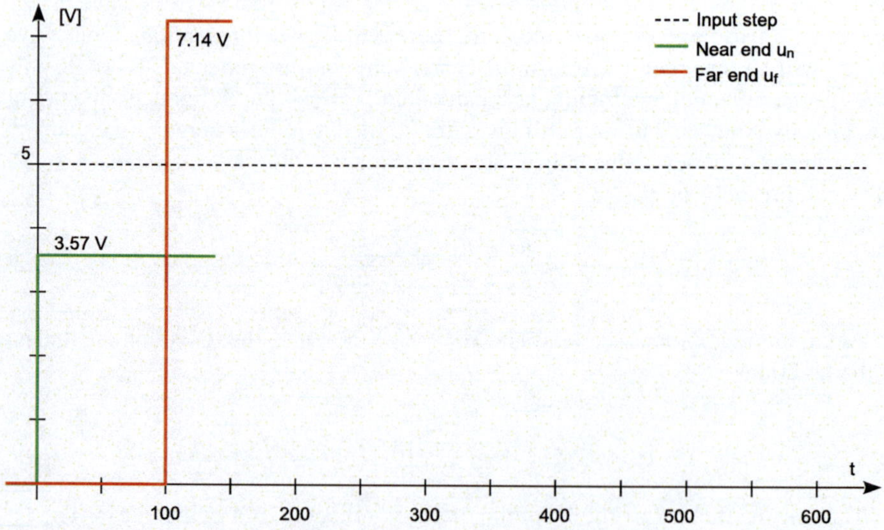

Fig. 5.9 Signal levels just after $t = 100$ ns

$$\gamma_n = \frac{Z_i - Z_0}{Z_i + Z_0} = \frac{20 - 50}{20 + 50} = -0.429$$

Hence, the reflected wave is $-0.429 \times (+3.57) = -1.53$ V. So, after the wavefront impacts the near end, a wave with amplitude -1.53 V will go back again towards the far end. The '−' sign indicates a phase shift of 180°, which means that the back-going waves will interact destructively with the incoming waves. The incoming wavefront left a voltage level of 7.14 V behind. The new wavefront going back to the far end will leave a voltage of 7.14 − 1.53 = 5.61 V behind. This is illustrated in Fig. 5.10 and in Fig. 5.11, we have updated our timing diagram; the near-end voltage goes up to 5.61 V.

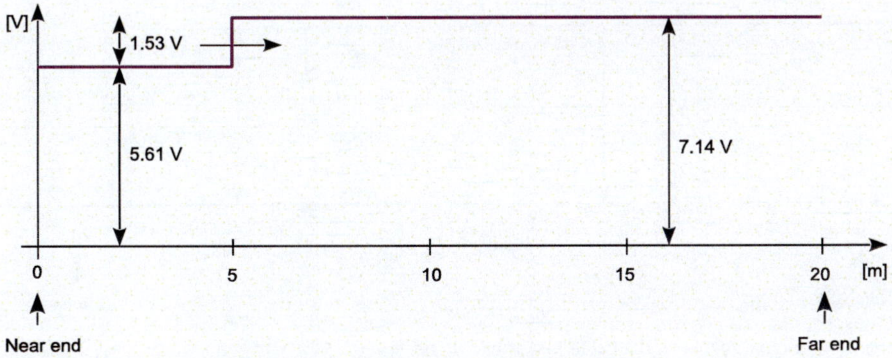

Fig. 5.10 The signal levels on the cable just after $t = 200$ ns

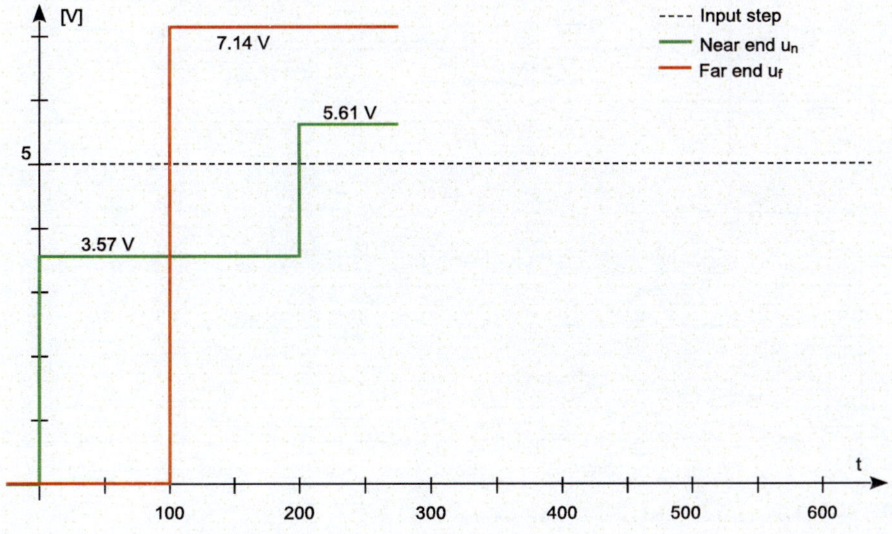

Fig. 5.11 The far- and near-end signals after 200 ns

After another 100 ns (t = 300 ns), the −1.53 V wavefront in Fig. 5.10 will hit the far end and since the reflection coefficient at the far end is (still) +1, all of the incoming wave is reflected; there will be a wavefront of −1.53 V going back to the near end. This will interact destructively with the incoming waves, leaving a voltage level of 5.61 – 1.53 = 4.08 V behind the wavefront, see Fig. 5.12. Hence, the far-end signal level will drop to 4.08 V, see Fig. 5.13.

At t = 400 ns, the −1.53 V wavefront returns to the near end where the reflection is −0.429 × (−1.53) = +0.65. The '+' sign indicates a constructive interaction with the incoming waves which means that the wavefront returning to the far end will

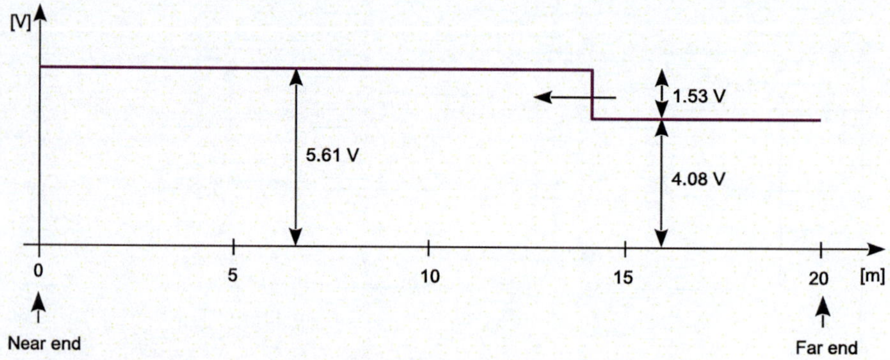

Fig. 5.12 Signal levels on the cable just after t = 300 ns

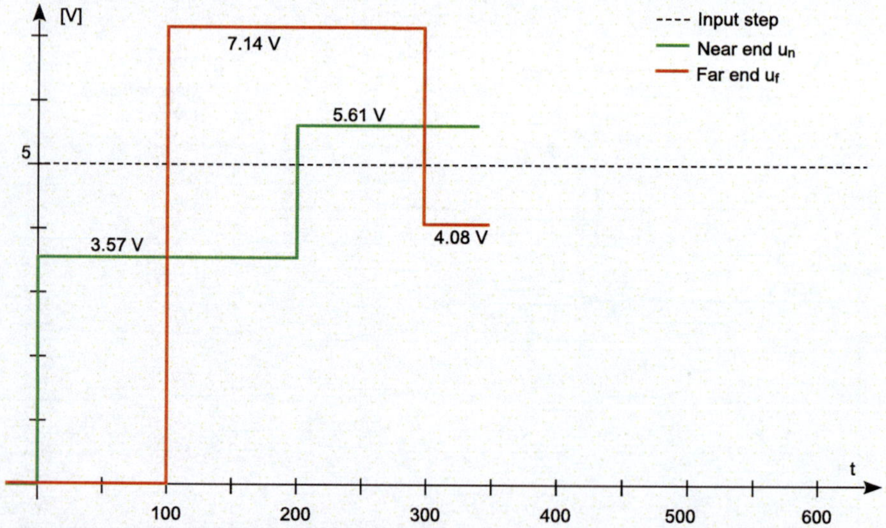

Fig. 5.13 Far- and near-end signals after 300 ns

leave behind a voltage level of 4.08 + 0.65 = 4.73 V, see Fig. 5.14, and this will also be the signal level at the near end, see the updated timing diagram in Fig. 5.15.

After 500 ns, the +0.65 V wavefront is reflected at the far end (+1) and a new +0.65 V wavefront goes back to the near end. The new +0.65 V wave interacts constructively with the incoming waves, leaving behind a voltage level of 4.73 + 0.65 = 5.38 V (= u_f), see Figs. 5.16 and 5.17.

At $t = 600$ ns, the 0.65 V wave returns to the near end and the reflected wave is −0.429 × 0.65 = −0.28 V, interacting destructively with the incoming waves, leaving 5.38 – 0.28 = 5.10 V behind (= u_n), see Figs. 5.18 and 5.19.

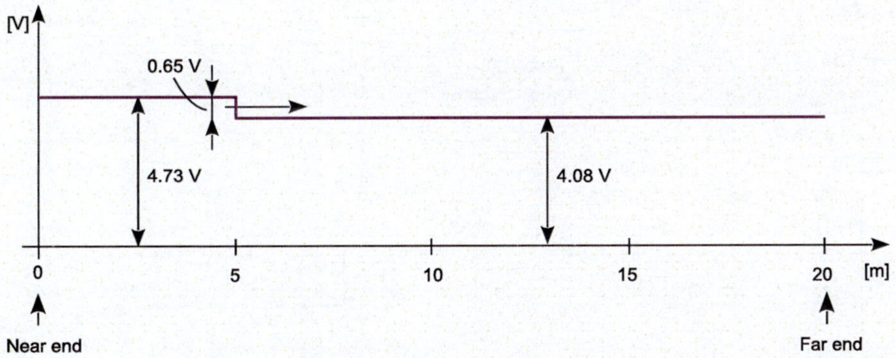

Fig. 5.14 Signal levels on the cable just after $t = 400$ ns

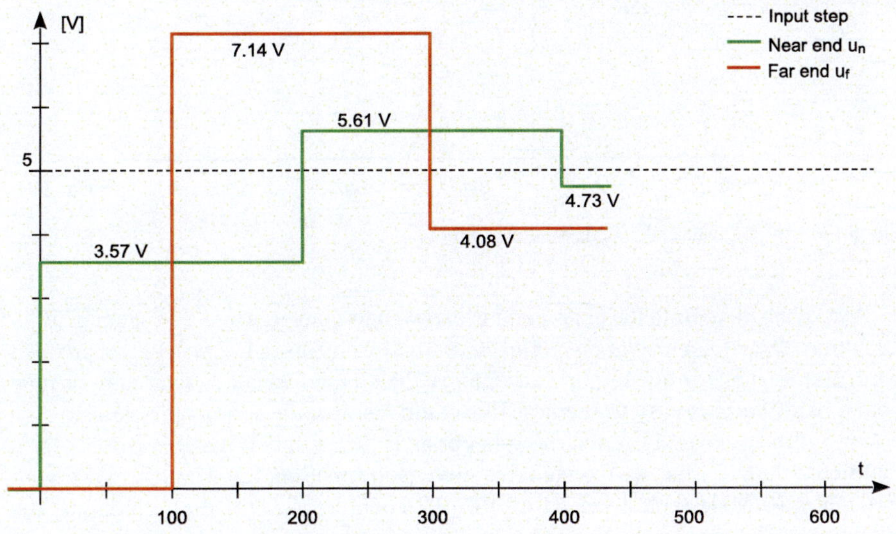

Fig. 5.15 Far- and near-end signals after 400 ns

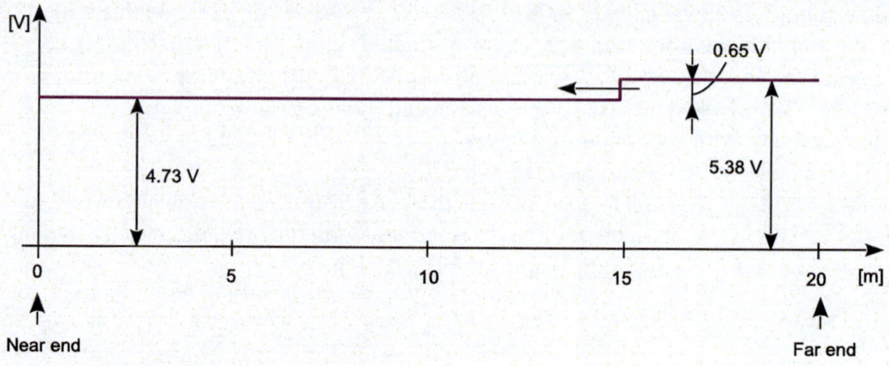

Fig. 5.16 Signal levels on the cable just after $t = 500$ ns

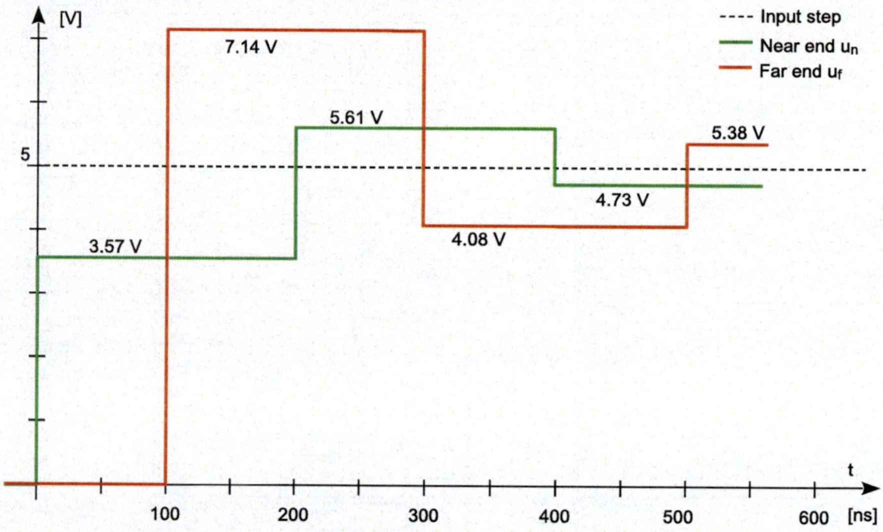

Fig. 5.17 Far- and near-end signals after 500 ns

We got the idea. Both the near- and far-end signals converge to +5 V, see Fig. 5.20. Of course, they do; it's an open circuit! Eventually, it is just a DC voltage on an open wire. But it takes a while! In this case, more than a microsecond. That may or may not be okay, but it is easy to see how this could cause some serious problems.

First, it takes over 1 μs for the reflections to peter out. What if we don't have that time? Suppose the source signal is not a step function, but a square wave with frequency 10 MHz, i.e., a period of 100 ns. Then, because of the reflections, the signal would be seriously distorted.

Second, suppose we have some kind of digital counting device at the far end (high-impedance input) with a trigger level of 5 V, then according to the timing diagram in Fig. 5.20, each pulse would generate multiple counts.

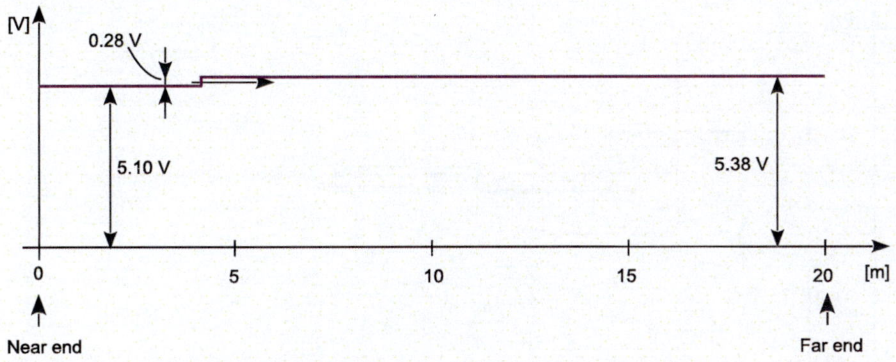

Fig. 5.18 Signal levels on the cable just after $t = 600$ ns

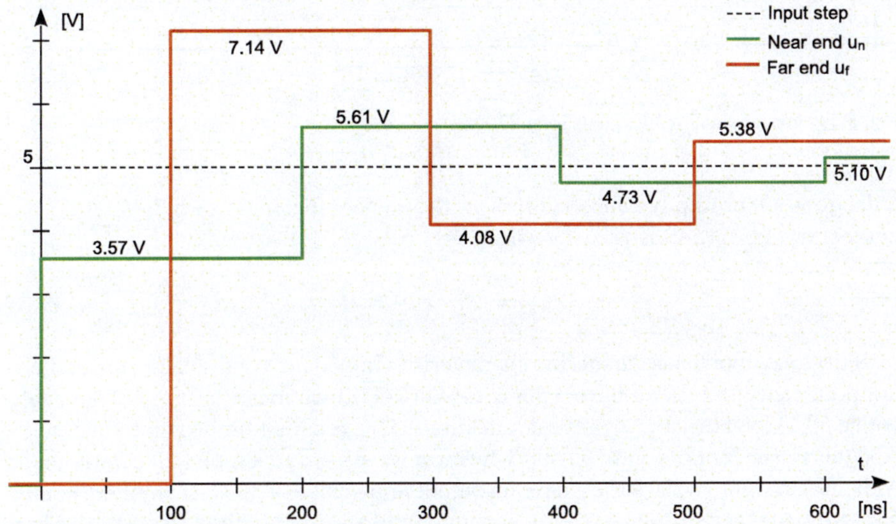

Fig. 5.19 Far- and near-end signals after 600 ns

Third, if this situation was allowed to occur in a serial digital network, like Ethernet or the automotive CAN network, it would corrupt all communication in the network.

5.3 Termination

Reflections in a transmission cable, like the ones we saw in Example 5.1, are almost always unwanted. Admittedly, there are some applications that capitalize on this phenomenon (see Sect. 5.5), but in most situations, pulse reflections are unwanted and need to be remedied. It is not that hard; Eq. (5.2) gives us the answer. The

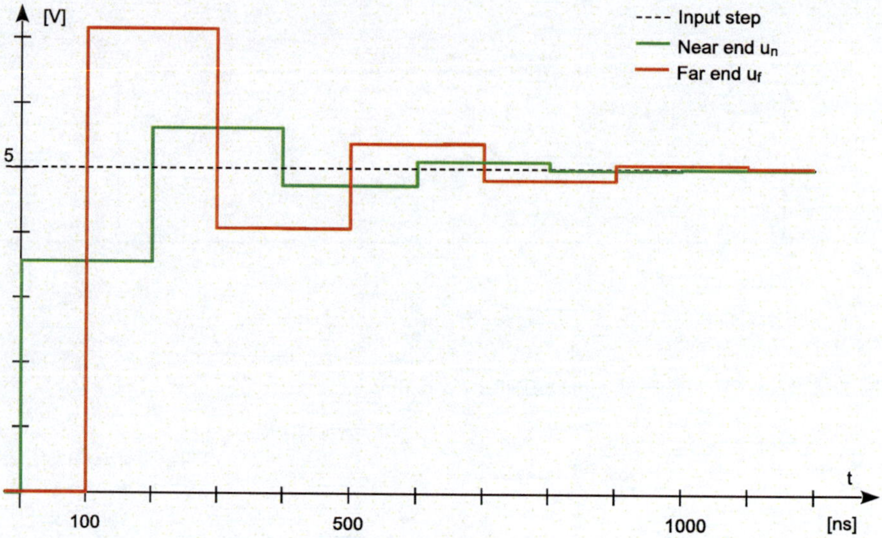

Fig. 5.20 Far- and near-end signals after 1200 ns

reflections are promptly cancelled if the reflection coefficient is zero. It is zero if the numerator in Eq. (5.2) is zero, i.e., if

$$Z_{\mathrm{load}} = Z_0 \Rightarrow \gamma = 0 \tag{5.3}$$

If the load impedance equals the characteristic impedance of the cable, then reflections are cancelled. If we have a 50-Ω cable (RG58) we must 'terminate' the cable with a 50-Ω load resistor to prevent reflections. This is called 'impedance matching'. And this is true for both ends. That is why most waveform generators, like the popular Agilent/Keysight 33220 series, have an output impedance of 50 Ω; if anything comes back from the far end, the waveform generator will absorb it. This is also why modern oscilloscopes always have a 50-Ω input option. The default input impedance of an oscilloscope is 1 MΩ, but it can always be changed to 50 Ω if you have a problem with reflections. Figure 5.21 illustrates an external 50-Ω terminator for an RG58 coax cable.

Don't you always have this problem? Of course, not. See Fig. 5.1. We had no problems there. The problem only occurs when you have 'long' cables or 'fast' signals. 'Long' and 'fast' are relative terms; a cable is 'long' compared to the signal's

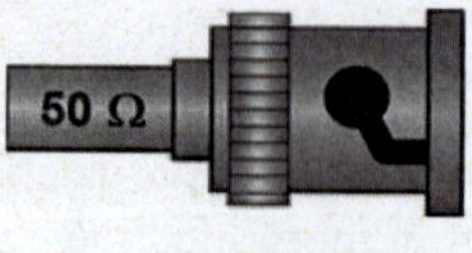

Fig. 5.21 50 Ω terminator

wavelength. You need to worry about cable termination if the cable is longer than about half the wavelength (for a sinusoidal signal). For a square signal, you terminate the cable if the signal rise time is shorter than the propagation time in the cable. (These are just rules of thumb. You must always assess the situation at hand.)

5.4 Splitting and Splicing

Sometimes one cable is not long enough and needs to be spliced, or maybe it needs to be split to feed multiple instruments. Splitting and splicing of high-speed transmission lines is a precarious task and we will investigate both here.

If you need to splice a cable, you should obviously be careful to use identical cables (preferably even from the same cable drum). However, sometimes you might have to splice an RG58 cable to another kind of cable (RG59, TP, …) which does not have the same characteristic impedance. If you connect two cables with different Z_0, there will be reflections at the the joint, see Fig. 5.22. To avoid that, you need to design a simple interface between the cables.

In Fig. 5.23, we want to splice an RG58 coax ($Z_1 = 50\ \Omega$) with an RG62 coax cable ($Z_2 = 93\ \Omega$). All we need is the simple resistor network in Fig. 5.23.

The resistor values are given by the expressions in Eq. (5.4):

$$R_1 = Z_1\sqrt{\frac{Z_2}{Z_2 - Z_1}} \quad R_2 = \sqrt{Z_2(Z_2 - Z_1)} \tag{5.4}$$

Fig. 5.22 Splicing two cables with different Z_0 will generate reflections at the joint

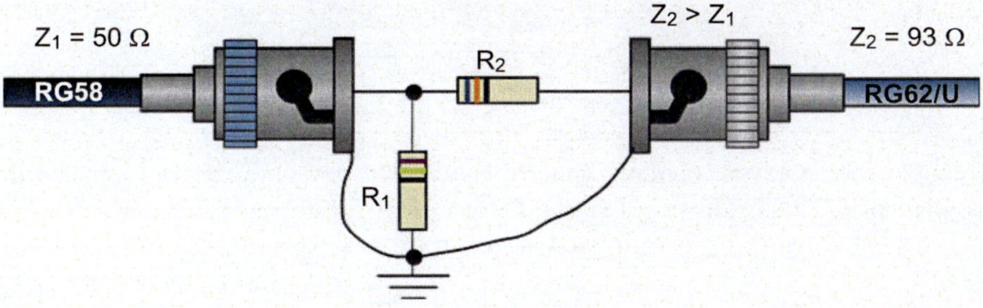

Fig. 5.23 Splicing two cables

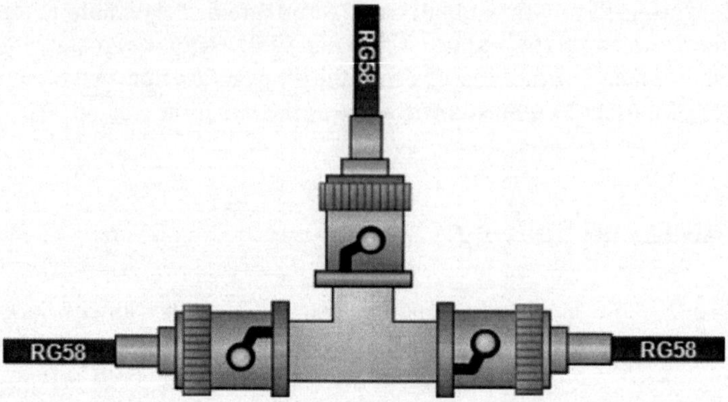

Fig. 5.24 Splitting a cable with a T-cross will cause reflections

Example 5.2 Find the resistor values for R_1 and R_2 if you need to splice an RG58 cable with an RG62 cable.

Solution $Z_1 = 50\ \Omega$ and $Z_2 = 93\ \Omega$:

$$R_1 = 50\sqrt{\frac{93}{93-50}} = 74\,\Omega \quad R_2 = \sqrt{93(93-50)} = 63\,\Omega$$

Another problem is splitting a transmission cable. Most people just use a 'T-cross' connector, see Fig. 5.24. This is not a good idea if you have 'fast' signals, because this splitting will cause reflections. The reason is that the incoming signal (from any direction) will see two cables, i.e., two 50-Ω impedances to ground, which corresponds to 25 Ω to ground, which is an impedance mismatch, and we will have reflections.

The trick is to insert a network so that the perceived impedance is always 50 Ω, regardless of where the signal comes from. The solution is illustrated in Fig. 5.25. This splitting is reflection-free since from any direction the total impedance is 16.7 + (16.7 + 50)//(16.7 + 50) = 16.7 + 66.7//67.7 = 16.7 + 33.3 = 50 Ω.

5.5 Attenuation

Because of the non-zero ohm resistance in any conductor, a propagating signal will be attenuated. This is illustrated in Fig. 5.26.

The attenuation factor is expressed in 'dB/m':

$$\alpha = \frac{20 \cdot \log \frac{u_x}{u_0}}{x} [\text{dB/m}] \tag{5.5}$$

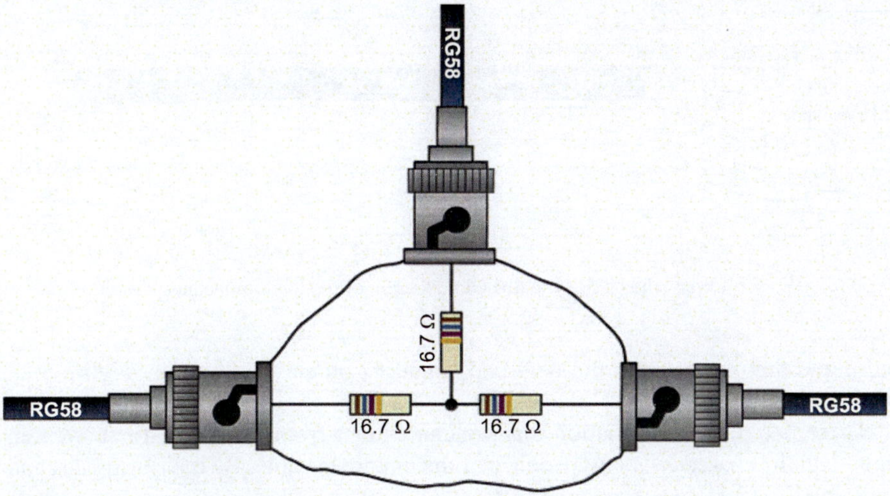

Fig. 5.25 Reflection-free splitting

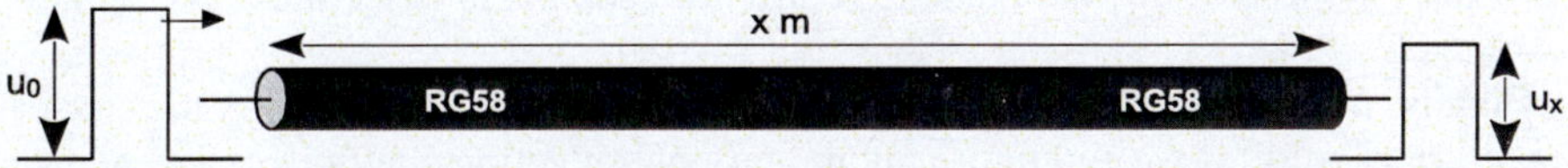

Fig. 5.26 The signal is attenuated in the cable

and hence

$$u_x = u_0 \cdot 10^{\alpha x/20} \tag{5.6}$$

There is one important aspect of the attenuation factor that you need to know; it is highly frequency dependent, $\alpha = \alpha(f)$. You will find an attenuation number in the data sheet but remember that it is for a specific frequency. For example, it could say '−0.1 dB/m @ 50 MHz'. That means that you can use that attenuation factor only for a 50 MHz *sinusoidal* signal. If you use anything else, you must measure the attenuation yourself (by first applying your signal to a cable with a known length.)

5.6 Time Domain Reflectometry

Time Domain Reflectometry, TDR, is an example of a measurement technique that takes advantage of impedance mismatching. Consider the setup in Fig. 5.27.

In Fig. 5.27, the cable type and the cable length are unknown, and so is the load impedance at the far end. Next, suppose the waveform generator sends out a 'short'

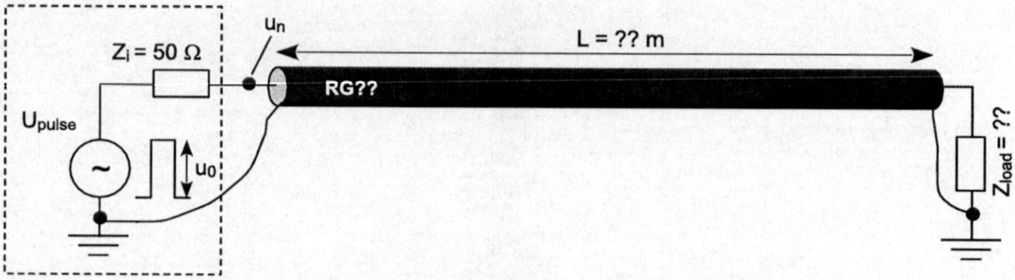

Fig. 5.27 We don't know the cable type, the cable length, or the load impedance

pulse and that we monitor the near-end signal u_n on an oscilloscope. Figure 5.28 illustrates u_n.

There is a lot of information that we can extract from Fig. 5.28. First, we can figure out the characteristic impedance of the unknown cable. By measuring u_{Z0} and applying Eq. (5.1), we can solve for Z_0.

And once we know Z_0, we can identify the cable and then we know the propagation velocity in the cable. And if we know the signal velocity, we can find the cable length, since the time it takes for the pulse to return corresponds to twice the cable length: $L = v \cdot t_{2L}/2$.

Finally, we can also figure out the far-end load impedance by studying the size of the returned pulse; if we can figure out the far-end reflection coefficient, we can use Eq. (5.2) and solve for Z_{load} ($= Z_2$). But this is a little precarious.

The returned pulse will be smaller than u_{Z0}, but it is important to understand that there are *two* reasons for that. It will lose amplitude partly because of the attenuation in the cable and partly because of the reflection against the far-end impedance (see Fig. 5.29).

If we send out a pulse of height u_{Z0}, then according to Eq. (5.6), $u_{Z0} \cdot 10^{\alpha L/20}$ will arrive at the far end. At the far end, there will be wave reflections and the size of the reflected wavefront will be $u_{Z0} \cdot 10^{\alpha L/20} \cdot \gamma_f$. On its way back to the near end, the pulse will again be attenuated, so the height of the pulse returning to the near end is

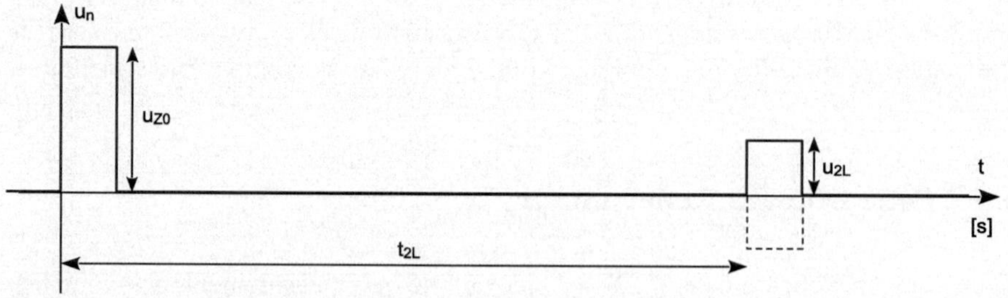

Fig. 5.28 A typical TDR response

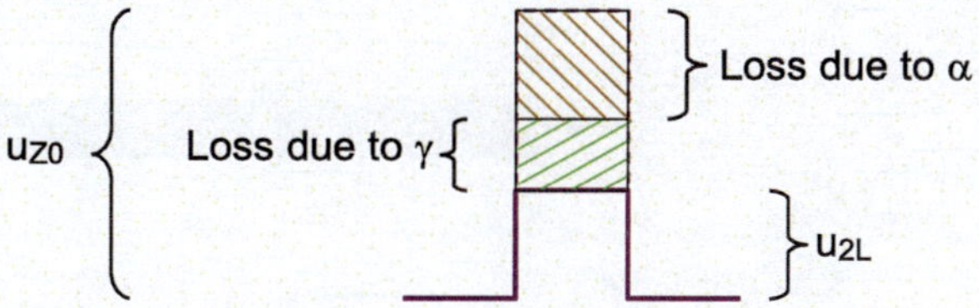

Fig. 5.29 There are two contributions to the signal loss

$$u_{2L} = \underbrace{\underbrace{\underbrace{\underbrace{u_{Z0}}_{\text{Sent out from near end}} \cdot 10^{\alpha L/20}}_{\text{Arriving at far end}} \cdot \gamma_{\mathrm{f}}}_{\text{Reflected back from far end}} \cdot 10^{\alpha L/20}}_{\text{Returns to near end}} = u_{Z0}\gamma_{\mathrm{f}} \cdot 10^{\alpha L/10} \tag{5.7}$$

And solving for γ_{f}:

$$\gamma_{\mathrm{f}} = \frac{u_{2L}}{u_{Z0}} \cdot 10^{-\alpha L/10} \tag{5.8}$$

Inserting this γ_{f} number into Eq. (5.2), we can solve for the load impedance. There is just one problem; we must know the attenuation coefficient in the cable for our probing pulse. There is no easy way to find α. Once you have identified the cable (from Z_0), you must find an identical cable of known length (open-ended) and send in your pulse and study the attenuation of the returned pulse.

This is the basic TDR theory. It can be used to find anyone of the parameters, Z_0, L, and γ, but in 'TDR applications, it is usually understood that it is used only to find L. For example, it is used to localize cable failures in long transmission lines or communication networks.

However, due to the versatility of the TDR technique, it has found applications in a wide range of areas. For example, a TDR-based technique for automatic monitoring of the water content in soil [1, 2] has been reported, landslide warning systems in Taiwan [3] and monitoring of rock mass response in underground mining [4] are other examples of TDR applications.

5.7 Solved Problems

Problem 5.1 A 10-m RG58 coax cable is used to transmit a 45-MHz sine signal. Does this cable need to be 50-Ω terminated?

Solution It needs to be terminated if the cable is longer than half the signal wavelength. The signal wavelength is

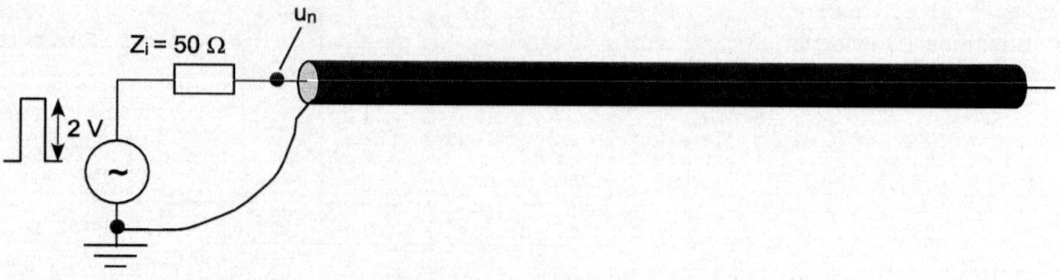

Fig. 5.30 A TDR experiment

$$\lambda = \frac{v}{f} = \frac{2 \cdot 10^8}{45 \cdot 10^6} = 4.4\,\mathrm{m}$$

Half the signal wavelength is 2.2 m which is shorter than the cable length. Yes, this cable needs to be 50-Ω terminated.

Problem 5.2 A CAN bus runs at a bit rate of 5 Mbits/s and the bits' risetime is 10 ns. How long network can you have before you need to terminate the ends of the bus[1]?

Solution If we assume the same velocity factor in a CAN bus cable as in an RG58 coax cable, the signal travels $2{\cdot}10^8 \times 10{\cdot}10^{-9} = 2$ m during the rise time. If the network is longer than that, it needs to be terminated (at both ends).

Problem 5.3 In Example 5.2, we got the resistor values 74 and 63 Ω. These are not 'standard' resistor values. How would you solve this if you only have access to resistors from the E12 series?

Solution The E12 series comprises the numbers 10, 12, 15, 22, 27, 33, 39, 47, 56, 68, and 82 (multiplied by any multiple of 10). Then we use the following combinations to get the desired resistances:

$$74\,\Omega = 47 + 27 \qquad 63\,\Omega = 82//270 = \frac{82 \cdot 270}{82 + 270} = 63$$

To get 74 Ω, we connect a 47-Ω resistor in series with a 27-Ω resistor. To get 63 Ω, we connect an 82-Ω resistor in parallel with a 270-Ω resistor.[2]

Problem 5.4 Consider the TDR experiment in Fig. 5.30. Figure 5.31 illustrates the near-end signal.

The propagation speed in the cable is $0.7c_0$.

---

[1] The characteristic impedance of a CAN bus cable is typically 120 Ω, so 'termination' in this case would be a 120-Ω resistor.

[2] Google "resistor e12 series online combination" and you will find an online tool for how to combine standard resistor values to any other resistance.

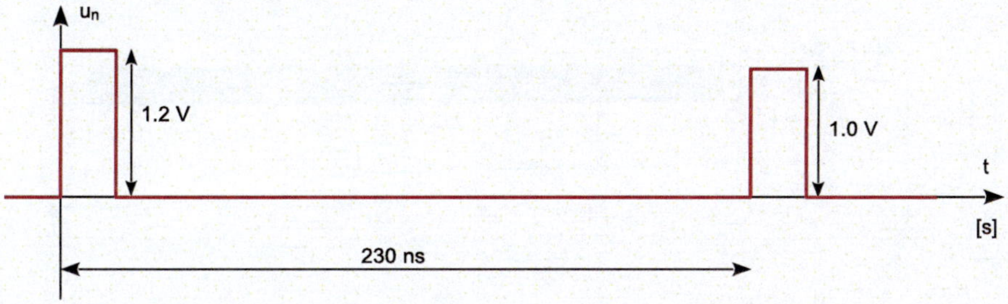

Fig. 5.31 The near-end signal

(a) Determine the characteristic impedance of the cable. (b) What is the cable length? (c) What is the attenuation in the cable? and (d) Draw the corresponding diagram if the far-end is short-circuited to ground.

Solution (a) Equation (5.1) gives $U_{Z0} = \dfrac{Z_0}{Z_0 + 50} \cdot 2 = 1.2 \Rightarrow 2Z_0 = 1.2Z_0 + 60$

$$\Rightarrow 0.8Z_0 = 60 \Rightarrow Z_0 = \underline{75\,\Omega}$$

(b) $L = vt/2 = 0.7 \cdot 3 \cdot 10^8 \cdot 230 \cdot 10^{-9}/2 = \underline{24.15\text{ m}}$

(c) $\alpha = \dfrac{20\log\frac{u_x}{u_0}}{2\cdot L} = \dfrac{20\log\frac{1}{1.2}}{2\cdot 24.15} = \underline{-0.033\text{ dB/m}}$

(d) Short-circuit to ground $\Rightarrow \gamma_f = -1$ (see Fig. 5.32).

Problem 5.5 Consider the experiment in Fig. 5.33.

This is what we know about the cable: It is 30 m long, has a characteristic impedance of 50 Ω, and, for the pulse-type signal in Fig. 5.33, the attenuation is −0.02 dB/m.

Determine the load impedance if the near-end signal looks as in Fig. 5.34.

Solution $u_x = 2 \cdot 10^{-0.02\cdot 30/20}\gamma_f \cdot 10^{-0.02\cdot 30/20} = 1.74\gamma_f = -1 \Rightarrow \gamma_f = -0.574$

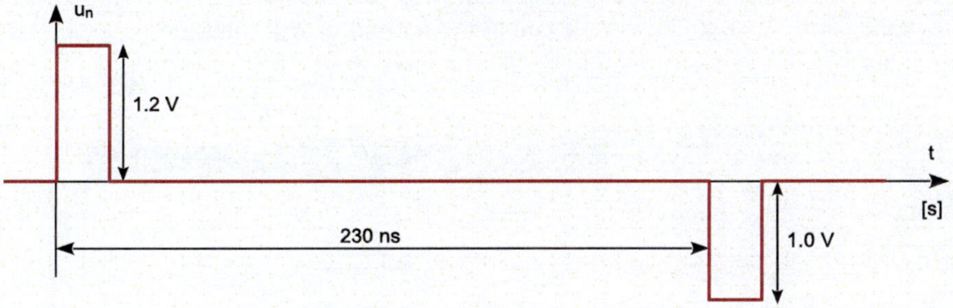

Fig. 5.32 The near-end signal if the far end is short-circuited

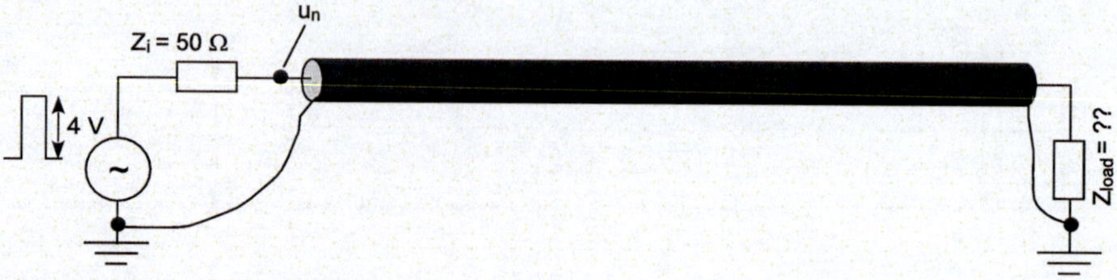

Fig. 5.33 The TDR experiment

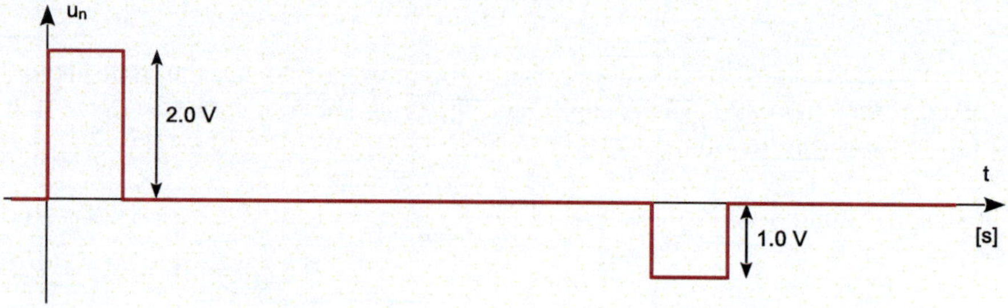

Fig. 5.34 The near-end signal

$$\gamma_f = \frac{Z_{\text{load}} - 50}{Z_{\text{load}} + 50} = -0.574 \Rightarrow Z_{\text{load}} - 50 = -0.574 \cdot Z_{\text{load}} - 28.7$$

$$1.574 \cdot Z_{\text{load}} = 21.3 \Rightarrow \underline{Z_{\text{load}} = 13.5\ \Omega}$$

Problem 5.6 Consider the experiment in Fig. 5.35 (the input signal is a step function).

Plot the near-end signal if Z_{load} is (a) an open end, (b) a short-circuit to ground, (c) a 50-Ω resistor, (d) a capacitor, and (e) an inductor. (Disregard attenuation in this problem.)

Solution First, let's figure out what the step response of u_{n} is. Since we have an RG58 cable, it has a characteristic impedance of 50 Ω; at $t = 0$, 2 V will propagate down the cable. If anything is reflected at the far end, it will return to the near end

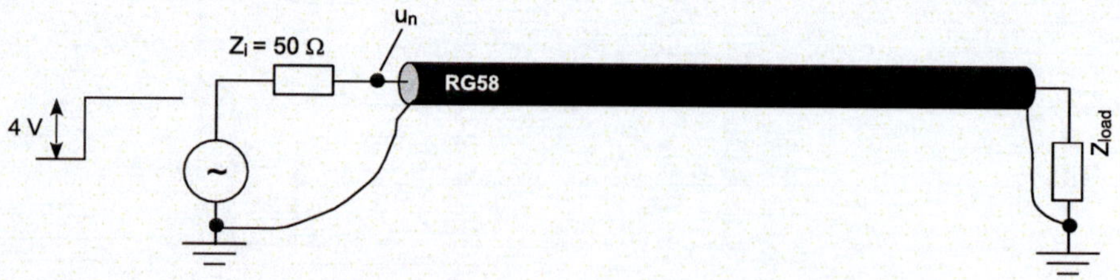

Fig. 5.35 A TDR experiment

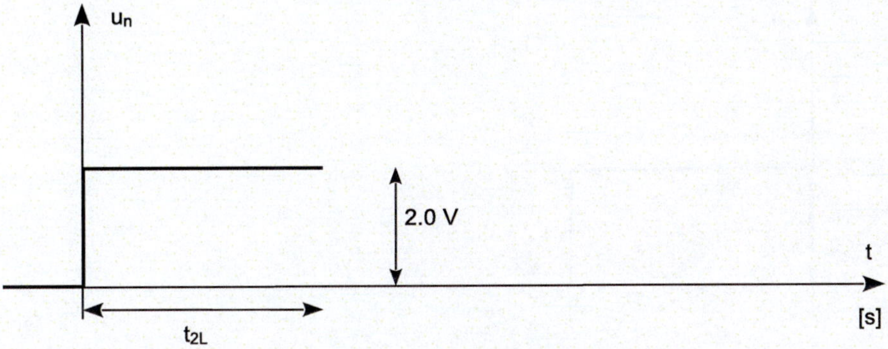

Fig. 5.36 The first t_{2L} seconds are independent of the load

after $t_{2L} = 2L/v$ seconds. So, whatever Z_{load} is, the near-end signal will always look the same for the first t_{2L} seconds, see Fig. 5.36.

If the far end is open, $\gamma_f = +1$ and all the incoming +2 V is reflected, interacting constructively with the incoming waves, leaving a voltage of $2 + 2 = 4$ V behind. When it reaches the near end, u_n will = 4 V, see Fig. 5.37.

If the far end load is a short-circuit, the reflection coefficient is –1, and the reflected 2 V will interact destructively with the incoming waves and since $2 - 2 = 0$, the voltage level on the cable is cancelled. At t_{2L}, $u_n = 0$, see Fig. 5.38.

If $Z_{load} = 50\ \Omega$, the far-end reflection coefficient is 0 and nothing is returned, see Fig. 5.39.

If the load is a capacitor, it will initially be uncharged and as long as it is uncharged, current can pass right through it (like a short-circuit). But soon (sooner rather than later) it will be fully charged and then it will stop conducting current and act like an open circuit. Hence, at first it will act like Fig. 5.38, but soon it will act as in Fig. 5.37. Figure 5.40 illustrates the capacitor case.

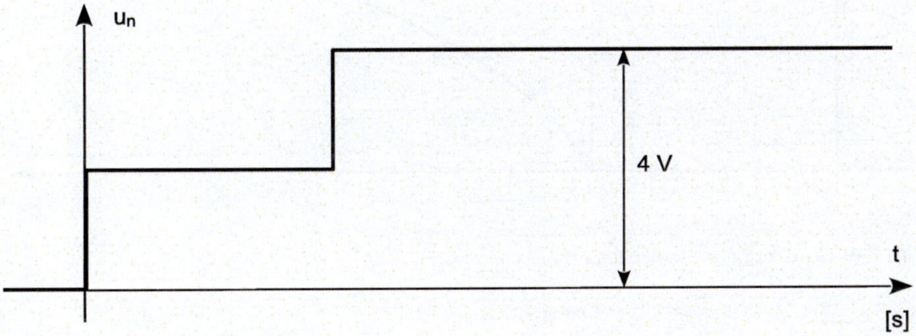

Fig. 5.37 u_n if $Z_{load} = \infty$

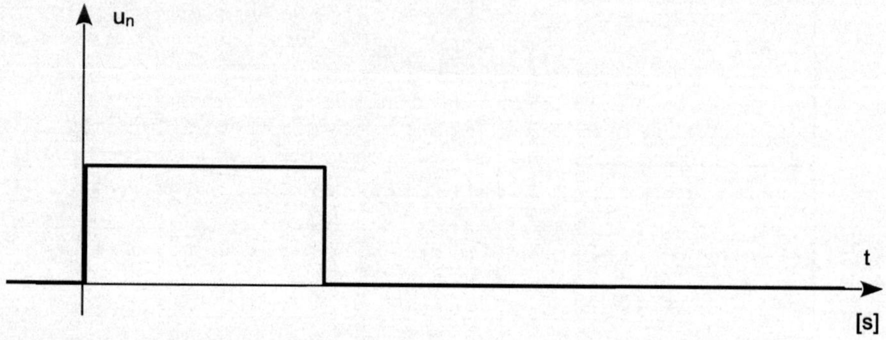

Fig. 5.38 u_n if $Z_{\text{load}} = 0$

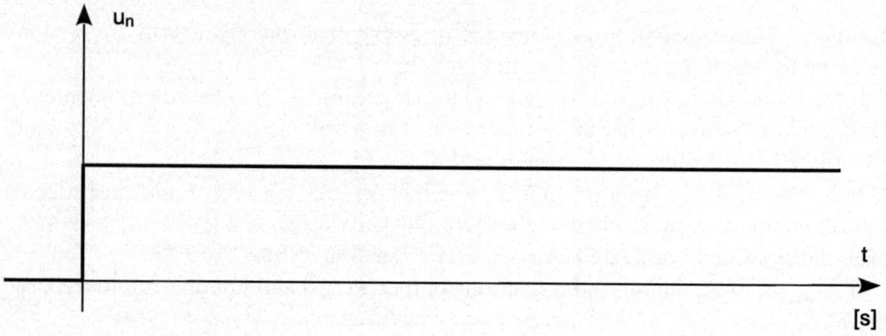

Fig. 5.39 u_n if $Z_{\text{load}} = 50\ \Omega$

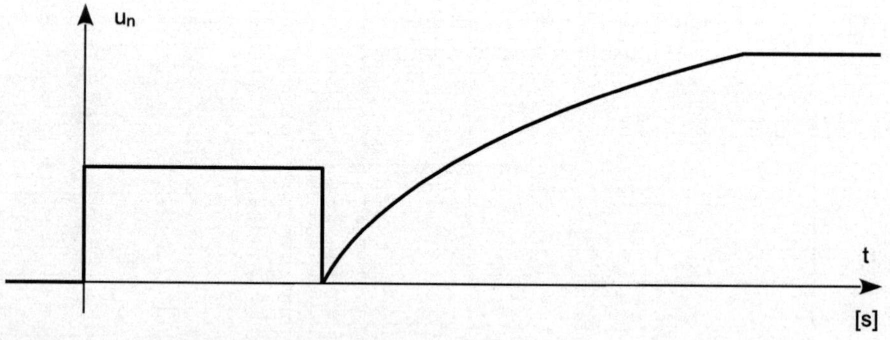

Fig. 5.40 u_n if $Z_{\text{load}} =$ a capacitor

On the other hand, if the load is an inductor, it will initially act as an open circuit (Fig. 5.37), but once it understands that it is just a DC signal, it will act as a short-circuit (Fig. 5.38). Figure 5.41 illustrates the inductor case.

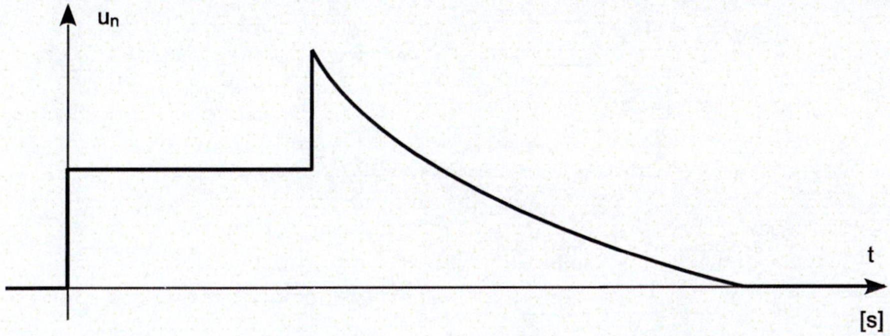

Fig. 5.41 u_n if Z_{load} = an inductor

Fig. 5.42 Splice the cables

Problem 5.7 In an experiment, a TP signal cable needs to be spliced to an RG58 coax cable, see Fig. 5.42. How would you do that if the characteristic impedance of the TP cable is 75 Ω?

Solution Eq. (5.4) gives

$$R_1 = 50\sqrt{\frac{75}{75-50}} = 87\,\Omega = 100\,\Omega\,//\,680\,\Omega$$

$$R_2 = \sqrt{75(75-50)} = 43\,\Omega = 10\,\Omega + 33\,\Omega$$

Above we have assumed that we only have access to E12 series resistor values. The reflection-free splicing is illustrated in Fig. 5.43.

Problem 5.8 Suggest another way to do the splitting in Fig. 5.25.

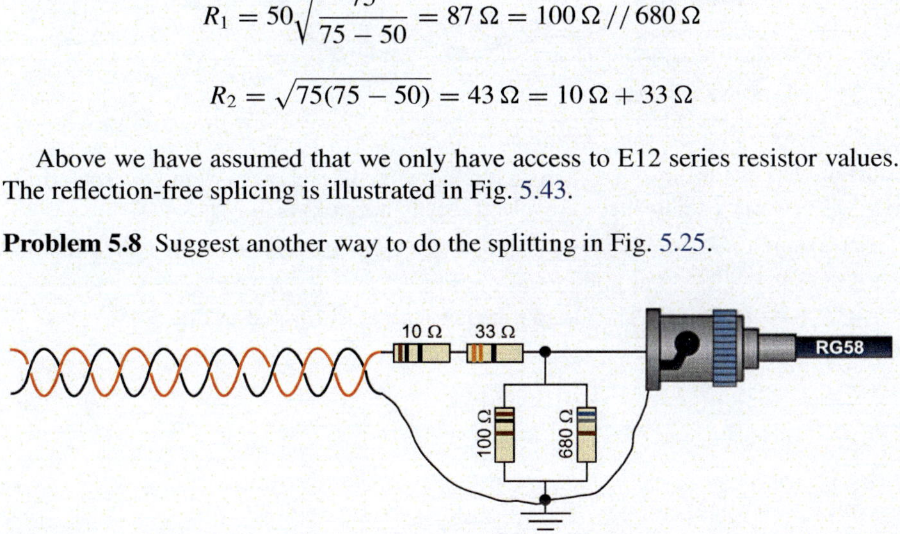

Fig. 5.43 Reflection-free splicing

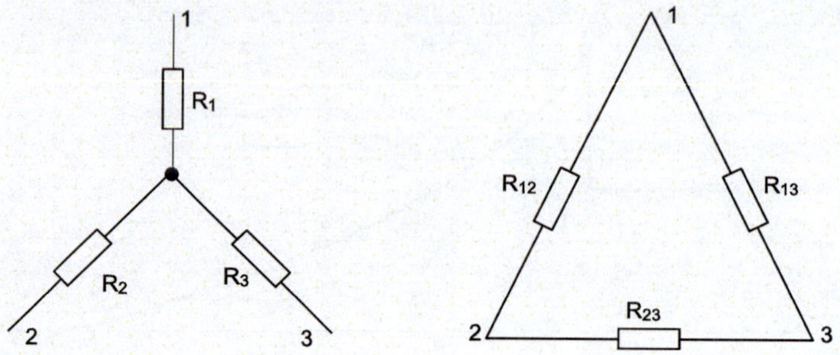

Fig. 5.44 Y and Δ networks

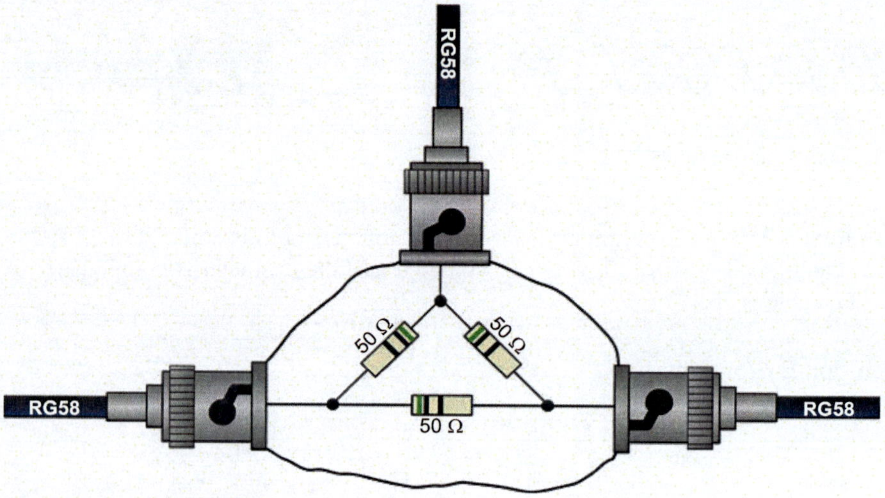

Fig. 5.45 Alternative splitting

Solution The circuit in Fig. 5.25 is a 'Y' network ('wye'). This can be transferred to a 'Δ' network ('delta') (see any electricity handbook) (see Fig. 5.44).

For example, $R_{12} = R_1 R_2 \left(\frac{1}{R_1} + \frac{1}{R_2} + \frac{1}{R_3} \right)$, but if $R_1 = R_2 = R_3$, then this is reduced to $R_{12} = 3R_1$. For our network in Fig. 5.25, that would give us $R_{12} = R_{23} = R_{13} = 3 \times 16.7 = 50\ \Omega$. The alternative network is illustrated in Fig. 5.45.

References

1. Wraith, J.M., et al. 2005. Spatially characterizing apparent electrical conductivity and water content of surface soils with time domain reflectometry. *Computers and Electronics in Agriculture* 46 (1–3): 239–261.
2. Walker, J.P., G.R. Willgoose, and J.D. Kalma. 2004. In situ measurement of soil moisture: A comparison of techniques. *Journal of Hydrology* 293 (1–4): 85–99.
3. Su, M.-B., I.-H. Chen, and C.-H. Liao. 2009. Using TDR cables and GPS for landslide monitoring in high mountain area. *Journal of Geotechnical and Geoenvironmental Engineering* 135 (8): 1113–1121.
4. O'Connor, K.M., and L.V. Wade. 1994. Applications of time domain reflectometry in the mining industry. In *Proceedings of the symposium and workshop on TDR in environmental, infrastructure and mining applications, Northwestern University, Evanston, IL, USA, Sep 1994.*

Chapter 6
Probes

Abstract This chapter emphasizes the need for probes in electrical measurements. A 'probe' is used to minimize the measurement system's interference with the measurement object (observing without interfering). Passive probes are introduced in Sect. 6.2, active probes in Sect. 6.3, and current probes in Sect. 6.4.

6.1 Introduction

Consider the simple network in Fig. 6.1.

It is easy to see that the potential at point A is 2.5 V. But, of course, we don't really *know* that until we *measure* it. A common DMM (or an oscilloscope) has an input impedance of 1 MΩ. If we connect that over the 200-kΩ resistor, see Fig. 6.2, then the voltage meter's impedance will be in parallel with the 200-kΩ resistor, and the total impedance will be 200//1000 = 167 kΩ, see Fig. 6.2.

By trying to *measure* the voltage, we *change* the system. In this example, the result is off by almost 10%! System interference is inevitable; you can (almost) never measure something without disturbing the system. The important thing here is to first be aware of that, and second, to do what you can to minimize your instruments' interference.

In the previous example, it is obvious that the problem is the 'high' source impedance (or the 'low' instrument impedance); the source's and the instrument's impedances are of the same order. Obviously, the remedy is to increase the impedance of the instrument so that it is 'much higher' than the source impedance.

That is exactly what a 'probe' does; it increases the impedance of the instrument to minimize the instrument's interference with the system. In the previous DC example, the probe would just be a simple high-Ω resistor in series with the voltage meter. In Fig. 6.3, we have connected a 9-MΩ resistor in series, which means that the total impedance is now 200//10000 = 196 kΩ and the potential in point A is now 2.47 V; our new instrument only causes a disturbance corresponding to 1%. That is quite an improvement.

L. Bengtsson, *Electrical Measurement Techniques*,
https://doi.org/10.1007/978-981-99-8187-8_6

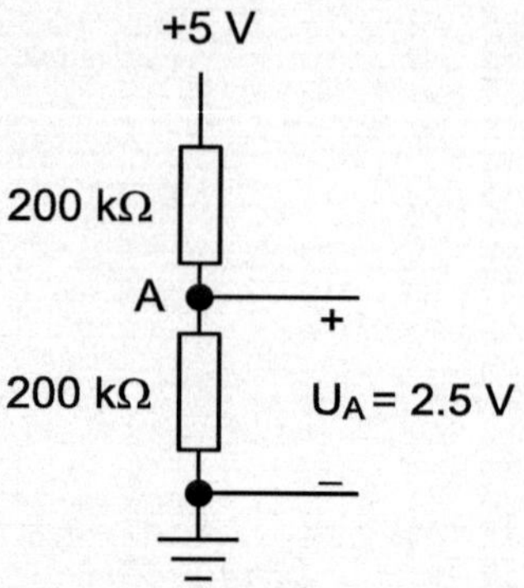

Fig. 6.1 Simple DC measurement

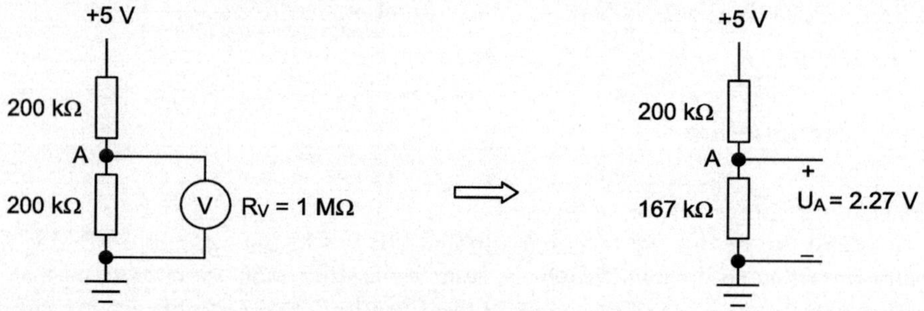

Fig. 6.2 Connecting the instrument will *change* the impedance

Fig. 6.3 Using a probe

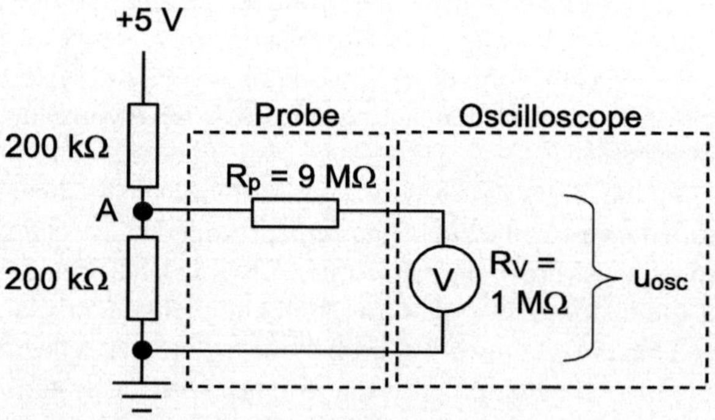

Of course, it could be argued that this will cause an error in the measurement; the voltage meter in Fig. 6.3 will only measure one-tenth of this (0.247 V), but that is easily compensated for (just multiply by a factor of 10).

In this example, it was easy and straightforward to design a probe. However, probes are mostly used with oscilloscopes measuring AC signals, and that makes things a little more complicated. The main objective here is to design probes for oscilloscopes.

6.2 Passive Probes

First, we need an accurate model of the oscilloscope's input. An oscilloscope input is always marked with a resistance and a capacitance, see Fig. 6.4. That refers to the input impedance values as illustrated in the right-hand side of Fig. 6.4; the input is modeled as a resistor and a capacitor in parallel. (The 'capacitor' is not a capacitor; it is the inherent capacitance of the input.)

Hence, the input impedance is

$$Z_1 = X_{C1}//R_1 = \frac{\frac{1}{\mathrm{j}\omega C_1} R_1}{\frac{1}{\mathrm{j}\omega C_1} + R_1} = \frac{R_1}{1 + \mathrm{j}\omega R_1 C_1} \tag{6.1}$$

In Eq. (6.1), we see the problem; Z_1 depends on the frequency.

In Fig. 6.3, we multiplied the voltage meter reading by a factor of 10 to get the right answer. That is acceptable, if we can *always* multiply by the same constant (= 10) to get the right answer, but if we did the same trick in Fig. 6.4, the portion of the input voltage that falls over R_1 would depend on the frequency; we would have to calculate a new multiplication factor for each new frequency, and that would make the trick next to useless. So, for AC signals and oscilloscopes, we need to be a little more creative. Figure 6.5 illustrates our 'creative' solution.

In Fig. 6.5, the 'probe' is the R_2/C_2 circuit, and this is what we need to make the fraction of u_{in} that ends up over R_1 independent of the input signal's frequency. That may look unlikely at first sight, since we have added one more component that has a frequency-dependent impedance (C_2), but we will prove it shortly. First, we consider the DC case. If the input signal is DC, then we can disregard the capacitors and we are back to the same problem as in Fig. 6.3. The most common probes are 'X10' probes, meaning that only one-tenth of the input signal ends up over R_1 and that's what we are aiming for here. Well, for that to be true for DC signals, we must have $R_2 = 9\ \text{M}\Omega$, just like in Fig. 6.3.

So, we already know the value of R_2. The trick now is to, if possible, select a value for C_2 such that the fraction of the voltage u_{in} that falls over R_1 is always one-tenth

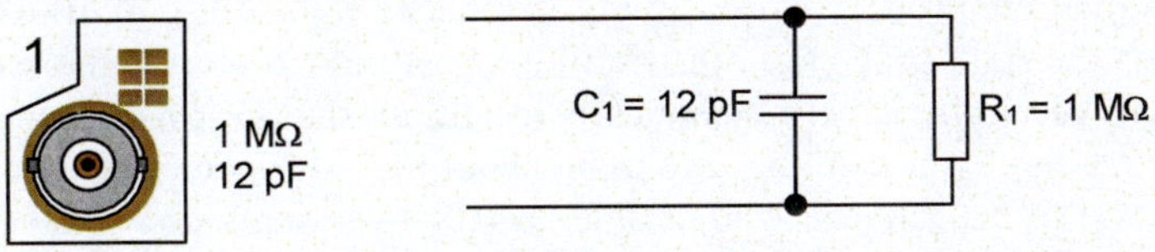

Fig. 6.4 Oscilloscope input

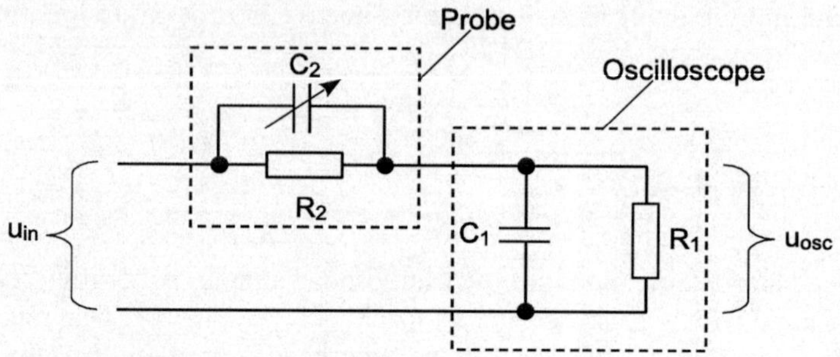

Fig. 6.5 Oscilloscope with probe ($R_1 = 1\ \mathrm{M\Omega}$)

of u_{in}, regardless of the input signal's frequency. If we can't find such a capacitance, the whole idea will be rejected. Let's see what u_{osc} is

$$u_{\mathrm{osc}} = \frac{\frac{R_1}{1+\mathrm{j}\omega R_1 C_1}}{\underbrace{\frac{R_1}{1+\mathrm{j}\omega R_1 C_1} + \frac{R_2}{1+\mathrm{j}\omega R_2 C_2}}_{\delta}} \times u_{\mathrm{in}} \tag{6.2}$$

Equation (6.2) does not look very promising; there are a lot of ωs in Eq. (6.2) and we want δ to be independent of ω; we want $\delta = 1/10$. Well, that is possible; if we make $R_1 C_1 = R_2 C_2$, we can cancel all the $1 + \mathrm{j}\omega R_x C_x$ denominators and then:

$$u_{\mathrm{osc}} = \frac{R_1}{R_1 + R_2} \times u_{\mathrm{in}} = \frac{1}{1+9} \times u_{\mathrm{in}} = \frac{1}{10} \times u_{\mathrm{in}} \tag{6.3}$$

which is exactly what we are looking for. So, if

$$C_2 = \frac{R_1}{R_2} C_1 = \frac{1}{9} \times 12 = 1.33\,\mathrm{pF} \tag{6.4}$$

we have a *frequency-independent voltage divider* in Fig. 6.5, and that's what a *passive* probe is. Figure 6.6 illustrates a typical oscilloscope probe.

The advantage of an oscilloscope probe is that your instrument's 'disturbance' on the system is significantly reduced (which is paramount), but the disadvantage is that since only a fraction of the input voltage (one-tenth) ends up over the oscilloscope (the rest is over the probe), the sensitivity is reduced by a factor of 10. However, if you insert the right probe (i.e., the oscilloscope vendor's probe), the scope will automatically recognize it and automatically recalibrate the vertical scale to give you the correct reading (you don't have to think about the 1/10 factor). As you can see in Fig. 6.5, the C_2 capacitance is adjustable, usually by a small screw on the probe's head, (see oscilloscope manual for the procedure), but modern oscilloscopes have an 'auto-calibration' option for probes (in some menu somewhere…).

Apart from not disturbing the voltage at the measurement point, there is one more thing you need to consider. We will illustrate that by an example.

Example 6.1 In Fig. 6.7, two systems are cascaded, one with bandwidth 100 MHz and one with bandwidth 120 MHz. What is the overall bandwidth of this system?

Solution On an exam, a lot of students would answer '100 MHz', arguing that in a cascaded system, the component with the narrowest bandwidth defines the overall bandwidth. That is wrong (and earns you zero credits on an exam). It is a lot worse than that. Go back to Chap. 1, Fig. 1.4, and Eqs. (1.7) and (1.8), and you will see why. The overall risetime of the system in Fig. 6.7 is

$$t_{\text{total}} = \sqrt{\left(\frac{0.35}{0.12}\right)^2 + \left(\frac{0.35}{0.10}\right)^2} = 4.56\,\text{ns}$$

which means that the overall bandwidth is

$$B_{\text{total}} = \frac{0.35}{4.56} = 77\,\text{MHz}$$

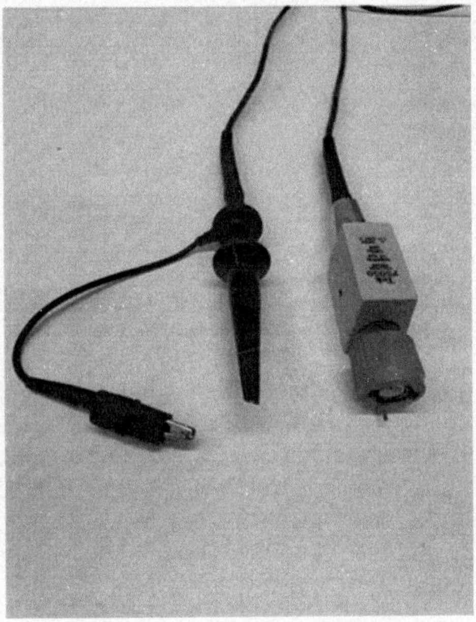

Fig. 6.6 A 'passive' oscilloscope probe

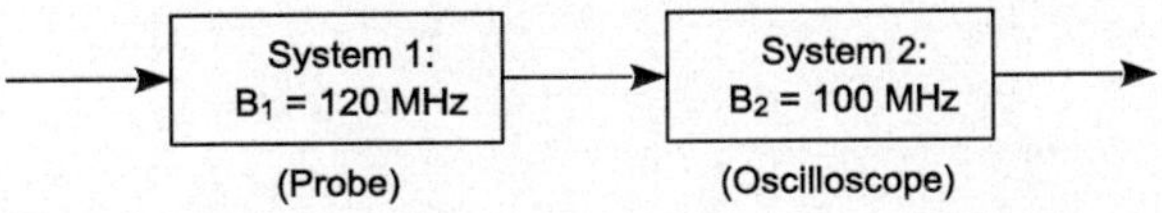

Fig. 6.7 Two cascaded systems

We learn two things from this example. First, we learn that the overall bandwidth of a cascaded system of components is always less than the bandwidth of the system with the narrowest bandwidth. Second, notice that by adding the probe to the oscilloscope above, we significantly reduced the system's bandwidth. We went from 100 MHz without the probe to 77 MHz, with the probe. So, we learned one more thing:

The probe's bandwidth must be significantly larger than the oscilloscope's bandwidth (or the overall bandwidth will be reduced).

6.3 Active Probes

If high input impedance is what we want, then we can easily find a better solution; feed the input signal to the input of a op amp in a voltage follower circuit, see Fig. 6.8.

The probe in Fig. 6.8 is an 'active' probe; it is 'active' because it needs external power to work. Anyway, this is obviously the perfect probe! The input of an op amp is extremely large so it would not induce any disturbance at the measurement point, and it also does not reduce the sensitivity like the passive probe does. So, if this probe is so perfect, why don't we always use it? Why do we bother with passive probes at all? The probe in Fig. 6.8 is so perfect, it is almost too good to be true.... And when something sounds too good to be true, it usually is. Active probes are no exception. The problem with active probes is that they are *very* expensive. An active probe could easily be €10,000. Why are they so expensive? It looks like a simple enough circuit. The answer is in the previous example. We concluded that the probe's bandwidth must be much larger than the oscilloscope's bandwidth. If you have a 500 MHz oscilloscope, you need maybe a 5 GHz probe. So, even if the circuit in Fig. 6.8 looks simple and innocent enough, if you scale it up to the GHz range (and above), it becomes a very complicated (and expensive) design. So, unless you have very deep pockets, you will be stuck with passive probes.

Here is a common question about active probes: If 'active' means that it needs external power, does that mean that I need to supply power from an external DC power supply?

No, you don't. Again, if you use the right probe, the oscilloscope will recognize it, and provide the necessary power automatically through the copper pads on the oscilloscope input, see Fig. 6.9. (So, for God's sake, don't buy the wrong active probe to your oscilloscope! If the scope doesn't recognize it, it is a lot of money down the drain.)

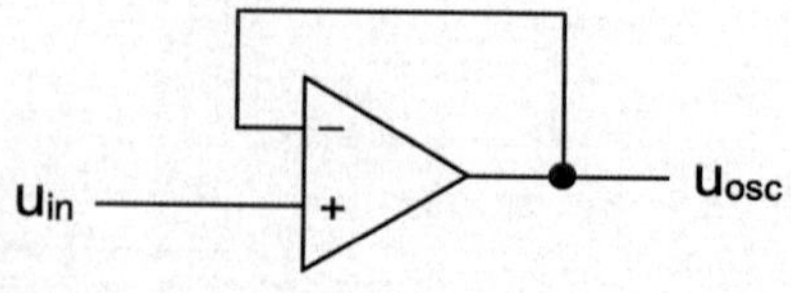

Fig. 6.8 An 'active' probe

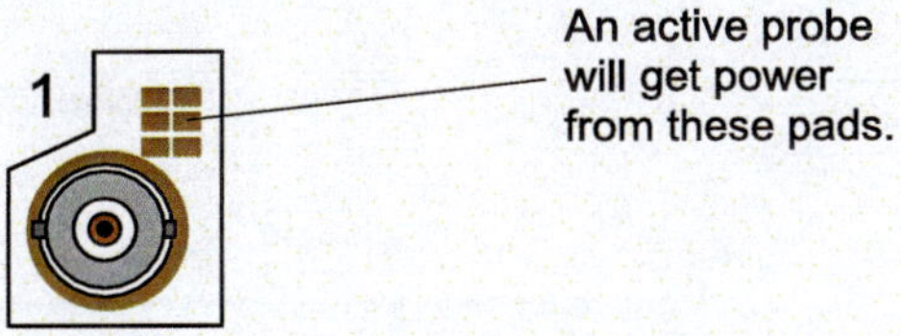

Fig. 6.9 An active probe is powered from the oscilloscope

6.4 Current Probes

Oscilloscopes are *voltage meters*; they inherently measure voltage. You could of course measure current by measuring the voltage across some reference resistor, but the vertical scale on your scope would still be voltage [V]. However, you can measure current with an oscilloscope (with amps [A] on the vertical axis) if you buy a *current* probe.

Current probes for oscilloscopes are non-contact probes; the probe clamps an iron core around the wire-under-test and the induced magnetic flux in the iron core is proportional to the current. Figure 6.10 illustrates the principle.

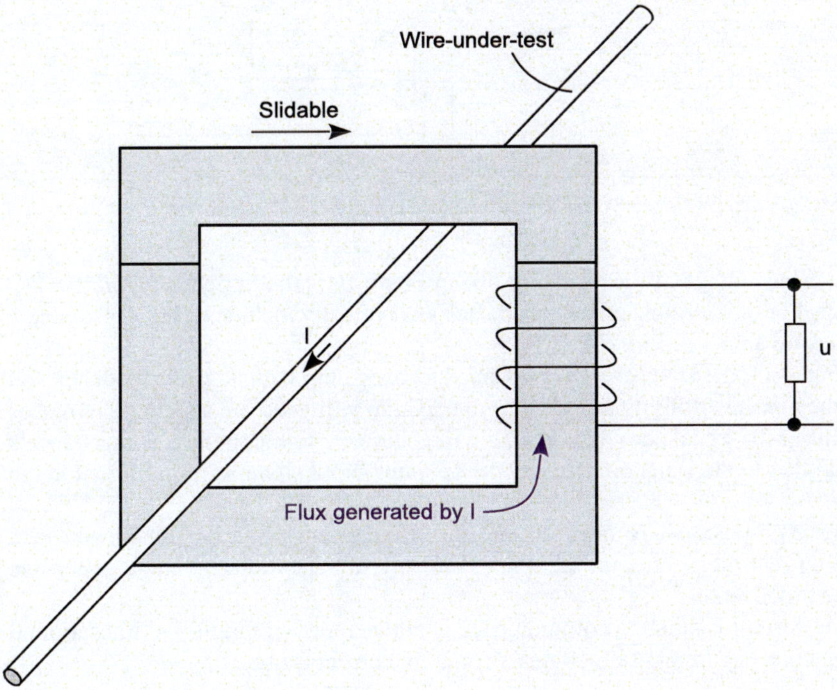

Fig. 6.10 The current probe principle

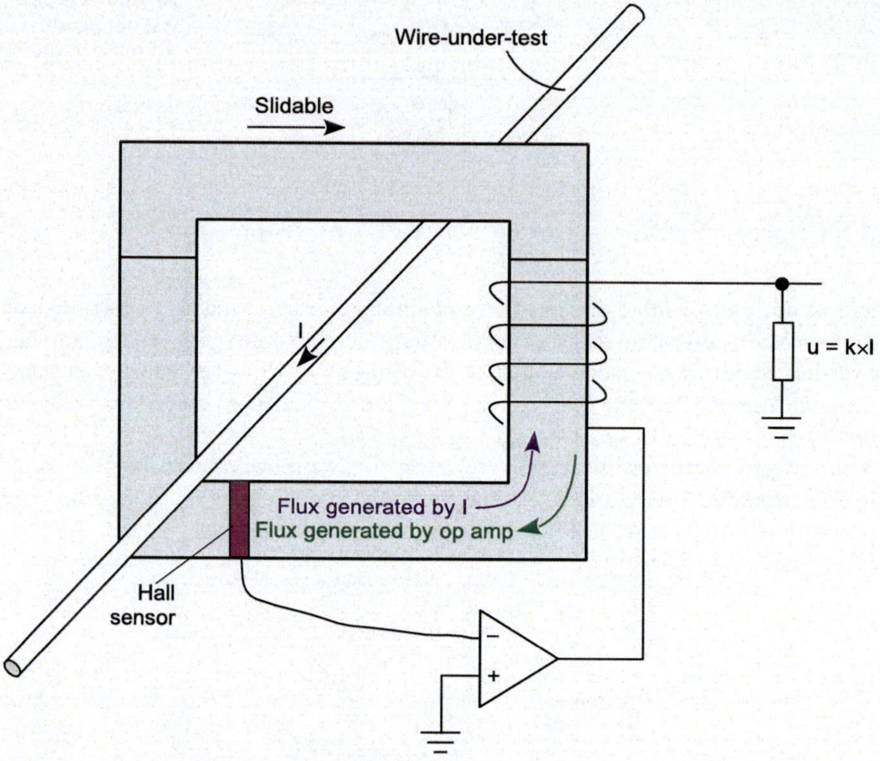

Fig. 6.11 Current probe with op amp feedback

The disadvantage of the current probe in Fig. 6.10 is that it doesn't work for DC currents. For that reason, current probe designs typically include a Hall sensor and a feedback op amp, see Fig. 6.11.

In Fig. 6.11, the Hall sensor signal is fed to an op amp with negative feedback and since the non-inverting input is at 0 V, the op amp will generate a signal that makes the Hall sensor signal zero; the op amp will generate a signal in the coil that cancels the total flux in the iron core. Hence, the op amp output signal is proportional to the current.

Current probes are expensive too, and the price depends on the bandwidth you need for the current measurement; the higher the current bandwidth the more expensive will the probe be.

If the current is small, a common trick is to twist the wire multiple times around the iron core. Figure 6.12 illustrates a common current probe.

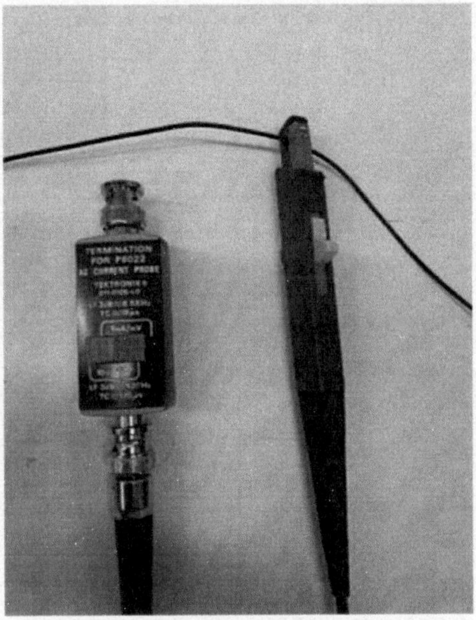

Fig. 6.12 Current probe

6.5 Solved Problems

Problem 6.1 Design a $\times 10$ probe for the oscilloscope in Fig. 6.13.

Solution Fig. 6.14 illustrates the probe. If we first consider the DC case (ignore the capacitors), we can find R_2:

$$\frac{3}{3+R_2} = \frac{1}{10} \Rightarrow \underline{R_2 = 27\,\mathrm{M\Omega}}$$

Equation (6.4) gives us $C_2 = R_1C_1/R_2 = 3 \cdot \frac{40}{27} = \underline{4.44\,\mathrm{pF}}$.

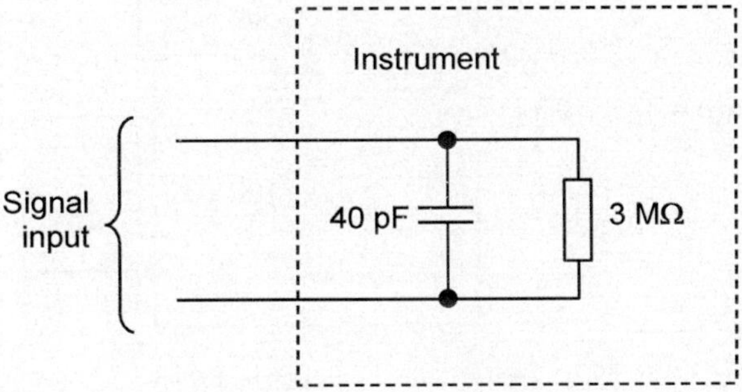

Fig. 6.13 Instrument input

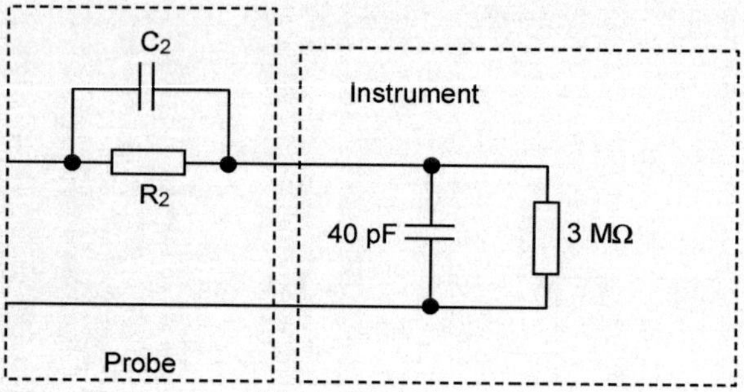

Fig. 6.14 The probe

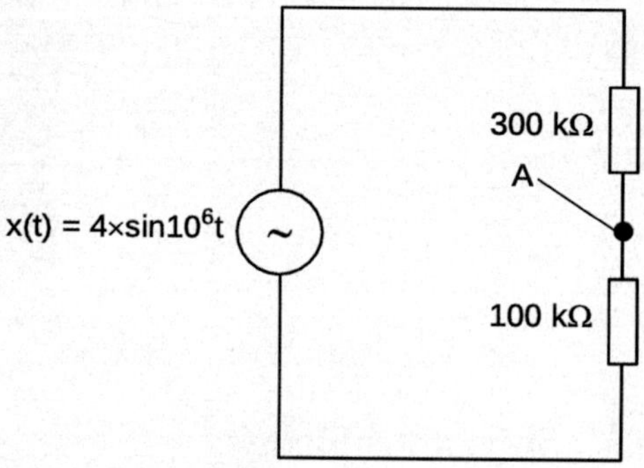

Fig. 6.15 Find the 'true' potential at A

Problem 6.2a What is the voltage at point A in Fig. 6.15?

Solution One-fourth of $x(t)$ falls over the 100-kΩ resistor: $\underline{u_A(t) = 1 \times \sin 10^6 t \text{ V}}$.

Problem 6.2b What signal will the oscilloscope in Fig. 6.16 measure?

Solution 100//1000 = 90.9 kΩ. $X_C = -j/\omega C = -j/10^6 \cdot 15 \cdot 10^{-12} = -j66.7\,\text{k}\Omega$

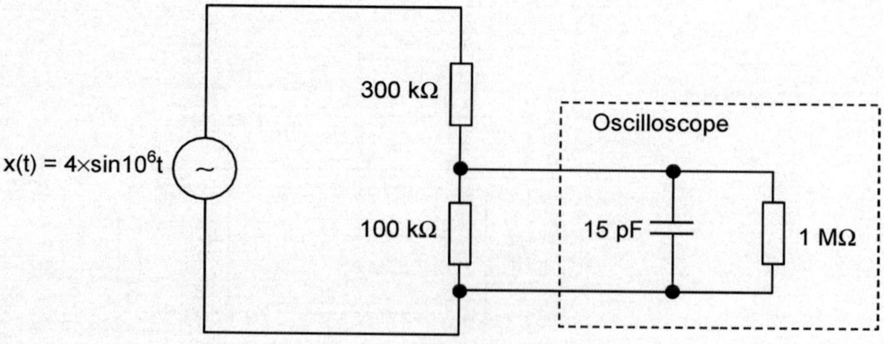

Fig. 6.16 What signal will the oscilloscope display?

$$Z = 90.9// - \mathrm{j}66.7 = \frac{-\mathrm{j}66.7 \cdot 90.9}{90.9 - \mathrm{j}66.7} = \frac{6063 \cdot \mathrm{e}^{-\mathrm{j}90^\circ}}{112.7 \cdot \mathrm{e}^{-\mathrm{j}36.3^\circ}} = 53.8 \cdot \mathrm{e}^{-\mathrm{j}53.7^\circ}$$
$$= 31.8 - \mathrm{j}43.5\,\mathrm{k}\Omega$$

$$u_{\mathrm{A}}(t) = \frac{Z}{Z + 300} \cdot x(t) = \frac{53.8 \cdot \mathrm{e}^{-\mathrm{j}53.7^\circ}}{31.8 - \mathrm{j}43.4 + 300} \cdot 4 \cdot \mathrm{e}^{\mathrm{j}10^6 t} = \frac{53.8 \cdot \mathrm{e}^{-\mathrm{j}53.7^\circ}}{334.6 \cdot \mathrm{e}^{-\mathrm{j}7.5^\circ}} \cdot 4 \cdot \mathrm{e}^{\mathrm{j}10^6 t}$$
$$= \underline{0.64 \cdot \mathrm{e}^{\mathrm{j}\left(10^6 t - 46.2^\circ\right)}}$$

Compared to the 'true' value, we have an amplitude error of 36% and a phase error of 46.2°.

Problem 6.2c Design a × 10 probe for the oscilloscope in Fig. 6.16.

Solution $R_2 = 9\,\mathrm{M}\Omega$, $C_2 = 15/9 = 1.67\,\mathrm{pF}$.

Problem 6.2d What will the oscilloscope measure if we use the probe at point A?

Solution Fig. 6.17 illustrates the oscilloscope with the probe.

First, we find the impedances Z_1 and Z_2 (see Eq. (6.1)):

$$Z_1 = \frac{1\,\mathrm{M}}{1 + \mathrm{j} \cdot 10^6 \cdot 10^6 \cdot 15 \cdot 10^{-12}} = 66.5 \cdot \mathrm{e}^{-\mathrm{j}86.2^\circ}\,\mathrm{k}\Omega = 4.4 - \mathrm{j}66.4\,\mathrm{k}\Omega$$

$$Z_2 = \frac{9\,\mathrm{M}}{1 + \mathrm{j} \cdot 10^6 \cdot 9 \cdot 10^6 \cdot 1.67 \cdot 10^{-12}} = 598.7 \cdot \mathrm{e}^{-\mathrm{j}86.2^\circ}\,\mathrm{k}\Omega = 39.7 - \mathrm{j}597.4\,\mathrm{k}\Omega$$

$$Z_1 + Z_2 = 44.1 - \mathrm{j}663.8\,\mathrm{k}\Omega = 665.3 \cdot \mathrm{e}^{-\mathrm{j}86.2^\circ}\,\mathrm{k}\Omega$$

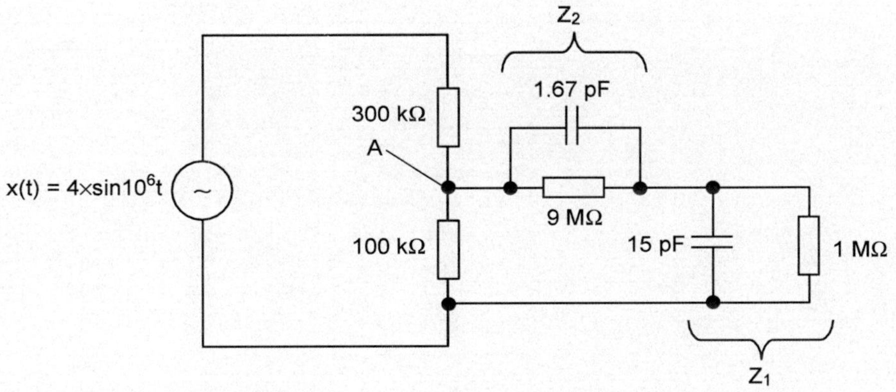

Fig. 6.17 Measuring with probe

$$Z_A = (Z_1 + Z_2)//100\,\text{k}\Omega = \frac{100 \cdot 665.3 \cdot e^{-j86.2^\circ}}{144.1 - j663.8} = \frac{66530 \cdot e^{-j86.2^\circ}}{679.3 \cdot e^{-j77.8^\circ}}$$
$$= 97.9 \cdot e^{-j8.4^\circ}\,\text{k}\Omega = 96.8 - j14.3\,\text{k}\Omega$$

$$u_A(t) = \frac{Z_A}{Z_A + 300} \cdot x(t) = \frac{97.9 \cdot e^{-j8.4^\circ}}{396.8 - j14.3} \cdot 4 \cdot e^{j10^6 t} = \frac{97.9 \cdot e^{-j8.4^\circ}}{397.1 \cdot e^{-j2.1^\circ}} \cdot 4 \cdot e^{j10^6 t}$$
$$= 0.99 \cdot e^{-j(10^6 t - 6.3^\circ)} = \underline{0.99 \times \sin(10^6 t - 6.3^\circ)\,\text{V}}$$

We now have a 1% error in the amplitude and 6.3-degree error in the phase.

Chapter 7
Transform Theory

Abstract This long and important chapter introduces a mathematical tool called *transforms.* This is probably the most important mathematical operation used in electrical measurements. It transforms a signal in time space to frequency space, and this is extremely common (and useful) to understand and analyze your measurement signal. The focus in this chapter is the *understanding* of transforms and it starts with understanding exactly what is meant by the *frequency* (and later we must understand why the frequency can be a complex number). Transform theory is by most students perceived to be 'hard' and the main reason for that is that there appears to be so many different transform expressions; depending on the nature of the (time) signal, it is necessary to use different mathematical expressions, but they all really do the same thing (i.e., transfer a time signal to frequency space). Because there are so many different expressions, this chapter tries to organize them for you (see Table 7.6). Several different transforms are introduced; the Fourier transform, the discrete Fourier transform, the Fast Fourier transform, the Laplace transform, and the z transform, but remember they all do the same thing; they take your signal from time space to frequency space. The main objective of this chapter is to help the reader *understand* transforms and see how they are related. This chapter also introduces the Bode plot and defines LTI systems (linear and time-invariant system).

7.1 Introduction

You should consider this to be one of the most important chapters in this book. 'Transform theory' is what we use to transfer a signal from 'time space' to 'frequency space'. It will become obvious that we can learn so much more about any signal (or system) if we leave time space and go to frequency space. Our primary objective here is to do a 'frequency analysis' and the tool(s) we need to do that is called a 'transform'. There is not just one transform, there are several different ones, depending on the signal we are looking at. The signal could be 'analog' or it could be 'digital'. In this context, 'digital' means 'sampled' (the mathematical term is 'discrete'). We will also introduce a few new frequency variables. For example, we will have both

L. Bengtsson, *Electrical Measurement Techniques*,
https://doi.org/10.1007/978-981-99-8187-8_7

'non-complex' and 'complex' frequencies. That gives us four transforms already, see Table 7.1.

We will fill the gaps in this table as we learn about the transform tools in this chapter.

But we need to start from the beginning. We should start by properly defining exactly what we mean by 'frequency'. I'm sure you have a clear idea about what a signal's 'frequency' is, but our definition of 'frequency' in this context is very strict: If a signal has frequency f, it is a *harmonic* function (sinusoidal) with period $T = 1/f$. Hence, the signal in Fig. 7.1 has frequency f, but not the signal in Fig. 7.2. (Well, it does have the frequency f, but also at lot of other frequencies, see Example 7.1.)

Hence, when we say 'frequency' we mean the frequency of a harmonic signal.

We also need to define the word 'analysis', but first we will present a fundamental theorem.

Theorem *All signals can be expressed as a sum of cosines*:

$$\begin{aligned} x(t) &= a_0 + a_1 \cos(\omega_1 t + \varphi_1) + a_2 \cos(\omega_2 t + \varphi_2) + \ldots \\ &= a_0 + \sum_k a_k \cos(\omega_k t + \varphi_k) \end{aligned} \tag{7.1}$$

Table 7.1 The transform map

	Non-complex frequency	Complex frequency
Analog		
Sampled		

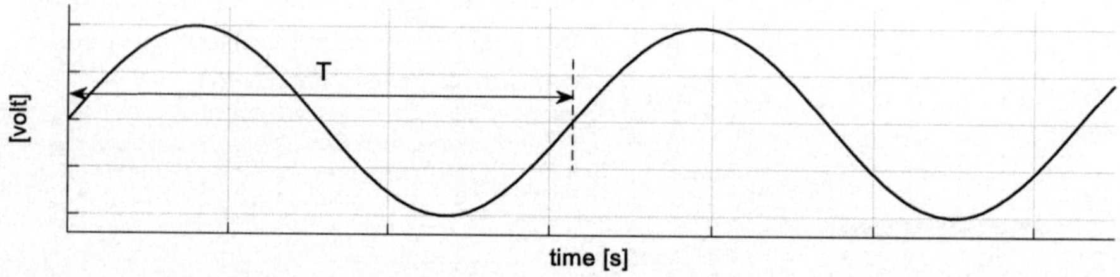

Fig. 7.1 A signal with frequency $f = 1/T$

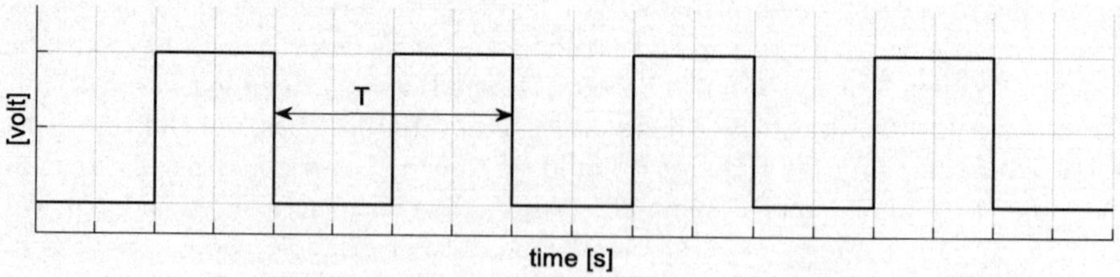

Fig. 7.2 This signal has frequency $f = 1/T$ and many more (see Example 7.1)

The domain of k depends on the signal; if $x(t)$ is periodic with period T, then the domain of k is all positive integers and $\omega \in k \cdot \omega_0$, where

$$\omega_0 = \frac{2\pi}{T} \Rightarrow \omega_0 T = 2\pi \tag{7.2}$$

On the other hand, if $x(t)$ is not periodic, then k can be any number and $\omega \in \mathbb{R}$ (all real numbers). When $x(t)$ is not periodic, Eq. (7.1) doesn't make sense. We will come back to that later.

Now we can define 'analysis'. In general, 'analysis' means 'decompose into constituents'. When we do 'frequency analysis' of a signal, we express it as in Eq. (7.1), i.e., we write it as a linear combination of cosines (the cosines are the 'base vectors'). If $x(t)$ is periodic, we already know the frequencies (they are $k \cdot \omega_0$); we only need to find all the amplitudes a_k and phase angles φ_k. The tool we use to find the amplitudes and the phase angles is called the *Fourier transform*.

7.2 The Fourier Transform

The Fourier transform takes our signal and produces an expression that is a function of the frequency, $X(\omega)$ (Fig. 7.3); this is a complex function and the magnitude of $X(\omega)$ gives us the amplitude of the cosine with frequency ω, and the argument (arg $X(\omega)$) of $X(\omega)$ gives us the phase angle.

Depending on the signal, we can identify three different cases that we need to treat separately.

7.2.1 *Case 1: Signal is Periodic*

In the first case, the signal is periodic, the period is T and, in that case, only frequencies which are a multiple of ω_0 can exist in Eq. (7.1). For that reason, we write the Fourier transform $X(\omega)$ as $X(\omega) = X(k{\cdot}\omega_0) = X(k)$, and we define it as

$$X(k) = \frac{1}{T}\int_0^T x(t) \cdot e^{-j\omega t} dt = \frac{1}{T}\int_0^T x(t) \cdot e^{-jk\omega_0 t} dt \tag{7.3}$$

Example 7.1 Express the square wave in Fig. 7.4 as a sum of cosines.

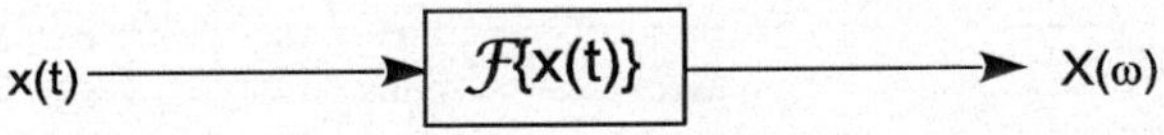

Fig. 7.3 The Fourier transform operator

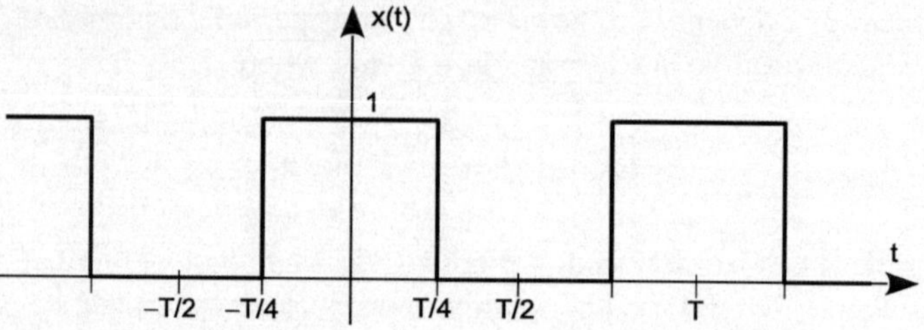

Fig. 7.4 We will analyze a square signal

Solution First, we find the Fourier transform expression. In Eq. (7.3), we can integrate over any period. Since the signal is symmetric, we choose to integrate over a symmetric interval:

$$X(k) = \frac{1}{T}\int_{-T/2}^{T/2} x(t)\mathrm{e}^{-\mathrm{j}k\omega_0 t}dt = \frac{1}{T}\int_{-T/4}^{T/4} 1\cdot \mathrm{e}^{-\mathrm{j}k\omega_0 t}dt = -\frac{1}{\mathrm{j}k\omega_0 T}\left[\mathrm{e}^{-\mathrm{j}k\omega_0 t}\right]_{-T/4}^{T/4}$$
$$= -\frac{1}{\mathrm{j}k2\pi}\left(\mathrm{e}^{-\mathrm{j}k\omega_0 \frac{T}{4}} - \mathrm{e}^{\mathrm{j}k\omega_0 \frac{T}{4}}\right) = \frac{1}{k\pi}\cdot\frac{1}{2\mathrm{j}}\left(\mathrm{e}^{\mathrm{j}k\pi/2} - \mathrm{e}^{-\mathrm{j}k\pi/2}\right) = \frac{1}{k\pi}\sin\frac{k\pi}{2} \tag{7.4}$$

In Eq. (7.4), we can immediately see that $X(k) = 0$ if k is an even number ($\neq 0$). If k is odd, then $\sin k\pi/2 = \pm 1$ and $X(1) = 1/\pi$, $X(3) = -1/3\pi$, $X(5) = 1/5\pi$, etc. For $k = 0$, we need to find the limit when $k \to 0$. We use l'Hospital's rule to find the limit:

$$\lim_{k\to 0}\frac{\sin k\frac{\pi}{2}}{k\pi} = \lim_{k\to 0}\frac{\frac{\pi}{2}\cos k\frac{\pi}{2}}{\pi} = \frac{1}{2}$$

Hence, we get the following Fourier transform:

$$X(k) = \begin{cases} \frac{1}{2} & \text{if } k = 0 \\ \pm\frac{1}{k\pi} & \text{if } k \text{ is odd} \\ 0 & \text{if } k \text{ is even} \end{cases} \tag{7.5}$$

Notice in Eq. (7.5) that k is both positive *and* negative, which suggests negative frequencies. But of course, frequencies cannot be negative; the negative k values are only a consequence of Euler's formulas for sine and cosine:

$$a_k\cos(\omega_k t + \varphi_k) = a_k\frac{1}{2}\left(\mathrm{e}^{\mathrm{j}(\omega_k t+\varphi_k)} + \mathrm{e}^{-\mathrm{j}(\omega_k t+\varphi_k)}\right)$$
$$= \underbrace{\frac{1}{2}a_k\cdot \mathrm{e}^{\mathrm{j}\varphi_k}\cdot\mathrm{e}^{\mathrm{j}\omega_k t}}_{X(k)} + \underbrace{\frac{1}{2}a_k\cdot \mathrm{e}^{-\mathrm{j}\varphi_k}\cdot\mathrm{e}^{-\mathrm{j}\omega_k t}}_{X(-k)} \tag{7.6}$$

Euler's formula for cosine assigns half the signal energy to the 'positive' frequency and the other half to the 'negative' frequency. Except for $k = 0$, of course, where $+0$ and -0 coincide, and all the energy is in $k = 0$ ($k = 0$ is the 'DC' part of the signal). From Eq. (7.6), we can see that

$$|X(k)| = \frac{1}{2}a_k \Rightarrow a_k = 2 \cdot |X(k)| \quad k \neq 0 \tag{7.7a}$$

$$\varphi_k = \arg X(k) \tag{7.7b}$$

$$a_0 = |X(0)| \tag{7.7c}$$

Expression (7.7) is the link between the Fourier transform and the sum of cosines in Eq. (7.1). In our example, we get

$$\begin{aligned} a_0 &= |X(0)| = \frac{1}{2} \\ X(1) &= \frac{1}{\pi} = \frac{1}{\pi}\mathrm{e}^{\mathrm{j}0} \Rightarrow a_1 = 2 \cdot \frac{1}{\pi} \qquad \varphi_1 = 0 \\ X(3) &= -\frac{1}{3\pi} = \frac{1}{3\pi}\mathrm{e}^{\mathrm{j}\pi} \Rightarrow a_3 = 2 \cdot \frac{1}{3\pi} \qquad \varphi_3 = \pi \\ X(5) &= \frac{1}{5\pi} = \frac{1}{5\pi}\mathrm{e}^{\mathrm{j}0} \Rightarrow a_5 = 2 \cdot \frac{1}{5\pi} \qquad \varphi_5 = 0 \end{aligned}$$

Since $\cos(\alpha + \pi) = -\cos(\alpha)$, we can now use Expression (7.1) to write $x(t)$ as

$$x(t) = \frac{1}{2} + \frac{2}{\pi}\left(\cos\omega_0 t - \frac{1}{3}\cos 3\omega_0 t + \frac{1}{5}\cos 5\omega_0 t - \ldots..\right) \tag{7.8}$$

In Fig. 7.5, we have plotted the amplitude spectrum of $x(t)$ and in Fig. 7.6 we have plotted the different components and the sum of the first four terms in Eq. (7.8) to compare it with the original square signal.

We can learn a few important things from this example. First, notice that we use k as the frequency variable. Get used to that! The frequency is a multiple of ω_0 and it is very important that you get used expressing frequency in terms of k. $\omega_0 = 2\pi/T$

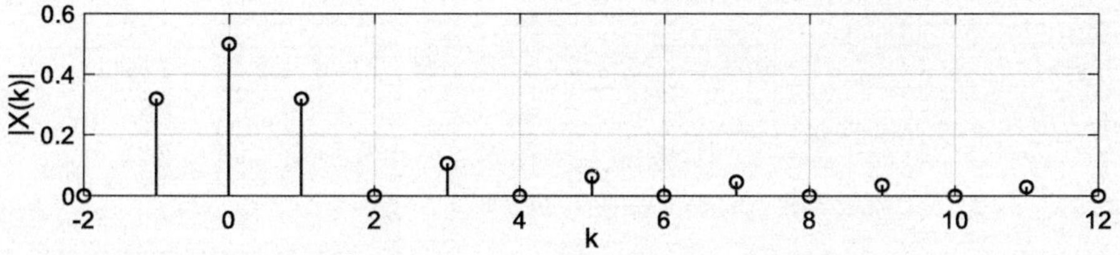

Fig. 7.5 The amplitude spectrum of $x(t)$

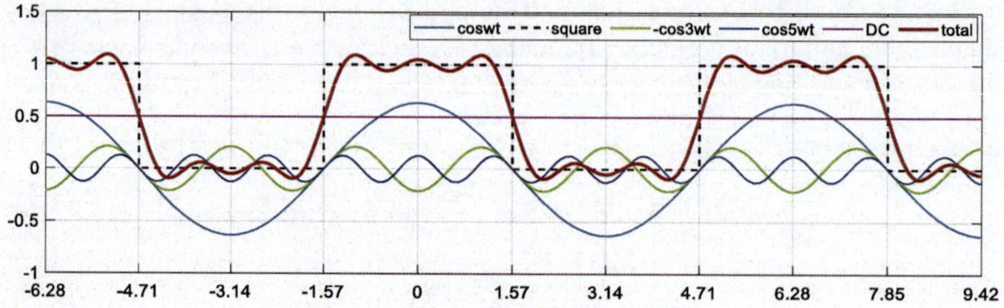

Fig. 7.6 The different frequency components of $x(t)$

or $f_0 = 1/T$. So, for example, when we say that the frequency is '3', we mean that it is $3 \times f_0$. Here is an alternative way to think of k: We know the period T of the signal and k tells us how many periods of a cosine that fits in T, see Fig. 7.7.

Second, notice that we only plotted the amplitude spectrum in Fig. 7.5; we didn't plot the phase spectrum. We could have done that too, but we didn't. This is typical; in most cases, we only care about the distribution of amplitudes in a signal (see Chap. 8 about spectral analysis).

Let's return to Eq. (7.3). The Fourier transform expression has a very important feature; the amplitude spectrum is *symmetric*, i.e.,

$$X(-k) = X^*(k) \Rightarrow |X(-k)| = |X(k)| \tag{7.9}$$

(In Eq. (7.3), it doesn't matter if we change sign of k or j.) As we will see later, this is a feature that characterizes all Fourier transforms.

Now we can update our transform map in Table 7.1 with our first transform.

In Table 7.2, we had to split the analog/non-complex frequency cell into two, because we are not done with this signal category yet.

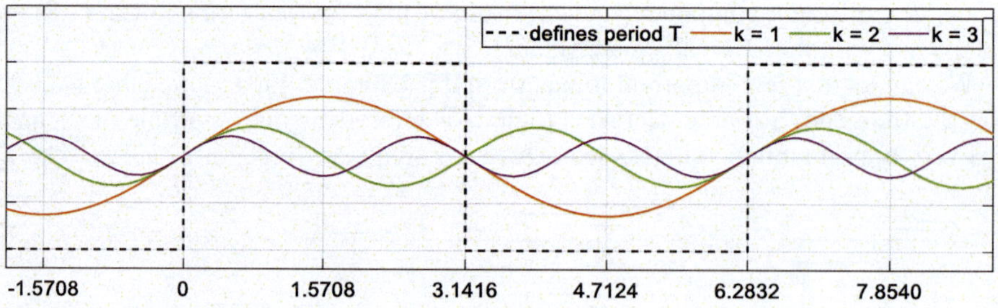

Fig. 7.7 k is our frequency variable

Table 7.2 The transform map

	Non-complex frequency	Complex frequency
Analog	$X(k) = \frac{1}{T}\int_{0}^{T} x(t)\mathrm{e}^{-\mathrm{j}k\omega_0 t}dt$	
Sampled		

7.2.2 *Case 2: Signal is Non-Periodic, But 'Time-Limited'*

Another name for a 'non-periodic, time-limited' signal is a 'transient', see Fig. 7.8.

A transient is characterized by the fact that its *energy* is limited. (A periodic signal has infinite energy, but limited *power*.) Since a transient doesn't have a period T, we cannot define an ω_0 and the frequencies of the cosines in Eq. (7.1) cannot be predicted with some simple formula. As a matter of fact, when $x(t)$ is a transient, all frequencies are allowed and $\omega \in \mathbb{R}$ (all real numbers). That means that Expression (7.1) is no longer meaningful, and when we describe a transient, we only present its amplitude spectrum (which is now a *continuous* function). Fourier transform Expression (7.3) must be manipulated to allow any frequency:

$$X(\omega) = \int_{-\infty}^{+\infty} x(t) \cdot \mathrm{e}^{-\mathrm{j}\omega t}dt \tag{7.10}$$

Example 7.2 Plot the magnitude spectrum of the transient in Fig. 7.9. What is the signal's bandwidth?

Solution First we need to find the Fourier transform:

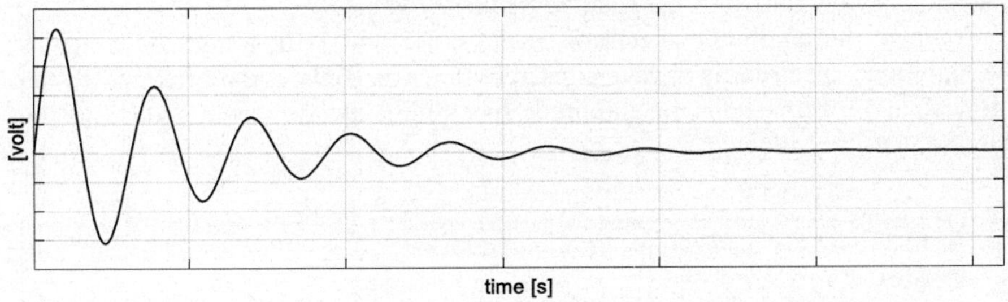

Fig. 7.8 A transient

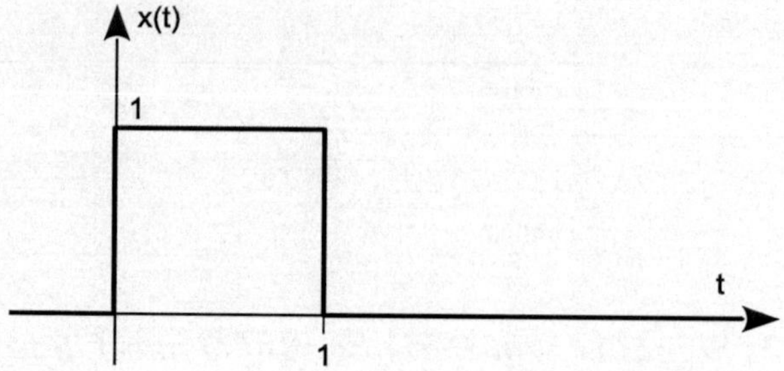

Fig. 7.9 A transient

$$X(\omega)=\int_0^1 1\cdot e^{-j\omega t}dt=-\frac{1}{j\omega}\cdot\left[e^{-j\omega t}\right]_0^1=-\frac{1}{\omega}\cdot\frac{1}{j}\left(e^{-j\omega}-1\right)=$$

$$=-\frac{2}{\omega}\cdot\underbrace{\frac{1}{2j}\left(e^{-j\omega/2}-e^{j\omega/2}\right)}_{-\sin\omega/2}\cdot e^{-j\omega/2}=\frac{2}{\omega}\cdot\sin\frac{\omega}{2}\cdot e^{-j\omega/2}=\text{sinc}\frac{\omega}{2}\cdot e^{-j\omega/2}\quad(7.11)$$

$$\Rightarrow |X(\omega)|=|\text{sinc}\omega/2|$$

The amplitude spectrum is plotted in Fig. 7.10.

In Fig. 7.10, we can see that the signal's bandwidth is infinite, but most of its energy is in the interval 0… 2π rad/s.

Notice in Eq. (7.10) that the Fourier transform for a transient is also symmetric:

$$X(-\omega)=X^*(\omega)\Rightarrow|X(-\omega)|=|X(\omega)|\quad(7.12)$$

So, if we also plotted the amplitude spectrum in Fig. 7.10 for negative frequencies, it would just be a mirror of the positive frequency values.

Compare the amplitude spectrums in Figs. 7.5 and 7.10; for periodic signals, the amplitude spectrum is always a discrete function (only certain frequencies are allowed) and for transients the amplitude spectrum is always a continuous function (any frequency is allowed).

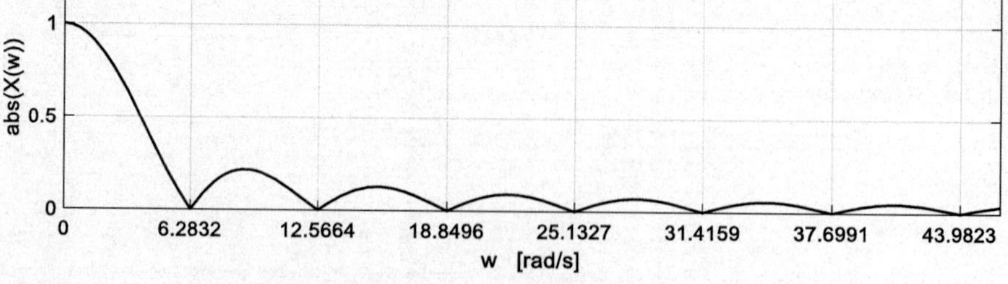

Fig. 7.10 The amplitude spectrum is now a continuous function

Table 7.3 The transform map

		Non-complex frequency	Complex frequency
Analog	Periodic	$X(k) = \frac{1}{T}\int_{0}^{T} x(t)\mathrm{e}^{-\mathrm{j}k\omega_0 t}\,dt$	
	Transient	$X(\omega) = \int_{-\infty}^{\infty} x(t)\mathrm{e}^{-\mathrm{j}\omega t}\,dt$	
Sampled			

Let's update our transform map with our new Fourier transform (Table 7.3).

7.2.3 *Case 3: Signal is Non-Periodic and Infinite*

Figure 7.11 illustrates a non-periodic, infinite signal.

This signal is a little bit of a headache because we don't have a transform formula for it. It is not periodic, and it is not a transient. It is not even deterministic, i.e., we don't have a closed-form expression for it to put into a formula. So, what do we do? If you think about it, this must be what most 'real-life' signals look like, and there are plenty of situations where we need to know the amplitude spectrum of this kind of signals (to find the signal bandwidth, for example).

The answer is that we must *sample* it. 'Sampling' means recording its signal value at regular time intervals, see Fig. 7.12.

T_S is the sampling time interval, the time between each sample, and the inverse is the *sampling rate*:

$$f_S = \frac{1}{T_S}[\mathrm{S/s}] \tag{7.13}$$

(The unit is 'Samples per second'.) Sampling is a precarious operation; there is a strict rule that you must always follow:

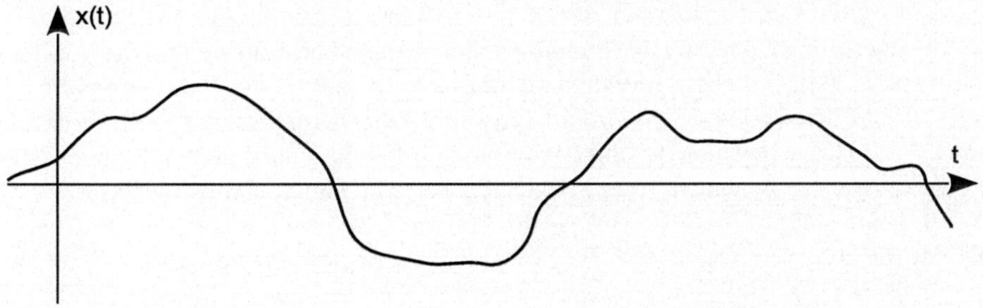

Fig. 7.11 A non-periodic and infinite signal

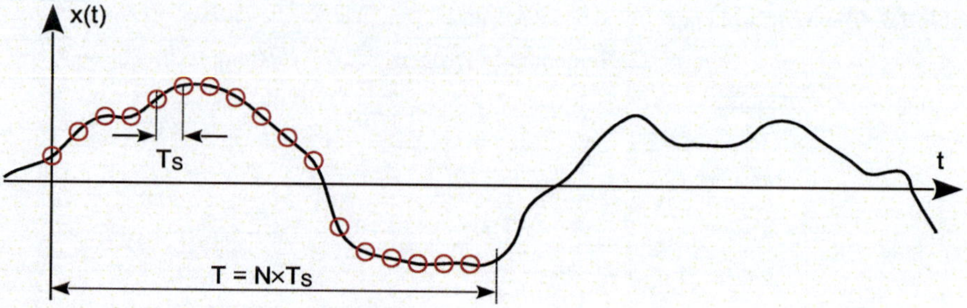

Fig. 7.12 Sampling a signal

Theorem *When you sample a signal with a maximum frequency of* f_{max} *(= the signal bandwidth), the sampling rate must exceed* $2{\cdot}f_{max}$*:*

$$f_S > 2 \cdot f_{\max} \Rightarrow f_{\max} < \frac{f_S}{2} \tag{7.14}$$

This is the *sampling theorem*, sometimes also called the *Nyquist* sampling theorem (or the *Shannon* sampling theorem[1]). Notice the use of '$>$' and '$<$' in Eq. (7.14) and *not* '$\geq$' and '$\leq$'. The sampling rate must be *greater than* $2{\cdot}f_{\max}$. Not *greater than or equal*! This is important and we will talk a lot more about this later. A consequence of the sampling is that the signal becomes *discrete*: $x(t)$ becomes $x(n)$:

$$x(t) \rightarrow x(n \cdot T_S) = x(n) \tag{7.15}$$

$x(n)$ should be interpreted as $x(n{\cdot}T_S)$. Also, we will usually write 'x_n' instead of $x(n)$ (laziness wins).

Another thing we need to decide is when to stop sampling. The signal is infinite, but we can't just sample forever. Sooner or later, we must stop sampling and do something with our samples (like finding the Fourier transform). Hence, we take N samples and then we stop (temporarily) to do some calculations on our samples. When we stop sampling, we have observed the signal for a time duration T:

$$T = N \times T_S \tag{7.16}$$

The question is: What do these samples represent and how do we find the Fourier transform? Well, the answer is that we must adjust one of the Expressions (7.3) or (7.10) to discrete time. (We don't have any other expressions for the Fourier transform.) But, which one? Using Expression (7.3) would indicate that we consider our N samples to be exactly one period of a periodic signal and using Expression (7.10) would suggest that we consider the N samples to be a transient, equal to zero outside the observed time $T = N{\cdot}T_S$. So, whichever expression we use, it will not be

[1] Neither Shannon nor Nyquist'discovered' the sampling theorem; Edmund Whittaker published it already in 1915, but Shannon and Nyquist are usually credited for it. Fair or not, that is how it is.

completely correct, but again, these are all the expressions we have. And, as it turns out, it doesn't matter which one we choose, the result will be the same! However, we will arrive at the result a little faster (and a little more elegantly) if we use Expression (7.3): We consider our N samples to be exactly one period of a periodic signal with period $T = N{\cdot}T_S$. We denote our samples in Fig. 7.12 as $\{x_0, x_1, x_2, x_3, x_4, ..x_{N-1}\}$. To find the Fourier transform of the samples, we adjust Eq. (7.3) for discrete time: $t \to n{\cdot}T_S$.

$$X(k) = \frac{1}{T}\int_0^T x(t)\cdot \mathrm{e}^{-\mathrm{j}k\omega_0 t} = dt\ \{t \to nT_\mathrm{S}\} = \frac{1}{N}\sum_{n=0}^{N-1} x_n \cdot \mathrm{e}^{-\mathrm{j}k\omega_0 nT_\mathrm{S}} \tag{7.17}$$

But, $\omega_0 T_S = \omega_0 T/N = 2\pi/N$, so

$$X(k) = \underbrace{\frac{1}{N}}_{\text{Drop it!}}\sum_{n=0}^{N-1} x_n \cdot \mathrm{e}^{-\mathrm{j}k\frac{2\pi}{N}\cdot n} = \sum_{n=0}^{N-1} x_n \cdot \mathrm{e}^{-\mathrm{j}k\frac{2\pi}{N}\cdot n} = \mathrm{DFT} \tag{7.18}$$

Expression (7.18) is called the *Discrete Fourier Transform*, or just DFT. Notice that we dropped the 1/N factor above; it carries no information (since it only scales each $X(k)$ value by the same factor), and we drop it because our samples will of course be processed by a computer algorithm with a real-time constraint and multiplying by 1/N is just a waste of processor time that doesn't add any information. (Our $X(k)$ samples will be N times 'too large', though. We need to keep that in mind when we use a computer to find the Fourier transform. See Example 7.4.)

Equation (7.18) requires exactly N (complex) multiplications for each $X(k)$, and there will be exactly N $X(k)$s to calculate (we will explain why there are exactly N values later), so the DFT requires N^2 complex multiplications. This is a time-consuming task that puts a limit on the real-time sampling rate. However, analyzing Eq. (7.18) in more detail reveals that a lot of the multiplications are identical and/or 'symmetrical'.

When a computer computes the discrete Fourier transform, it takes advantage of the symmetries in Eq. (7.18) by using an algorithm known as the *Fast Fourier transform* or the 'FFT algorithm'. It was first presented by Cooley and Tukey in 1965 [1], and it reduces the number of complex multiplications to only $N{\cdot}\log_2 N$. Anyway, the details of that algorithm are not important to us (let the computer worry about that). In this context, 'DFT' and 'FFT' mean the same thing and will be used interchangeably.

Before we present any examples of how to use the DFT, we make some observations. First, the Fourier transform Expression (7.18) is still symmetric ($X(-k) = X^*(k)$), so we still have symmetric amplitude spectra. However, the DFT expression has a *new property* that we have not seen in our earlier expressions, and this new property is all-important! To see this property, we find the DFT of $X(k + N)$:

$$X(k+N) = \sum x_n e^{-j(k+N)\frac{2\pi}{N}\cdot n} = \sum x_n e^{-jk\frac{2\pi}{N}n} \cdot \underbrace{e^{-j2\pi n}}_{=1} = \sum x_n e^{-jk\frac{2\pi}{N}n} \\ = X(k) \tag{7.19}$$

We can easily see that Eq. (7.19) holds for any multiple of N: $X(k+mN) = X(k)$, which proves that the DFT is periodic.

If You Take N Samples of X(t), X(k) Becomes Periodic with Period N!

(Which explains why we only need to calculate exactly N $X(k)$ values.)

Notice also that we *still use k as our frequency variable*! If the frequency is '3', it still means that the cosine has a frequency such that its period fits three times in T. The only difference is that T is now the 'observation time' defined in Fig. 7.12. The frequency of the 'fundamental' is $f_0 = 1/T$, and all other signal frequencies must (still) be a multiple if this frequency:

$$f = k \cdot f_0 = k\frac{1}{T} = k\frac{1}{NT_S} = k\frac{f_S}{N} = k \cdot \Delta f \tag{7.20}$$

where $\Delta f = f_S/N$ is the frequency *resolution*. Notice that the frequency resolution is *improved* if we *reduce* the sampling rate (or take more samples). The sampling rate and the number of samples determine how small frequency differences we can resolve in the amplitude spectrum. A high sampling rate gives us a high resolution in the time domain, but a low resolution in the frequency domain.

7.2.4 FFT Outputs

If you use an FFT algorithm to compute the DFT, it will output exactly one period of $X(k)$; $X(0)$ .... $X(N-1)$. These samples need to be treated 'carefully'. First, remember the sampling theorem: only frequencies $< f_S/2$ are 'legit'. Are all N DFT samples produced by the FFT algorithm legit? Well, yes and no. First, according to Eq. (7.20), the frequencies are $k \cdot f_S/N$. Inserting this into the sampling theorem condition gives us

$$f = k \cdot \frac{f_S}{N} < \frac{f_S}{2} \Rightarrow k < \frac{N}{2} \tag{7.21}$$

Hence, you could argue that only the first half of the N $X(k)$ samples are 'legit'; the second half has frequencies that violate the sampling theorem. On the other hand, we have the symmetric and periodic properties of the DFT. Assume that $N/2 \le k < N-1$, i.e., k represents a frequency that violates the sampling theorem. Then,

$$X(k) \underbrace{=}_{\substack{\text{Due to}\\\text{periodicity}}} X(k-N) \underbrace{=}_{\substack{\text{Due to}\\\text{symmetry}}} X^*(N-k) \tag{7.22}$$

Suppose $N = 16$ and $k = 10$. Then $X(10) = X(-6) = X^*(6)$. Hence, you could argue that the second half of the N DFT samples *is* legit; they are just the 'negative' half (and will always be the complex conjugates of the 'positive' half). Figure 7.13 illustrates some of these important aspects of the DFT.

Example 7.3 Figure 7.14 illustrates a sinusoidal function (not sampled). Without doing any actual calculations, predict its Fourier transform.

Solution There is only one cosine in this signal, and it defines the period T. That means that only $X(1)$ and $X(-1)$ are $\neq 0$. Equation (7.7) gives us immediately that the magnitudes are 0.5. Since it is a sine, and not a cosine, it has a phase shift of −90°:

$$X(1) = 0.5 \cdot e^{-j90^\circ} = -0.5j \Rightarrow X(-1) = X^*(1) = 0.5j$$

$X(k) = 0$ for all $k \neq \pm 1$.

Notice how easy it was to find the Fourier transform once you understand what it represents.

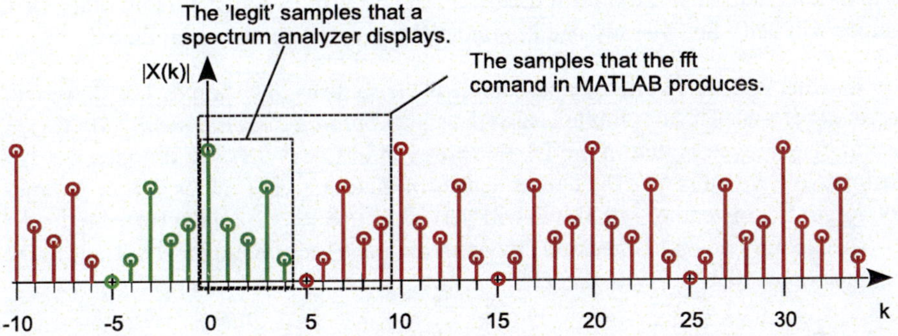

Fig. 7.13 A DFT spectrum when $N = 10$; notice the periodicity and the symmetry

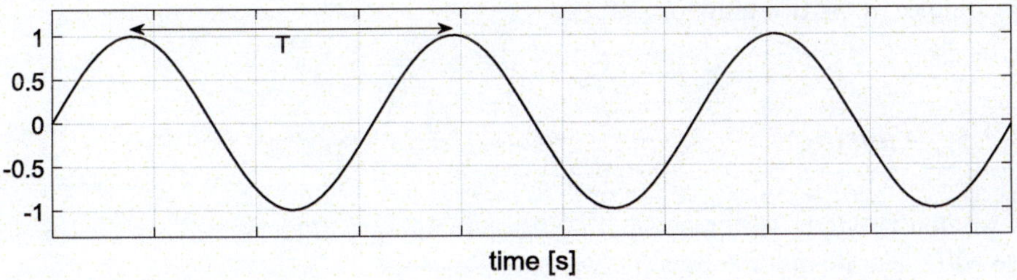

Fig. 7.14 A sinusoidal function

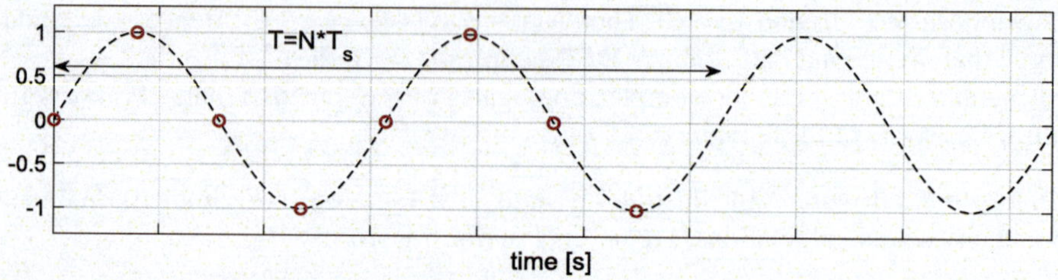

Fig. 7.15 Taking eight samples of a sinusoidal signal

Example 7.4 In Fig. 7.15, we have sampled the signal in Fig. 7.14. What would MATLAB (or any other signal processing software) produce if we used the *fft* command on these samples?

Solution It is important to understand that T is no longer determined by the signal's period; it is determined by the (total) sampling time $N{\cdot}T_S$ as indicated in Fig. 7.15. Since our signal's period fits exactly two times in this time T, only $X(2)$ will be $\neq 0$. From the symmetry property, we also know that $X(-2)$ would be $= X^*(2)$, but the *fft* algorithm doesn't produce, $X(-2)$, it only produces $X(k)$ for $k = 0\ldots7$. However, the periodic property of the DFT implies that $X(-2) = X(-2+8) = X(6)$ which indeed is a number that the *fft* algorithm produces. Hence $X(6) = X^*(2) \neq 0$, all other DFT values will be 0. So, we only need to find $X(2)$, everything else is predicted.

So, what is $X(2)$? Well, the phase angle is the same as in Example 7.3, so we will again get a '−j'. The amplitude is also the same, *but that does not mean that X(2) =* −0.5j! The reason is that in the DFT expression in Eq. (7.18), we dropped the 1/N factor! Now we must face the consequences of that; our $X(k)$ values will be N times too large! So, instead of −0.5j, we will get −0.5j × 8 = −4j (and $X(6)$ = 4j).

We encourage you to check this by applying the *fft* command in MATLAB to the eight samples [0,1,0,−1,0,1,0,−1].

Finally, we should update our transform map, see Table 7.4. Notice in Table 7.4 that we use the same frequency variable k in two transforms, and that they mean the same thing. Almost….! Make sure you understand the subtle difference between the k variables in the two transforms.

7.2.5 *Aliasing*

If we don't comply with the sampling theorem, there will be consequences, which we will illustrate in the following two examples.

Example 7.5 Plot the amplitude spectrum of the signal $x(t) = \sin 2\pi 1000t$, if it is sampled at a rate of f_S = 10 kS/s.

Table 7.4 The transform map

	Non-complex frequency		Complex frequency
Analog	Periodic	$X(k) = \frac{1}{T}\int_0^T x(t)\mathrm{e}^{-\mathrm{j}k\omega_0 t}dt$	
	Transient	$X(\omega) = \int_{-\infty}^{\infty} x(t)\mathrm{e}^{-\mathrm{j}\omega t}dt$	
Sampled		$X(k) = \sum_{n=0}^{N-1} x_n\mathrm{e}^{-\mathrm{j}k\frac{2\pi}{N}n}$	

Solution This is easy; there is only one frequency (= 1000 Hz), so we will have a spectrum peak at $f = 1000$ Hz. But, *since we are sampling*, there will be other peaks too, due to the symmetry and periodic properties. First, due to symmetry, we will have an equally large peak for $f = -1000$ Hz, and then these two peaks will repeat for every multiple of the sampling frequency, i.e., we will have peaks at (m·10,000 ± 1000) Hz, see Fig. 7.16.

Due to the symmetry and periodic properties of the DFT, there will be a lot of 'peaks' in the amplitude spectrum (infinitely many), but since we only look in the 'allowed' range 0 …f_S/2, that is not a problem; our 1000 Hz signal shows up correctly in this region. The other peaks are called 'aliasing' peaks and in this example, they don't cause us any trouble, but they will cause you trouble if you don't comply with the sampling theorem. The next example will illustrate that.

Example 7.6 Plot the amplitude spectrum of the signal $x(t) = \sin 2\pi 9000t$, if it is sampled at a rate of $f_S = 10$ kS/s.

Solution Just as easy; 'real' peak at 9000 Hz, 'symmetry' peak at −9000 Hz and then the 'periodic' peaks at (m·10,000 ± 9000) Hz. That will give us peaks at frequencies …−9000, + 1000, + 11,000, + 21,000 ….. and −1000, 9000, 19,000, 29,000 ….. *Exactly at the same positions as in the previous example*!! (Fig. 7.17).

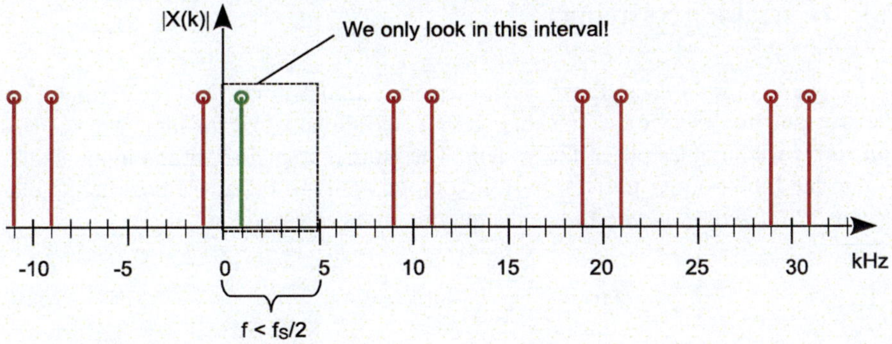

Fig. 7.16 Amplitude spectrum of a 1000 Hz sinusoidal signal sampled at 10 kS/s

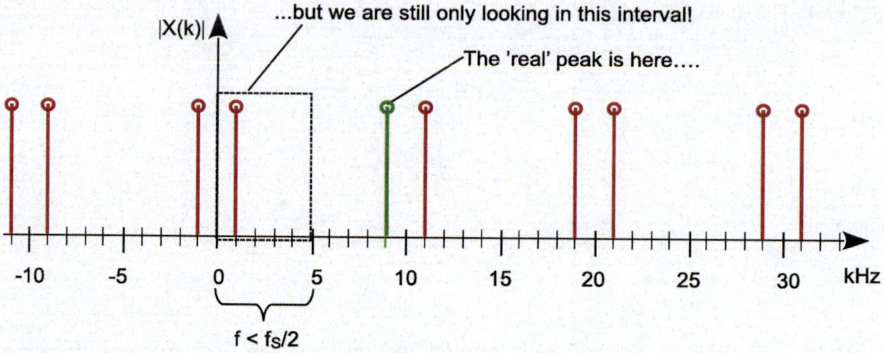

Fig. 7.17 Same amplitude spectrum as in Fig. 7.16

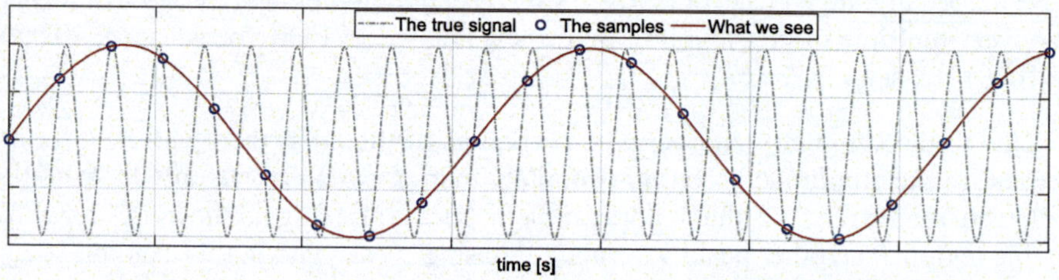

Fig. 7.18 Aliasing in the time domain; sampling violates the Nyquist theorem

Since we still only look in the Nyquist interval ($f < f_S/2$), we will only see the 1000 Hz peak and draw the erroneous conclusion that we are measuring a sinusoidal with frequency 1000 Hz. We have *aliasing* in our amplitude spectrum, which will happen when you don't comply with the sampling theorem. Figure 7.18 illustrates what aliasing looks like in the time domain.

7.3 Describing Systems

So far, we have only used transforms to describe signals, but it is very common to also use them to describe systems. In Fig. 7.19, the signal $x(t)$ is the input to a system and $y(t)$ is what comes out of the system. The Fourier transform of the input signal is $X(\omega)$ and the Fourier transform of the output signal is $Y(\omega)$. We now define the system's *transfer function* $H(\omega)$ as the quotient between $Y(\omega)$ and $X(\omega)$:

$$H(\omega) = \frac{Y(\omega)}{X(\omega)} = |H(\omega)| \cdot \mathrm{e}^{\mathrm{j}\varphi(\omega)} \tag{7.23}$$

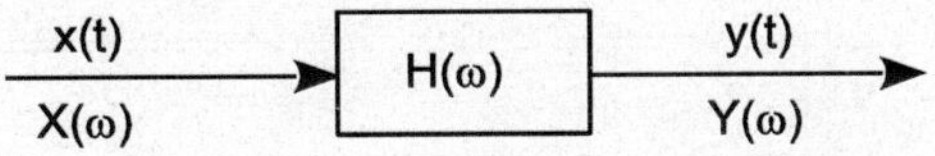

Fig. 7.19 Signals and systems are described with transforms

$H(\omega)$ is in general a complex function where $|H(\omega)|$ is the *amplification* diagram (or 'gain' diagram) and $\varphi(\omega)$ is the *phase* diagram. The amplification diagram tells you what happens to the amplitudes of the cosines in $x(t)$ and the phase diagram tells you what happens to their phase angles. An example will make this clear.

Example 7.7 If the signal $x(t) = 4\cos(20000t + 25°)$ is the input to the system in Fig. 7.20, what will the output be?

Solution First we find $y(t)$:

$$y(t) = \frac{X_C}{X_C + R} \cdot x(t)$$

$$= \frac{\frac{1}{\mathrm{j}\omega C}}{\frac{1}{\mathrm{j}\omega C} + R} x(t) = \frac{1}{1 + \mathrm{j}\omega RC} x(t) \Rightarrow Y(\omega) = \frac{1}{1 + \mathrm{j}\omega RC} X(\omega)$$

$$\Rightarrow H(\omega) = \frac{Y(\omega)}{X(\omega)} = \frac{1}{1 + \mathrm{j}\omega RC} = \frac{1 \cdot \mathrm{e}^{\mathrm{j}\cdot 0}}{\sqrt{1^2 + (\omega RC)^2} \cdot \mathrm{e}^{\mathrm{j}\tan^{-1}\omega RC/1}}$$

$$= \underbrace{\frac{1}{\sqrt{1 + R^2C^2\omega^2}}}_{|H(\omega)|} \cdot \mathrm{e}^{\overbrace{-\mathrm{j}\tan^{-1} RC\omega}^{\varphi(\omega)}}$$

Since $|H(\omega)|$ is a first-order polynomial in ω, the system in Fig. 7.20 is a first-order system. It is also a *lowpass* system, since $|H(0)| = 1$, and $|H(\omega)| \to 0$, when $\omega \to \infty$. In Fig. 7.21, we have plotted the *Bode diagram* of the system, i.e., $|H(\omega)|$ and $\varphi(\omega)$ for $R = 1$ kΩ and $C = 100$ nF. From this plot, we can see that a signal with frequency 20 krad/s will be attenuated by a factor of 0.45 and the phase angle will be shifted by −63 degrees. Hence, the output $y(t)$ in Fig. 7.20 is

$$y(t) = 0.45 \cdot 4\cos(20000t + 25° - 63°) = 1.8\cos(20000t - 38°)$$

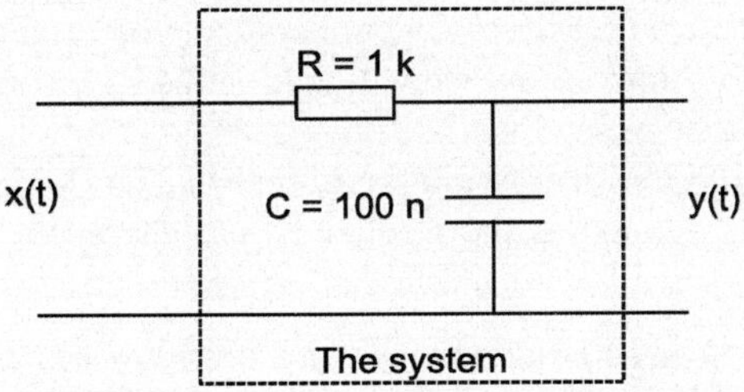

Fig. 7.20 A first-order system

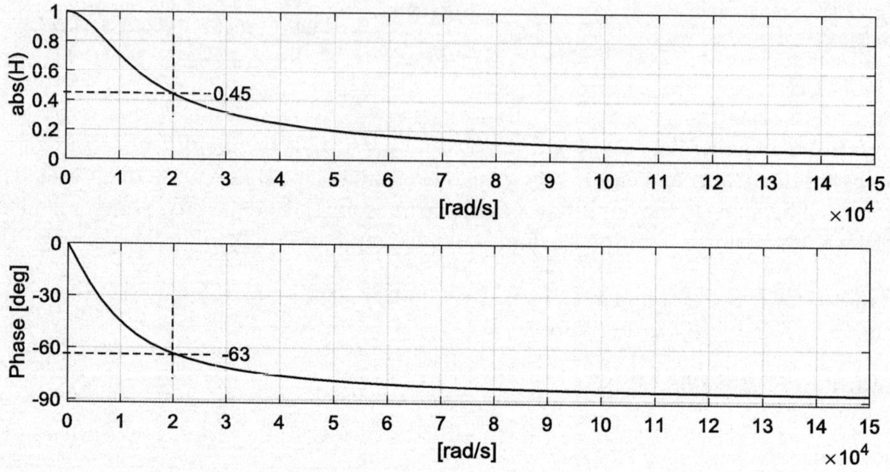

Fig. 7.21 The Bode plot

Notice in Example 7.7 that we mix radians and degrees in the cosine argument. Mathematically this is of course wrong, but it is common practice, simply because it is easier to imagine the size of an angle in degrees than in radians. You just have to keep that in mind and be careful when you use your calculator.

Another detail worth pointing out in this example is that the system changed the amplitude and the phase of the signal, but it didn't change the frequency. This is what characterizes *linear and time-invariant* systems (LTI), and we will only treat LTI systems in this book. That is the most common restraint in signal processing textbooks.[2]

The Bode plot in Fig. 7.21 is what characterizes any system; if you know the transfer function, you know the Bode plot and then you know everything you need to know about the system; with 'everything' we mean that you can predict the output for any signal input. As a matter of fact, systems are characterized by the $|H(\omega)|$ diagram (which we will call the *amplification* diagram or *gain* diagram) and there are, in general, six different types of systems: Lowpass, highpass, bandpass, stop-band, resonance, and notch systems. Their characteristic amplification diagrams are illustrated in Fig. 7.22 on the next page.

Most systems are 'by nature' lowpass, like amplifiers, instruments, and transmission lines (they are typically also first-order systems). In general, to get anything else than a first-order lowpass system, you need to design a 'filter' (see Chaps. 9 and 10). For that reason, we will take a closer look at the amplification diagram of a lowpass system, see Fig. 7.23.

First, in most diagrams, both axes are logarithmic. Second, the system's *bandwidth* is defined as the frequency ω_B where the amplification has decreased by -3 dB

[2] Look for textbooks about 'Non-linear systems' or 'Adaptive systems' if you want to go beyond the LTI restriction.

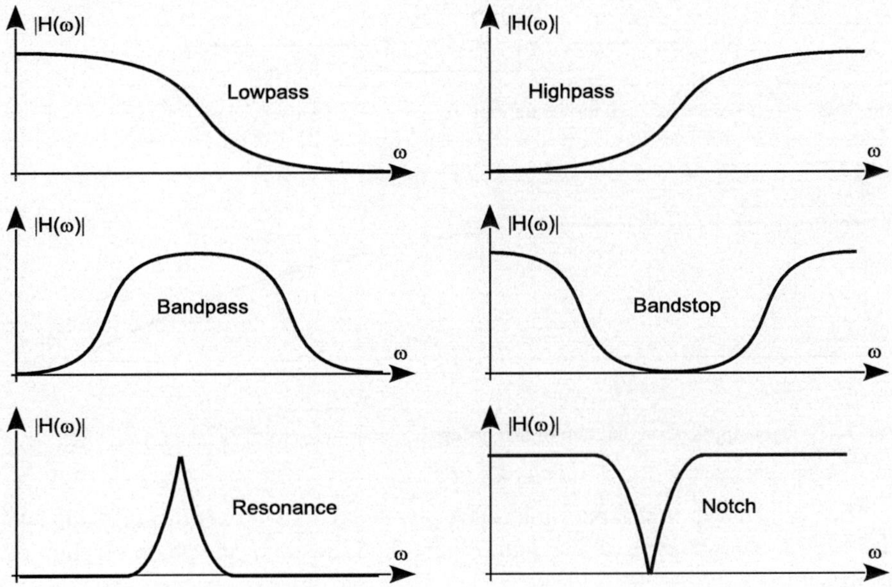

Fig. 7.22 Six different types of systems

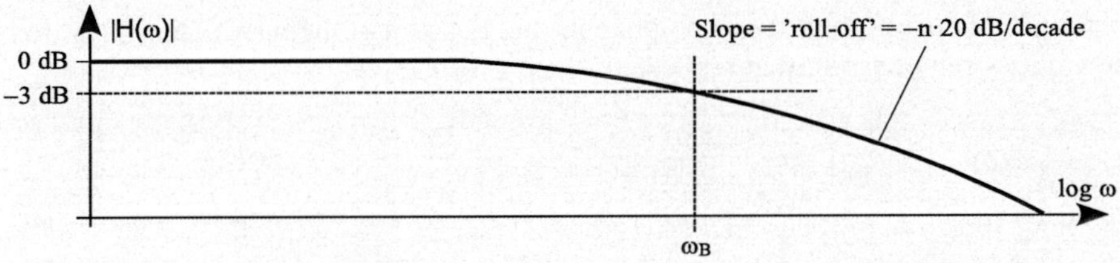

Fig. 7.23 A lowpass system

(compared to the amplification at $\omega = 0$). Third, in the stopband ($\omega > \omega_B$), the amplification drops as $n{\cdot}20$ dB/decade, where n is the system order. Hence, you can figure out the system order by looking at the stopband roll-off.

7.3.1 Distortion-Free Systems

Consider the signal $x(t) = \sin t + 0.33 \sin 3t$. This signal is passed through the lowpass system in Fig. 7.24 with the amplification diagram in Fig. 7.25. $x(t)$ has frequencies 1 and 3 rad/s and according to the amplification diagram, neither of the amplitudes are affected by the system (since the amplification = 1 for both signals.) Does that mean that $y(t) = x(t)$?

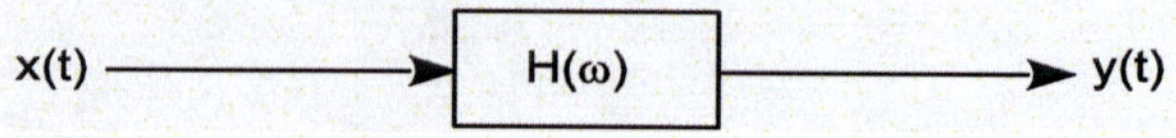

Fig. 7.24 *x(t)* is passed through the system *H(ω)*.

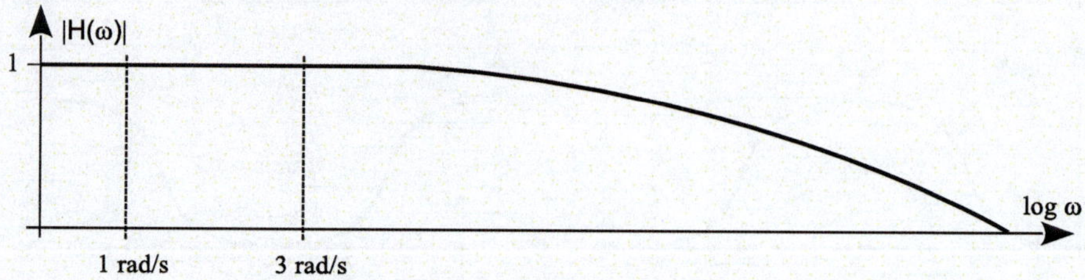

Fig. 7.25 The amplification diagram of the system

The right answer to that question is: *We don't know*! To answer that question, we also need to know the phase diagram. Figure 7.26 illustrates the phase diagram of the system.

Now we can answer the question: The answer is 'No, $y(t)$ will not equal $x(t)$'. Here is the reason.

Figure 7.27 illustrates the signal $x(t)$ and its two components and Fig. 7.28 illustrates what happens to the two components and the sum of them (which is $y(t)$) after they have been phase shifted $-90°$ and $-120°$, respectively.

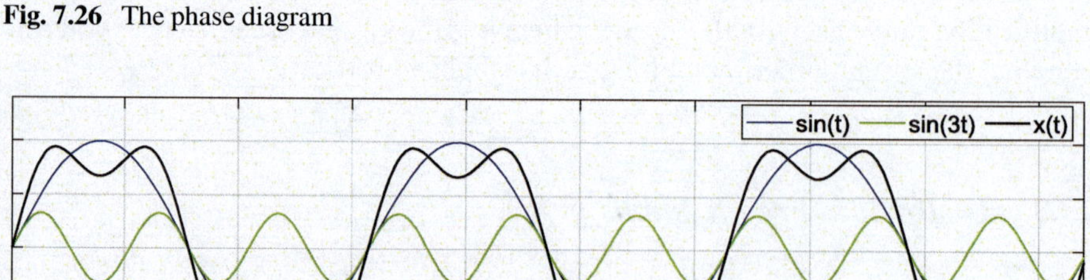

Fig. 7.26 The phase diagram

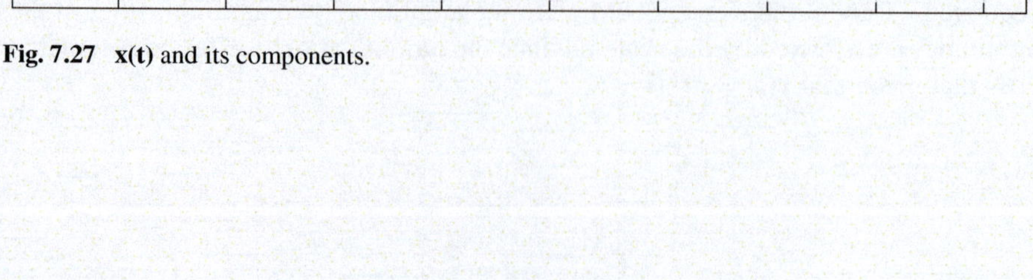

Fig. 7.27 **x(t)** and its components.

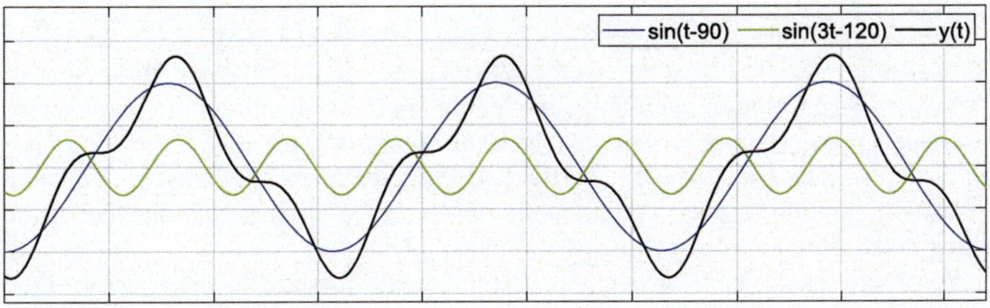

Fig. 7.28 $y(t)$ and its components.

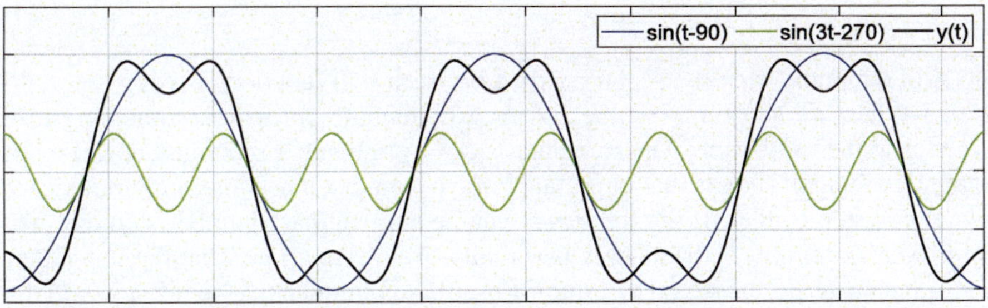

Fig. 7.29 $y(t)$ and its components.

From Fig. 7.28, we can see that even though the amplitudes are not affected, we still have distortion because of the phase diagram. So, what is wrong with the phase diagram in Fig. 7.26? To have a distortions-free passage, the phase diagram must be *linear*, i.e., $\varphi(\omega) = k{\cdot}\omega$, and we can see immediately that the phase diagram in Fig. 7.26 is not linear and distortion is expected.

For a distortion-free passage of $x(t)$, the phase shift for $\omega = 3$ rad/s must be three times higher than the phase shift for $\omega = 1$ rad/s. Figure 7.29 illustrates the system output when sin3t is phase shifted $-270°$ $(= 3 \times (-90°) = 3 \times \varphi(1))$.

So, distortion-free systems are characterized by linear phase diagrams; however, remember that it only needs to be linear in the passband ($\omega < \omega_B$); we don't care what happens in the stopband since these signals are attenuated anyway.

7.4 Complex Frequencies

Now that we know something about systems, and how to describe them, we will take a new look at our frequency variables. From experience, I know that wrapping your head around all the frequency variables in transform theory is the hardest part. So far, we have introduced f, ω, and k, but I'm sorry to say, we are only half-way through.

The Fourier transform expressions in Eqs. (7.3), (7.10), and (7.18) are really *scalar products*, where x is the vector we 'analyze' and $e^{-j\omega t}$ is the 'base vector' (base *function*). Compare with what you learned in linear algebra; to find the vector *component along the x-axis, you take the scalar product of the vector and the x base vector*: $v_x = \langle \overline{v} | \overline{e}_x \rangle = \left(v_x, v_y\right) \cdot (1, 0) = v_x$. Well, that is exactly what we do when we use the Fourier transform (*signals are vectors!*). We do it to find the size of the signal in the different 'cosine directions' in Eq. (7.1).

You could argue against that, saying: 'But if the cosines are the base vectors, how come we don't have cosines in the Fourier transform expressions?' Well, *we do*! We have just used Euler's formula for cosine:

$$Ae^{j\omega t} = A\cos\omega t(+jA\sin\omega t) \tag{7.24}$$

(And remember that the scalar product between two functions is $\langle f, g\rangle = \int f \cdot g^* dt$.) When we substitute cosine for an exponential expression, something else happens at the same time: The frequency goes from being a real number (ω) to an imaginary number ($j\omega$). Is that important? Maybe not, but it becomes important if we take it one step further: If the frequency can be an imaginary number, could it also be a *complex* number? The answer is actually 'Yes'! We know what the imaginary part (ω) represents (the signal's harmonic oscillation frequency), but what would the real part of the frequency represent? We will assign the letter s to our new complex frequency variable:

$$s = \sigma + j\omega \tag{7.25}$$

We will later come back to what physical property the real part (σ) represents. First, let's update our transform expression with our new frequency. We only need to adjust Expression (7.10) (it will become clear later why we don't worry about Eq. (7.3)). In Eq. (7.10), we substitute $j\omega$ for s:

$$X(\omega) \to X(\sigma + j\omega) = \int x(t)e^{-(\sigma + j\omega)t}dt = \int x(t)e^{-st}dt = X(s) \tag{7.26}$$

$X(s)$ is the *Laplace* transform (and notice that the Fourier transform is just a special case of the Laplace transform, when $\sigma = 0$ and $s = j\omega$).

Now, let's first figure out what the real part of the new frequency variable represents. First, we substitute $j\omega$ for s in Eq. (7.24):

$$Ae^{j\omega t} \to Ae^{st} = Ae^{(\sigma + j\omega)t} = Ae^{\sigma t} \cdot e^{j\omega t} = \underbrace{Ae^{\sigma t}}_{\text{Amplitude!}} \cos\omega t \tag{7.27}$$

Look at Eq. (7.27). By introducing a real part in the frequency, we can also represent harmonic functions with exponentially decaying/increasing amplitudes! (With the Fourier transform we can only process harmonic signals with constant amplitudes.) From Eq. (7.27), we can also see that if $\sigma > 0$, we have a harmonic signal with exponentially growing amplitude! We *never* want that![3] It would drive our signal output to the power supply limit (and we would suddenly have a non-linear, malfunctioning system). Hence, we (almost) always want to make sure that we have signals where the real part of the frequency is ≤ 0.

However, our main concern here is not signals; it is systems.

7.4.1 Laplace Representation of Systems

We previously derived the expression $H(\omega) = 1/(1 + \mathrm{j}RC\omega)$ for the transfer function of the system in Fig. 7.20. We can, and usually do, express the transfer function using the complex frequency s, just substitute $\mathrm{j}\omega$ for s:

$$H(s) = \frac{1}{1 + RCs} \tag{7.28}$$

In Chap. 9, we will investigate the details of filters and how to describe and design them, but Eq. (7.28) is a typical filter equation; in general, a filter transfer function is a quotient between two polynomials of s, and the filter 'order' is determined by the highest polynomial order. Right now, we are trying to understand the Laplace transform, and to have something to work with we will use the following second-order filter:

$$H(s) = \frac{4s}{s^2 + 2s + 2} \tag{7.29}$$

First, let's find the Bode plot to see what kind of filter we are dealing with. Substitute s for $\mathrm{j}\omega$ to get the Fourier transform:

$$\begin{aligned} H(\omega) &= \frac{4\mathrm{j}\omega}{-\omega^2 + 2\mathrm{j}\omega + 2} = \frac{4\omega \mathrm{e}^{\mathrm{j}90^\circ}}{\sqrt{\left(2-\omega^2\right)^2 + 4\omega^2} \cdot \mathrm{e}^{\mathrm{j}\tan^{-1} 2\omega/(2-\omega^2)}} = \\ &= \frac{4\omega}{\sqrt{\left(2-\omega^2\right)^2 + 4\omega^2}} \cdot \mathrm{e}^{\mathrm{j}\left(90^\circ - \tan^{-1} 2\omega/(2-\omega^2)\right)} = |H(\omega)| \cdot \mathrm{e}^{\mathrm{j}\varphi(\omega)} \end{aligned} \tag{7.30}$$

Figure 7.30 illustrates the Bode diagram of this system. From Fig. 7.30, we can see that we are dealing with a bandpass system with a peak amplification around 1 rad/s. Also, the discontinuity in the phase diagram is a not real; it is because the MATLAB arctan function returns values in the range $-\pi/2 \ldots \pi/2$ only.

Speaking of MATLAB, there is a faster way to get the Bode plot of a given system. If we define the numerator and denominator as $b = [4,0]$ and $a = [1,2,2]$,

---

[3] Unless you are designing an oscillator.

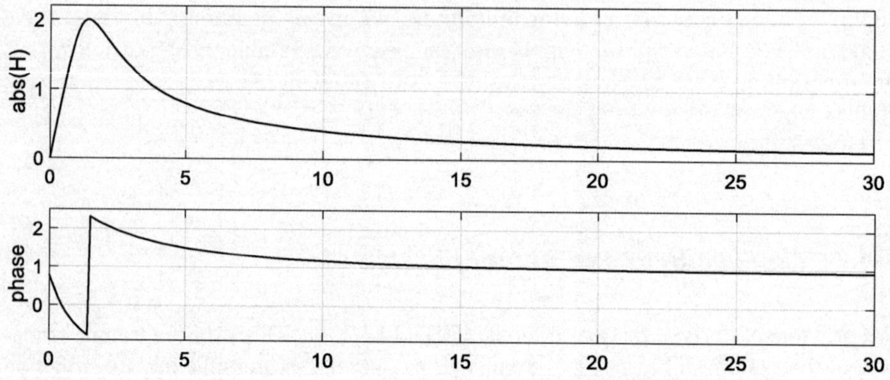

Fig. 7.30 The Bode plot

the *freqs*(*b*,*a*) command will immediately plot the Bode diagram (with logarithmic axes and no arctan folding), see Fig. 7.31.

Next, we substitute s for $\sigma + j\omega$, in Eq. (7.29):

$$H(\sigma,\omega) = \frac{4(\sigma + j\omega)}{(\sigma + j\omega)^2 + 2(\sigma + j\omega) + 2} \tag{7.31}$$

Let's look at the magnitude function of this expression.

$$|H(\sigma,\omega)| = \frac{4\sqrt{\sigma^2 + \omega^2}}{\sqrt{\left(\sigma^2 - \omega^2 + 2\sigma + 2\right)^2 + 4\omega^2(\sigma + 1)^2}} \tag{7.32}$$

Since this is a function of two variables, the amplification diagram of a system represented by the Laplace transform will be a 3D graph, see Fig. 7.32. In Fig. 7.32,

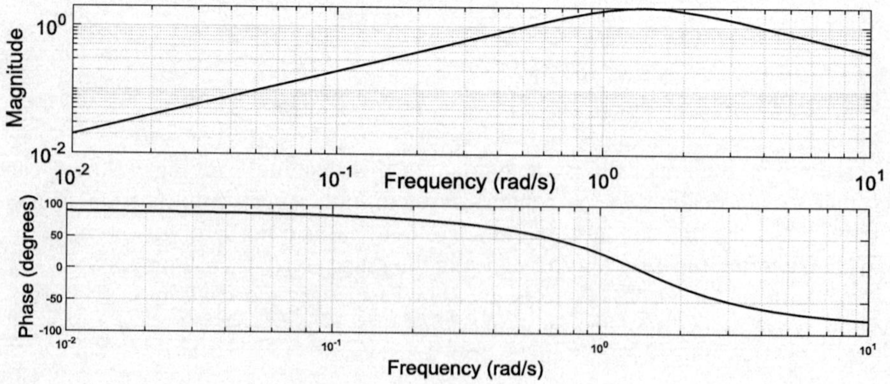

Fig. 7.31 Using the *freqs* command in MATLAB

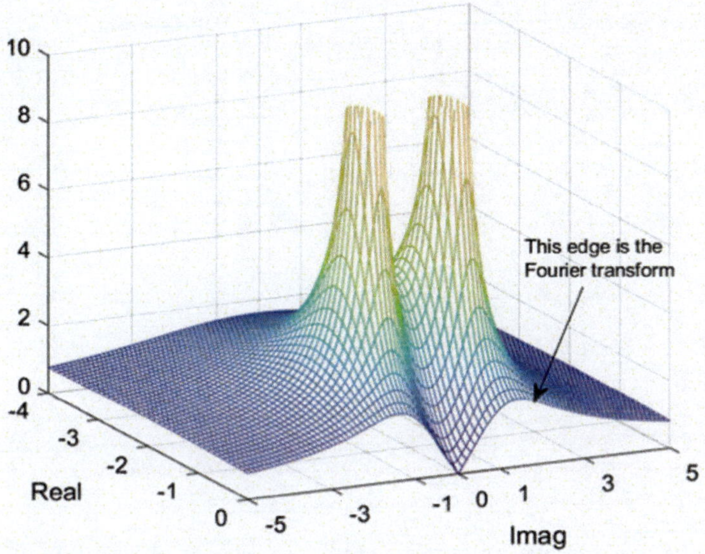

Fig. 7.32 The amplification diagram of the system in Eq. (7.29) is a 3D graph

'Real' is the σ-axis and 'Imag' is the jω-axis. Notice most of all in Fig. 7.32 that we have 'cut' the graph along $\sigma = 0$; the edge line in the cut corresponds to the Fourier transform's amplification diagram. Compare the edge line with the top diagram in Fig. 7.30.

We will comment more on this 3D graph in a minute, but first we must introduce the concept of a system's 'poles' and 'zeros'.

The roots of the numerator are the 'zeros' and the roots of the denominator are the 'poles'. Our system in Eq. (7.29) has one zero only; $s = 0$. The poles are

$$s^2 + 2s + 2 = s^2 + 2s + 1 + 1 = (s + 1)^2 + 1 = 0 \Rightarrow s = -1 \pm \mathrm{j} \qquad (7.33)$$

In Fig. 7.33, we have marked the poles ('X') and zeros ('O') in the s plane (using the *pzplot* command in MATLAB). Compare Figs. 7.32 and 7.33; the poles in Fig. 7.33 coincide with the 'poles' in Fig. 7.32 and the zero in Fig. 7.33 coincides with the (0,0)-coordinate where the amplification diagram touches the zero plane. Now we understand the names 'poles' and 'zeros'. When you get used to this kind of representation of systems, you will be able to immediately identify the pole-zero diagram in Fig. 7.33 as a bandpass system.

So far, we have presented three different ways to represent a system: the transfer function, the Bode plot, and the pole-zero diagram. They all say the same thing and it is important that you learn to transfer smoothly between the different representations. In Chap. 9, we will introduce a few more ways to represent systems, but this is all we need for now.

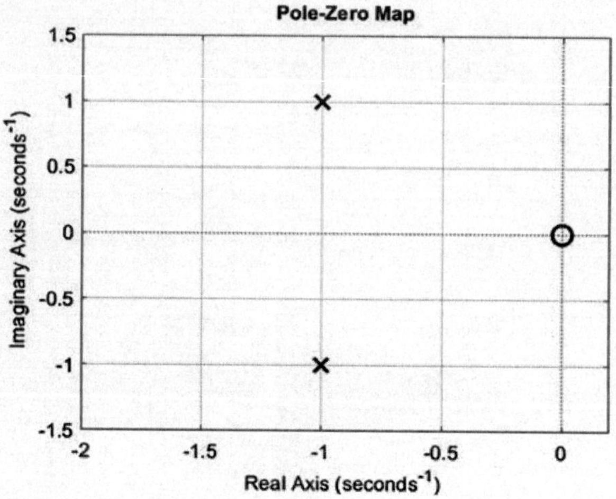

Fig. 7.33 Poles and zeros in the s plane

Hopefully, you can also see the benefit of introducing the complex frequency variable s; it provides so much more information about the system than when we only use the 'one-dimensional' frequency variable $\mathrm{j}\omega$.

It is time to update our transform map, see Table 7.5.

From Table 7.5, it is obvious where we must go next; we have a void in our transform map. We need to introduce a complex frequency variable for the sampled case and find a discrete-time correspondence to the Laplace transform.

But before we do that, we should look at one more aspect of systems and their pole-zero diagram. We do that in the following example.

Example 7.8 A system's *impulse response* $h(t)$ is the system's output when the input is an impulse, see Fig. 7.34. By 'impulse' we mean a Dirac impulse:

$$\delta(t) = \begin{cases} 0 \text{ if } t \neq 0 \\ \int\limits_{-\infty}^{\infty} \delta(t) dt = 1 \end{cases} \tag{7.34}$$

Table 7.5 The transform map

		Non-complex frequency	Complex frequency
Analog	Periodic	$X(k) = \frac{1}{T}\int\limits_{0}^{T} x(t)\mathrm{e}^{-\mathrm{j}k\omega_0 t} dt$	$X(s) = \int\limits_{0}^{\infty} x(t)\mathrm{e}^{-st} dt$
	Transient	$X(\omega) = \int\limits_{-\infty}^{\infty} x(t)\mathrm{e}^{-\mathrm{j}\omega t} dt$	
Sampled		$X(k) = \sum\limits_{n=0}^{N-1} x_n \mathrm{e}^{-\mathrm{j}k\frac{2\pi}{N}n}$	

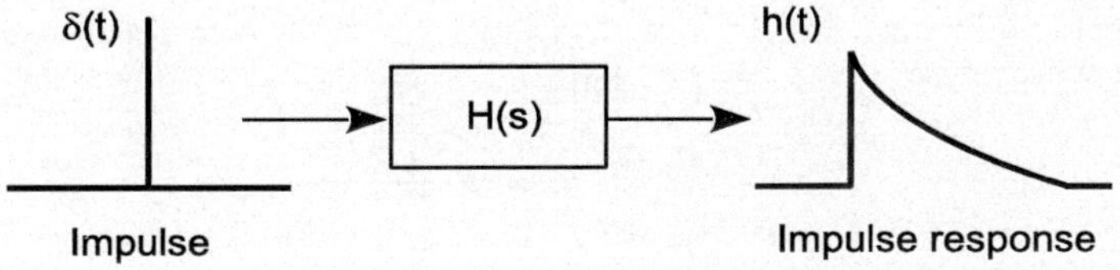

Fig. 7.34 The impulse response

The impulse response is a very important characteristic of a system; the Laplace transform of the impulse response is the transfer function:

$$H(s) = \int_0^\infty h(t)\mathrm{e}^{-st}dt \Rightarrow h(t) = H^{-1}(s) \tag{7.35}$$

Suppose that the impulse response for some system is $h(t) = \mathrm{e}^{-at}$. What is the domain of a that guarantees a stable system? What does that imply for the system's poles?

Solution Obviously, we must have $a > 0$ to have a stable system; if $a \leq 0$, just a short impulse input would generate an output that would never 'die'. A 'healthy' system is characterized by 'limited input' must generate a 'limited output'. To see what that implies for the system's poles, we need to find the poles. And to find the poles we need the transfer function:

$$H(s) = \int_0^\infty h(t)\mathrm{e}^{-st}dt = \int_0^\infty \mathrm{e}^{-at}\mathrm{e}^{-st}dt = \int_0^\infty \mathrm{e}^{-(a+s)t}dt = -\frac{1}{a+s}\left[\mathrm{e}^{-(a+s)t}\right]_0^\infty$$
$$= -\frac{1}{a+s}(0-1) = \frac{1}{s+a} \tag{7.36}$$

The system has no zeros, just one pole in $s_p = -a$. Since we already restricted a to be > 0, obviously this pole can only be in the 'negative' half of the s plane (the left half), see Fig. 7.35.

The conclusion in the previous example is true in general; a system's poles must be in the left half of the s plane to be stable. (There is no such restriction for the zeros.)

7.4.2 The z Transform

The objective here is to find the Laplace transform correspondence in discrete-time space (for a sampled signal). Before we do that, let's talk about what properties we would expect to find in such a transform.

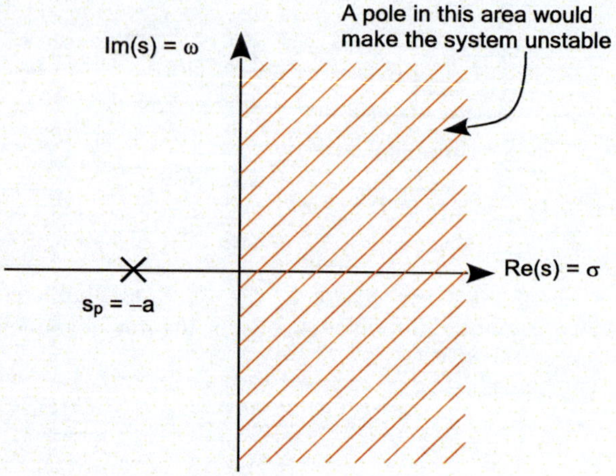

Fig. 7.35 Poles must be in the left half of the *s* plane

In the Laplace transform, we find the Fourier transform by setting $\sigma = 0$, i.e., the Fourier transform is along the jω-axis in the *s* plane. We already know that that can't be true for the discrete-time 'Laplace transform', since we have already concluded that the discrete Fourier transform of a sampled signal is *periodic*! Then it can't be on a straight line, because that would require an infinite number of poles and zeros. The only thing that would account for the periodicity of the discrete-time Fourier transform is if the Fourier transform *is on a circle*! When we transfer from the *s* plane to the corresponding space for sampled signals, we must make sure that the jω-axis is transferred to a circle. How do we do that? That's easy. We use the following transfer trick:

$$z = e^{sT_S} \tag{7.37}$$

(T_S is the sampling interval time.) We call the new space the '*z* space' and the 'Laplace transform' for sampled signals is called the *z transform*. Let's take a closer look at Eq. (7.37):

$$z = e^{sT_S} = e^{(\sigma + j\omega)T_S} = e^{\sigma T_S} \cdot e^{j\omega T_S} = |z| \cdot e^{j\Omega} \tag{7.38}$$

where

$$|z| = e^{\sigma T_S} \tag{7.39a}$$

$$\Omega = \omega T_S \tag{7.39b}$$

z is illustrated in Fig. 7.36: $e^{\sigma T_S}$ is the 'length' of *z* and ωT_S is the 'angle'.

To find the Fourier transform in this new space, we set $\sigma = 0$, and from Eq. (7.39a) we conclude that *the Fourier transform is on the unit circle in z space* ($|z| = 1$). This

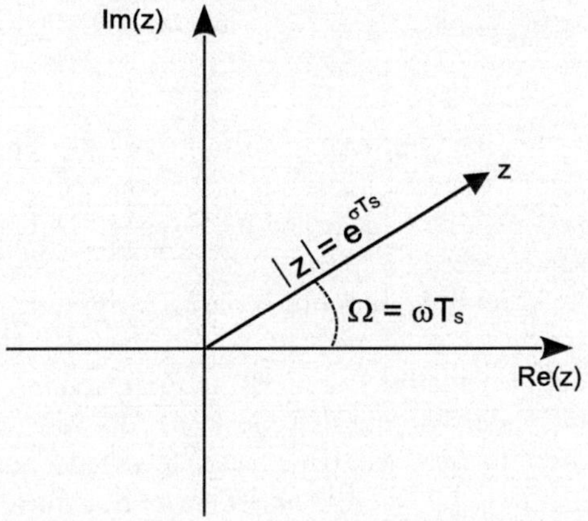

Fig. 7.36 The z frequency

will make it periodic with period $\Omega = 2\pi$. That corresponds to a period in frequency equal to f_S:

$$\Omega = \omega T_S = \omega \frac{2\pi}{\omega_S} = 2\pi \frac{\omega}{\omega_S} = 2\pi \frac{f}{f_S} \tag{7.40}$$

(Ω is the 'normalized' frequency.) Hence, the z frequency variable reflects the periodicity of the Fourier transform in sampled signals. Let's find the transforms that correspond to the Laplace and Fourier transforms for sampled signals. First, we discretize the time in the Laplace transform, and then we substitute e^{sT_s} for z:

$$\begin{aligned} X(s) &= \int_0^\infty x(t)e^{-st}dt = \{t \to nT_s\} = \sum_{n=0}^{\infty} x_n e^{-snT_S} = \sum_{n=0}^{\infty} x_n \left(e^{sT_S}\right)^{-n} \\ &= \sum_{n=0}^{\infty} x_n z^{-n} = X(z) \end{aligned} \tag{7.41}$$

This is the *z transform* and corresponds to the Laplace transform in continuous time. We will take a closer look at it in a minute, but let's first also derive the Fourier transform. Setting $\sigma = 0$ and substituting s for $j\omega$ in Eq. (7.41) give us:

$$\sum_{n=0}^{\infty} x_n e^{-j\omega n T_S} = \sum_{n=0}^{\infty} x_n e^{-j\Omega n} = X(\Omega) \tag{7.42}$$

Equation (7.42) is the Fourier transform for discrete-time signals. (Not to be confused with the discrete Fourier transform, the DFT! We'll talk more about that later.) We have requested that it should be periodic with period $\Omega = 2\pi$. Let's check that:

$$X(\Omega+2\pi)=\sum x_n \mathrm{e}^{-\mathrm{j}(\Omega+2\pi)n}=\sum x_n \mathrm{e}^{-\mathrm{j}\Omega n}\cdot\underbrace{\mathrm{e}^{-\mathrm{j}2\pi n}}_{=1}=\sum x_n \mathrm{e}^{-\mathrm{j}\Omega n}=X(\Omega)$$

So, the Fourier transform is periodic with a period $\Omega=2\pi$, which according to Eq. (7.40) corresponds to a frequency of $f=f_S$. Figure 7.37 illustrates the relation between the s space and the z space. (Notice that $\omega=0$ and $\omega=\omega_S$ ends up in the same place in z space.)

You might be a little confused right now. We just derived the Fourier transform for sampled signals, but didn't we do that already in Sect. 7.2.3, Eq. (7.18)? No, we didn't actually! Eq. (7.18) is the *discrete Fourier transform*. It is a very common misunderstanding that the word 'discrete' in 'discrete Fourier transform' refers to discrete *time*, i.e., to the fact that we have sampled the signal. It does not! Expression (7.42) is the *discrete-time* Fourier transform.

Expression (7.18) is the *discrete Fourier transform*, i.e., it is a discrete version of Eq. (7.42), where 'discrete' means that we only calculate Eq. (7.42) for certain Ω-values (you could say that we sample $X(\Omega)$). The DFT is an adaption to the 'real-world' situation if you like. Look at Eq. (7.42); the domain of Ω is $\mathbb{R}$; all Ω-values are allowed, and we don't have time to calculate $X(\Omega)$ for that many Ω-values. As if that was not enough; the Fourier transform in Eq. (7.42) *sums forever*! In a 'real-world' application, we must stop sampling at some point (after N samples) and due to real-time constraints, we want to calculate $X(\Omega)$ for just enough Ω-values; no more and no less than necessary. So how many $X(\Omega)$-values are 'enough'? Well, by definition, it is enough when we have enough $X(\Omega)$-values to be able to reproduce x_n by an inverse Fourier transform, i.e., when $X^{-1}(\Omega)=x_n$. And, if we take N samples of a signal, we must have (at least) N $X(\Omega)$-values to get the signal back when we do an inverse transform.

If we have N samples and calculate Eq. (7.42) for exactly N Ω-values, the distance between Ω-values must be $2\pi/N$ and

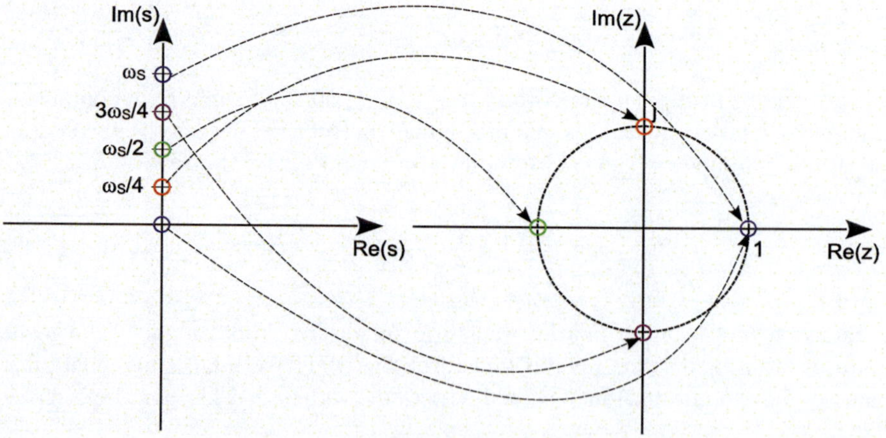

Fig. 7.37 From s space to z space

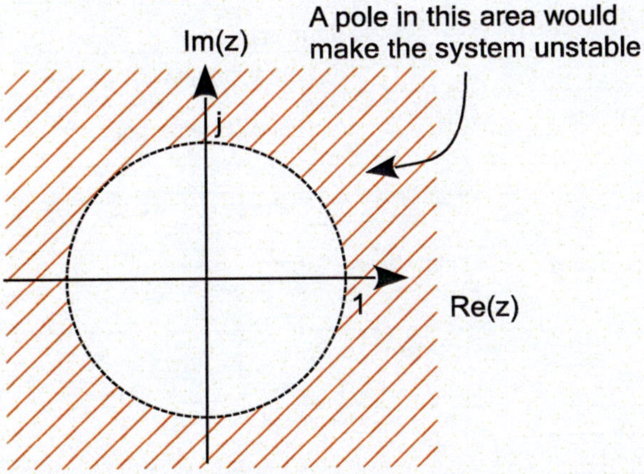

Fig. 7.38 Poles must be within the unit circle

$$\Omega = k \cdot \frac{2\pi}{N} \tag{7.43}$$

Inserting that into Eq. (7.42) gives us

$$X(\Omega) = X\left(k \cdot \frac{2\pi}{N}\right) = X(k) = \sum_{n=0}^{N-1} x_n e^{-jk\frac{2\pi}{N}n} \tag{7.44}$$

which is exactly the DFT expression in Eq. (7.18). The DFT is a discrete (sampled) version of the Fourier transform in Eq. (7.42). That's why it is 'discrete', not because it processes discrete signals.

Finally, let's also see what happens to our 'poles restraint' in Fig. 7.35. For a continuous-time system to be stable, all poles must be in the left half of the s plane, i.e., $\sigma < 0$. How does that translate to discrete time? Well, inserting $\sigma < 0$ in Eq. (7.39a) implies that $|z| < 1$, i.e., in the z plane, all poles must be within the unit circle for the system to be stable, see Fig. 7.38.

Time to update (to complete!) our transform map, see Table 7.6.

Admittedly, there are a lot of transforms, but hopefully, after reading this chapter you can see how they fit together and Table 7.6 may help you to 'organize' them. We will have plenty of reasons to use transforms later in this book. In Fig. 7.39, we have summarized the three different transforms for sampled signals and their relationship.

7.5 Solved Problems

Problem 7.1 Plot the amplitude spectrum of the signal $x(t)$ in Fig. 7.40.

Solution First we need to find the Fourier transform. The signal is periodic with period $T = 0.1$ s. In the time interval $0 \ldots 0.1$ s, the signal equation is $x(t) = 1 - 10t$

Table 7.6 The complete transform map

		Non-complex frequency	Complex frequency
Analog	Periodic	$X(k) = \frac{1}{T}\int\limits_0^T x(t)e^{-jk\omega_0 t}dt$	$X(s) = \int\limits_0^\infty x(t)e^{-st}dt$
	Transient	$X(\omega) = \int\limits_{-\infty}^{\infty} x(t)e^{-j\omega t}dt$	
Sampled	DFT	$X(k) = \sum\limits_{n=0}^{N-1} x_n e^{-jk\frac{2\pi}{N}n}$	$X(z) = \sum\limits_{n=0}^{\infty} x_n z^{-n}$
	Fourier	$X(\Omega) = \sum\limits_{n=0}^{\infty} x_n e^{-j\Omega n}$	

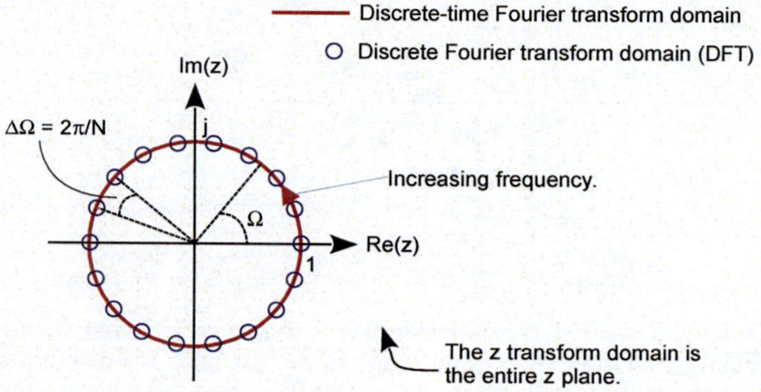

Fig. 7.39 Transform domains for sampled signals

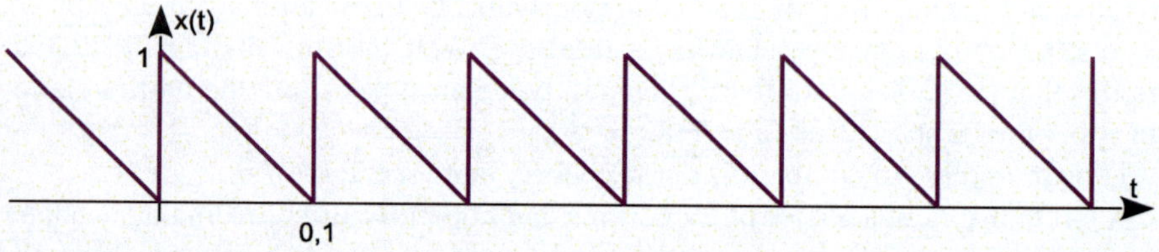

Fig. 7.40 A sawtooth signal

$= 1 - t/T$. The Fourier transform is

$$X(k) = \frac{1}{T}\int_0^T \left(1 - \frac{1}{T}t\right) \cdot \mathrm{e}^{-\mathrm{j}k\omega_0 t}\,dt$$

$$= \frac{1}{T}\left\{-\frac{1}{\mathrm{j}k\omega_0}\left[\left(1 - \frac{1}{T}t\right) \cdot \mathrm{e}^{-\mathrm{j}k\omega_0 t}\right]_0^T - \frac{1}{\mathrm{j}k\omega_0 T}\int_0^T \mathrm{e}^{-\mathrm{j}k\omega_0 t}\,dt\right\}$$

$$= \frac{1}{T}\left\{\frac{1}{\mathrm{j}k\omega_0} + \frac{1}{\mathrm{j}^2k^2\omega_0^2 T}\left[\mathrm{e}^{-\mathrm{j}k\omega_0 t}\right]_0^T\right\} = \frac{1}{T}\left\{\frac{1}{\mathrm{j}k\omega_0} - \underbrace{\frac{1}{k^2\omega_0^2 T}\left(\underbrace{\mathrm{e}^{-\mathrm{j}k2\pi}}_{=1} - 1\right)}_{=0 \text{ if } k\neq 0}\right\}$$

$$= \underline{\frac{1}{\mathrm{j}k2\pi}} \qquad k \neq 0$$

We need to treat the case $k = 0$, separately. Since $k = 0$ represents the DC component in the signal, we can easily see in Fig. 7.40 that $X(0)$ must be $= 0.5$. But let's calculate it anyway:

$$X(0) = \frac{1}{T}\int_0^T \left(1 - \frac{1}{T}t\right)dt = \frac{1}{T}\left[t - \frac{1}{2T}t^2\right]_0^T = \frac{1}{T}\left(T - \frac{T}{2}\right) = \frac{1}{2}$$

Hence:

$$|X(k)| = \begin{cases} \frac{1}{2} & k = 0 \\ \frac{1}{k2\pi} & k \neq 0 \end{cases}$$

Figure 7.41 illustrates the amplitude spectrum.

Problem 7.2 Prove that the signal $x(t - t_0)$ has the same amplitude spectrum as the signal $x(t)$.

Solution To prove this, we do a substitution: $\tau = t - t_0$. The Fourier transform of $x(t - t_0)$ is

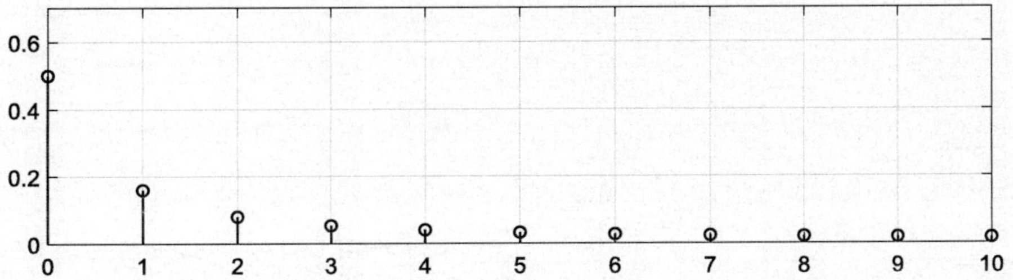

Fig. 7.41 Amplitude spectrum of sawtooth

$$\int x(t-t_0)\mathrm{e}^{-\mathrm{j}\omega t}dt = \int x(\tau)\mathrm{e}^{-\mathrm{j}\omega(\tau+t_0)}d\tau = \int x(\tau)\mathrm{e}^{-\mathrm{j}\omega\tau}\cdot \mathrm{e}^{-\mathrm{j}\omega t_0}d\tau$$
$$= \mathrm{e}^{-\mathrm{j}\omega t_0}\int x(\tau)\mathrm{e}^{-\mathrm{j}\omega\tau}d\tau = \mathrm{e}^{-\mathrm{j}\omega t_0}X(\omega) = \mathrm{e}^{-\mathrm{j}\omega t_0}|X(\omega)|\cdot \mathrm{e}^{-\mathrm{j}\varphi(\omega)}$$
$$= |X(\omega)|\cdot \mathrm{e}^{-\mathrm{j}(\varphi(\omega)+\omega t_0)}$$

A time shift only adds a phase angle ωt_0 to the phase diagram. The amplitude spectrum is unaffected by time shifts in the signal.

Problem 7.3 How does the bandwidth of the signal in Example 7.2 depend on the pulse width?

Solution We set the pulse width to a (instead of 1). According to the previous example, we can shift the pulse anywhere we want to, it doesn't affect the amplitude spectrum. Hence, we assume it is between $-a/2$ and $a/2$:

$$\int_{-a/2}^{a/2} \mathrm{e}^{-\mathrm{j}\omega t}dt = -\frac{1}{\mathrm{j}\omega}\left[\mathrm{e}^{-\mathrm{j}\omega t}\right]_{-a/2}^{a/2} = -\frac{2a}{\omega a}\cdot\frac{1}{2\mathrm{j}}\left(\mathrm{e}^{-\mathrm{j}\omega a/2} - \mathrm{e}^{\mathrm{j}\omega a/2}\right)$$
$$= a\cdot\frac{\sin\omega a/2}{\omega a/2} = a\cdot \mathrm{sinc}\frac{\omega a}{2} \Rightarrow |X(\omega)| = a\cdot\left|\mathrm{sinc}\frac{\omega a}{2}\right|$$

This expression is plotted in Fig. 7.42 and it is obvious that the bandwidth increases when the pulse width decreases. This is true in general; 'short in time, wide in frequency' (and vice versa).

Problem 7.4 Some FFT software produced the following output for a 16-sample input:

24 0 0 $-5+2\mathrm{j}$ 0 $30+18\mathrm{j}$ 0 0 0 0 0 $30-18\mathrm{j}$ 0 $-5-2\mathrm{j}$ 0 0

The 16 samples were sampled at 100 kS/s. Write down an expression for the analog signal that was sampled. (No aliasing occurred.)

Solution We have $X(0) = 24$, hence $a_0 = 24/16 = 1.5$. $X(3) = -5 + 2\mathrm{j}$ and $X(5) = 30 + 18\mathrm{j}$. Use Eq. (7.7) to find the corresponding amplitudes and phase angles (and divide the amplitudes by $N = 16$):

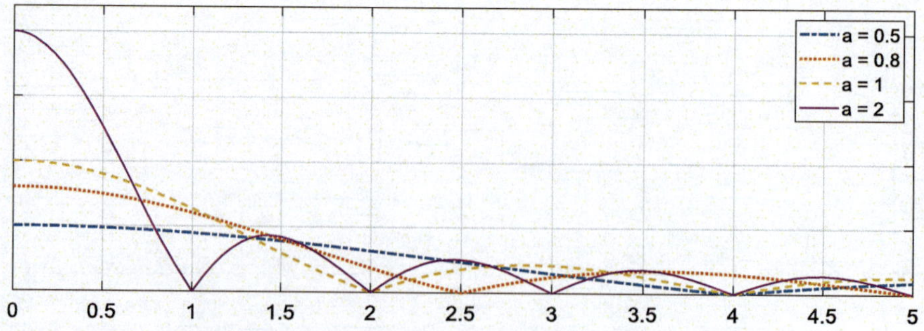

Fig. 7.42 The bandwidth increases when the pulse width decreases

$$X(3) = -5 + 2\mathrm{j} = 5.38\mathrm{e}^{\mathrm{j}158.2^\circ} \Rightarrow a_3 = 2 \cdot 5.38 \cdot \frac{1}{16} = 0.67 \qquad \varphi_3 = 158.2^\circ$$

$$X(5) = 30 + 18\mathrm{j} = 35.98\mathrm{e}^{\mathrm{j}31.0^\circ} \Rightarrow a_5 = 2 \cdot 35.98 \cdot \frac{1}{16} = 4.50 \qquad \varphi_5 = 31.0^\circ$$

Frequencies are

$$k \cdot \frac{f_S}{N} = k \cdot \frac{100}{16}\ \text{kHz} = k \cdot 6.25\ \text{kHz} \Rightarrow f_3 = 3 \cdot 6.25 = 18.75\ \text{kHz},\ f_5 = 31.25\ \text{kHz}$$

$$\underline{x(t) = 1.50 + 0.67\cos(2\pi 18750t + 158.2^\circ) + 4.50\cos(2\pi 31250t + 31^\circ)\ \text{V}}$$

Problem 7.5 If you sample the signal $x(t) = 10\sin(2\pi 200000t - 45^\circ)$, with a sampling rate of 75 kS/s, what frequency would you see?

Solution Signal frequency is 200 kHz which is $> f_S/2 = 37.5$ kHz, and there will be aliasing. We need to find the aliasing frequency that ends up in the 'observation' interval 0 ... 37.5 kHz, because that is the frequency we will 'see'. First, we subtract 75 from 200 repeatedly: 200, 125, 50, −25, −100... None of these falls into the observation area. Next, *add* 75 repeatedly to −200: −125, −50, 25, 100, ... Obviously, 25 kHz is the aliasing frequency that ends up in our observation interval. Answer: We will see a frequency of 25 kHz.

Problem 7.6 If $x(t)$ has the Laplace transform $X(s)$, what is the Laplace transform of $x'(t)$?

Solution Integrating by parts:

$$\int_0^\infty x'(t)\mathrm{e}^{-st}dt = \left[x(t)\mathrm{e}^{-st}\right]_0^\infty + s\int_0^\infty x(t)\mathrm{e}^{-st}dt = 0 - x(0) + sX(s)$$
$$= \underline{sX(s) - x(0)}$$

Problem 7.7 If $x(t)$ has the Laplace transform $X(s)$, what is the Laplace transform of the primitive function of $x(t)$?

Solution Setting $y(t) = \int_0^t x(\tau)d\tau$, then $y'(t) = x(t)$ and $y(0) = 0$. According to Example 7.6, we have that the Laplace transform of $y'(t)$ is

$$\mathcal{L}(y'(t)) = sY(s) - y(0) = sY(s) \Rightarrow Y(s) = \frac{\mathcal{L}(y'(t))}{s} = \frac{\mathcal{L}(x(t))}{s}$$
$$= \frac{X(s)}{s} = \underline{\frac{1}{s}X(s)}$$

Problem 7.8 What is the Laplace transform of a 'step function', see Fig. 7.43.

Solution $\int_0^\infty 1 \cdot \mathrm{e}^{-st}dt = -\frac{1}{s}\left[\mathrm{e}^{-st}\right]_0^\infty = -\frac{1}{s}(0-1) = \frac{1}{s}.$

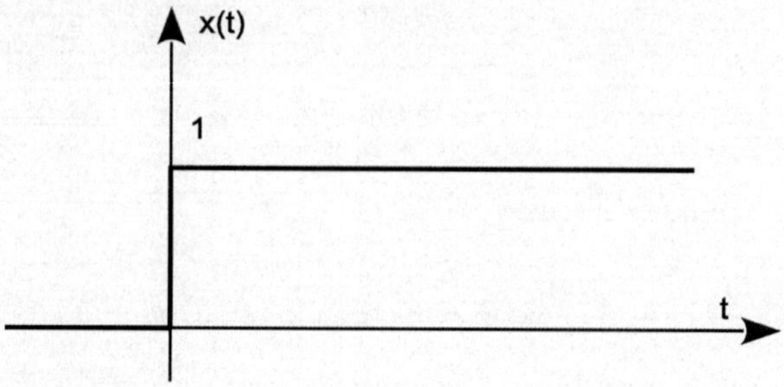

Fig. 7.43 A 'step function' (sometimes called a 'Heaviside' function)

Problem 7.9 What is the Laplace transform of $x(t) = \mathrm{e}^{-at}$? ($x(t) = 0$ for $t < 0$.)

Solution $\int_0^\infty \mathrm{e}^{-at} \cdot \mathrm{e}^{-st} dt = \int_0^\infty \mathrm{e}^{-(s+a)t} dt = -\frac{1}{s+a}\left[\mathrm{e}^{-(s+a)t}\right]_0^\infty = \frac{1}{s+a}$.

Problem 7.10 If the z transform of $x(n)$ is $X(z)$, then what is the z transform of $x(n - n_0)$?

Solution Inserting $x(n - n_0)$ into Eq. (7.41) gives us

$$\sum_{n=0}^{\infty} x_{n-n_0} z^{-n} = \left\{ \begin{array}{l} m = n - n_0 \\ n = m + n_0 \end{array} \right\} =$$

$$\sum_{m=-n_0}^{\infty} x_m z^{-m-n_0} = \{\text{assuming } x_n = 0 \text{ if } n < 0\}$$

$$= \sum_{m=0}^{\infty} x_m z^{-m} z^{-n_0} = \underline{X(z) \cdot z^{-n_0}}$$

Problem 7.11 If $x(t)$ has the Laplace transform $X(s)$, what is the Laplace transform of the delayed signal $x(t - t_0)$?

Solution Substituting τ for $t - t_0$ and that $x(t) = 0$ for $t < 0$:

$$\mathcal{L}(x(t - t_0)) = \int_0^\infty x(t - t_0)\mathrm{e}^{-st} dt = \left\{ \begin{array}{l} \tau = t - t_0 \\ d\tau = dt \\ t = \tau + t_0 \end{array} \right\} = \int_{-t_0}^{\infty} x(\tau)\mathrm{e}^{-s(\tau+t_0)} d\tau$$

$$= \mathrm{e}^{-st_0} \int_0^\infty x(\tau)\mathrm{e}^{-s\tau} d\tau = \mathrm{e}^{-st_0} X(s)$$

If $x(t)$ has the Laplace transform $X(s)$, then $x(t - t_0)$ has the Laplace transform $\underline{\mathrm{e}^{-st_0} X(s)}$.

Reference

1. Cooley, J.W., and J.W. Tukey. 1965. An algorithm for the machine calculation of complex Fourier series. *Mathematics of computation* 19 (90): 297–301.

Chapter 8
Spectrum Analyzers

Abstract Spectrum analyzers are one of the most common instruments used in a measurement laboratory (often integrated into the oscilloscope) and it is an imperative skill of a scientist to be able to handle a spectrum analyzer and to interpret its output. A spectrum analyzer is the most basic application of transform theory, and this chapter relies heavily on the previous chapter. Spectrum analyzers can be analog (non-sampling) or digital (sampling). Both kinds of analyzers are treated in this chapter, but the focus is on sampling systems based on the fast Fourier transform. Fundamental spectrum concepts like spectral leakage and windows, resolution bandwidth, and heterodyne analyzers are highlighted and illustrated by examples.

8.1 Introduction

In general, a 'spectrum analyzer' produces the Fourier transform of a signal. However, in most situations, it is understood that only the magnitude ($|X(\omega)|$) of the Fourier transform is of interest. Some spectrum analyzers produce the phase diagram too, but here we will consider spectrum analyzers that produce the magnitude diagram only.

A spectrum analyzer can be 'digital' or 'analog'. We will treat 'analog' spectrum analyzers in Sect. 8.4. Digital spectrum analyzers sample the signal and calculate the discrete Fourier transform (Expression (7.18)), but of course, they use the FFT algorithm to speed up the math (see Sect. 7.2). We know from the previous chapter that Fourier transform spectra can be corrupted by aliasing, so here we will assume that the signal is sampled with respect to the sampling theorem. Instead, we will concentrate on other aspects of the Fourier transform spectrum.

Our starting point is Expression (7.20), that specifies the resolution of the FFT spectrum:

$$f = k \cdot \frac{f_S}{N} = k \cdot \Delta f \tag{8.1}$$

where Δf is the resolution of the FFT spectrum ($|X(k)|$).

L. Bengtsson, *Electrical Measurement Techniques*,
https://doi.org/10.1007/978-981-99-8187-8_8

Consider the analog signal in Fig. 8.1. To produce the FFT, we must sample it, and we must stop sampling it at some point; we take N samples of the signal at a rate of f_S. If we would sample the signal in the interval indicated in Fig. 8.1, the signal's period would fit *exactly* five times in this interval, and that would correspond to a 'frequency' of $k = 5$ in Eq. (8.1). The FFT spectrum would produce a single peak for $k = 5$, see Fig. 8.2; k represents the number of periods of the signal that fits in the sampling interval.

The spectrum in Fig. 8.2 looks nice and clean, but if you think about it, we were quite lucky in our sampling; we sampled *exactly* five periods of the signal! What are the odds of that in a 'real' situation? In a real situation, we would not know the signal frequency and it is more likely to look as in Fig. 8.3.

In Fig. 8.3, the sampling interval is not an exact multiple of the signal's frequency; it corresponds to $k \approx 4.8$. But the FFT only outputs frequencies for integer values of k. So, in what FFT channel will we see this signal? Well, the answer is that in this case, which is the most common case, the signal will be smeared out over *all* frequencies (over all k values), see Fig. 8.4. However, the k frequency closest to 4.8

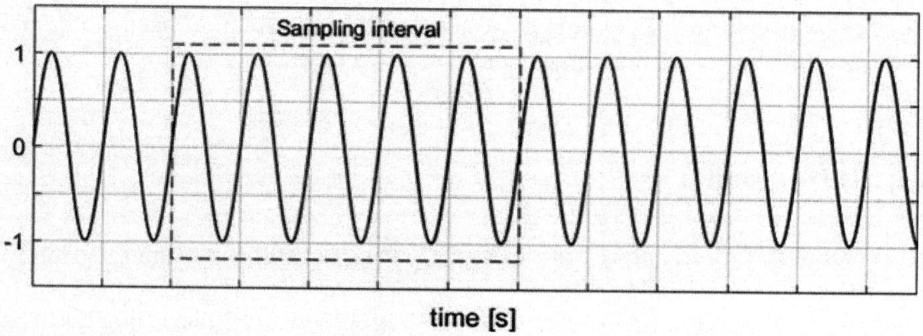

Fig. 8.1 Sampling a sinusoidal signal

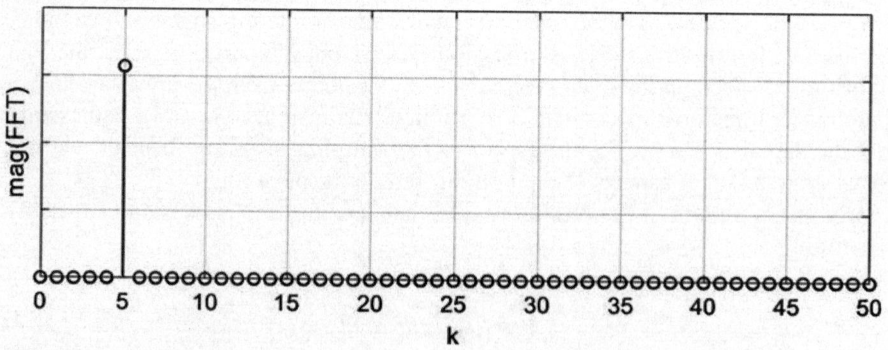

Fig. 8.2 The FFT spectrum

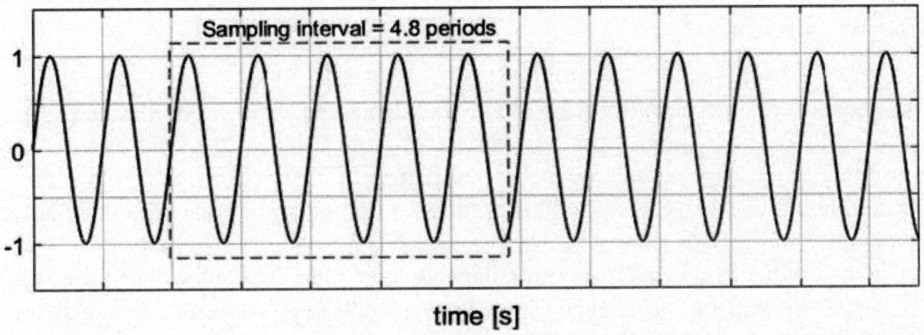

Fig. 8.3 'Real' case: The sampling interval is not an exact multiple of the signal's period

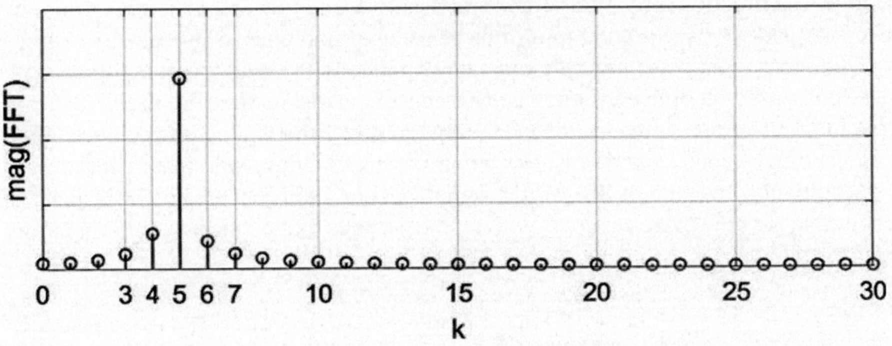

Fig. 8.4 Spectral leakage

(= 5) will have the largest value and the other frequencies' magnitudes will drop off with the distance to 4.8, see Fig. 8.4.

This phenomenon, the 'smearing out' of the signal's energy over a wide range of frequencies in the FFT spectrum, is called 'spectral leakage'. If we only learn that the k frequency represents a signal period that fits k times in the sampling interval, then we could (erroneously) conclude from the spectrum in Fig. 8.4 that the original signal consists of many sinusoidal signals with a wide range of frequencies, so it is important to understand that multiple peaks in an FFT spectrum can be (and often are) caused by spectral leakage.

Apart from introducing an uncertainty in the signal's frequency, the biggest disadvantage of leakage is that the frequency peak is 'broadened', which reduces our ability to resolve signals with close frequencies in the spectrum. If that is the problem, then we need to reduce the leakage.

8.2 Windows

The reason for the spectral leakage is obvious from Fig. 8.3; we don't sample an exact multiple of the signal's period. Remember from Chap. 7, the DFT expression assumes that we have sampled exactly one period of a periodic signal. If we take the sampled interval in Fig. 8.3 and repeat it, we get the signal that the DFT really represents, see Fig. 8.5.

The periodic signal will have singularities, and then it is no wonder that the FFT spectrum will be 'broadened'. The reason for the spectrum broadening is the discontinuities in the periodic signal in Fig. 8.5; if we want to remedy the spectral broadening, we must remedy the discontinuities.

We remedy the discontinuities by applying a 'window' to the sampled data. A window is a function that is zero (or almost) at both ends. There are a lot of window functions and we have plotted four of the most common ones in Fig. 8.6.

That means that if we multiply our sampled data with a window function, it will be forced to zero at both ends and that will cancel the discontinuities. In Fig. 8.7, we have multiplied our samples with a Hanning window function, and it is obvious that the resulting periodic function is now a continuous function with no singularities.

Admittedly, the window distorts the signal, but our main concern here was spectral resolution, not amplitude accuracy. In Fig. 8.8, we have plotted the new FFT spectrum after the windowing.

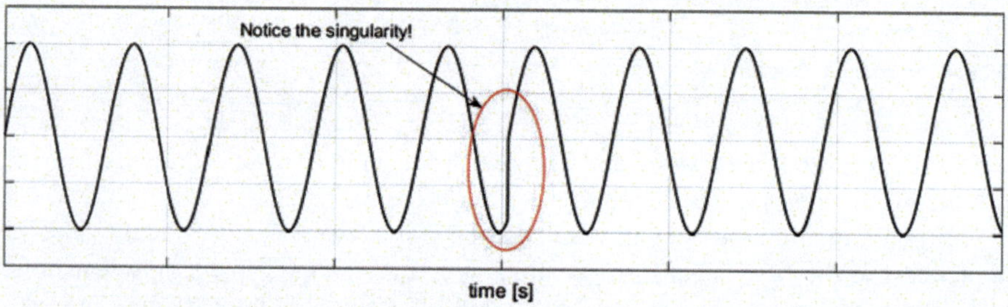

Fig. 8.5 The periodic signal has a singularity

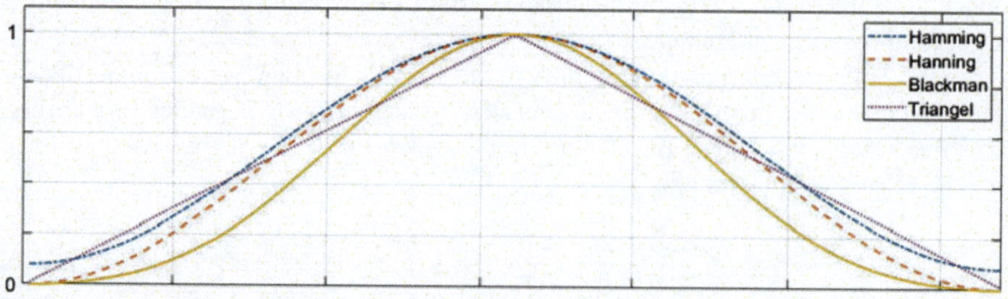

Fig. 8.6 Some window functions

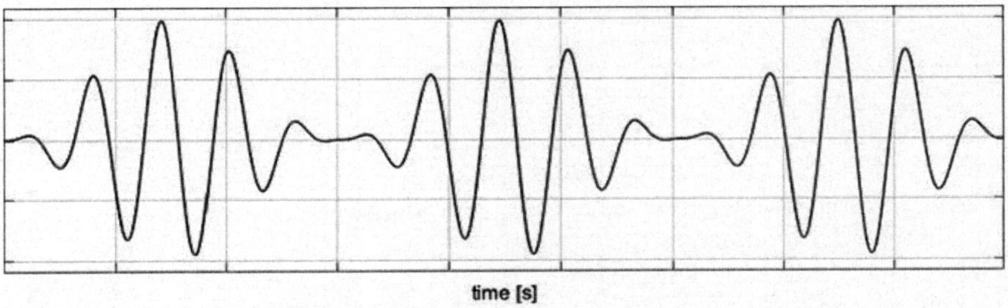

Fig. 8.7 Applying a Hanning window: The periodic signal has no discontinuities

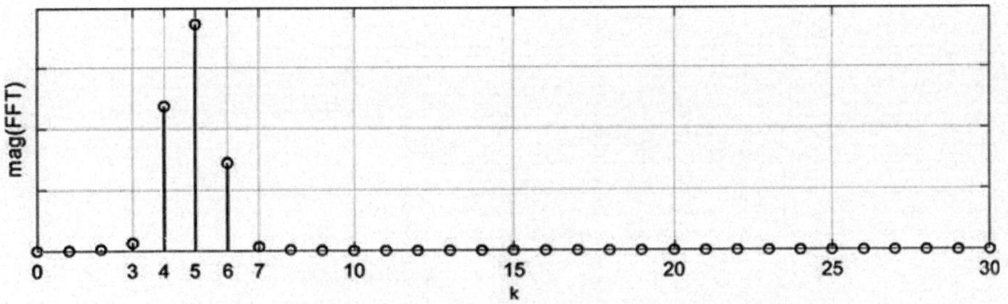

Fig. 8.8 The FFT spectrum after a Hanning window has been applied to the signal

If we compare the spectrum in Fig. 8.8 with the spectrum in Fig. 8.4, we can see that the spectrum has been narrowed, it is spread out over three k-numbers only (in Fig. 8.4 it is spread over 8–10 k-numbers).

Modern oscilloscopes that have an FFT analyzer also have a set of window functions that you can choose from. Different windows work best for different situations, so you just try them all and see which one works best for your signal. Windows are particularly useful when you need to resolve spectrum peaks where a 'small' peak is drowned in the leakage from a 'big' peak. This is illustrated in the next example.

Example 8.1 Figure 8.9 illustrates the amplitude spectrum of the signal $x(t) = 10 \sin 100t + 0.5 \sin 104t$. ($f_S = 100$ S/s, $N = 501$.)

As we can see in Fig. 8.9, the 104 rad/s signal is hard to discern; it drowns in the leakage from the 100 rad/s signal. Try resolving the peaks by applying a window to the sampled data.

Solution Fig. 8.10 illustrates what the spectrum looks like after a Bartlett window has been applied to the data. Comparing it with Fig. 8.9, we can see that it is now easier to discern the small peak.

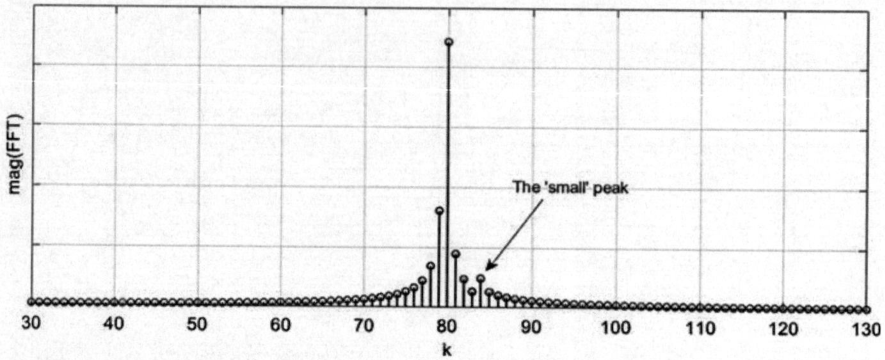

Fig. 8.9 The 'small' peak drowns in the 'big' peak's leakage

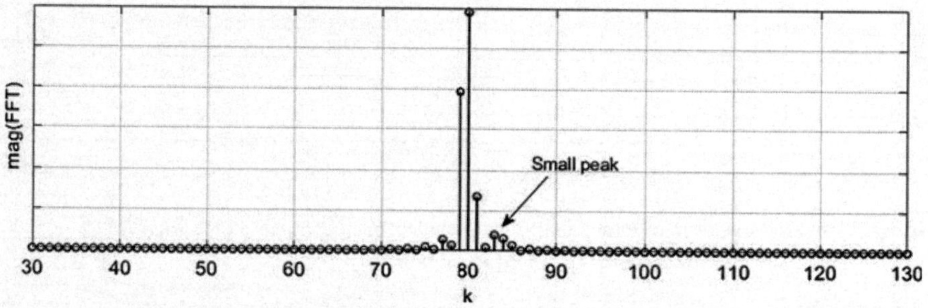

Fig. 8.10 Spectral resolution has been improved by a Bartlett window

8.3 Resolution Bandwidth

8.3.1 Quantifying the Leakage

It is possible to quantify (and predict) the leakage exactly, and it is not that hard either. To explain how to do that, we need some numbers to work with. Let's assume that we take 128 samples of a sine signal with frequency 336 Hz at a rate of 10 kS/s. The total sampling time is $N{\cdot}T_S = 128/10\,\text{k} = 12.8$ ms. The signal period is 1/336 = 2.97 ms, and hence we expect a peak in the DFT spectrum at

$$k = \frac{12.8}{2.97} = 4.3 \tag{8.2}$$

Since the DFT spectrum only provides values for integer numbers of k, there will be leakage; the biggest peak will be at $k = 4$, the second biggest will be at $k = 5$, etc., see Fig. 8.11.

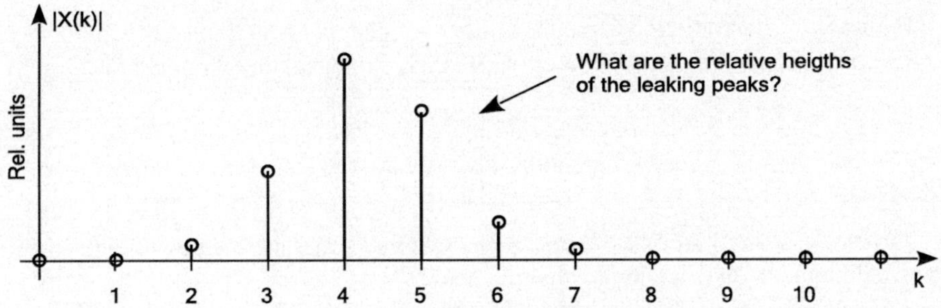

Fig. 8.11 Estimated spectrum: We will figure out the relative sizes of the peaks in the leakage

Our objective here is to calculate the relative heights of the peaks in Fig. 8.11. It is not that hard once you understand how to model each 'channel' (= each k). Admittedly, we have not covered filters yet, so you might need to read this section again after you have read Chapter 9, but we think this is understandable even without the filter theory in Chapter 9.

First, we will refer to each k frequency in the DFT spectrum as a 'channel', and we model each channel as a *narrow resonance filter* with center frequency $\Delta f = k \cdot f_S$. However, this resonance filter is *not* infinitely narrow; the frequency response function for each channel is a sinc function, see Fig. 8.12.

Notice in Fig. 8.12 that the period of the sinc function is 2π; the sinc function's zeros coincide exactly with all the other channels' center frequencies. If the signal frequency had been exactly $k = 4$, the channel 4 peak would have had height 1 (normalized), but since it is 0.3π to the right of channel 4, the channel 4 height will be

$$\text{Channel 4: } \frac{\sin 0.3\pi}{0.3\pi} = 0.86$$

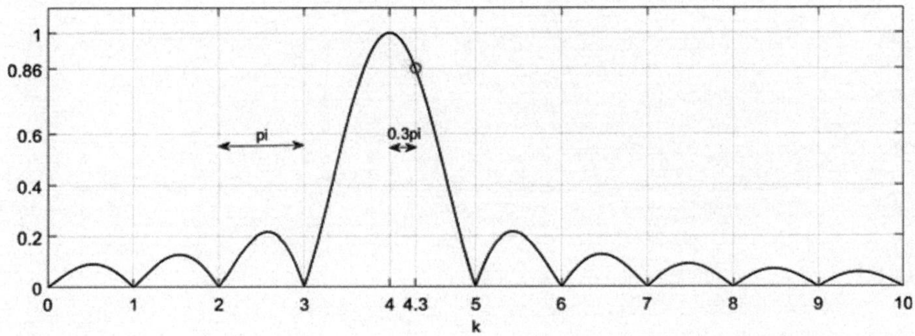

Fig. 8.12 The relative peak height in channel 4 will be 0.86

Figure 8.13 illustrates the resonance filter over channel 5, and since the frequency $k = 4.3$ is 0.7π away from the center of channel 5, the peak height in channel 5 will be

$$\text{Channel 5: } \frac{\sin 0.7\pi}{0.7\pi} = 0.37$$

We center a sinc function over each channel and calculate the peak height from the 'real' signal's distance to the channel center.

And once you accept the 'sinc resonance filter' model for each DFT channel, there is an easier way to do this; if you center a sinc function over the 'real' frequency, you can read the relative height in each channel from the sinc function's value over the channel, see Fig. 8.14. Compare Fig. 8.14 with Fig. 8.15 where the real DFT spectrum has been plotted.

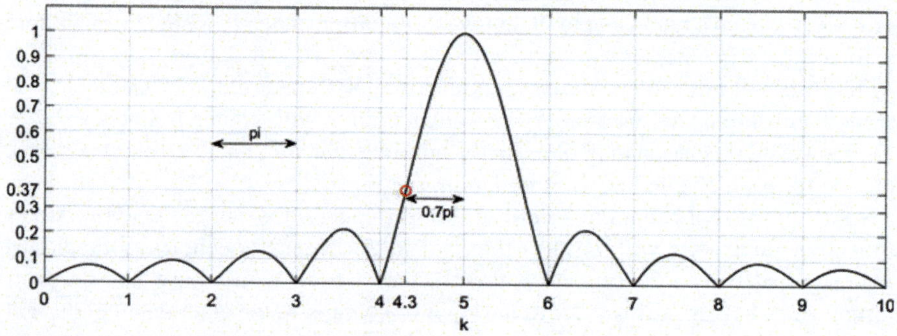

Fig. 8.13 The relative peak in channel 5 will be 0.37

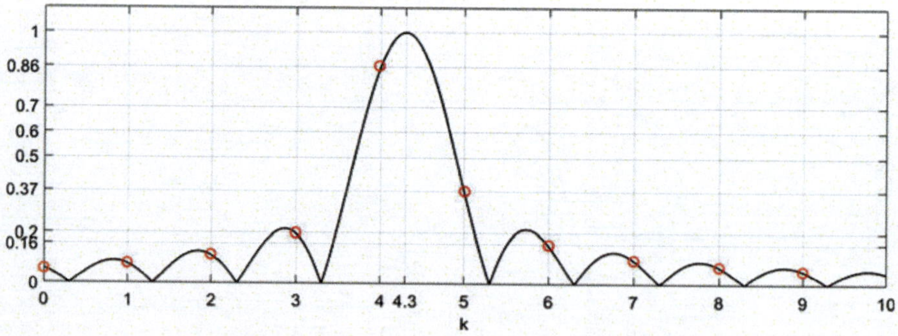

Fig. 8.14 Center the sinc function over $k = 4.3$

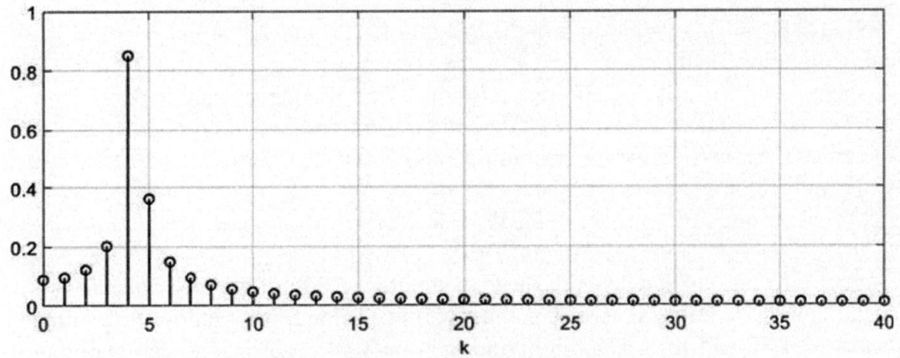

Fig. 8.15 The real DFT spectrum (compare with Fig. 8.14)

8.3.2 Resolution Bandwidth

Equation (8.1) defines the resolution Δf in the FFT spectrum. However, Δf is not the most common number used to describe the resolution in FFT spectra; the most common number is the *resolution bandwidth*, RBW. The RBW is almost the same as Δf but there is a subtle difference that we think is important to understand, and the RBW has an advantage over Δf; RBW is adjusted for different windows.

In Chap. 9, we will learn that a filter's bandwidth is defined as the '3 dB' limit, i.e., the frequency where the gain has decreased by –3 dB ($= 1/\sqrt{2} \approx 0.707$). Translated to our sinc resonance filters in the FFT spectrum, the bandwidth will be the distance between the upper and lower limits where the gain has decreased by –3 dB, see Fig. 8.16.

To find the exact bandwidth, we must solve the equation $\sin x/x = 1/\sqrt{2}$ which has the solutions $x = \pm 0.44\pi$, which means that the bandwidth is $2 \cdot 0.44\pi \approx 0.9\pi$ or $0.9 \cdot \Delta f$ (since the distance between each channel is π in terms of 'sinc angle' and Δf in terms of 'Hz').

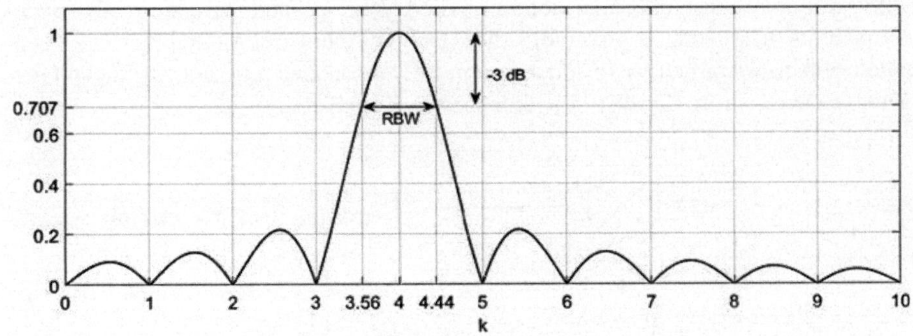

Fig. 8.16 Defining the resolution bandwidth

So, RBW $= 0.9 \cdot \Delta f$ which indicates that RBW and Δf are almost the same and it shouldn't really matter which one we use, but it does, because RBW can be compensated for the use of different windows. The 'frequency space' consequence of applying a window is that we change the resolution bandwidth in each channel. The general expression for the resolution bandwidth is

$$\mathrm{RBW} = w \cdot \Delta f \tag{8.3}$$

where w is a constant that depends on what window we use. If we have no window (= 'rectangular' window), then $w = 0.9$, but if we have a Hamming window, for example, then $w = 1.30$ and for a Hanning window $w = 1.44$. That means that applying a window *increases* the RBW which would indicate an impairment in resolution, but because the *leakage* is reduced, the overall resolution could still improve. However, if you don't suffer from leakage, you should not use a window because it deteriorates the resolution bandwidth.

8.4 Heterodyne Analyzers

It is not necessary to sample the signal to find it's amplitude spectrum, it is quite possible to design analog (non-sampling) hardware that produces $|X(\omega)|$. That hardware is based on the following trigonometric identity:

$$A\cos\alpha \cdot B\cos\beta = \frac{AB}{2}(\cos(\alpha+\beta) + \cos(\alpha-\beta)) \tag{8.4}$$

If we multiply two trigonometric functions, we get two new ones: One with the 'sum-of-angles' and one with the 'difference-of-angles'. That implies that if we multiply two sinusoidal *signals* with different frequencies, we will get two new sinusoidal signals, one with the sum frequency and one with the difference frequency, see Fig. 8.17.

Multiplying two signals is sometimes called 'mixing', and there are ready-made components that do that, for example, the AD633 circuit from Analog Devices. As a matter of fact, we only have to add a very narrow resonance filter and an AC voltage meter to the circuit in Fig. 8.17 to get a spectrum analyzer.

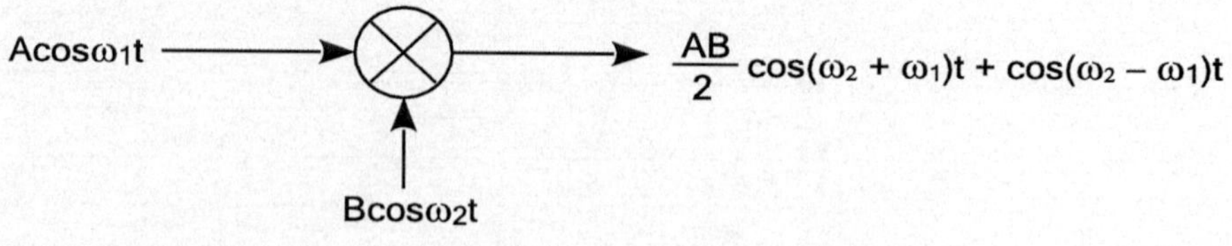

Fig. 8.17 'Mixing' two signals produces the sum and difference frequencies

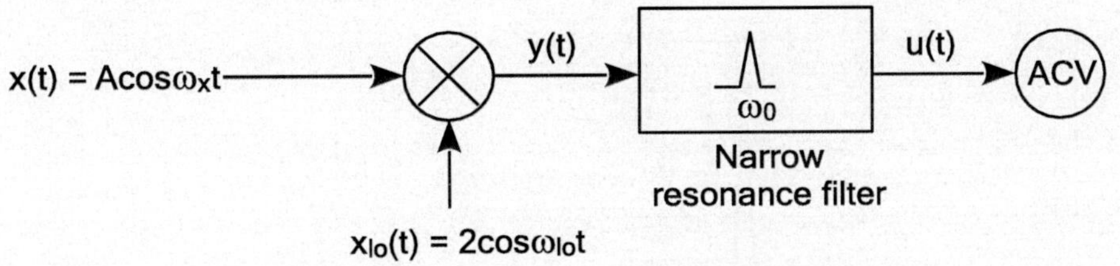

Fig. 8.18 An analog spectrum analyzer (a 'heterodyne' analyzer)

In Fig. 8.18, *x(t)* is the signal we want to analyze (i.e., find the magnitude of its Fourier transform) and 'lo' is short for 'local oscillator'. In Fig. 8.18, the signal $y(t)$ is

$$y(t) = \frac{2 \cdot A}{2}(\cos(\omega_{\text{lo}} + \omega_{\text{x}})t + \cos(\omega_{\text{lo}} - \omega_{\text{x}})t) = \\ A\cos(\omega_{\text{lo}} + \omega_{\text{x}})t + A\cos(\omega_{\text{lo}} - \omega_{\text{x}})t \tag{8.5}$$

Next, we *sweep* the local oscillator's frequency; we start on ω_0 and sweep continuously up to $\omega_0 + \omega_{\text{bw}}$. ('bw' is short for 'bandwidth'.) Also remember that the narrow resonance filter only allows signals with the exact frequency ω_0 to pass. That means that $u(t) \neq 0$ only if $\omega_{\text{lo}} + \omega_{\text{x}} = \omega_0$, or if $\omega_{\text{lo}} - \omega_{\text{x}} = \omega_0$, i.e., if

$$\text{Sum term: } \omega_{\text{lo}} + \omega_{\text{x}} = \omega_0 \Rightarrow \omega_{\text{lo}} = \omega_0 - \omega_{\text{x}} \tag{8.6a}$$

$$\text{Difference term: } \omega_{\text{lo}} - \omega_{\text{x}} = \omega_0 \Rightarrow \omega_{\text{lo}} = \omega_0 + \omega_{\text{x}} \tag{8.6b}$$

Since the local oscillator's frequency *starts* on ω_0 (and sweeps upward), the condition $\omega_{\text{lo}} = \omega_0 - \omega_{\text{x}}$ will *never* happen; the resonance filter will always cancel the sum signal in Eq. (8.5).

The difference term, however, *will* pass the resonance filter if only $\omega_{\text{lo}} = \omega_0 + \omega_{\text{x}} < \omega_0 + \omega_{\text{bw}}$, i.e., if $\omega_{\text{x}} < \omega_{\text{bw}}$; the signal $x(t)$ must be within the analyzer's bandwidth. When $\omega_{\text{lo}} = \omega_0 + \omega_{\text{x}}$, then $u(t) = A\cos\omega_0 t$, and the ACV meter will register its amplitude (or rms value). For all other frequencies, the ACV will read 0 V.

So, we sweep the local oscillator's frequency from ω_0 to $\omega_0 + \omega_{\text{bw}}$ and read the ACV. Figure 8.19 illustrates the ACV reading as a function of the local oscillator's frequency.

All we need to do is to rescale the frequency axis and we have the magnitude of the Fourier transform of $x(t)$. (Even better since we don't have the ½ factor on the vertical scale as the Fourier transform does.) This kind of frequency analyzers are called *heterodyne* analyzers and more expensive oscilloscopes have built-in heterodyne analyzers.

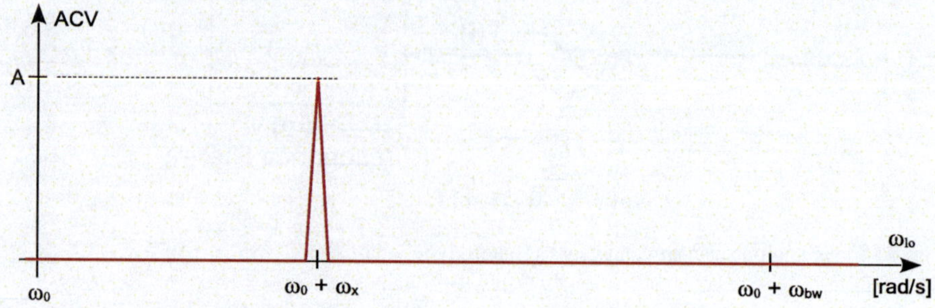

Fig. 8.19 The ACV reading as a function of the local oscillator frequency

The advantage of heterodyne analyzers is that they don't sample, so they are not limited by the sampling theorem (no aliasing); heterodyne analyzers typically have much higher bandwidths than FFT analyzers.

8.5 Solved Problems

Problem 8.1a A 500 Hz sine signal is sampled 256 times at a sampling rate of 5 kS/s. Find the relative sizes of the six largest peaks in the FFT spectrum (for $k < 128$).

Problem 8.1b How would the spectrum change if the sampling rate is changed to 5.12 kS/s?

Solution The frequency resolution is $\Delta f = 5000/256 = 19.53$ Hz, and the 500 Hz sine corresponds to a frequency $k = 500/19.53 = 25.6$.

That means that the 'biggest' peaks will be 26, 25, 27, 24, 28, and 23, in that order. The '26 peak' is only 0.4π from the sinc maximum, so its relative size will be

$$\text{Channel 26:} \quad \frac{\sin 0.4\pi}{0.4\pi} = 0.76$$

Channel 25 is 0.6π from the sinc maximum, and for the following channels we add one π for each channel:

$$\text{Channel 25:} \quad \frac{\sin 0.6\pi}{0.6\pi} = 0.50 \qquad \text{Channel 27:} \quad \frac{\sin 1.4\pi}{1.4\pi} = 0.22$$

$$\text{Channel 24:} \quad \frac{\sin 1.6\pi}{1.6\pi} = 0.19 \qquad \text{Channel 28:} \quad \frac{\sin 2.4\pi}{2.4\pi} = 0.13$$

$$\text{Channel 23:} \quad \frac{\sin 2.6\pi}{2.6\pi} = 0.12$$

In Fig. 8.20, we have plotted these six peaks together with the sinc function and in Fig. 8.21, we have plotted the real FFT spectrum for the sampled 500 Hz signal. If we compare Figs. 8.20 and 8.21, we can see that our prediction in Fig. 8.20 is correct.

If we change the sampling rate to 5.12 kS/s, then $\Delta f = 5120/256 = 20$ Hz exactly, and the 500 Hz sine corresponds to $k = 500/20 = 25$. That would give a single peak at $k = 25$ in the FFT spectrum, see Fig. 8.22.

Problem 8.2 Figure 8.23 illustrates the FFT spectrum of some unknown signal that was sampled at 10 kS/s and 100 samples were taken. What can you say about the signal?

Solution The spectrum is perfectly symmetric around $k = 20.5$; the signal that was sampled was a sinusoidal signal with frequency $f = 20.5 \cdot \Delta f = 20.5 \cdot 10{,}000/100 = \underline{2050 \text{ Hz}}$.

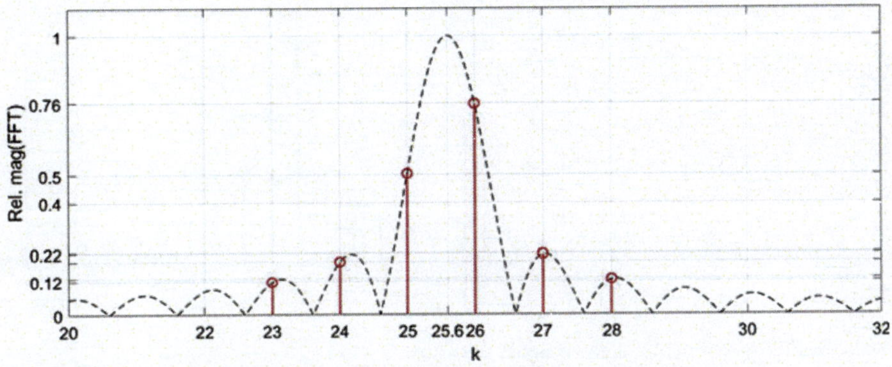

Fig. 8.20 Predicting the six biggest peaks in the FFT spectrum

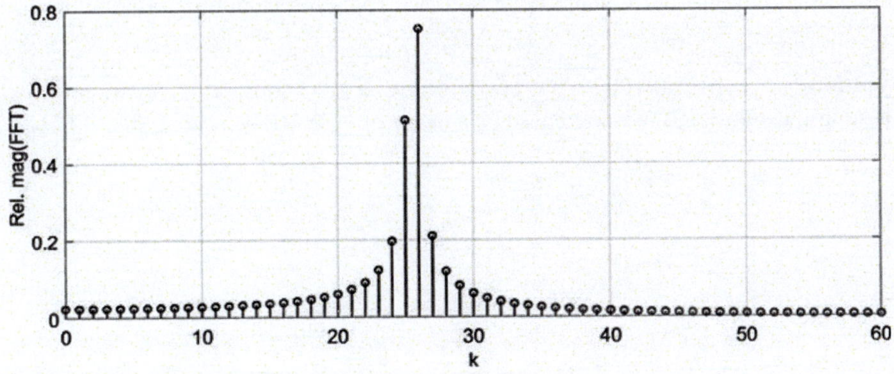

Fig. 8.21 The real spectrum (using the *fft* command in MATLAB)

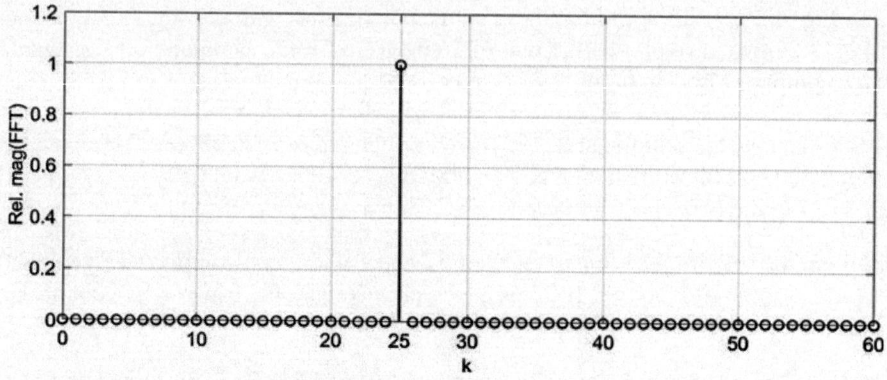

Fig. 8.22 FFT spectrum if the sampling rate is changed to 5.12 kS/s

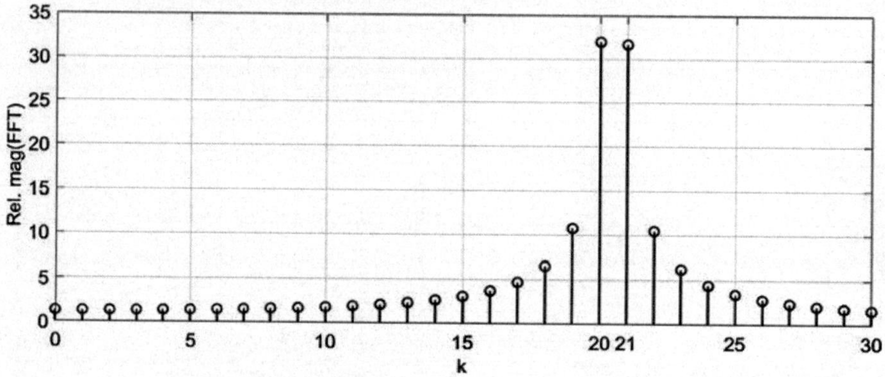

Fig. 8.23 FFT spectrum of unknown signal

Problem 8.3 What is the resolution bandwidth of the FFT analyzer in Problem 8.2? What would the resolution bandwidth be if we applied a Hamming window to the data?

Solution $\Delta f = 10{,}000/100 = 100$ Hz, so RBW $= 0.9 \cdot 100 = \underline{90 \text{ Hz}}$. If we apply a Hamming window, RBW is increased by a factor of 1.30: RBW $= 90 \cdot 1.30 = \underline{117 \text{ Hz}}$.

Chapter 9
Analog Filters

Abstract A filter is used to discriminate unwanted signals in a complex measurement signal. This chapter first introduces passive filters (RCL filters) of first and second order. Next, biquad filters, switched capacitor filters, and state-variable filters are introduced. The quality factor is defined (the Q factor) and different filter characteristics and filter models (Butterworth, Chebyshev, and elliptic models) are illustrated. Analog filters can be implemented using so-called Sallen–Key links. Section 9.6 demonstrates how a given filter characteristic can be transformed into any other filter characteristic. To process the signal in time space, it is necessary to introduce a mathematical operation called *convolution* (Sect. 9.7). Like transforms, convolution is usually considered to be hard to grasp by students, but this chapter emphasizes the *understanding* of convolution by using graphical examples.

9.1 Introduction

By 'analog' filters we mean filters that can be implemented in hardware, using analog electronics. (We will treat 'digital' filters in Chap. 10.) Filter theory depends heavily on the transform theory that we presented in Chap. 7, like Bode plots (Sect. 7.3) and pole-zero diagrams.

9.2 First-Order Filters

9.2.1 Passive Filters

A first-order passive filter is a simple voltage division between two impedances, see Fig. 9.1.

The impedances are either real or imaginary (they could be complex, but we limit the presentation here to non-complex impedances). The transfer function is simple enough:

L. Bengtsson, *Electrical Measurement Techniques*,
https://doi.org/10.1007/978-981-99-8187-8_9

$$y(t) = \frac{Z_2}{Z_1 + Z_2} x(t) \Rightarrow Y(\omega) = \frac{Z_2}{Z_1 + Z_2} X(\omega)$$
$$\Rightarrow H(\omega) = \frac{Y(\omega)}{X(\omega)} = \frac{Z_2}{Z_1 + Z_2} \tag{9.1}$$

In Example 7.7, we illustrated the case where $Z_1 = R$ and $Z_2 = 1/j\omega C$:

$$H(\omega) = \frac{1}{1 + j\omega RC} \Rightarrow H(s) = \frac{1}{1 + sRC} \tag{9.2}$$

This is a lowpass system with a pole in $s = -1/RC$. By replacing Z_1 and Z_2 in Eq. (9.1) with other combinations of R, $j\omega L$ or $1/j\omega C$, we can get highpass filters too.

The filter in Fig. 9.1 has a disadvantage; if it is cascaded with a second filter module, its output impedance will be connected in parallel with the next stage's input impedance and that might change the characteristics of the filter. This is easily avoided by inserting an op amp as a voltage follower, see Fig. 9.2, and once we have an op amp, we might as well take advantage of it and use it as a non-inverting amplifier to add an arbitrary amplification, see Fig. 9.3. Filters with op amps are called 'active' filters.

Fig. 9.1 First-order filter

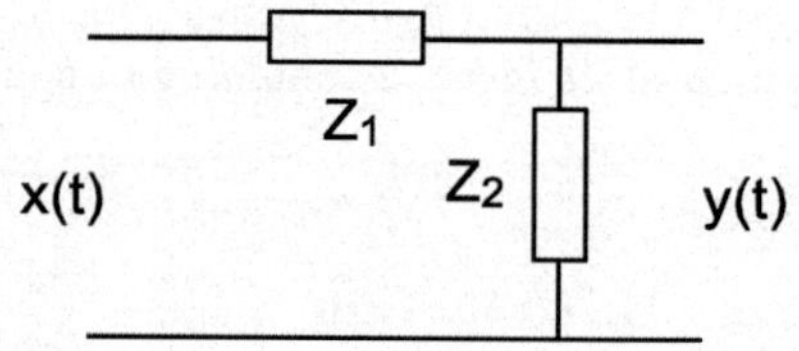

Fig. 9.2 First-order filter with voltage follower

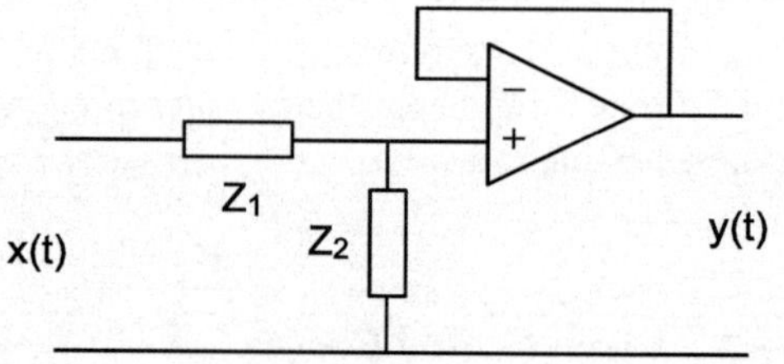

Fig. 9.3 First-order filter with amplification $(1 + R_2/R_1)$

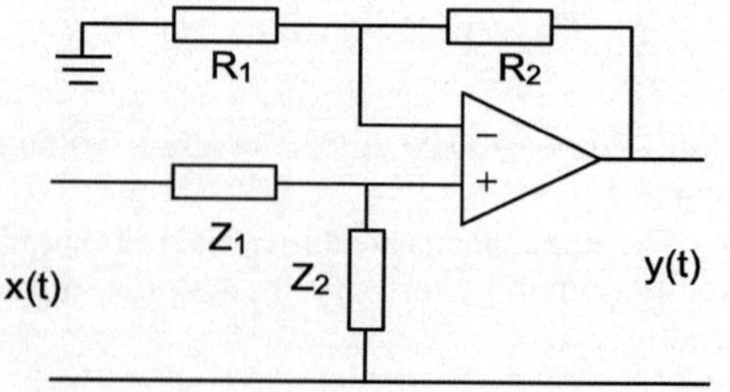

9.3 Second-Order Filters

9.3.1 *'Biquad'*

The general expression for a second-order filter is

$$H(s) = \frac{N(s)}{D(s)} = \frac{b_2 s^2 + b_1 s + b_0}{s^2 + a_1 s + a_0} \tag{9.3}$$

We know from Sect. 7.4 that the roots of the denominator polynomial represent the system's singular points (the 'poles') and they determine the system's resonance frequencies. If the filter in Eq. (9.3) has the poles $s = -\sigma_p \pm \mathrm{j}\omega_p$ (remember that the poles must be in the left half-plane in s space), then the denominator polynomial is

$$\begin{aligned} D(s) &= (s - s_{p1})(s - s_{p2}) = (s + \sigma_p - \mathrm{j}\omega_p)(s + \sigma_p + \mathrm{j}\omega_p) \\ &= s^2 + 2\sigma_p s + \sigma_p^2 + \omega_p^2 \end{aligned} \tag{9.4}$$

$\sqrt{\sigma_p^2 + \omega_p^2} = \omega_0$ is of course the poles' distance to the origin, and we define the system's 'quality factor' as

$$Q = \frac{\omega_0}{2 \cdot |\sigma_p|} \Rightarrow |\sigma_p| = \frac{\omega_0}{2Q} \tag{9.5}$$

Hence, Q is the poles' distance from the origin divided by their distance from the jω-axis, see Fig. 9.4.

This means that we can write the denominator polynomial in (9.3) as

$$D(s) = s^2 + \frac{\omega_0}{Q} s + \omega_0^2 \tag{9.6}$$

Inserting that into Eq. (9.3) gives us

$$H(s) = \frac{N(s)}{D(s)} = \frac{b_2 s^2 + b_1 s + b_0}{s^2 + \frac{\omega_0}{Q} s + \omega_0^2} \tag{9.7}$$

In Eq. (9.7), we have a second-order polynomial in both the numerator and the denominator. Since we have *two* quadratic polynomials, the filter is sometimes referred to as 'biquadratic' or just 'biquad'. It is the denominator polynomial (the poles) that determines the filter's resonance frequency (ω_0), but it is the numerator polynomial that determines the filter type (lowpass, highpass, bandpass, etc.). That gives us three important special cases.

Fig. 9.4 Defining Q and ω_0

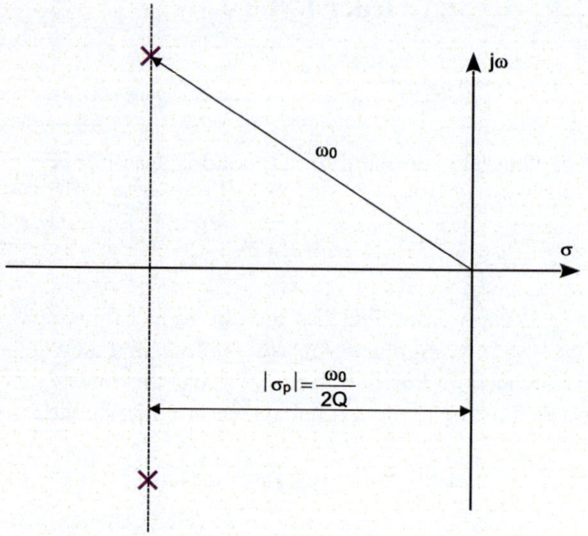

9.3.2 Lowpass: $b_2 = b_1 = 0$

With $b_2 = b_1 = 0$ we get the transfer function

$$H(s) = \frac{b_0}{s^2 + \frac{\omega_0}{Q}s + \omega_0^2} \Rightarrow |H(\omega)| = \frac{|b_0|}{\sqrt{\left(\frac{\omega_0}{Q}\omega\right)^2 + \left(\omega_0^2 - \omega^2\right)^2}} \tag{9.8}$$

In Fig. 9.5, we have plotted $|H(\omega)|$ for $b_0 = \omega_0 = 1$ for different quality factors. Notice first the lowpass characteristics; the amplification is =1 for low frequencies and =0 for high frequencies. Another thing to notice is the 'resonance' at $\omega = 1$ ($= \omega_0$) and that the peak gets higher and sharper with increasing Q. This is easy

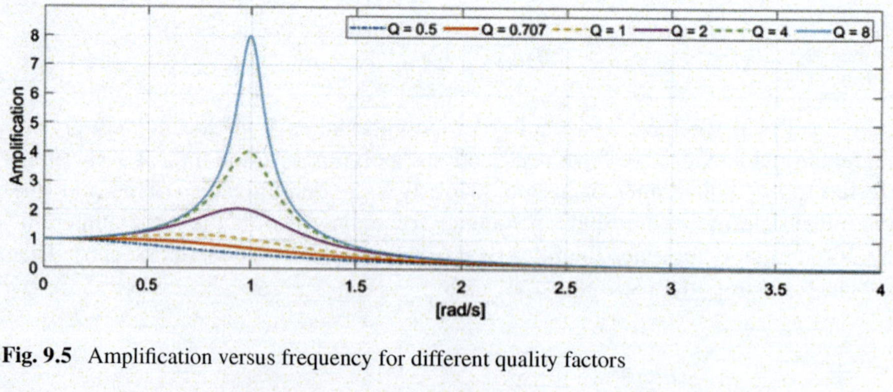

Fig. 9.5 Amplification versus frequency for different quality factors

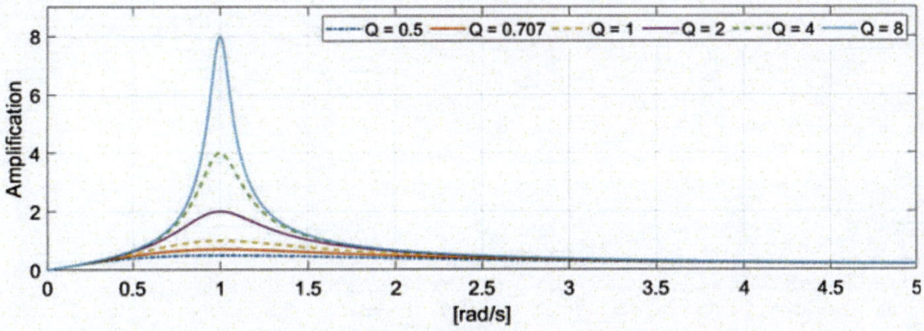

Fig. 9.6 Amplification diagram for bandpass filter

to understand from Fig. 9.4; when Q increases, the pole's distance to the jω-axis decreases and the closer the pole is to the jω-axis, the higher and sharper is the resonance peak (see also Fig. 7.32). The special case where $Q = 0.707$ $(1/\sqrt{2})$ generates no 'overshoot' and the system response is 'flat' (in the 'passband'; we will later refer to this case as the 'Butterworth filter'). For $Q > 0.707$, we have overshoots (the system is 'underdamped'), for $Q < 0.707$ the system is 'overdamped' and the system is 'critically damped' if $Q = 0.707$. (See also Sect. 18.5.2 about step response of second-order systems.)

9.3.3 Bandpass: $b_2 = b_0 = 0$

With $b_2 = b_0 = 0$ we get the transfer function

$$H(s) = \frac{b_1 s}{s^2 + \frac{\omega_0}{Q}s + \omega_0^2} \Rightarrow |H(\omega)| = \frac{|b_1\omega|}{\sqrt{\left(\frac{\omega_0}{Q}\omega\right)^2 + \left(\omega_0^2 - \omega^2\right)^2}} \tag{9.9}$$

In Fig. 9.6, we have plotted $|H(\omega)|$ for $b_1 = \omega_0 = 1$. Notice the bandpass characteristics; the amplification diagram goes to zero at both ends. The resonance peak is still at $\omega = \omega_0 = 1$ and the peak gets higher and sharper with increasing Q. In a lowpass filter, you try to keep the peak as low as possible, but in a resonance filter, the sharpness of the peak is intentional; the sharper the resonance peak is, the more selective is the filter.

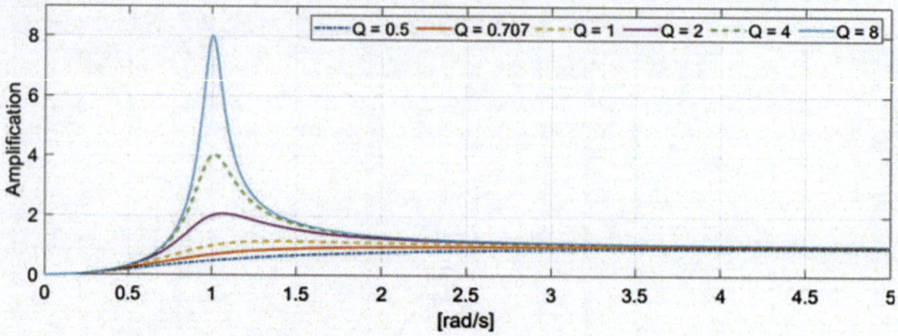

Fig. 9.7 Amplification diagram for a second-order highpass filter

9.3.4 Highpass: $b_1 = b_0 = 0$

With $b_1 = b_0 = 0$ we get the transfer function

$$H(s) = \frac{b_2 s^2}{s^2 + \frac{\omega_0}{Q}s + \omega_0^2} \Rightarrow |H(\omega)| = \frac{|b_2\omega^2|}{\sqrt{\left(\frac{\omega_0}{Q}\omega\right)^2 + \left(\omega_0^2 - \omega^2\right)^2}} \tag{9.10a}$$

In Fig. 9.7, we have plotted $|H(\omega)|$ for $b_2 = \omega_0 = 1$.

This is a highpass filter since amplification =0 for low frequencies and =1 for high frequencies.

9.4 Implementations

There are several different ways to implement analog filters in hardware and we will present the most common ones here.

9.4.1 The Double Integral Method

If we have a highpass filter as in Eq. (9.10a), then (see Fig. 9.8)

Fig. 9.8 Second-order highpass filter

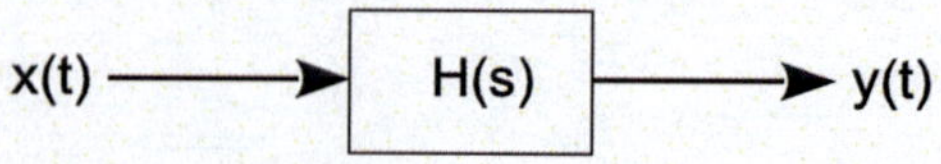

$$H(s) = \frac{b_2 s^2}{s^2 + \frac{\omega_0}{Q}s + \omega_0^2} = \frac{Y(s)}{X(s)} \tag{9.10b}$$

and $s^2Y(s) + \frac{\omega_0}{Q}sY(s) + \omega_0^2 Y(s) = b_2 s^2 X(s)$. Next, we rearrange this expression as follows:

$$Y(s) + \frac{1}{s}\frac{\omega_0}{Q}Y(s) + \frac{1}{s^2}\omega_0^2 Y(s) = b_2 X(s)$$

$$\Rightarrow Y(s) = b_2 X(s) - \frac{\omega_0}{s}\frac{1}{Q}Y(s) - \frac{\omega_0^2}{s^2}Y(s) \tag{9.11}$$

In Problem 7.7, we learned that if the function $f(t)$ has the Laplace transform $F(s)$, then the Laplace transform of the integral of $f(t)$ is $F(s)/s$. Hence, if we integrate the function $y(t)$ using an integrator with time constant $1/\omega_0$ (see Problem 9.1), the output is $\omega_0 Y(s)/s$ and if we integrate it *again*, the output will be $\omega_0^2 Y(s)/s^2$. That means that we can represent the system in Eq. (9.11) with the block model in Fig. 9.9.

Next, we rearrange the blocks in Fig. 9.9 as illustrated in Fig. 9.10. The highpass output signal is the signal right after the summator in Fig. 9.10. However, since an integration corresponds to a division by s in frequency space, the output after the first integrator is a bandpass filter (Eq. (9.9)) and the output after the second integrator is a lowpass filter (Eq. (9.8)); with a single design we can get three different filters!

The derivatives of a signal are called the 'state variables' of the signal and for that reason the filter(s) in Fig. 9.10 is called a 'state-variable' filter. Figure 9.11 illustrates how it is implemented with only three op amps.

This implementation is usually referred to as the 'KHN biquad' (from Kerwin–Huelsman–Newcomb) and you only need two design equations:

$$\omega_0 = \frac{1}{RC} \qquad \frac{R_3}{R_2} = 2Q - 1 \tag{9.12}$$

(The R_1 resistors' value doesn't matter; they are only part of the summation circuit.) Another usual name for this filter is 'UAF', which stands for *Universal Active Filter*. The UAF42 circuit from Burr-Brown is an example of such a circuit.

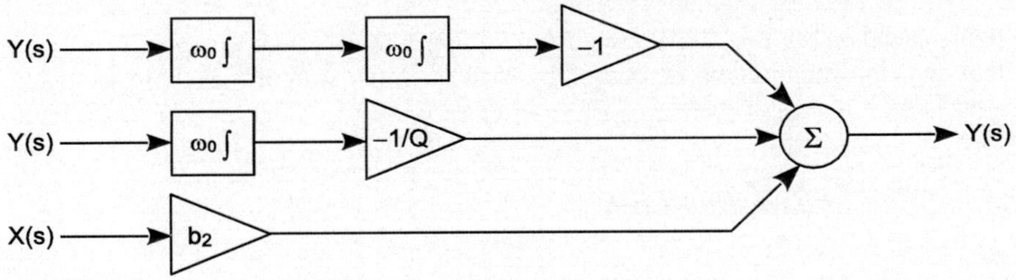

Fig. 9.9 Block diagram of Eq. (9.11)

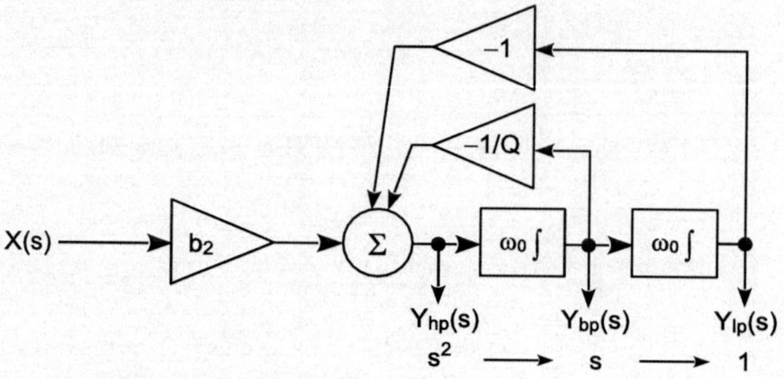

Fig. 9.10 Rearranging the blocks; a 'state-variable' filter

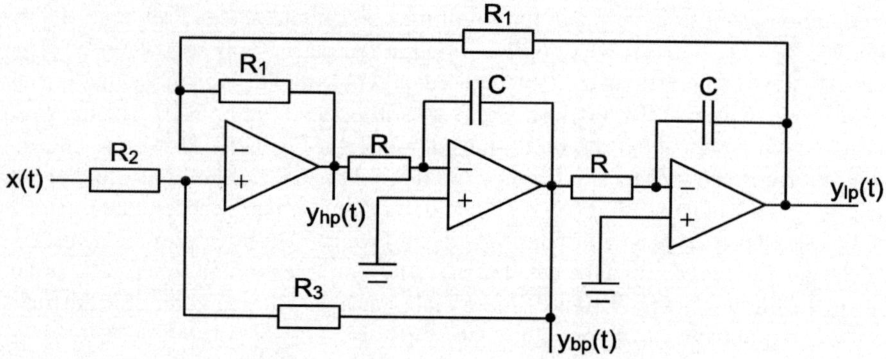

Fig. 9.11 A state-variable filter can be implemented with only three op amps

Some final comments about the quality factor Q. For a lowpass or a highpass filter, the Q number determines the 'steepness' of the filter, i.e., the width of the transition area from the passband to the stopband. For a bandpass filter, Q determines the 'selectiveness', i.e., how narrow it is:

$$Q = \frac{\omega_0}{\omega_u - \omega_l} \tag{9.13}$$

where ω_u and ω_l are the '3 dB' frequencies ('upper' and 'lower', respectively), i.e., where the amplification has decreased by 3 dB on either side of the resonance peak.

9.4.2 The Sallen–Key Link

The first-order lowpass RC filter is the most basic of all filters (see Example 7.7). By adding a voltage follower, it doesn't impose any load on the next step, and if we are adding a voltage follower, we might as well turn that into a non-inverting amplifier; that turns our filter into a first-order 'Sallen–Key' link, see Fig. 9.12.

The transfer function is still $1/(1 + j\omega RC)$, see Example 7.7, and the *cutoff* frequency (where the amplification is down by 3 dB) is

$$|H(\omega)| = \frac{1}{\sqrt{1+(\omega RC)^2}} = \frac{1}{\sqrt{2}} \Rightarrow \omega_c = 1/RC \quad (9.14)$$

and the DC amplification can be adjusted arbitrarily with the R_1 and R_2 resistors ($1 + R_2/R_1$). We can easily change it into a highpass filter by changing places with R and C.

Figure 9.13 illustrates a second-order lowpass Sallen–Key link.

The transfer function is (see Problem 9.3):

$$H(s) = \frac{1/R_1R_2C_1C_2}{s^2 + \frac{R_1+R_2}{R_1R_2C_1}s + 1/R_1R_2C_1C_2} \quad (9.15)$$

or, if we set $R_1 = R_2 = R$, then

$$H(s) = \frac{1/R^2C_1C_2}{s^2 + \frac{2}{RC_1}s + 1/R^2C_1C_2} \quad (9.16)$$

Comparing with Eq. (9.7), we can see that

$$\frac{2}{RC_1} = \frac{\omega_0}{Q} \quad (9.17)$$

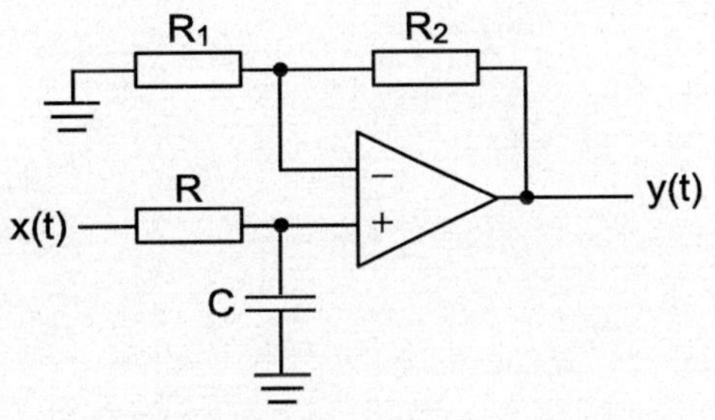

Fig. 9.12 First-order Sallen–Key link

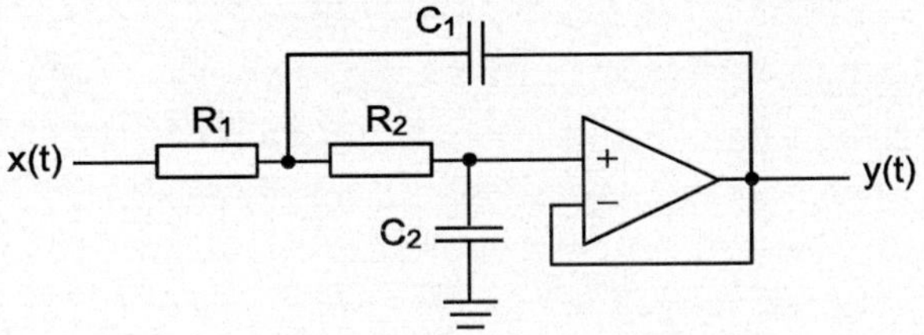

Fig. 9.13 Second-order Sallen–Key link

and

$$\omega_0^2 = \frac{1}{R^2 C_1 C_2} \tag{9.18}$$

Using Eqs. (9.17) and (9.18), we can implement any ω_0 and Q values we want to, by selecting the right R, C_1 and C_2 values. Also, by changing places with resistors and capacitors in Fig. 9.13, we get a highpass filter.

9.4.3 Switched Capacitors

As we have seen above, both the biquad and the Sallen–Key filters depend on passive components like resistors and capacitors. When implemented in integrated circuits, silicon area saving is paramount; capacitors don't require much area, but resistors do. For that reason, a technique to implement resistors using a capacitor has been developed. This technique is called the 'switched capacitor' technique and it replaces a resistor with a capacitor and two switches (the switches are of course two transistors). Figure 9.14 illustrates the switched capacitor resistor.

The switches are controlled by two clock signals that are 180° out of phase; when one switch is closed, the other one is open. The clock signal's period is $T = 1/f_0$ and the duty cycle is 50%. When switch S_1 is closed, the capacitor is charged by U_1, and when S_2 is closed, the capacitor is charged (or discharged) by U_2. Figures 9.15 and 9.16 illustrate the two situations.

Hence, over one clock period T, the change in charge over the capacitor is $\Delta Q = C(U_1 - U_2)$. By definition, current is the change of charge per time unit; the *average* current between the end points during one period is

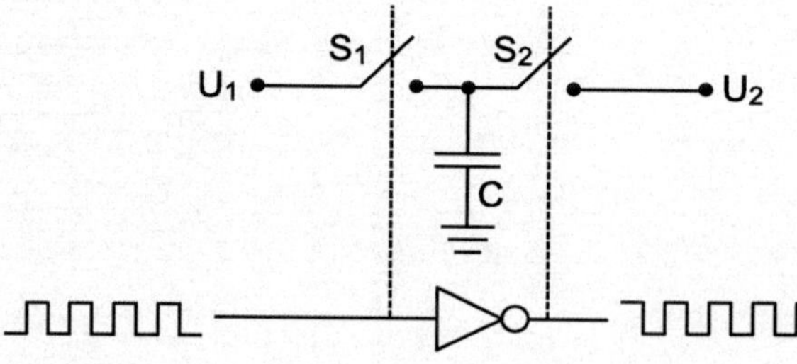

Fig. 9.14 A switched capacitor resistor

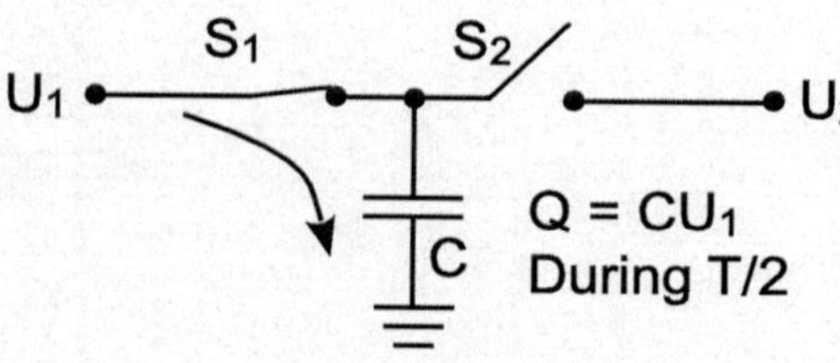

Fig. 9.15 S_1 is closed

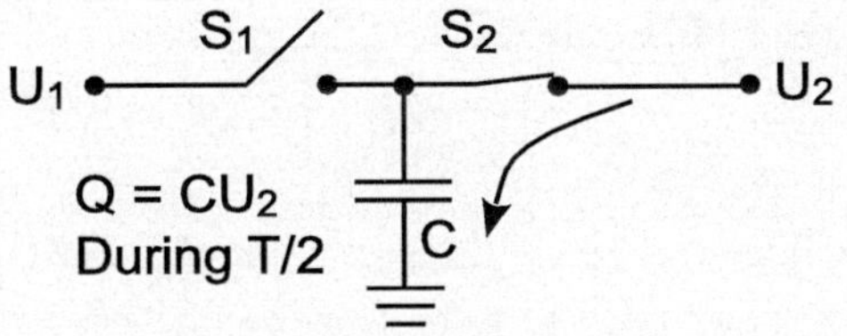

Fig. 9.16 S_2 is closed

$$i = \frac{\Delta Q}{T} = \frac{C(U_1 - U_2)}{1/f_0} \Rightarrow R = \frac{U_1 - U_2}{i} = \frac{1}{Cf_0} \tag{9.19}$$

From Eq. (9.19), we can see that this circuit corresponds to a resistance between the end points that depends on the capacitor and the clock frequency. By using a switched capacitor as the resistance in a filter design, we can control the filter parameters with the clock frequency. MAX7400 is an example of an integrated filter that uses switched capacitors.

9.4.4 More About Passive Filters

In Fig. 9.1, we only have two impedances, and that limits our range of filters to lowpass and highpass filters. If we introduce a third impedance, we can also design bandpass and bandstop filters (since they need to be second-order filters), see Fig. 9.17.

For example, Fig. 9.18 illustrates a bandpass filter. The transfer function is

$$\begin{aligned} H(s) &= \frac{X_L//X_C}{R + X_L//X_C} = \frac{\frac{sL\cdot 1/sC}{sL+1/sC}}{R + \frac{sL\cdot 1/sC}{sL+1/sC}} = \frac{sL \cdot 1/sC}{sLR + R/sC + sL \cdot 1/sC} \\ &= \frac{sL}{s^2LRC + sL + R} = \frac{s/RC}{s^2 + s/RC + 1/LC} \end{aligned} \tag{9.20}$$

And according to Eq. (9.9), this is a bandpass filter with $\omega_0 = 1/\sqrt{LC}$ and $Q = R\sqrt{C/L}$. By changing places with R and the LC network in Fig. 9.18, we get a bandstop filter.

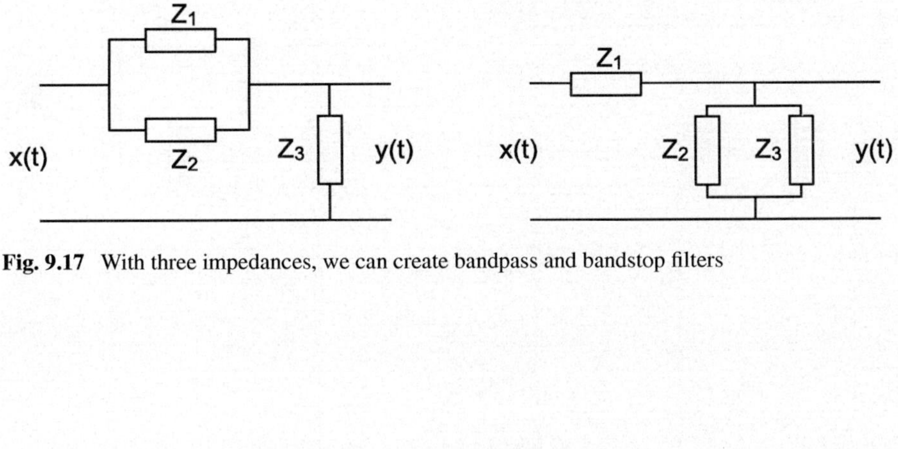

Fig. 9.17 With three impedances, we can create bandpass and bandstop filters

Fig. 9.18 A bandpass filter

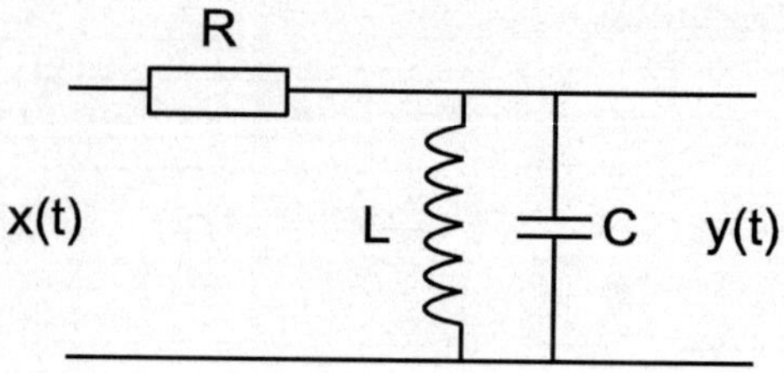

9.4.5 *Special Cases*

A special case of bandstop filters is the 'notch' filter, see Fig. 7.22, that is designed to block just one single frequency. For example, in a physics lab, the power line frequency, 50 or 60 Hz, is omnipresent in measurements and it can be blocked with a passive second-order filter called the 'Twin-T' notch filter, see Fig. 9.19. This filter has a zero on the imaginary axis at $\omega = 1/RC$.

9.5 Filter Models

Equation (9.3) is the general expression for a second-order filter, and it can certainly be extended to an arbitrary order filter. The filter coefficients (the a_i and b_i polynomial coefficients) are optimized for certain conditions:

In a 'Butterworth' filter, the amplification diagram (in the Bode plot) is as 'flat' as possible in the passband (no 'ripple').

Chebyshev and Cauer filters are more selective (the transition area from passband to stopband is narrower) at the expense of some passband ripple.

The Bessel filter is not that selective but has the advantage of a very linear phase diagram (which is what we need to minimize the signal distortion, see Sect. 7.3.1).

We will give a brief presentation of the Butterworth, Chebyshev, and Cauer filters here. In the following presentation, we will only treat the lowpass filter types; in Sect. 9.6, we will show you how to transform a lowpass filter to any other filter type.

Fig. 9.19 The Twin-T notch filter

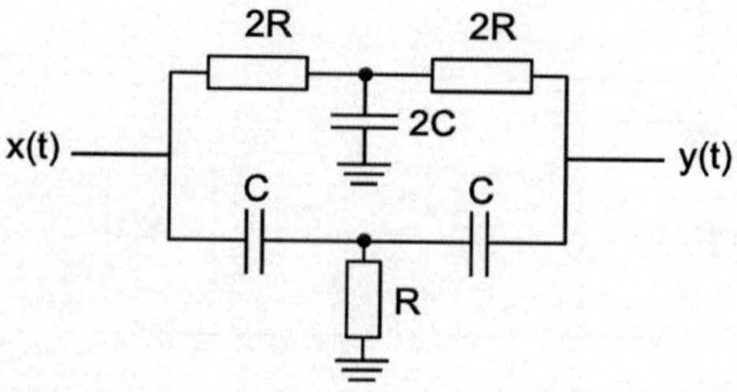

9.5.1 Butterworth

A Butterworth filter is an 'all-pole' filter, i.e., no zeros, and all the poles are on a perfect circle (semi-circle) in the s plane, see Fig. 9.20. If the filter order is odd, there is a pole on the negative σ-axis and the angle between the poles is π/n, where n is the filter order. In a Butterworth filter, the numerator polynomial $N(s) = 1$. Comparing Fig. 9.20 with Fig. 9.4, we can see that $|\sigma_p| = \omega_0 \cos\pi/4 = \omega_0/\sqrt{2}$. Equation (9.5) gives us Butterworth filter's quality factor:

$$Q = \frac{\omega_0}{2 \cdot \omega_0/\sqrt{2}} = \frac{1}{\sqrt{2}} \approx 0.707 \tag{9.21}$$

Comparing this with Fig. 9.5, we can see that this represents an amplification diagram with no overshoot ('critically' damped); the amplification diagram is 'maximally flat' in the passband, which is the hallmark of all Butterworth filters. This is a consequence of the fact that the poles are on a circle and the circle radius determines the resonance frequency ω_0.

In Table 9.1, you can see the denominator polynomials for all Butterworth filters up to the seventh order for $\omega_0 = 1$. (In Sect. 9.6 we will show you how to transfer them to other ω_0s.)

Figure 9.21 illustrates the amplification diagram for Butterworth filters of different orders; the higher the order, the more selective is the filter. Notice in Fig. 9.21 that the amplification diagrams have no 'ripple'; it declines monotonically.

Notice in Table 9.1 that higher order filters are written as a product of first- and second-order filters; they are typically implemented by cascading first- and second-order filters.

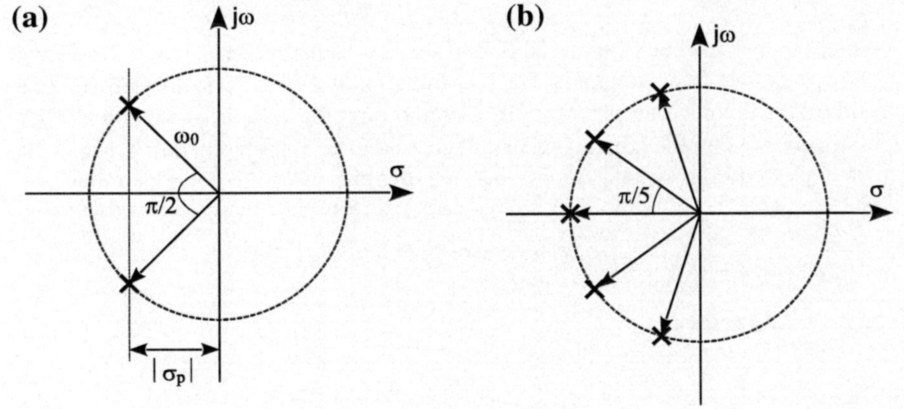

Fig. 9.20 **a** Second-order Butterworth filter. **b** Fifth-order Butterworth

Table 9.1 Butterworth filter polynomials

Order	Polynomial
1	$s+1$
2	$s^2+1.41s+1$
3	$(s+1)(s^2+s+1)$
4	$(s^2+0.765s+1)(s^2+1.848s+1)$
5	$(s+1)(s^2+0.618s+1)(s^2+1.618s+1)$
6	$(s^2+0.518s+1)(s^2+1.414s+1)(s^2+1.932s+1)$
7	$(s+1)(s^2+0.444s+1)(s^2+1.246s+1)(s^2+1.802s+1)$

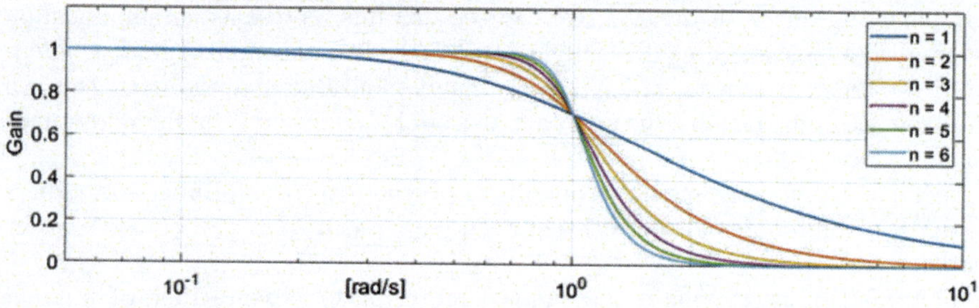

Fig. 9.21 Amplification diagram for Butterworth filters

To see exactly how that works, let's take a fourth-order filter as an example:

$$H(s)=\frac{1}{s^2+0.765s+1}\cdot\frac{1}{s^2+1.848s+1} \tag{9.22}$$

Since $\omega_0=1$ for both filters, the filters' quality factors are $1/0.765=1.307$ and $1/1.848=0.541$. If we compare these Q numbers with Fig. 9.5, we can see that the first filter is underdamped ($Q>0.707$) and the other one is overdamped ($Q<0.707$), but if we combine them, the overall gain diagram is perfectly flat. In Fig. 9.22, we have plotted both systems' gain diagrams together with the combined diagram. (Combined $Q=1.307{\cdot}0.541=0.707$.)

9.5.2 *Chebyshev*

A Chebyshev filter is more selective (steeper roll-off) than a Butterworth filter of the same order. It is still an 'all-pole' filter, and the greater selectiveness is achieved by placing the poles on an ellipse instead of a circle, see Fig. 9.23. The elliptic shape of the poles' location is the reason for the greater selectiveness, but it creates a 'ripple' in the passband. In Fig. 9.24, we have plotted the gain diagram of the filter in Fig. 9.23.

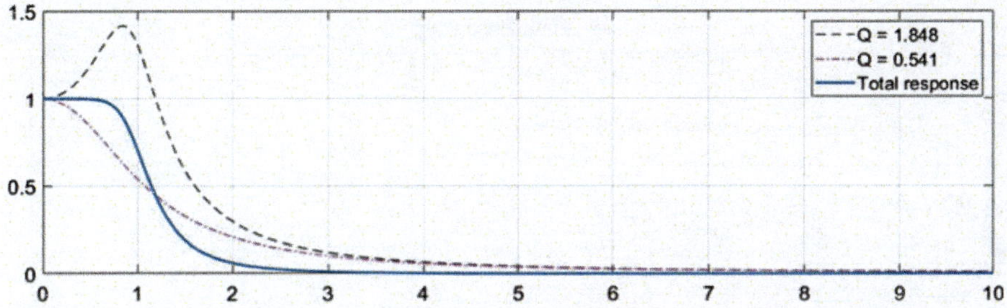

Fig. 9.22 A fourth-order filter = two cascaded second-order filters

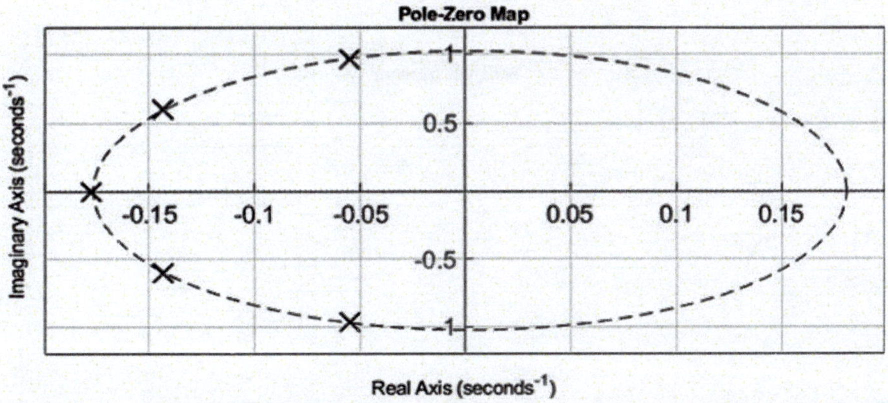

Fig. 9.23 A Chebyshev filter has all poles on an ellipse

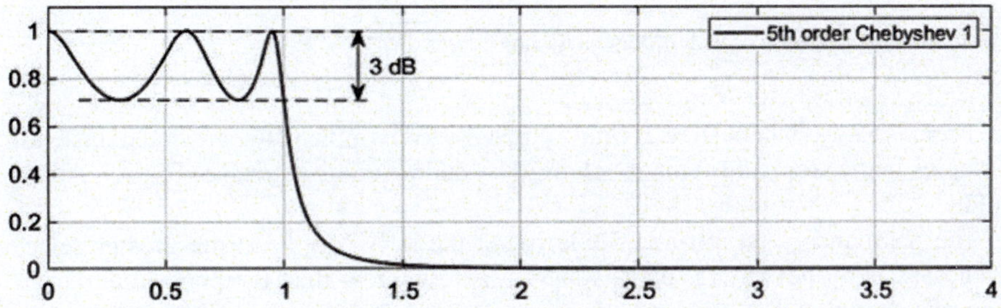

Fig. 9.24 Gain response of fifth-order Chebyshev 1 filter

The filter in Figs. 9.23 and 9.24 is a ‘Chebyshev 1’ filter and is characterized by its passband ripple. In Fig. 9.24, the ripple is 3 dB, but that is a design parameter that can be selected arbitrarily.

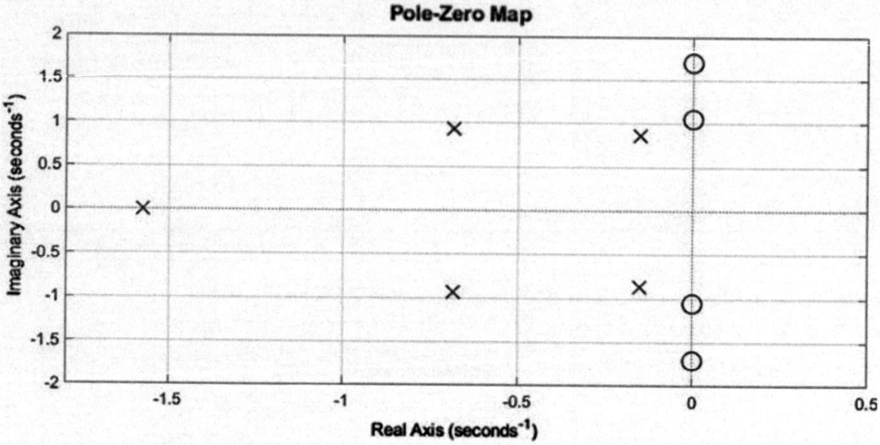

Fig. 9.25 Chebyshev 2 filters have zeros

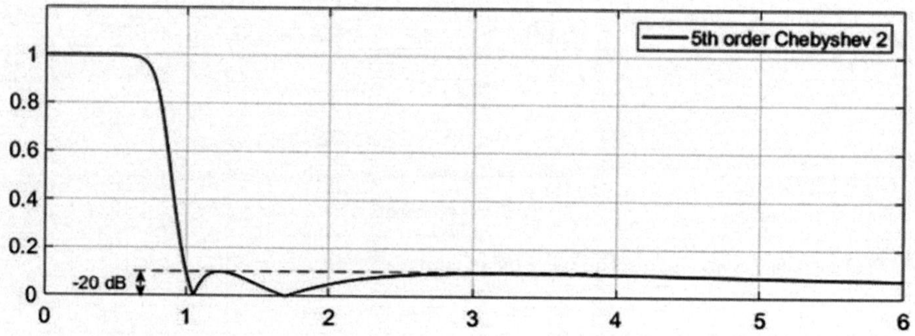

Fig. 9.26 Gain response of fifth-order Chebyshev 2 filter

There is also a 'Chebyshev 2' filter, which does not ripple in the passband; instead, it ripples in the stopband. This is achieved by adding some zeros, see Figs. 9.25 and 9.26.

The passband (or stopband) ripple is not the only 'cost' for the greater selectiveness; Chebyshev filters' phase diagram is less linear than the phase diagram of Butterworth filters, i.e., they are more prone to distort the signal.

9.5.3 Cauer

Chebyshev 1 filters ripple in the passband and Chebyshev 2 filters ripple in the stopband. What if we allowed ripple in both the passband and the stopband? Wouldn't that improve selectiveness even more? Yes, it would, and filters that ripple in both

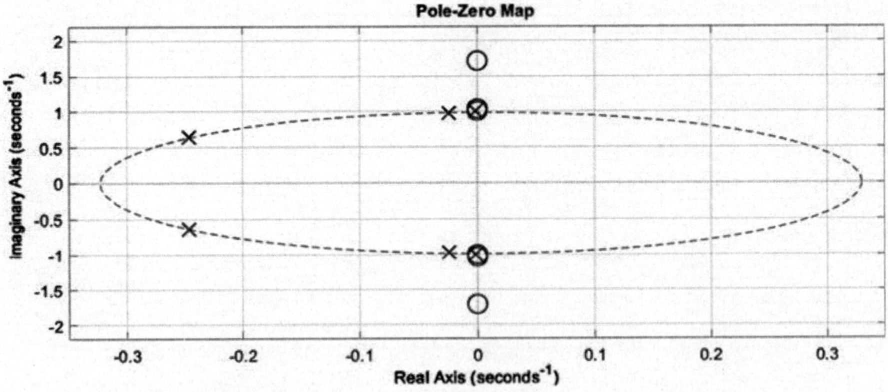

Fig. 9.27 Poles and zeros of a sixth-order Cauer filter

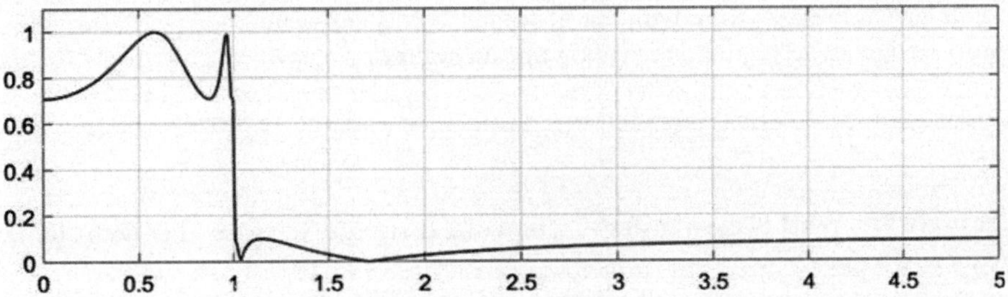

Fig. 9.28 Gain diagram of sixth-order Cauer filter (3 dB passband ripple, 20 dB stopband ripple)

passband and stopband are called 'Cauer' filters (or sometimes 'elliptic' filters). Figure 9.27 illustrates the poles'/zeros' location in the s plane for a sixth-order Cauer filter, and Fig. 9.28 illustrates the corresponding gain diagram.

In Fig. 9.29, we have plotted the gain diagram for all four filter models (third-order filters). Comparing them, we can clearly see that the Cauer filter is the most selective one but keep in mind that the 'cost' is ripple in both passband and stopband and that the phase diagram is less linear (higher degree of distortion).

9.6 Filter Transformations

So far, we have mostly treated lowpass filters (with cutoff frequency $\omega_0 = 1$). The reason is that that is what you start with and then you just 'transform' your filter to whatever type and frequency you want in your application. We will present these transformation equations here.

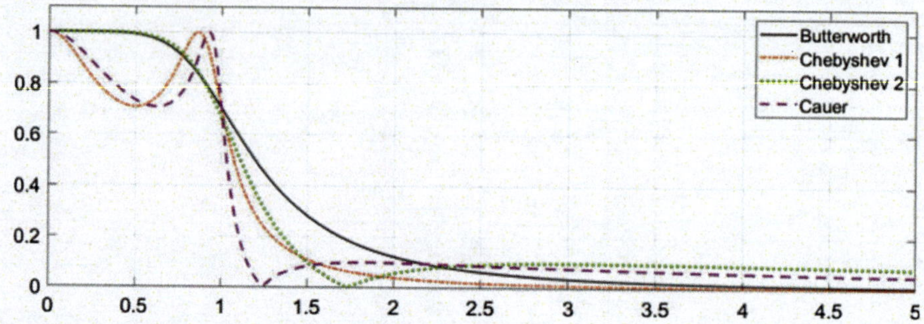

Fig. 9.29 Comparing filter models (passband ripple = 3 dB, stopband ripple = 20 dB)

9.6.1 Lowpass to Lowpass

If you have a lowpass filter with cutoff frequency ω_0 and want a lowpass filter with cutoff frequency ω_0', you do the following substitution:

$$s \to \frac{\omega_0}{\omega_0'} \cdot s \tag{9.23}$$

Example 9.1 What is the transfer function of a first-order lowpass filter with cutoff frequency 8 rad/s?

Solution The transfer function of a first-order lowpass filter with cutoff frequency = 1 is $1/(1 + s)$. Substituting $s/8$ for s gives us the new transfer function:

$$H(s) = \frac{1}{1 + s/8} = \frac{8}{s + 8}$$

The amplification diagram is plotted in Fig. 9.30.

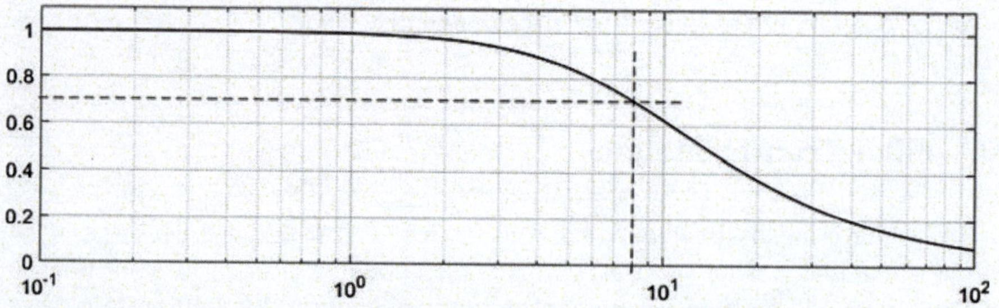

Fig. 9.30 A lowpass filter with cutoff frequency 8 rad/s

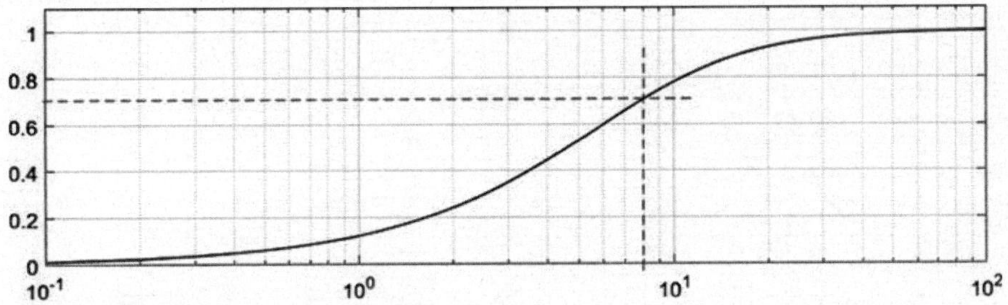

Fig. 9.31 A highpass filter with cutoff frequency 8 rad/s

9.6.2 Lowpass to Highpass

If you want to transform a lowpass filter with cutoff frequency ω_0 to a highpass filter with cutoff frequency ω_0', you do the substitution

$$s \rightarrow \frac{\omega_0 \omega_0'}{s} \tag{9.24}$$

Example 9.2 What is the transfer function of a first-order highpass filter with cutoff frequency 8 rad/s?

Solution Substituting $8/s$ for s: $H(s) = \frac{1}{8/s+1} = \frac{s}{s+8}$. Figure 9.31 illustrates the amplification diagram.

9.6.3 Lowpass to Bandpass

To transfer a lowpass filter with cutoff frequency ω_0 to a bandpass filter with the upper and lower cutoff frequencies ω_u and ω_l, you do the substitution:

$$s \rightarrow \frac{s^2 + \omega_l \omega_u}{s \cdot (\omega_u - \omega_l)} \cdot \omega_0 \tag{9.25}$$

Example 9.3 What is the transfer function of a second-order bandpass filter with upper and lower cutoff frequencies 25 and 20 rad/s, respectively?

Solution $\omega_u \omega_l = 25 \cdot 20 = 500$ and $\omega_u - \omega_l = 5$. The substitution we need to do is

$$s \rightarrow \frac{s^2 + 500}{5s} \Rightarrow H(s) = \frac{1}{\frac{s^2+500}{5s} + 1} = \frac{5s}{s^2 + 5s + 500}$$

Figure 9.32 illustrates the amplification diagram.

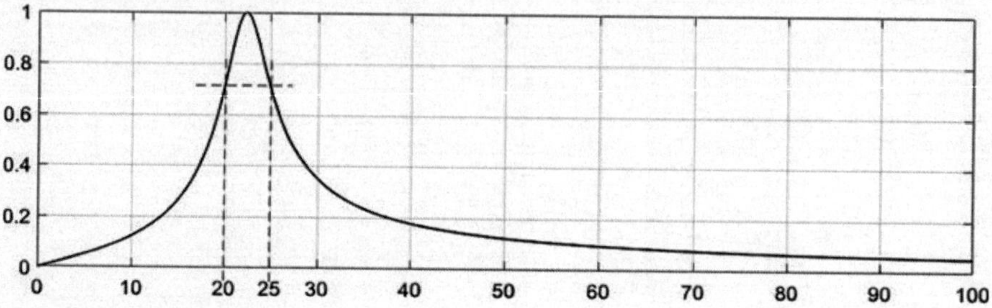

Fig. 9.32 A bandpass filter

9.6.4 Lowpass to Bandstop

You transform a lowpass filter with cutoff frequency ω_0 to a bandstop filter with upper and lower cutoff frequencies ω_u and ω_l, respectively, with the substitution

$$s \rightarrow \frac{s \cdot (\omega_u - \omega_l)}{s^2 + \omega_u \omega_l} \cdot \omega_0 \tag{9.26}$$

Example 9.4 What is the transfer function of a second-order bandstop filter with upper and lower cutoff frequencies 55 and 45 rad/s, respectively?

Solution $\omega_u \omega_l = 55 \cdot 45 = 2475$ and $\omega_u - \omega_l = 10$. The substitution we need to do is

$$s \rightarrow \frac{10s}{s^2 + 2475} \Rightarrow H(s) = \frac{1}{\frac{10s}{s^2+2475} + 1} = \frac{s^2 + 2475}{s^2 + 10s + 2475}$$

The amplification diagram of this transfer function is plotted in Fig. 9.33.

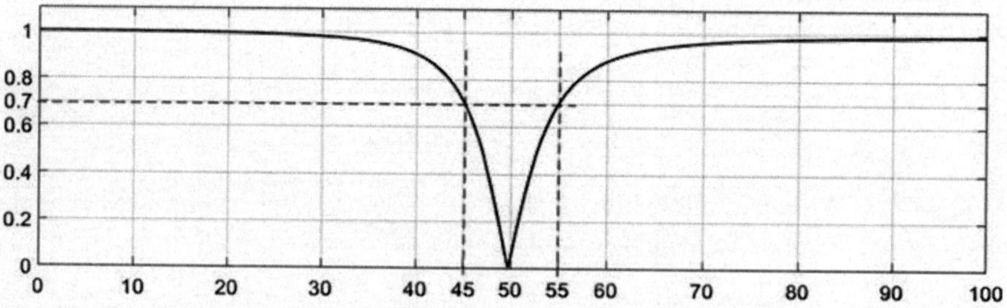

Fig. 9.33 A bandstop filter

9.7 Time Domain

In the frequency domain, we find the system's output by *multiplying* the input signal's Laplace transform (or the Fourier transform) with the system's transfer function, see Fig. 9.34. In the time domain, the system is represented by the system's impulse response (see Example 7.8), which is the inverse Laplace transform of the transfer function ($h(t) = H(s)^{-1}$).

It is a common misunderstanding that you should also multiply $h(t)$ by $x(t)$ to get the output $y(t)$; that is not the case! In time space, you must *convolve* the input signal with the impulse response. This is illustrated in Fig. 9.35 where the symbol $\otimes$ represents 'convolution'. Since this is a widely misunderstood concept, we will elucidate it in detail here.

9.7.1 Convolution

First, convolution is an integral:

$$y(t) = h(t) \otimes x(t) = \int_{-\infty}^{\infty} h(\tau)x(t-\tau)d\tau \tag{9.27}$$

This integral causes most students *a lot of* problems, so take a minute to really look at it. First, we have the *temporal variable* τ that we only use inside the integral; the integral output is still a function of t! Second, most people have trouble visualizing the function $x(t-\tau)$ and we will discuss that in detail in a minute, and third we integrate over *all* τs, where the two functions overlap. The output is the area of the 'product function' at each time t. The key to understanding convolution is to understand exactly *where* (in τ space) the function $x(t-\tau)$ is for every time t and exactly what the integral limits are (expressed in t).

So, before we do any convolution, let's look at the function $x(t-\tau)$ and what it looks like in τ space. First, in τ *space*, t is a *constant*! Let's consider the straight line $x(\tau) = 2\tau - 1$. This signal is plotted in Fig. 9.36. The signal $x(-\tau)$ is the 'mirror' of $x(\tau)$ around $\tau = 0$, see Fig. 9.37.

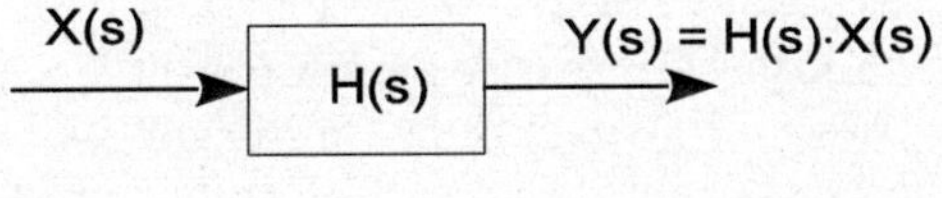

Fig. 9.34 Frequency space: Multiplication

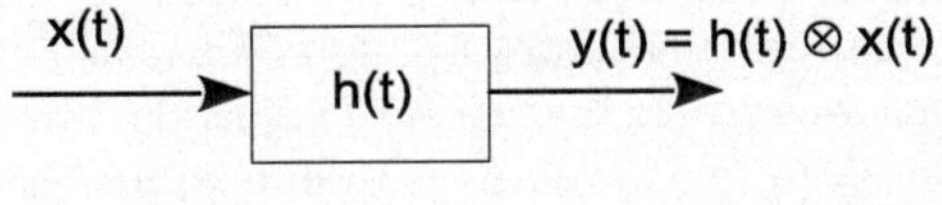

Fig. 9.35 Time space: Convolution

Fig. 9.36 $x(\tau) = 2\tau - 1$

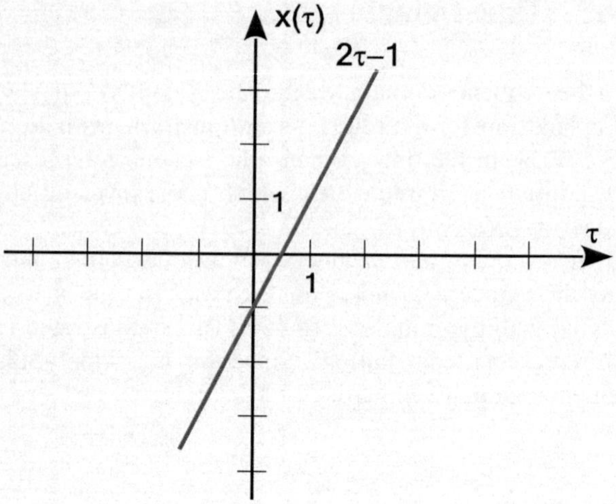

Fig. 9.37 $x(-\tau) = -2\tau - 1$

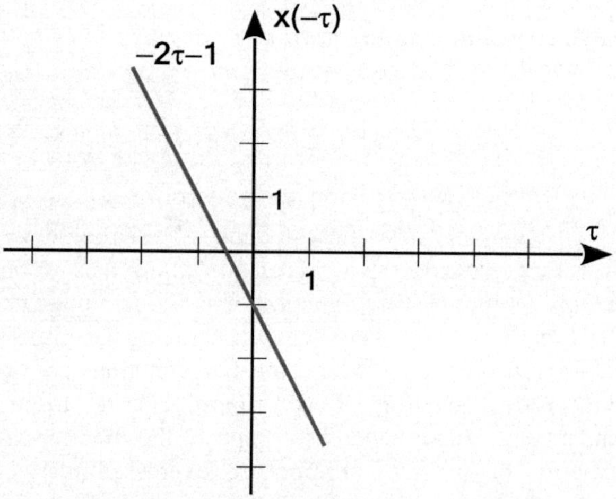

Next, we consider the function $x(\tau - t)$; remember that τ is our time variable here and t is a constant. In that case, $x(\tau - t)$ is a *delayed* copy of $x(\tau)$ (delayed by t). For example, if $t = 1$, then $x(\tau - 1) = 2(\tau - 1) - 1 = 2\tau - 3$, and if $t = 2$, then $x(\tau - 2) = 2(\tau - 2) - 1 = 2\tau - 5$, see Fig. 9.38.

And that brings us to the key question; what does $x(t - \tau)$ look like? Well, $x(t - \tau) = x(-(\tau - t))$, i.e., it is the 'mirror' of $x(t - \tau)$ around $\tau = t$. For example, if $t = 1$, then $x(1 - \tau) = 2(1 - \tau) - 1 = -2\tau + 1$, and if $t = 2$, then $x(2 - \tau) = 2(2 - \tau) - 1 = -2\tau + 3$, see Fig. 9.39.

In Figs. 9.40 and 9.41, we compare $x(\tau - t)$ and $x\ (t - \tau)$ for the same t values, and we can see that $x(t - \tau)$ is just the mirror image of $x(\tau - t)$ around $\tau = t$. We can draw two conclusions from this; first, as t increases, $x(t - \tau)$ moves to the *right*

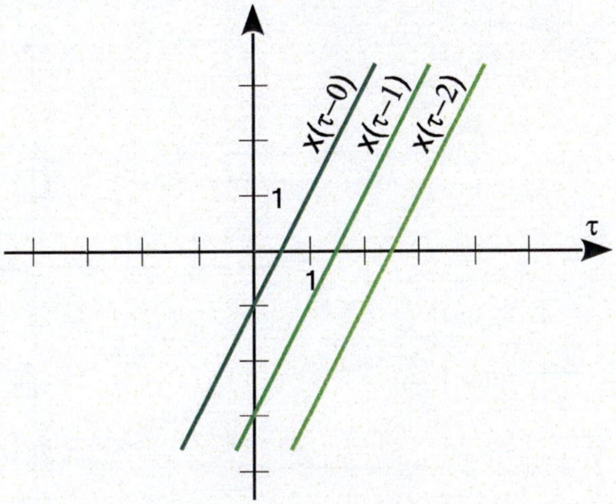

Fig. 9.38 $x(\tau - t)$ for $t = 0$, 1 and 2

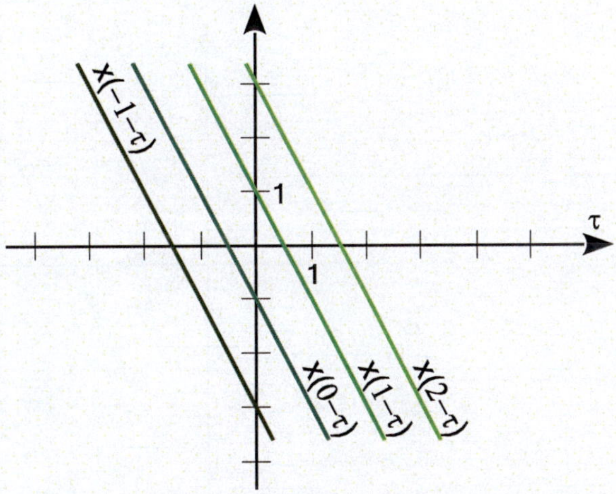

Fig. 9.39 $x(t - \tau)$ for $t =$ –1, 0, 1 and 2

along the τ-axis, and second, if $t < 0$, $x\ (t - \tau)$ moves to the left. $t = -\infty$ is to the far left and $t = +\infty$ is to the far right. Hence, when t goes from $-\infty$ to $+\infty$, the function $x(t - \tau)$ 'slides' from left to right along the τ-axis, see Fig. 9.42.

Now, let's go back to Eq. (9.27). $h(\tau)$ doesn't move with t (independent of t); for each time t, we must figure out where $x(t - \tau)$ is, multiply it with $h(\tau)$ and then integrate over all τ, i.e., over all τs where the two functions overlap. If you understand Fig. 9.42, you will know where $x(t - \tau)$ is, but that is only half the problem; the second part is what the integration limits are. We illustrate that with an example.

Example 9.5 Figure 9.43 illustrates $h(t)$ and Fig. 9.44 illustrates $x(t)$. What is the convolution of $h(t)$ and $x(t)$; find $y(t) = h(t) \otimes x(t)$.

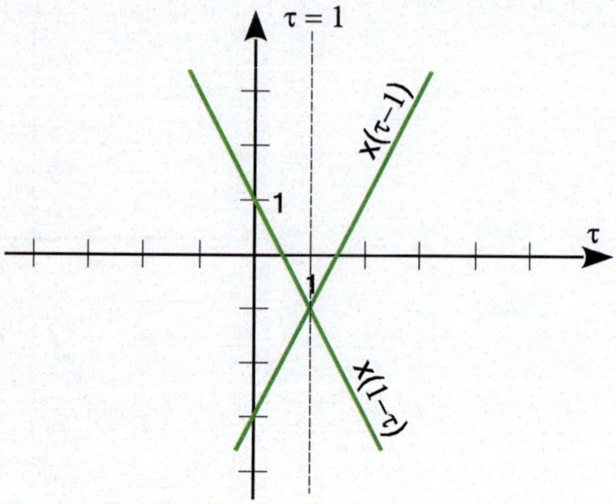

Fig. 9.40 $x(1-\tau)$ is $x(\tau-1)$ mirrored in $t = 1$

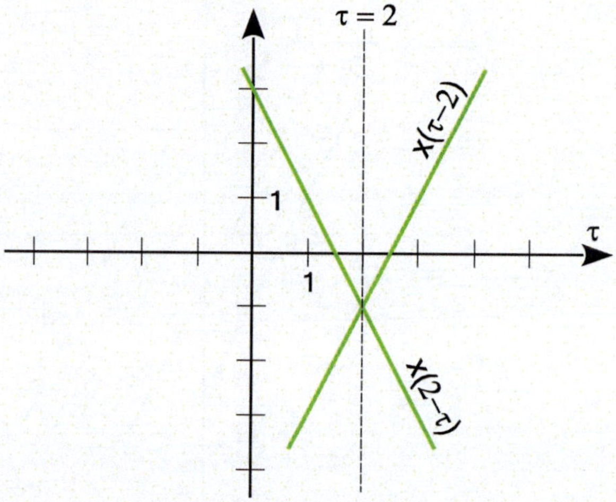

Fig. 9.41 $x(2-\tau)$ is $x(\tau-2)$ mirrored in $\tau = 2$

Solution First, we plot $x(t-\tau)$ for some different t values to make sure we understand what goes on, see Fig. 9.45. In Fig. 9.45, we can see that as $x(t-\tau)$ slides from left to right, there is no overlap between $x(t-\tau)$ and $h(\tau)$ until right after $t = 0$; hence the convolution $h(t) \otimes x(t) = 0$ for $t < 0$. Figure 9.46 illustrates the situation for $0 < t < 1$. From Fig. 9.46, we can see that the integration limits are $\tau = -1$ and $\tau = -(1-t) = t-1$. In this interval, $x(t-\tau) = 1$ and $h(\tau) = \tau + 1$. Now we can calculate the convolution expression (9.27) for $t \in [0, 1]$:

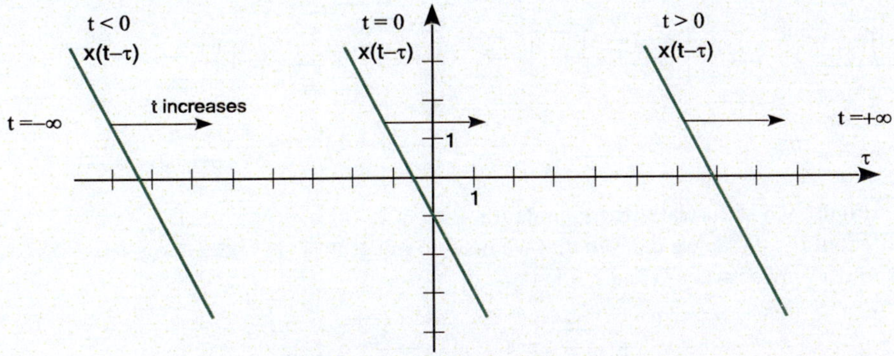

Fig. 9.42 $x(t-\tau)$ 'slides' from left to right as t increases

Fig. 9.43 $h(t)$

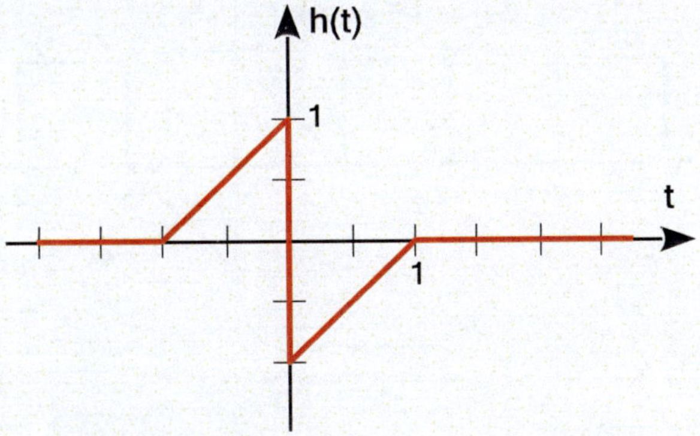

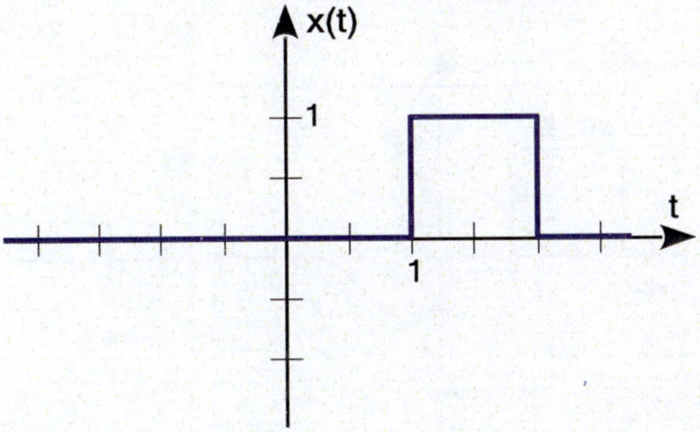

Fig. 9.44 $x(t)$

$$\int_{-1}^{t-1} (\tau + 1) \cdot 1 d\tau = \left[\frac{1}{2}\tau^2 + \tau\right]_{-1}^{t-1} = \frac{1}{2}(t-1)^2 + t - 1 - \left(\frac{1}{2} - 1\right)$$

$$= \frac{1}{2}\left(t^2 - 2t + 1\right) + t - \frac{1}{2} = \underline{\frac{1}{2}t^2} \quad \text{for } 0 \le t < 1$$

Figure 9.47 illustrates the signals for $1 < t < 2$.

From Fig. 9.47, we can see that we need to integrate $\tau + 1$ between $\tau = t - 2$ and 0, and $\tau - 1$, between 0 and $t - 1$:

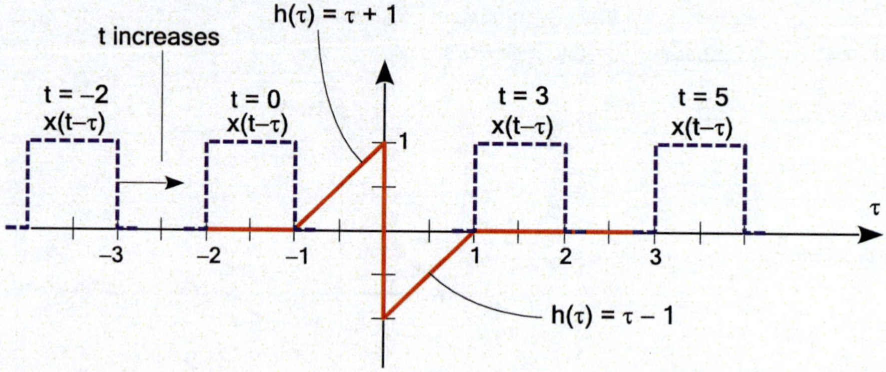

Fig. 9.45 $x(t - \tau)$ for different times t

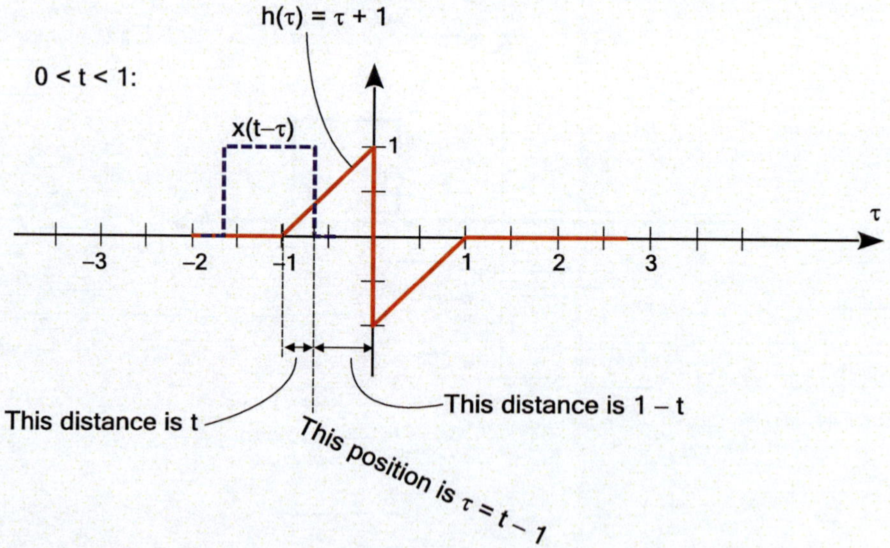

Fig. 9.46 $0 < t < 1$

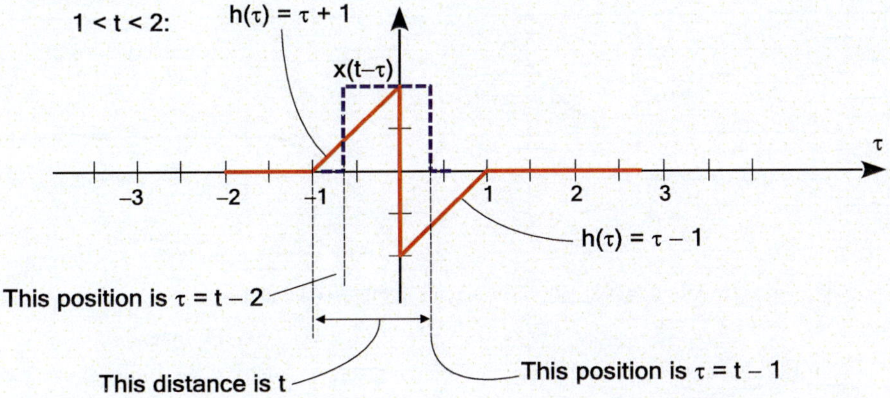

Fig. 9.47 $1 < t < 2$

$$\int_{t-2}^{0} (\tau+1)\cdot 1 d\tau + \int_{0}^{t-1} (\tau-1)\cdot 1 d\tau = \left[\frac{1}{2}\tau^2+\tau\right]_{t-2}^{0} + \left[\frac{1}{2}\tau^2-\tau\right]_{0}^{t-1}$$

$$= -\frac{1}{2}(t-2)^2 - t + 2 + \frac{1}{2}(t-1)^2 - t + 1 = \cdots = \underline{-t+1.5} \quad \text{for } 1 \le t < 2$$

Figure 9.48 illustrates the signals for $2 < t < 3$.

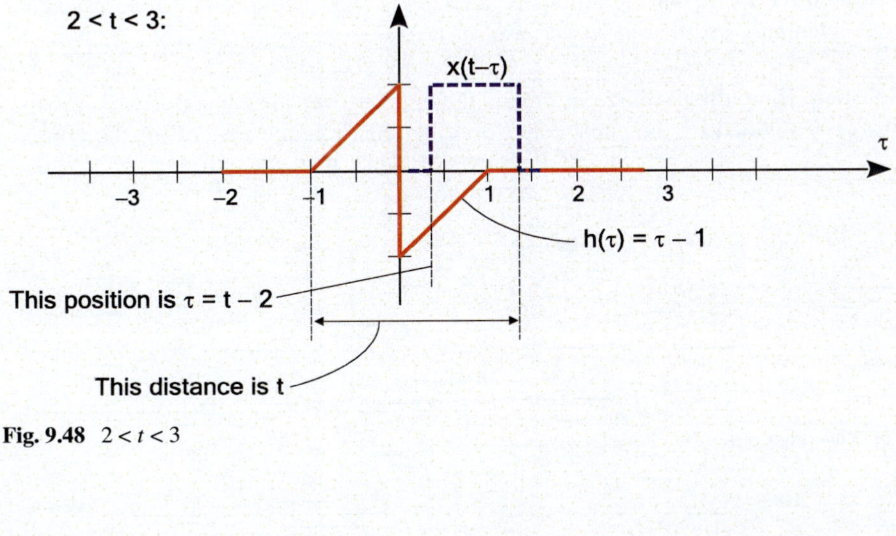

Fig. 9.48 $2 < t < 3$

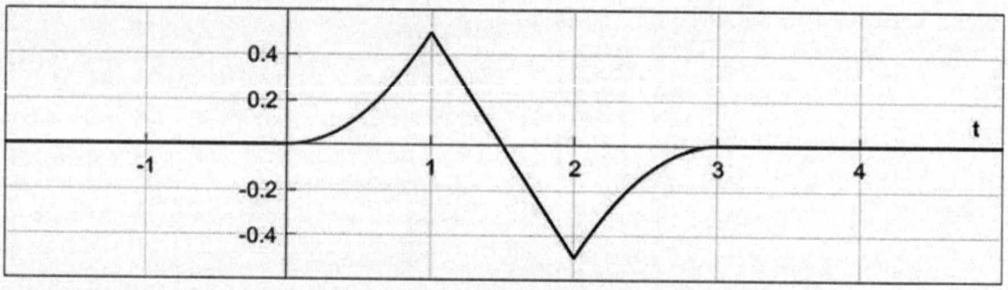

Fig. 9.49 The convolution $h(t) \otimes x(t)$ of the signals in Figs. 9.43 and 9.44

From Fig. 9.48, it is clear that we should integrate $\tau - 1$ over the interval $t - 2 < \tau < 1$:

$$\int_{t-2}^{1} (\tau - 1) \cdot 1 d\tau = \left[\frac{1}{2}\tau^2 - \tau\right]_{t-2}^{1} = \frac{1}{2} - 1 - \left(\frac{1}{2}(t-2)^2 - (t-2)\right) =$$

$$\cdots = \underline{-\frac{1}{2}t^2 + 3t - 4.5} \quad \text{for } 2 \le t < 3$$

In Fig. 9.49, we have plotted the convolution expressions for all ts.

Convolution can really challenge your patience, but remember, there are *two* keys; understanding how $x(t - \tau)$ moves in τ space and figuring out what the integration limits are. You need to do a couple of convolution integrals on your own before the penny drops.

9.8 Solved Problems

Problem 9.1 Prove that the simple RC filter in Fig. 9.50, under certain circumstances, is an integrator, i.e., that $u_y \sim \int u_x dt$.

Solution The output voltage u_y equals the voltage over the capacitor which is the charge Q divided by the capacitance and the charge is the integral of the current i: $u_y = u_C = \frac{Q}{C} = \frac{1}{C}\int idt$. If $R >> X_C = 1/\omega C$, then the current $i \approx u_x/R$ and

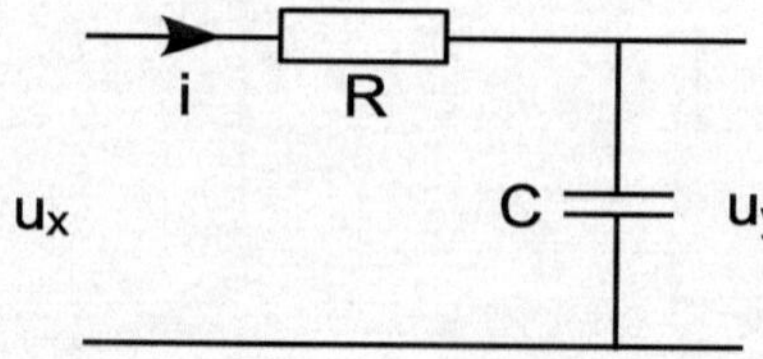

Fig. 9.50 First-order RC filter

$u_y = \frac{1}{C}\int \frac{u_x}{R}dt = \frac{1}{RC}\int u_x dt$. Hence, the first-order lowpass filter in Fig. 9.50 is an integrator if $R >> 1/\omega C$, i.e., if $RC = \tau \gg \frac{1}{\omega} = \frac{T}{2\pi}$.

Conclusion: The lowpass filter acts as an integrator if the filter time constant is much greater than the signal period.

Problem 9.2 Consider the filter $H(s) = \frac{12.5s}{s^2+12.5s+625}$. **a** What kind of filter is this? **b** What is ω_0? **c** What is Q? **d** What is the maximum amplification? **e** Plot the amplification diagram.

Solution First, $H(s) = 0$ for $s = 0$ and $s = \infty$, so it is a bandpass filter. Second, we can rewrite the transfer function as $H(s) = \frac{12.5s}{s^2+\frac{25}{2}s+25^2} \Rightarrow H(\omega) = \frac{12.5j\omega}{12.5j\omega+25^2-\omega^2} \Rightarrow$ $|H(\omega)| = \frac{|12.5\omega|}{\sqrt{12.5^2\omega^2+(25^2-\omega^2)^2}}$, and hence $\omega_0 = 25$ rad/s and $Q = 2$. We can also see that $|H(\omega)|$ has its maximum value when $\omega = 25$ rad/s and $|H(\omega = 25)| = 1$. The amplification diagram is plotted in Fig. 9.51.

Problem 9.3 Derive the transfer function of the Sallen–Key link in Fig. 9.13 (Eq. (9.15)).

Solution Referring to Fig. 9.52, the potential at point B is

$$U_B = \frac{1/sC_2}{R_2 + 1/sC_2}U_A = \frac{U_A}{1 + sR_2C_2} \Rightarrow U_A = U_B(1 + sR_2C_2) \tag{9.28}$$

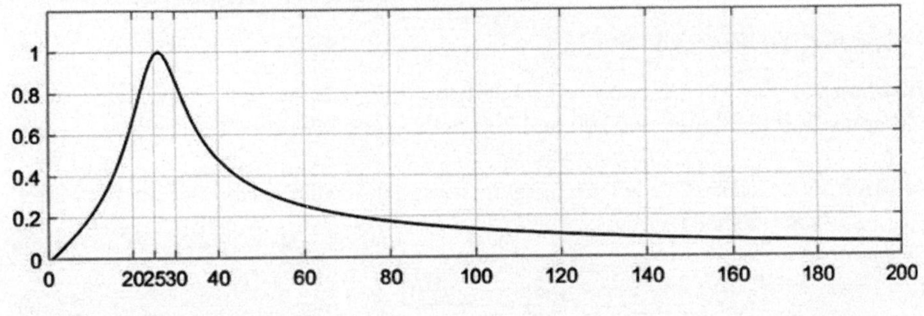

Fig. 9.51 Bandpass filter

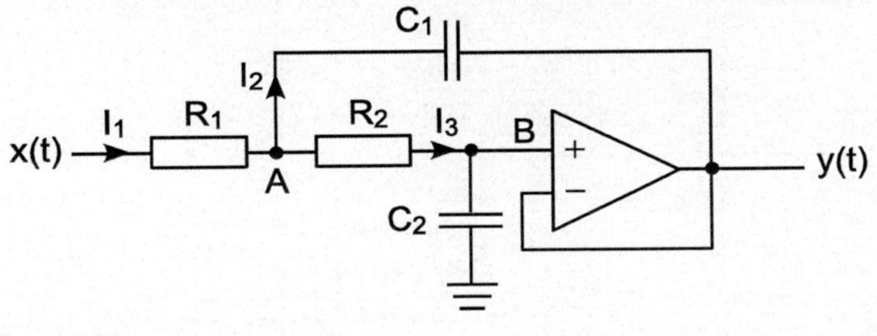

Fig. 9.52 The Sallen–Key link (second order)

At point A, $i_1 = i_2 + i_3$:

$$\frac{X(s) - U_A}{R_1} = \frac{U_A - Y(s)}{1/sC_1} + \frac{U_A - U_B}{R_2} = sC_1(U_A - Y(s)) + \frac{1}{R_2}(U_A - U_B)$$
$$\Rightarrow X(s) = U_A + sR_1C_1(U_A - Y(s)) + \frac{R_1}{R_2}(U_A - U_B) \tag{9.29}$$

Next, we insert Eq. (9.28) into Eq. (9.29):

$$X(s) = U_B(1 + sR_2C_2) + sR_1C_1(U_B(1 + sR_2C_2) - Y(s)) + \frac{R_1}{R_2}(U_B(1 + sR_2C_2) - U_B)$$

Since the op amp has negative feedback, the potential at point B follows $y(t)$ and $U_B = Y(s)$:

$$\begin{aligned} X(s) &= Y(s)\left((1 + sR_2C_2) + sR_1C_1((1 + sR_2C_2) - 1) + \frac{R_1}{R_2}((1 + sR_2C_2) - 1)\right) \\ &= Y(s)\big(1 + sR_2C_2 + s^2R_1R_2C_1C_2 + sR_1C_2\big) \end{aligned}$$

The transfer function is $H(s) = Y(s)/X(s)$:

$$H(s) = \frac{1}{s^2R_1R_2C_1C_2 + s(R_1 + R_2)C_2 + 1} = \frac{1/R_1R_2C_1C_2}{s^2 + s(R_1 + R_2)/R_1R_2C_1 + 1/R_1R_2C_1C_2}$$

And we have Expression (9.15).

Problem 9.4 Design a second-order bandstop filter, Chebyshev 1 type, with lower and upper cutoff frequencies 100 and 200 rad/s, respectively.

Solution From Table 9.2, we get the transfer function of a first-order lowpass filter with cutoff frequency 1 rad/s:

$$H(s) = \frac{1.024}{s + 1.024} \tag{9.30}$$

To transform this into a bandstop filter, we use the substitution in Eq. (9.26):

$$s \rightarrow \frac{s(200 - 100)}{s^2 + 200 \cdot 100} = \frac{100s}{s^2 + 20000}$$

Table 9.2 Chebyshev 1 filter polynomials (passband ripple = 3 dB)

Order	Polynomial
1	$1.024/(s + 1.024)$
2	$0.5012/(s^2 + 0.6449s + 0.7079)$
3	$0.2506/(s^3 + 0.5972s^2 + 0.9283s + 0.2506)$

Inserted into Eq. (9.30):

$$H(s) = \frac{1.024}{\frac{100s}{s^2+20000} + 1.024} = \frac{1.024(s^2 + 20000)}{100s + 1.024(s^2 + 20000)} = \frac{s^2 + 20000}{s^2 + 97.7s + 20000}$$

The amplification diagram of this filter is illustrated in Fig. 9.53.

Problem 9.5 A filter has an impulse response as illustrated in Fig. 9.54. What is the output $y(t)$ if Fig. 9.55 represents the input signal $x(t)$?

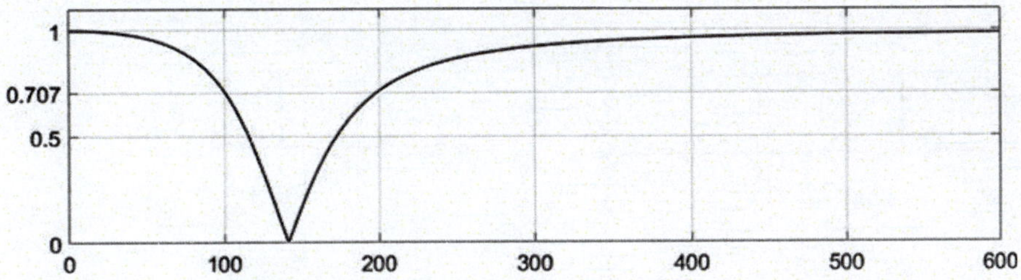

Fig. 9.53 Bandstop filter

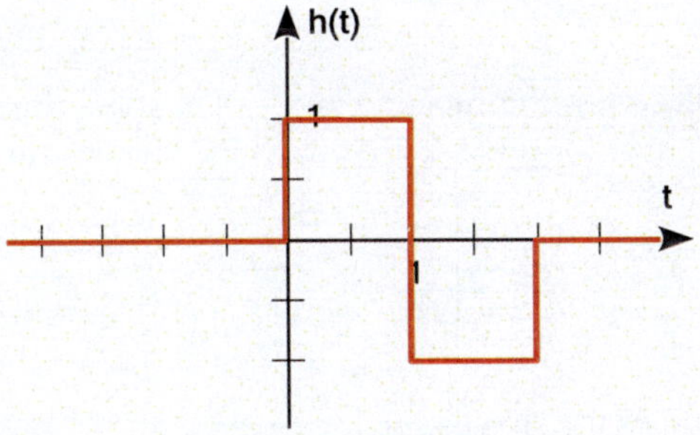

Fig. 9.54 Impulse response

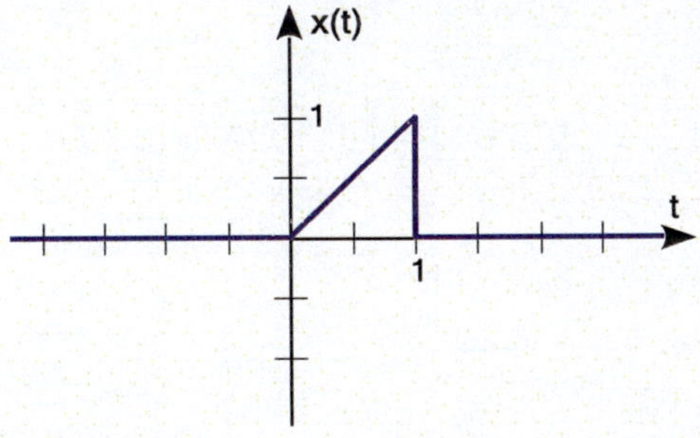

Fig. 9.55 Input signal $x(t)$

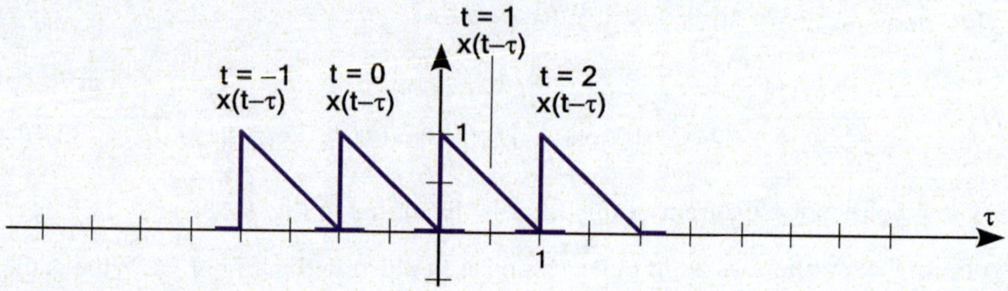

Fig. 9.56 $x(t-\tau)$ for different t

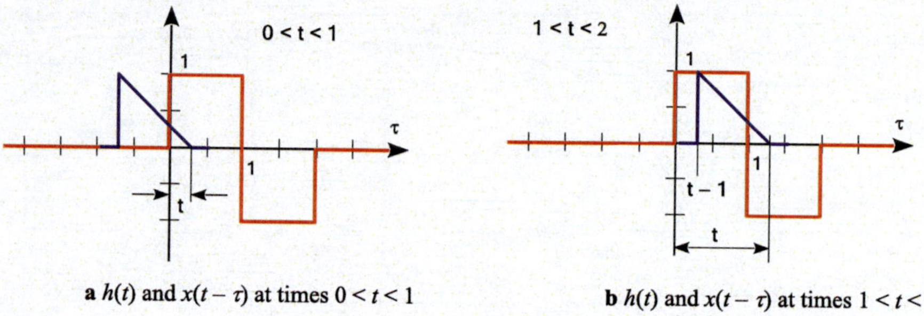

a $h(t)$ and $x(t-\tau)$ at times $0 < t < 1$ **b** $h(t)$ and $x(t-\tau)$ at times $1 < t < 2$

Fig. 9.57 **a** $h(t)$ and $x(t-\tau)$ at times $0 < t < 1$. **b** $h(t)$ and $x(t-\tau)$ at times $1 < t < 2$.

Solution Fig. 9.56 illustrates the function $x(t-\tau)$ at some times t, and Fig. 9.57a illustrates $h(t)$ and $x(t-\tau)$ for times $0 < t < 1$. Since $x(t) = t$ (in Fig. 9.55), $x(t-\tau) = t-\tau$. Hence:

$$\int_0^t 1\cdot(t-\tau)d\tau = \left[t\tau - \frac{1}{2}\tau^2\right]_0^t = t^2 - \frac{1}{2}t^2 = \underline{\frac{1}{2}t^2}$$

Figure 9.57b illustrates the signals for $1 < t < 2$:

$$\int_{t-1}^{1} 1\cdot(t-\tau)d\tau + \int_{1}^{t}(-1)\cdot(t-\tau)d\tau = \left[t\tau - \frac{1}{2}\tau^2\right]_{t-1}^{1} - \left[t\tau - \frac{1}{2}\tau^2\right]_{1}^{t}$$

$$= t - \frac{1}{2} - \left(t(t-1) - \frac{1}{2}(t-1)^2\right) - \left(t^2 - \frac{1}{2}t^2 - t + \frac{1}{2}\right) = \cdots = \underline{-t^2 + 2t - \frac{1}{2}}$$

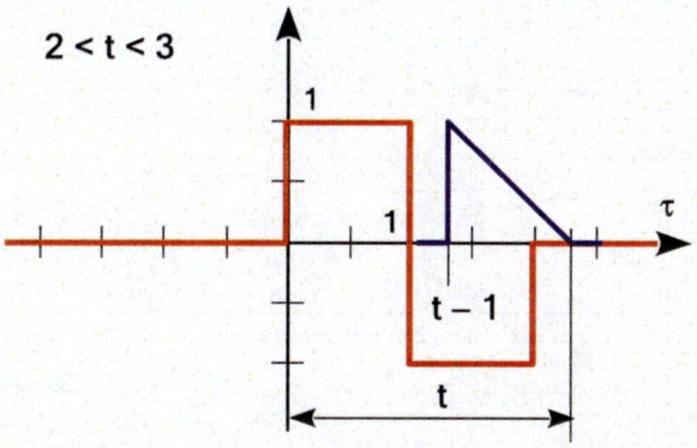

Fig. 9.58 $h(t)$ and $x(t-\tau)$ at times $2 < t < 3$

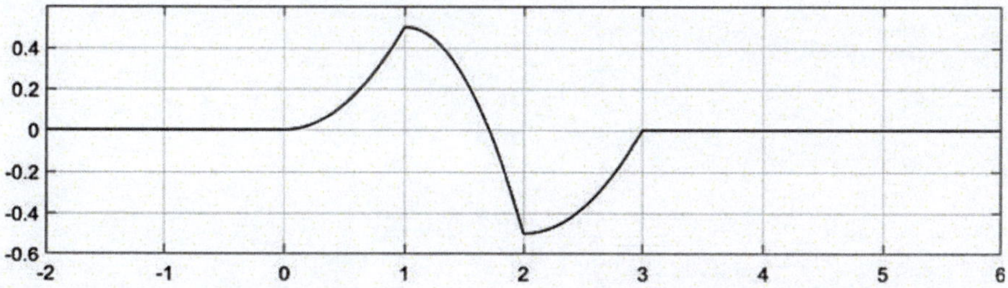

Fig. 9.59 The convolution $h(t) \otimes x(t)$

Figure 9.58 illustrates the signals for $2 < t < 3$:

$$\int_{t-1}^{2} (-1)(t-\tau)d\tau = \int_{t-1}^{2} (\tau - t)\,d\tau = \left[\frac{1}{2}\tau^2 - t\tau\right]_{t-1}^{2}$$

$$= 2 - 2t - \left(\frac{1}{2}(t-1)^2 - t(t-1)\right) = \cdots = \underline{\frac{1}{2}t^2 - 2t + \frac{3}{2}}$$

The convolution signal is illustrated in Fig. 9.59.

Chapter 10
Digital Filters

Abstract In this chapter *digital* filters are introduced. While an analog filter is implemented in hardware, a digital filter is implemented in software; it is a computer algorithm. Hence, this chapter works with sampled signals and the objective is to do the same thing with sampled signals as we did with analog signals (non-sampled signals) in Chap. 9, but instead of using hardware, software algorithms are presented that do the filtering. First, discrete-time convolution is defined, but the focus is on the FIR and IIR filters, which are the two most common digital filter algorithms for sampled signals. The reader will learn two design techniques for digital filters; the inverse Fourier transform method and the bilinear transformation method.

10.1 Introduction

In the previous chapter we looked at analog filters which are implemented in hardware. In this chapter, we will demonstrate how we can achieve the same signal processing results using computer algorithms. These computer algorithms are based on samples (from an ADC, see Chap. 11) and are called 'digital filters'. Compared to analog filters, digital filters have some advantages but certainly also some disadvantages and we will carefully point out the pros and cons of digital filters in this chapter. The objective here is to be able to design any (?) filter specified from an amplification diagram.

This chapter will depend heavily on our results from the z transform Sect. 7.4.2 and we will also refer to solved Problem 7.10.

Figure 10.1 illustrates our general model of an analog filter.

In Sect. 9.8 we learned that in the time domain, the output from this filter is the convolution between $h(t)$ and $x(t)$, Eq. (9.27):

$$y(t) = \int_{-\infty}^{\infty} h(\tau)x(t-\tau)d\tau \qquad (9.27) \tag{10.1}$$

To get a digital filter, we need to sample the signal, i.e., $t \rightarrow nT_S$:

L. Bengtsson, *Electrical Measurement Techniques*,
https://doi.org/10.1007/978-981-99-8187-8_10

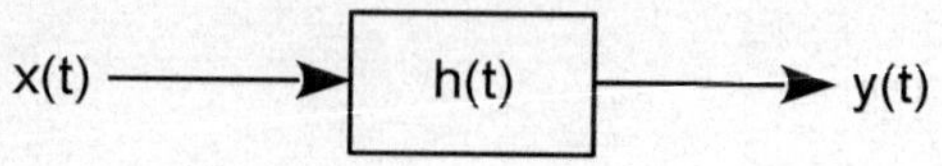

Fig. 10.1 Signal model

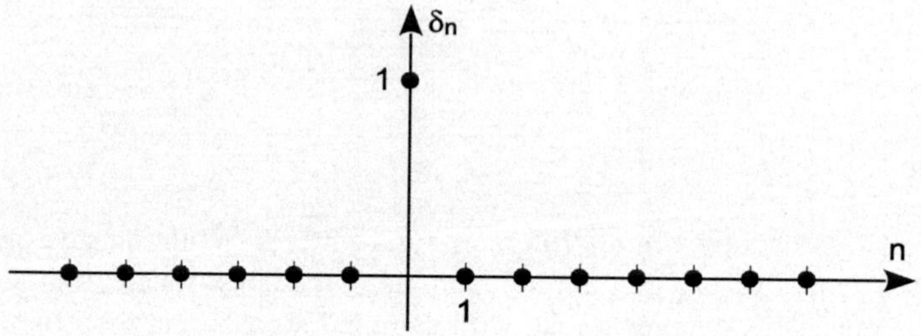

Fig. 10.2 The discrete-time Dirac impulse

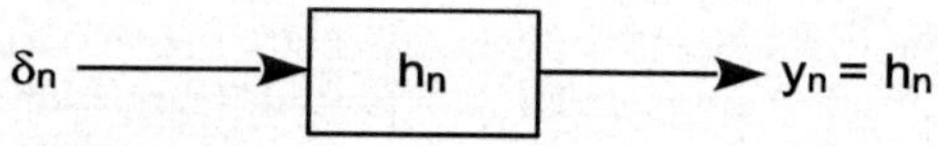

Fig. 10.3 Impulse and impulse response

$$y(t) \rightarrow y(nT_S) = y(n) = y_n = \sum_{i=-\infty}^{+\infty} h_i x_{n-i} \tag{10.2}$$

In Eq. (10.2), h_n is still the impulse response; the system's output when the input is an impulse. Figures 10.2 and 10.3 illustrate the Dirac impulse in discrete time.

First, we will assume that all our filters are 'causal'. That means that there can't be any output signal *before* there is an input signal. In Fig. 10.2, the impulse appears at time $n = 0$. If the system is casual, then h_n must $= 0$ if $n < 0$. (We could have non-causal digital filters, just not in real time. However, we limit the scope here to include only the causal filters.) That means that the summation in Eq. (10.2) should start at $i = 0$:

$$y_n = \sum_{i=0}^{\infty} h_i x_{n-i} \tag{10.3}$$

Second, if we write out Eq. (10.3) explicitly, then

$$y_n = h_0 x_n + h_1 x_{n-1} + h_2 x_{n-2} + h_3 x_{n-3} \ldots \tag{10.4}$$

x_n is the 'latest' sample and x_{n-1} is the second last sample, etc. Notice that we 'time reverse' the samples, just like we did in analog convolution (see Fig. 9.45). Table 10.1 illustrates the case where we have a time series of five samples $x = \{x_0, x_1, x_2, x_3, x_4\}$ and four impulse response coefficients.

Table 10.1 Convolution in discrete time

$x = [x_0\ x_1\ x_2\ x_3\ x_4]$	h_0	h_1	h_2	h_3	y_n
$x_4\ x_3\ x_2\ x_1$	$h_0\,x_0$				$y_0 = h_0\,x_0$
$x_4\ x_3\ x_2$	$h_0\,x_1$	$h_1\,x_0$			$y_1 = h_0\,x_1 + h_1\,x_0$
$x_4\ x_3$	$h_0\,x_2$	$h_1\,x_1$	$h_2\,x_0$		$y_2 = h_0\,x_2 + h_1\,x_1 + h_2\,x_0$
x_4	$h_0\,x_3$	$h_1\,x_2$	$h_2\,x_1$	$h_3\,x_0$	$y_3 = h_0\,x_3 + h_1\,x_2 + h_2\,x_1 + h_3\,x_0$
	$h_0\,x_4$	$h_1\,x_3$	$h_2\,x_2$	$h_3\,x_1$	$y_4 = h_0\,x_4 + h_1\,x_3 + h_2\,x_2 + h_3\,x_1$
		$h_1\,x_4$	$h_2\,x_3$	$h_3\,x_2$	$y_5 = h_1\,x_4 + h_2\,x_3 + h_3\,x_2$
			$h_2\,x_4$	$h_3\,x_3$	$y_6 = h_2\,x_4 + h_3\,x_3$
				$h_3\,x_4$	$y_7 = h_3\,x_4$
					$y_8 = 0$

In general, there are two kinds of digital filters. They all use input samples to calculate the next output sample, but some filters also use previous *output samples* in the algorithm to produce the next output sample. Just like analog filters, digital filters are represented by a transfer function that is a quotient between two polynomials; of course, for a digital filter it is a polynomial in z:

$$H(z) = \frac{B(z)}{A(z)} = \frac{b_0 + b_1 z^{-1} + b_2 z^{-2} + \ldots}{1 + a_1 z^{-1} + a_2 z^{-2} + \ldots} \tag{10.5}$$

The filters that don't use previous output samples in the next output sample algorithm are characterized by having $A(z) = 1$, and these are the filters we will start with.

10.2 FIR Filters

If $A(z) = 1$, then

$$H(z) = b_0 + b_1 z^{-1} + b_2 z^{-2} + \ldots \tag{10.6}$$

Also, remember that the transfer function is by definition the quotient between the output and input signals' transforms (the z transforms in this case):

$$\begin{aligned} H(z) &= \frac{Y(z)}{X(z)} = b_0 + b_1 z^{-1} + b_2 z^{-2} + \cdots \Rightarrow \\ Y(z) &= X(z)\big(b_0 + b_1 z^{-1} + b_2 z^{-2} + \ldots\big) = \\ &= b_0 X(z) + b_1 X(z) z^{-1} + b_2 X(z) z^{-2} + \ldots \end{aligned} \tag{10.7}$$

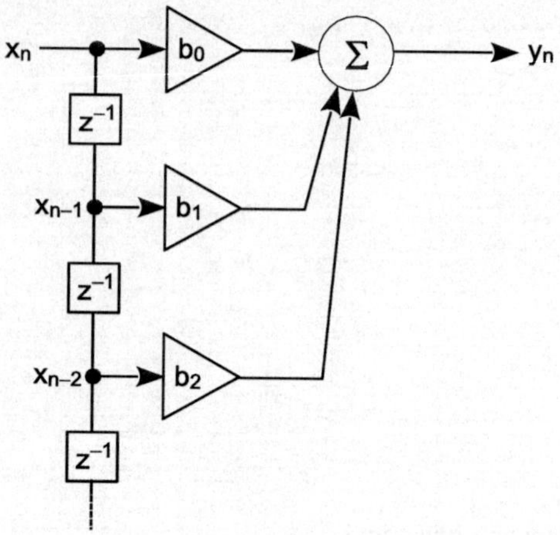

Fig. 10.4 The block diagram of a FIR filter

If we take the inverse z transform of both sides of Eq. (10.7) we get, remembering from Problem 7.10 that the z transform of x_{n-n_0} is $X(z)z^{-n_0}$:

$$y_n = b_0 x_n + b_1 x_{n-1} + b_2 x_{n-2} + \ldots \tag{10.8}$$

Comparing with Eq. (10.4), we can see that in this case, the filter coefficients b_i are also the impulse response coefficients. This means that the number of impulse response coefficients is limited by the number of filter coefficients, which is a finite number; this kind of filters are called 'FIR' filters; *Finite Impulse Response* filters. Figure 10.4 illustrates the *block diagram* of an FIR filter.

Notice in Fig. 10.4 that each sample must be 'saved' and pushed downwards in the delay chain. Notice also that each 'delay box' is represented by 'z^{-1}'; this represents the delay of one sample (in frequency space). (In time space it would be T_S.)

Also, each 'branch' in the delay chain is called a 'tap' and filters are sometimes characterized by the number of taps used: An n-tap filter has n delay taps.

Example 10.1 Plot the frequency response of the following FIR filter:

$$y_n = 0.25x_n + 0.25x_{n-1} + 0.25x_{n-2} + 0.25x_{n-3}$$

Solution This filter produces the average of four samples; we expect a lowpass behavior. To find the transfer function, we must first take the z transform of both sides:

$$Y(z) = \frac{1}{4}\left(X(z) + X(z)z^{-1} + X(z)z^{-2} + X(z)z^{-3}\right) =$$

$$= \frac{1}{4}\left(1 + z^{-1} + z^{-2} + z^{-3}\right)X(z) \Rightarrow$$

$$H(z) = \frac{Y(z)}{X(z)} = \frac{1}{4}\left(1 + z^{-1} + z^{-2} + z^{-3}\right) = \frac{1}{4}\left(z^{1.5} + z^{0.5} + z^{-0.5} + z^{-1.5}\right) \cdot z^{-1.5}$$

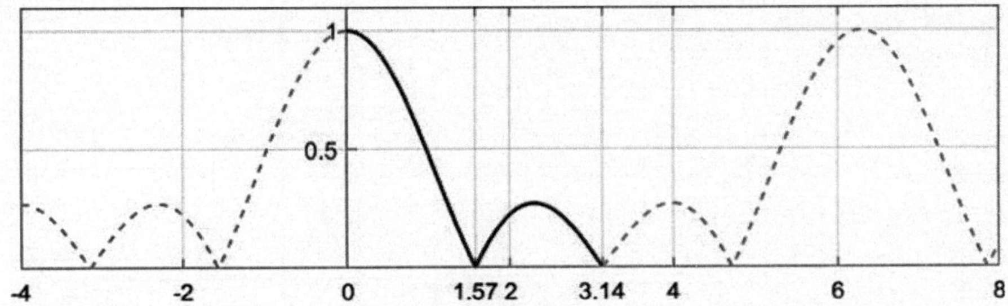

Fig. 10.5 The frequency response of the averaging filter

We get the Fourier transform by setting $z = e^{j\Omega}$:

$$H(\Omega) = H(z)|_{z=e^{j\Omega}} = \frac{1}{4}\left(e^{j1.5\Omega} + e^{j0.5\Omega} + e^{-j0.5\Omega} + e^{-j1.5\Omega}\right) \cdot e^{-j1.5\Omega} =$$

$$= \frac{1}{4} \cdot 2(\cos 0.5\Omega + \cos 1.5\Omega) \cdot e^{-j1.5\Omega} \Rightarrow \underline{|H(\Omega)| = \frac{1}{2} \cdot |\cos 0.5\Omega + \cos 1.5\Omega|}$$

The amplification diagram is plotted in Fig. 10.5, and as we expected, it has a lowpass characteristic.

In Sect. 7.4 we learned about poles and zeros of a system; the roots of the numerator polynomial in Eq. (10.5) are the 'zeros', and the roots of the denominator polynomial are the 'poles'. Obviously, a FIR filter doesn't have any poles; that makes it inherently stable. The averaging FIR filter in Example 10.1 has three zeros:

$$1 + z^{-1} + z^{-2} + z^{-3} = 0 \quad \Rightarrow z_1 = -1 \quad z_{2,3} = \pm j$$

These zeros are illustrated in Fig. 10.6. Compare this diagram to the frequency response in Fig. 10.5 and remember that the Fourier transform is on the unit circle in the z plane. Figure 10.6 indicates that we should have a zero response for 'frequencies' $\Omega = \pi/2$ and π, which is confirmed by the amplification diagram in Fig. 10.5.

10.3 IIR Filters

Let's see what happens if the denominator polynomial in Eq. (10.5) is $\neq 1$. For example, if $B(z) = 1$ and $A(z) = 1 + a_1 z^{-1}$ then

$$H(z) = \frac{1}{1 + a_1 z^{-1}} = \frac{Y(z)}{X(x)} \Rightarrow Y(z)\left(1 + a_1 z^{-1}\right) = X(z)$$

$$Y(z) + a_1 Y(z) z^{-1} = X(z) \Rightarrow y_n = -a_1 y_{n-1} + x_n$$

Let's see what the impulse response is: $x_n = \delta_n \Rightarrow y_n = h_n$

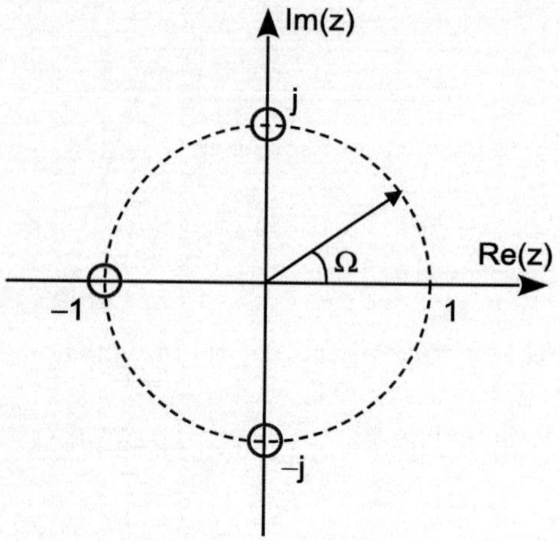

Fig. 10.6 Zeros of FIR filter in Example 10.1

$$
\begin{aligned}
y_0 &= -a_1 y_{n-1} + \delta_0 = 0 + 1 = 1 = h_0 \\
h_1 &= -a_1 h_0 + \delta_1 = -a_1 1 + 0 = -a_1 \\
h_2 &= -a_1 h_1 = a_1^2 \\
h_3 &= -a_1 h_2 = (-a_1)^3 \Rightarrow h_n = (-a_1)^n
\end{aligned}
$$

From this simple example we can see that the number of impulse response coefficients is infinite; when $A(z) \neq 1$, we have an *Infinite Impulse Response* filter, an 'IIR' filter. Since the nth impulse response coefficient is $(-a_1)^n$, we conclude that the filter is unstable if $|a_1| \geq 0$. The filter pole is

$$1 + a_1 z^{-1} = 0 \Rightarrow z_p = -a_1$$

Figure 10.7 illustrates the pole's location for the case where $|a_1| > 1$ and $|a_1| < 1$.
From Fig. 10.7 we can see that the filter is unstable if the pole is outside the unit circle; for digital filters all poles must be within the unit circle. This conclusion is general and consistent with our results in Sect. 7.4.2 (Fig. 7.38).

Equation (10.5) represents the general expression for a second-order IIR filter:

$$H(z) = \frac{b_0 + b_1 z^{-1} + b_2 z^{-2}}{1 + a_1 z^{-1} + a_2 z^{-2}} = \frac{Y(z)}{X(z)} \tag{10.9}$$

Taking the inverse z transform of Eq. (10.9) gives us the *difference equation* for the output sample:

$$y_n = -a_1 y_{n-1} - a_2 y_{n-2} + b_0 x_n + b_1 x_{n-1} + b_2 x_{n-2} \tag{10.10}$$

Figure 10.8 illustrates the block diagram.

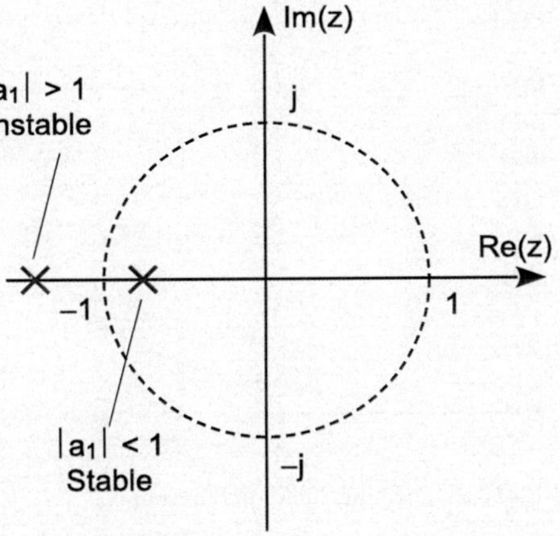

Fig. 10.7 Stable if poles are inside the unit circle

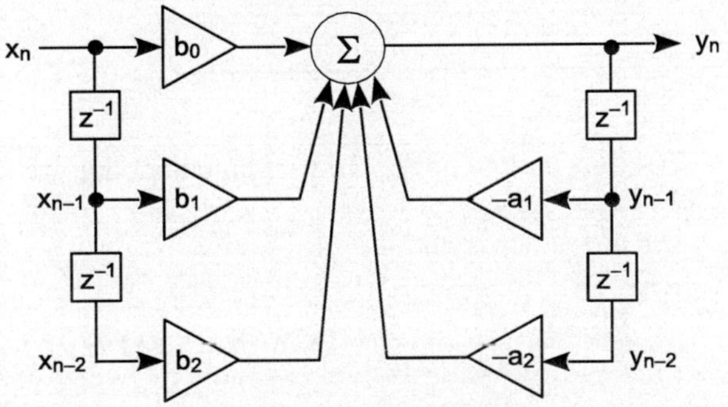

Fig. 10.8 Block diagram of second-order IIR filter

Example 10.2 Consider the filter $(z) = \frac{1-z^{-2}}{1+0.81z^{-2}}$.

a Where are the poles and the zeros? **b** What is the Fourier transform? **c** Find and plot the amplification diagram. **d** What is the difference equation? **e** Draw the block diagram. **f** What is the impulse response? **g** What is the output if the input is $x = [2, -1, 0.5]$?

Solution We can re-write the transfer function as follows:

$$H(z) = \frac{z^2 - 1}{z^2 + 0.81} = \frac{(z+1)(z-1)}{(z+0.9\mathrm{j})(z-0.9\mathrm{j})}$$

and it is obvious that we have two zeros (±1) and two poles (±0.9j). To find the Fourier transform, we replace z with $\mathrm{e}^{\mathrm{j}\Omega}$:

$$H(\Omega) = \frac{\mathrm{e}^{\mathrm{j}2\Omega} - 1}{\underline{\mathrm{e}^{\mathrm{j}2\Omega} + 0.81}} = \frac{\left(\mathrm{e}^{\mathrm{j}\Omega} - \mathrm{e}^{-\mathrm{j}\Omega}\right)\cdot \mathrm{e}^{\mathrm{j}\Omega}}{\cos 2\Omega + \mathrm{j}\sin 2\Omega + 0.81} = \frac{2\mathrm{j}\sin\Omega \cdot \mathrm{e}^{\mathrm{j}\Omega}}{(0.81 + \cos 2\Omega) + \mathrm{j}\sin 2\Omega} =$$

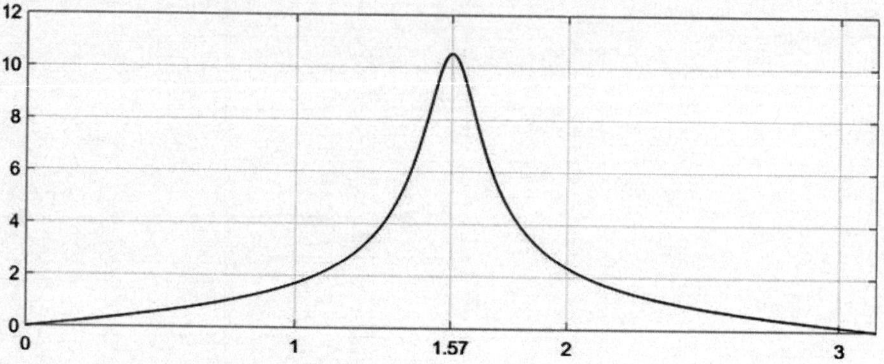

Fig. 10.9 The amplification diagram

$$= \underbrace{\frac{2 \cdot |\sin \Omega|}{\sqrt{(0.81 + \cos 2\Omega)^2 + (\sin 2\Omega)^2}}}_{=|H(\Omega)|} \cdot \frac{e^{j(\Omega + \pi/2)}}{e^{j\tan^{-1}(\sin 2\Omega/(0.81 + \cos 2\Omega))}}$$

$|H(\Omega)|$ is plotted in Fig. 10.9. (Remember that we have a pole in $z = +0.9j$, i.e., at $\Omega = \pi/2$.)

The difference equation:

$$\frac{Y(z)}{X(z)} = \frac{1 - z^{-2}}{1 + 0.81z^{-2}} \Rightarrow Y(z) + 0.81Y(z)z^{-2} = X(z) - X(z)z^{-2} \Rightarrow$$
$$y_n = -0.81y_{n-2} + x_n - x_{n-2}$$

Figure 10.10 illustrates the block diagram.

We get the impulse response by setting $x_n = \delta_n$:

$$h_0 = \delta_0 = 1 \quad h_1 = 0 \quad h_2 = -0.81 \cdot 1 - \delta_0 = -1.81 \quad h_3 = 0$$
$$h_4 = -0.81 \cdot (-1.81) = 1.47 \quad h_5 = 0 \quad h_6 = -0.81 \cdot 1.47 = -1.19 \quad h_7 = 0 \ldots.$$

The impulse response is plotted in Fig. 10.11.

Fig. 10.10 The block diagram

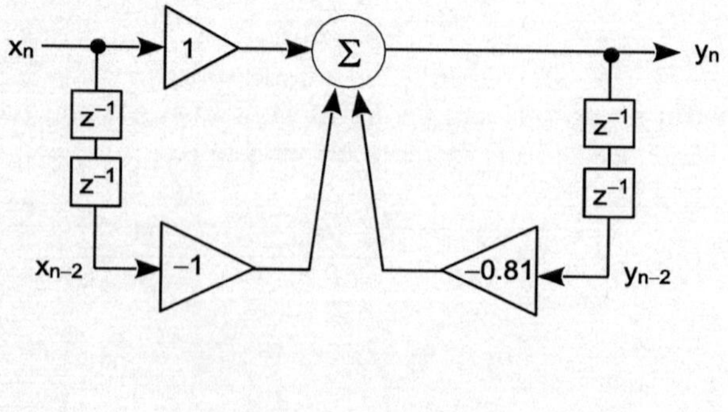

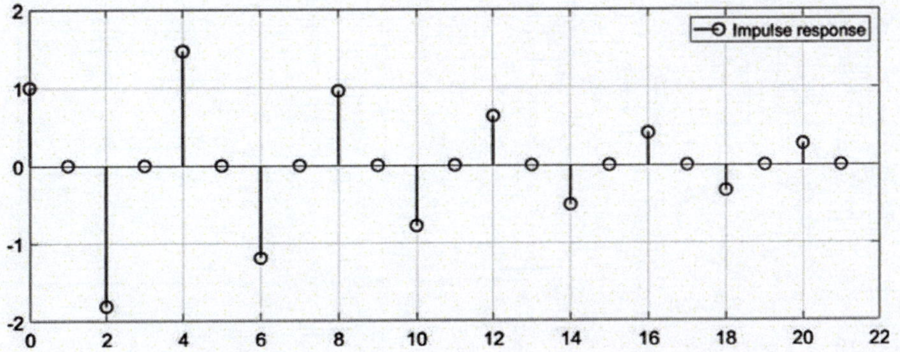

Fig. 10.11 The impulse response

To find the output for the input $x = [2, -1, 0.5]$, we could use a convolution table as in table 10.1, but here we take the opportunity to illustrate discrete-time convolution graphically, see Fig. 10.12.

In Fig. 10.12, we have plotted the signal x_{n-i}, for $n = -1$, see Eq. (10.3), and as time (n) increases, x_{n-i} slides right, and at each time (n) we stop and multiply all overlaps between x_{n-i} and h_i and then we sum all the products. Compare Fig. 10.12 to Fig. 9.42 in Chap. 9. Figure 10.13 illustrates the resulting output from the filter.

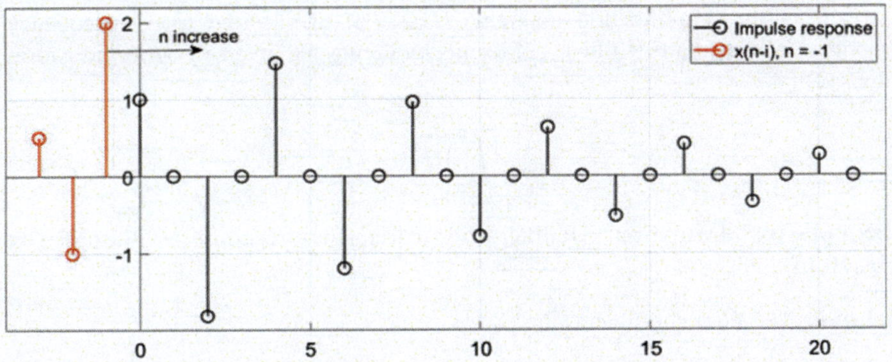

Fig. 10.12 Discrete time convolution

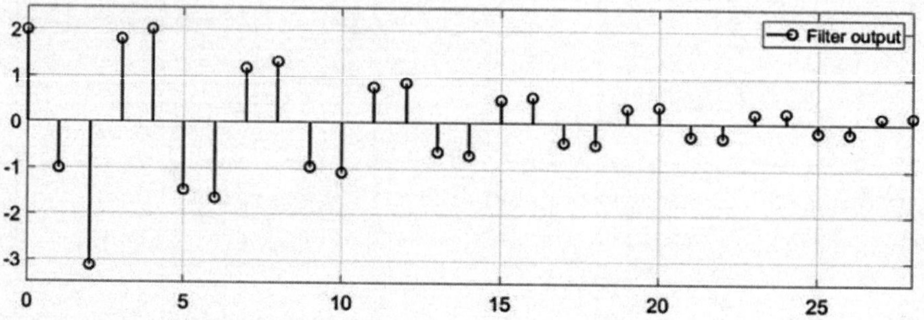

Fig. 10.13 Filter output

10.4 Designing Digital Filters

10.4.1 FIR Filters: The Inverse Fourier Transform Method

Now that we understand how IIR and FIR filters work, we need to learn how to design them. There are several methods to design digital filters, but here we will only learn one method for FIR filters and one method for IIR filters. We start with FIR filter design.

We use the *inverse Fourier transform* method to design FIR filters. This method starts from the amplification diagram $|H(\Omega)|$ of the desired filter's frequency response. We get the FIR filter coefficients by taking the inverse Fourier transform of$|H(\Omega)|$:

$$b_k = \frac{1}{2\pi} \int_{-\pi}^{\pi} |H(\Omega)| \cdot e^{jk\Omega} \tag{10.11}$$

Example 10.3 Design an FIR filter with a frequency response as illustrated in Fig. 10.14.

Solution The filter coefficients are

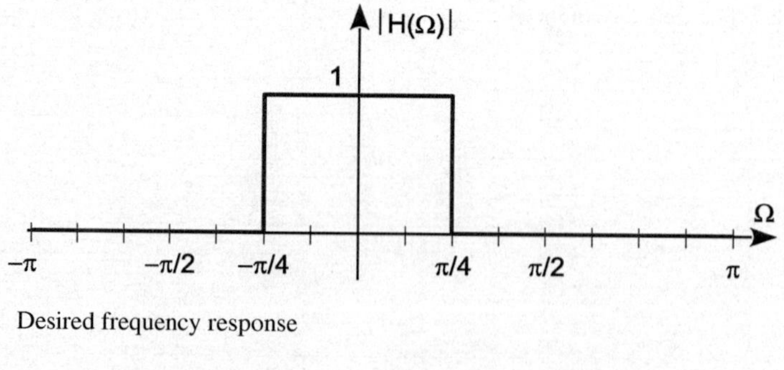

Fig. 10.14 Desired frequency response

$$b_k = \frac{1}{2\pi}\int_{-\pi/4}^{\pi/4} 1\cdot \mathrm{e}^{\mathrm{j}k\Omega} d\Omega = \frac{1}{2\pi \mathrm{j}k}\left[\mathrm{e}^{\mathrm{j}k\Omega}\right]_{-\pi/4}^{\pi/4}$$
$$= \frac{1}{\pi k}\cdot\frac{1}{2\mathrm{j}}\left(\mathrm{e}^{\mathrm{j}k\frac{\pi}{4}} - \mathrm{e}^{-\mathrm{j}k\frac{\pi}{4}}\right) = \frac{1}{\pi k}\sin\frac{k\pi}{4}$$

Table 10.2 contains the filter coefficient values for the first eleven coefficients.

Notice two things in Table 10.2. First, we have truncated the table since we must limit the number of coefficients. Second, there are 'negative-time' filter coefficients. That means that we must 'shift' the coefficients right (delay) to get a causal filter. Below are the difference equations for the 3-, 9- and 11-'tap' filters ($b_{\pm 4} = 0$):

$$y_n = 0.225x_n + 0.250x_{n-1} + 0.225x_{n-2}$$
$$y_n = 0.075x_n + 0.159x_{n-1} + 0.225x_{n-2} + 0.250x_{n-3} + 0.225x_{n-4} + \\ + 0.159x_{n-5} + 0.075x_{n-6}$$
$$y_n = -0.045x_n + 0.075x_{n-2} + 0.159x_{n-3} + 0.225x_{n-4} + 0.250x_{n-5} + \\ + 0.225x_{n-6} + 0.159x_{n-7} + 0.075x_{n-8} - 0.045x_{n-10}$$

Figure 10.15 illustrates the frequency response of these three filters (and a 21-tap filter).

From Fig. 10.15 it is obvious that the more taps we implement, the closer is the frequency response to the ideal response.

In general, IIR filters are more 'computational efficient' than FIR filters; they get the job done with less taps (less 'multiply-and-add' operations). So why use FIR filters at all? Well, we have already seen one reason; the lack of poles makes them inherently stable. But that is not the most important reason. The most important reason is stated in the following theorem:

Table 10.2 Filter coefficients

b_{-5}	b_{-4}	b_{-3}	b_{-2}	b_{-1}	b_0	b_1	b_2	b_3	b_4	b_5
−0.045	0.000	0.075	0.159	0.225	0.250	0.225	0.159	0.075	0.000	−0.045

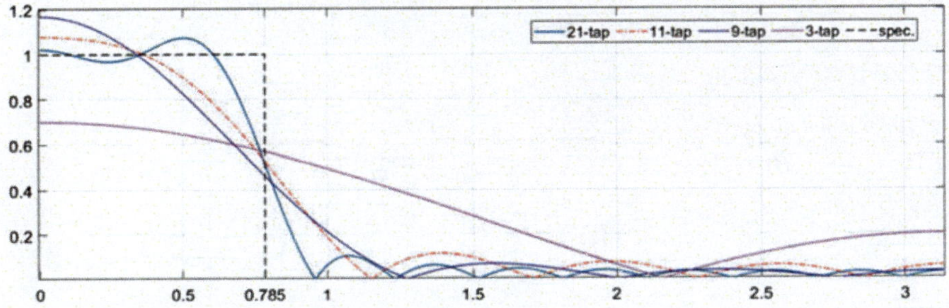

Fig. 10.15 The frequency response of the three FIR filters

Theorem FIR filters with a symmetric set of filter coefficients have linear phase diagrams.

Proof We will not give general proof here, only a 'convincing' example. 'Symmetric' filter coefficients means that if we have an n-tap filter, then $b_0 = b_{n-1}$, $b_1 = b_{n-2}$, etc. Let's take a symmetric 3-tap filter:

$$y_n = b_0 x_n + b_1 x_{n-1} + b_2 x_{n-2} = b_0 x_n + b_1 x_{n-1} + b_0 x_{n-2} \Rightarrow$$

$$Y(z) = X(z)\left(b_0 + b_1 z^{-1} + b_0 z^{-2}\right) \Rightarrow H(z) = \left(b_0 z^1 + b_1 + b_0 z^{-1}\right) \cdot z^{-1} \Rightarrow$$

$$H(\Omega) = \left(b_0 \mathrm{e}^{\mathrm{j}\Omega} + b_1 + b_0 \mathrm{e}^{-\mathrm{j}\Omega}\right) \cdot \mathrm{e}^{-\mathrm{j}\Omega} = (b_1 + 2b_0 \cos\Omega) \cdot \mathrm{e}^{-\mathrm{j}\Omega}$$

(Remember that filters with linear phase diagrams don't distort the signal.) And we can see that the phase function is $\varphi(\Omega) = -\Omega$, i.e., it is a linear function in Ω. This will hold true for any number of taps as long as the coefficients are 'symmetric', and it is not restricted to odd numbers of taps. See Example 10.1, where we had $n = 4$ and symmetric taps. The phase function was $\varphi(\Omega) = -1.5\Omega$.

10.4.2 IIR Filters: The Bilinear Transformation Method

When we design IIR filters, we start from an analog filter and then we try to mimic its frequency response in discrete time. Let's take the simple first-order RC filter in Fig. 10.16 as an example.

The transfer function is

$$H(s) = \frac{1}{1 + sRC} \Rightarrow \tag{10.12}$$

$$H(\Omega) = \frac{1}{1 + \mathrm{j}\omega RC} \Rightarrow |H(\Omega)| = \frac{1}{\sqrt{\left(1 + (\omega RC)^2\right)}} \tag{10.13}$$

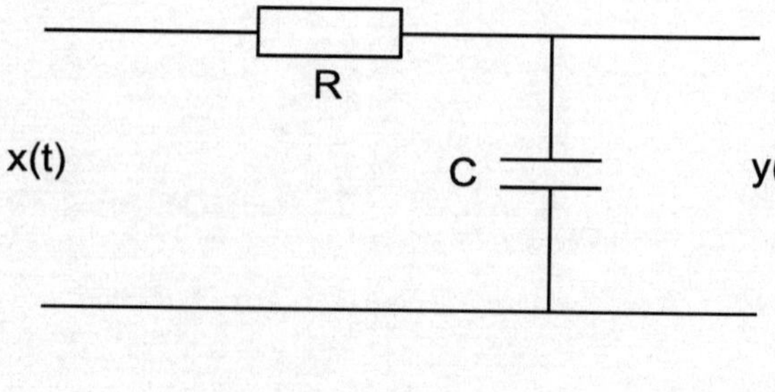

Fig. 10.16 First order RC filter

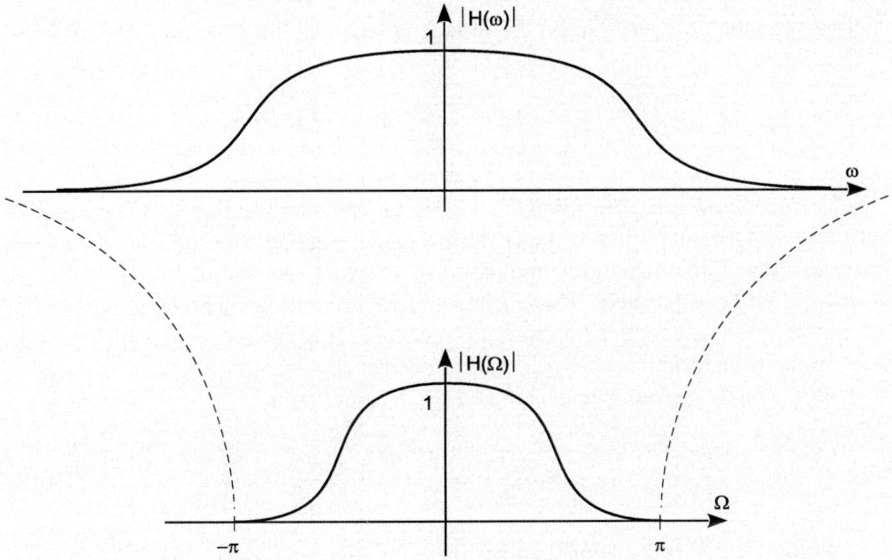

Fig. 10.17 The ω axis must be squeezed in between $-\pi$ and $+\pi$

This is a lowpass filter and its amplification diagram is illustrated in the top diagram in Fig. 10.17.

To mimic this behavior in discrete time, we must shoehorn the entire ω axis into the range $\pm\pi$ on the Ω axis, see Fig. 10.17.

We are looking for a transformation that transforms $-\infty$ to $-\pi$, 0 to 0 and $+\infty$ to $+\pi$. Do we know any such function? Yes, we do! The $\tan^{-1}\omega$ has the range $\pm\pi/2$, hence, if we just multiply it by 2, we will have our transformation formula. However, for reasons that we will explain later, we will multiply ω by k ($\tan^{-1}k\omega$ has the same range):

$$\Omega = 2 \cdot \tan^{-1} k\omega \tag{10.14}$$

Equation (10.14) is a transformation between the Fourier transforms in continuous and discrete time. Since we prefer working in the z and s spaces, we need to translate this expression into a transformation from s space to z space. We solve for $k\omega$ and multiply both sides by j:

$$\begin{aligned} \mathrm{j}k\omega &= \mathrm{j}\tan\frac{\Omega}{2} = \frac{\mathrm{j}\sin\Omega/2}{\cos\Omega/2} = \frac{\frac{1}{2}\left(\mathrm{e}^{\mathrm{j}\Omega/2} - \mathrm{e}^{-\mathrm{j}\Omega/2}\right)}{\frac{1}{2}\left(\mathrm{e}^{\mathrm{j}\Omega/2} + \mathrm{e}^{-\mathrm{j}\Omega/2}\right)} = \\ &= \frac{\mathrm{e}^{\mathrm{j}\Omega/2}\left(1 - \mathrm{e}^{-\mathrm{j}\Omega}\right)}{\mathrm{e}^{\mathrm{j}\Omega/2}\left(1 + \mathrm{e}^{-\mathrm{j}\Omega}\right)} = \frac{1 - \mathrm{e}^{-\mathrm{j}\Omega}}{1 + \mathrm{e}^{-\mathrm{j}\Omega}} \end{aligned}$$

Now we substitute s for $j\omega$ on the left-hand side and z for $e^{j\Omega}$ on the right-hand side:

$$ks = \frac{1 - z^{-1}}{1 + z^{-1}} \Rightarrow s = \frac{1}{k} \cdot \frac{1 - z^{-1}}{1 + z^{-1}}$$

All we need now is to determine the constant k. If we look at Fig. 10.17, we realize that there will be some distortion of the frequency response in the transformation from s to z space. In most digital applications we try to 'oversample', i.e., we stay as far away from f_S as possible. That means that we are more concerned about having the right frequency response for 'low' frequencies. This is what we use the constant k for; in the transformation we prioritize a correct representation of the low end of the ω axis. For 'low' frequencies, $\tan \Omega/2 \approx \Omega/2$, and then

$$j\omega k \approx j\frac{\Omega}{2} \Rightarrow k = \frac{\Omega}{2\omega} = \frac{\omega T_S}{2\omega} = \frac{T_S}{2} \Rightarrow s = \frac{2}{T_S} \cdot \frac{1 - z^{-1}}{1 + z^{-1}} \tag{10.15}$$

Equation (10.15) is the *bilinear transformation* that we use to 'convert' an analog transfer function into a corresponding digital IIR filter.

Example 10.4 Assuming in Eq. (10.12) that $RC = 0.01$, use bilinear transformation to design a corresponding digital IIR filter. The sampling rate of the digital filter is 1 kS/s. Plot the frequency response for both the analog and the digital filters and compare their cutoff frequencies.

Solution

$$H(z) = \frac{1}{1 + \frac{2}{T_S} \cdot \frac{1-z^{-1}}{1+z^{-1}} RC} = \frac{1 + z^{-1}}{1 + z^{-1} + \frac{2RC}{T_S}(1 - z^{-1})} =$$

$$= \frac{1 + z^{-1}}{\left(1 + \frac{2RC}{T_S}\right) + \left(1 - \frac{2RC}{T_S}\right) z^{-1}} = \frac{1 + z^{-1}}{21 - 19z^{-1}} = \frac{0.048 + 0.048z^{-1}}{1 - 0.905z^{-1}} = \frac{Y(z)}{X(z)}$$

$$\Rightarrow Y(z) - 0.905Y(z)z^{-1} = 0.048X(z) + 0.048X(z)z^{-1}$$

$$\underline{y_n = 0.905y_{n-1} + 0.048x_n + 0.048x_{n-1}}$$

The Fourier transform is $H(\Omega) = \frac{0.048+0.048e^{-j\Omega}}{1-0.905e^{-j\Omega}}$. In Fig. 10.18, we have plotted the amplification diagram of this filter with absolute frequencies on the x-axis, and in Fig. 10.19, we have plotted the frequency response of the original analog filter. Notice in Figs. 10.18 and 10.19 that the cutoff frequency is the same (100 rad/s). This is expected; from Eq. (10.14) we predict the 'analog' frequency of 100 rad/s to be transformed to

$$\Omega = 2\tan^{-1}\frac{T_S}{2}\omega = 2\tan^{-1}\frac{0.001}{2}100 = 0.1\,\text{rad} \Rightarrow 0.1\frac{1000}{2\pi} = 15.9\,\text{Hz} = 100\,\text{rad/s}$$

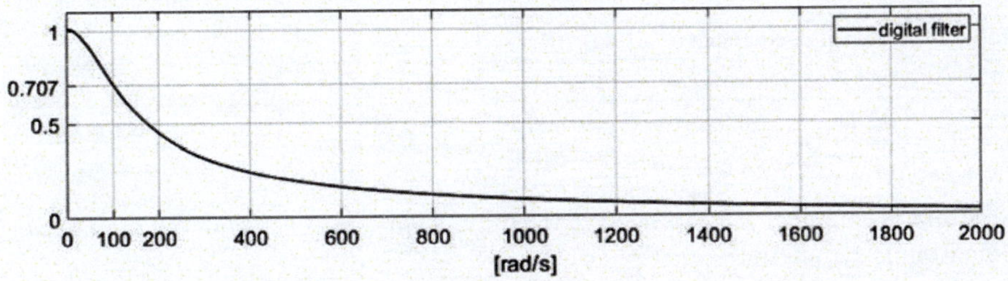

Fig. 10.18 Digital filter from bilinear transformation

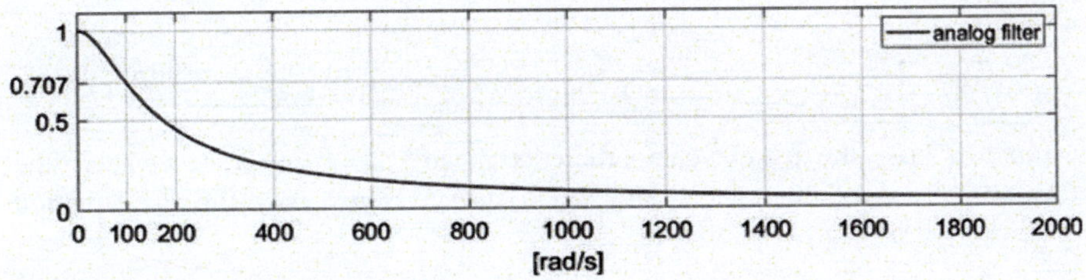

Fig. 10.19 Analog 'model' filter

Notice also how well the shape of the analog filter transfers to the digital filter's frequency response.

10.5 Solved Problems

Problem 10.1 Design a FIR filter with a frequency response as in Fig. 10.20.

Use only the first nine non-zero filter coefficients. Plot the frequency response of the resulting filter.

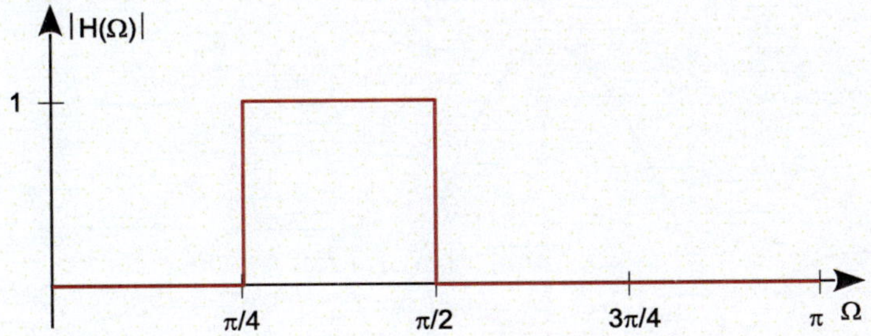

Fig. 10.20 A bandpass filter

Table 10.3 FIR filter coefficients

k	0	± 1	± 2	± 3	± 4	± 5
b_k	0.250	0.093	−0.159	−0.181	0	0.109

Solution Remember that we need to integrate from $-\pi$ to $+\pi$ and that the Fourier transform is symmetric:

$$b_k = \frac{1}{2\pi}\left(\int_{-\pi/2}^{-\pi/4} 1\cdot e^{jk\omega}d\omega + \int_{\pi/4}^{\pi/2} 1\cdot e^{jk\omega}d\omega\right) = \frac{1}{2\pi}\cdot\frac{1}{jk}\left(\left[e^{jk\omega}\right]_{-\pi/2}^{-\pi/4} + \left[e^{jk\omega}\right]_{\pi/4}^{\pi/2}\right) =$$

$$= \frac{1}{\pi k}\cdot\frac{1}{2j}\left(e^{-jk\pi/4} - e^{-jk\pi/2} + e^{jk\pi/2} - e^{jk\pi/4}\right) = \underline{\frac{1}{\pi k}\left(\sin k\frac{\pi}{2} - \sin k\frac{\pi}{4}\right)}$$

Table 10.3 lists the first eleven coefficients. Since $b_{\pm 4} = 0$ and since we need "the first nine non-zero coefficients" we must also use $b_{\pm 5}$. Hence, the difference equation is

$$y_n = 0.109x_n - 0.181x_{n-2} - 0.159x_{n-3} + 0.093x_{n-4} + 0.25x_{n-5} + 0.093x_{n-6} -$$

$$-0.159x_{n-7} - 0.181x_{n-8} + 0.109x_{n-10}$$

The frequency response is plotted in Fig. 10.21.

Problem 10.2 First, plot the frequency response of the analog filter $H(s) = s/(s+10)$. What type of filter is it and what is the cutoff frequency? Next, use bilinear transformation to design the corresponding IIR filter using a sampling rate of 100 S/s. Plot the frequency response of this filter. What is the cutoff frequency of the IIR filter?

Solution The Fourier transform is

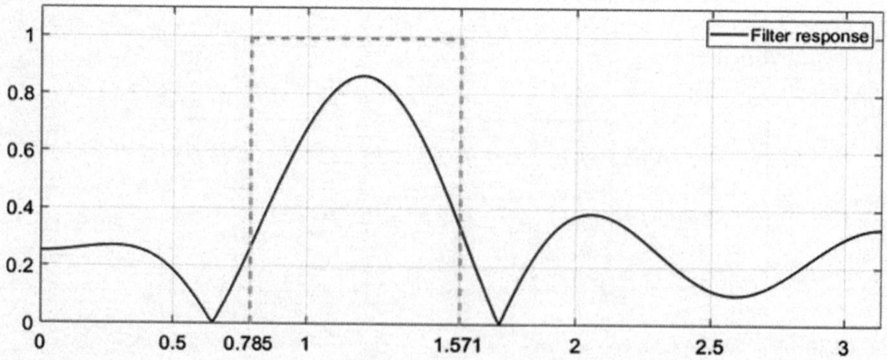

Fig. 10.21 The frequency response

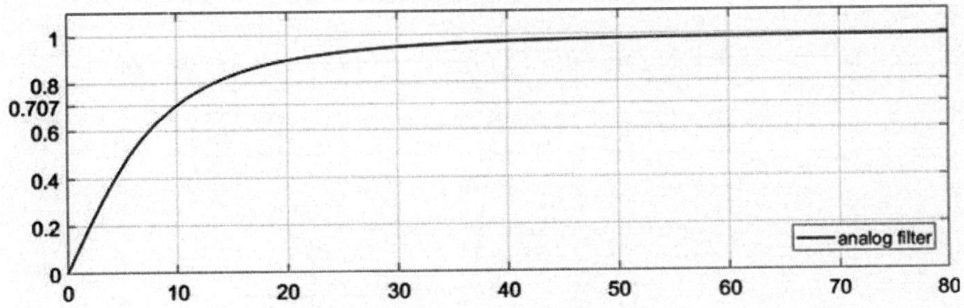

Fig. 10.22 Frequency response of analog filter

$$H(\omega) = \frac{\mathrm{j}\omega}{\mathrm{j}\omega + 10} = \frac{|\omega|}{\sqrt{\omega^2 + 10^2}} \cdot \mathrm{e}^{\mathrm{j}(90^\circ - \tan^{-1}\omega/10)} = |H(\omega)| \cdot \mathrm{e}^{\mathrm{j}\varphi(\omega)}$$

The frequency response is plotted in Fig. 10.22; it is a highpass filter with a cutoff frequency of 10 rad/s.

Since $2/T_S = 2/(1/100) = 200$, the bilinear transformation to z space is

$$H(z) = \frac{200 \cdot \frac{1-z^{-1}}{1+z^{-1}}}{200 \cdot \frac{1-z^{-1}}{1+z^{-1}} + 10} = \frac{1 - z^{-1}}{1 - z^{-1} + 0.05\left(1 + z^{-1}\right)} = \frac{1 - z^{-1}}{1.05 - 0.95z^{-1}} =$$

$$= \frac{0.952 - 0.952z^{-1}}{1 - 0.905z^{-1}} = \frac{Y(z)}{X(z)} \Rightarrow$$

$$\Rightarrow Y(z) - 0.905Y(z)z^{-1} = 0.952X(z) - 0.952X(z)z^{-1} \Rightarrow$$

$$\Rightarrow y_n = 0.905y_{n-1} + 0.952x_n - 0.952x_{n-1}$$

The frequency response is plotted in Fig. 10.23.

The cutoff frequency of 10 rad/s in analog space has been transformed to 0.1 rad in z space. This is expected since

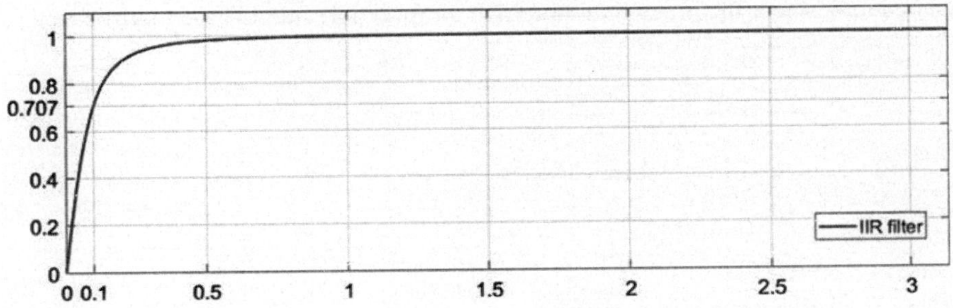

Fig. 10.23 Frequency response of the IIR filter

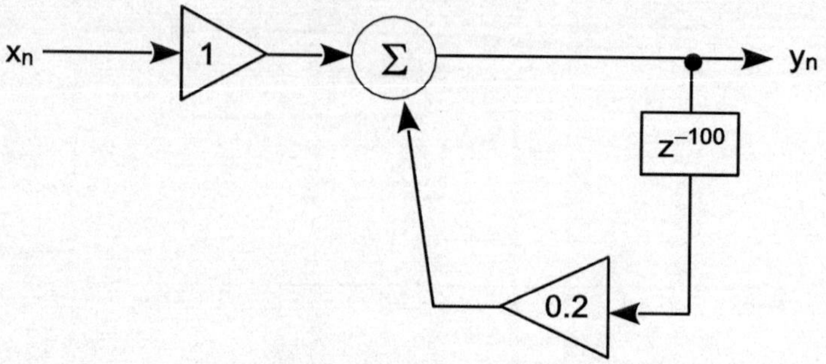

Fig. 10.24 IIR filter

$$\Omega = 2\tan^{-1}\frac{T_S}{2}\omega = 2\tan^{-1}\frac{10}{200} = 0.1\,\text{rad}$$

This corresponds to a frequency of $\frac{0.1}{2\pi}100 = 1.6\,\text{Hz} \; = 10\,\text{rad/s}$.

Problem 10.3 Look at the IIR filter in Fig. 10.24. What does this filter do? What is the application of such a filter?

Solution The difference equation is $y_n = 0.2y_{n-100} + x_n$ and the transfer function is

$$H(z) = \frac{1}{1 - 0.2z^{-100}} \quad \text{Poles: } 1 - 0.2z^{-100} = 0 \quad \Rightarrow z^{100} = 0.2 \quad \Rightarrow$$

$$\left(A \cdot e^{j\varphi}\right)^{100} = 0.2 \cdot e^{j(0° \pm n360°)} \quad \Rightarrow A = 0.2^{1/100} = 0.98 \quad \varphi = \pm n \cdot 3.6°$$

The poles' location in z space is illustrated in Fig. 10.25; there are 100 poles and only the first 10 poles are marked in Fig. 10.25. The frequency response is illustrated in Fig. 10.26.

From the look of the frequency response, this kind of filter is sometimes called a 'comb' filter. What does it do? It adds an attenuated delayed output sample to the present output sample; that will generate an 'echo' effect. This has obvious audio applications where the input is the microphone, and the output is the loudspeaker.

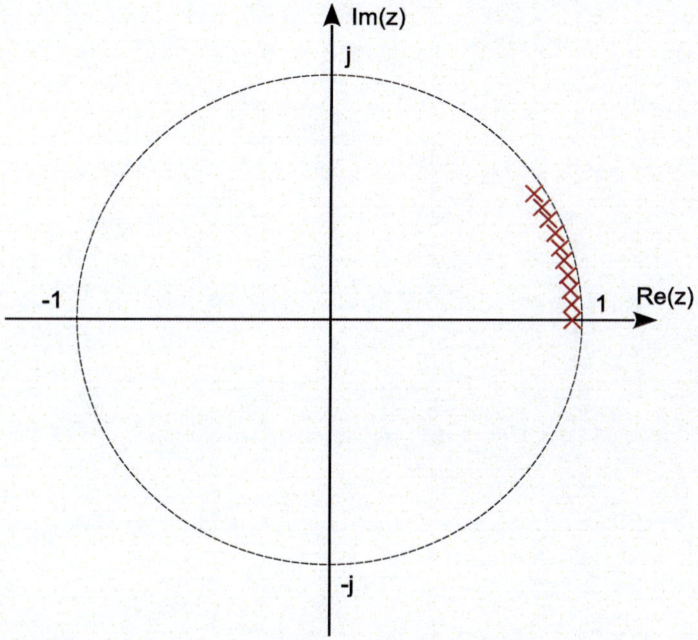

Fig. 10.25 Pole chart; first ten poles (of 100)

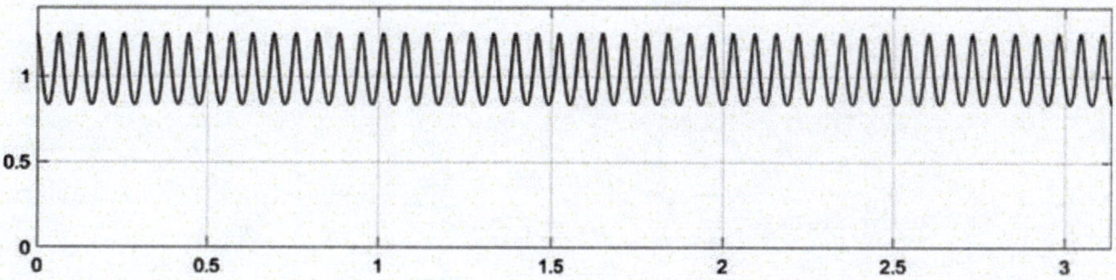

Fig. 10.26 Frequency response of ‘comb’ filter

Chapter 11
ADCs and Sampling

Abstract Almost all real signals are analog by nature and almost all measurement systems are digital. That means that in most measurement systems, the signal must be converted from the analog world to the digital world and a basic understanding of this process is paramount. First, quantization and quantization noise are discussed in general and then a few different analog-to-digital converter techniques (ADCs) are presented. The first one is the successive approximation ADC (SAR) followed by the flash ADC, the pipeline ADC, and the dual slope ADC. Level-crossing ADCs and the sigma-delta ADCs are also presented (but they are less common in the physics lab). The theory also includes the concept of an equivalent number of bits (ENOB) and 'dithering'. This chapter also digs a little deeper into the sampling process; the benefit of oversampling and how to achieve extreme sampling rates (time-interleaved sampling and equivalent-time sampling).

11.1 Introduction

We have previously 'sampled' to 'discretize' an analog signal, but we never got into the details of how this is implemented in hardware. The sampling unit is a central component in any measurement system, but the sampling itself is only part of the secret. The 'sample' value of an analog signal can have any value; its range is all real numbers ($\in \mathbb{R}$). Since the 'end station' of most samples is 'some kind of computer' (digital device), the sample must also be 'digitized' (we will call it 'quantized'). The quantization is done by an *analog-to-digital converter* (ADC) and there are only a handful techniques to implement an ADC in electronics. Finally, we will discuss some 'advanced' aspects on sampling and quantization theory.

L. Bengtsson, *Electrical Measurement Techniques*,
https://doi.org/10.1007/978-981-99-8187-8_11

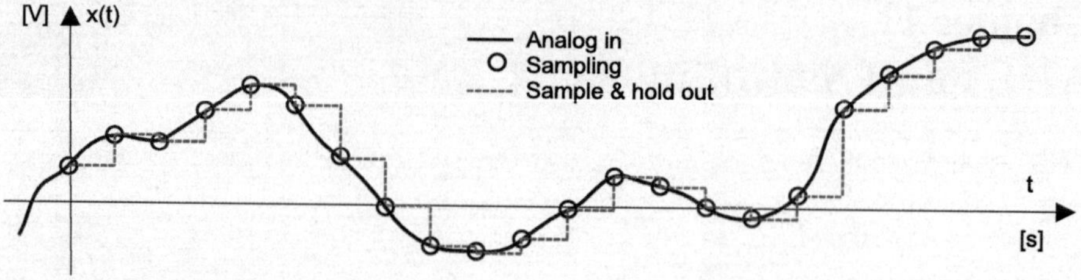

Fig. 11.1 Sampling

11.2 Sampling

For the quantization process to work (the analog-to-digital conversion), the sample value must be constant during the entire conversion process. This is illustrated in Fig. 11.1; the dotted line represents the input value to the ADC. It is the responsibility of the sample and hold unit to take a sample and hold it constant until the quantization is completed.

Figure 11.2 illustrates a sample and hold unit. The S&H unit consists of three components: A voltage follower, a capacitor, and a switch. The switch is controlled by the sampling clock signal. It is 'closed' during the 'positive' period of the sampling clock and during this time the capacitor is charged to the input signal level. The high input impedance of the op-amp ensures that it doesn't discharge during the 'negative' period. Hence, the output of the voltage follower will be the dotted line in Fig. 11.1.

11.3 Quantization and Quantization Noise

An ADC takes an analog voltage (the sample) and converts it into an integer (a binary integer) that is proportional to the sample voltage. An ADC is characterized by two parameters. The most important parameter is the number of bits, n, in the digital (binary) output integer. The second parameter is the reference voltage U_{ref}. Figure 11.3 illustrates an n-bit ADC.

The ADC divides the reference voltage into 2^n equidistant levels (usually). If, for example, $U_{\text{ref}} = 3.3$ V and $n = 8$, we will have 256 levels, and the distance between each level is

$$\Delta U = \frac{U_{\text{ref}}}{2^n} = \frac{3.3}{2^8} = 12.89\,\text{mV} \tag{11.1}$$

ΔU is the *resolution* of the ADC.[1] Figure 11.4 illustrates how the reference voltage is divided into 256 levels (2^n levels in the general case).

[1] Sometimes just the number of bits (n) is used for the resolution: "The ADC has a resolution of 12 bits."

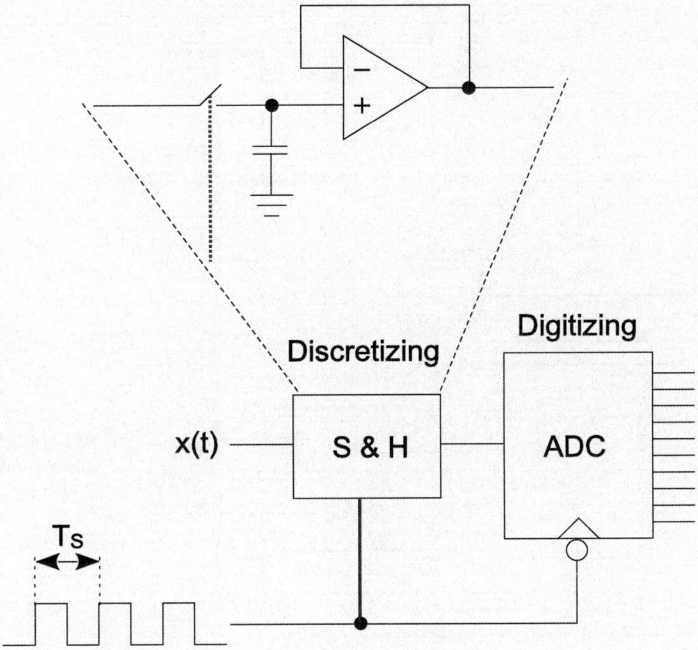

Fig. 11.2 Sample and hold unit

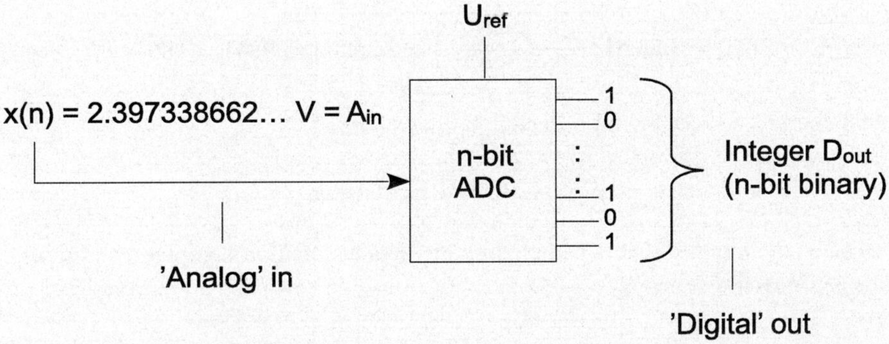

Fig. 11.3 Analog-to-digital conversion

Notice that there are 256 levels, numbered from 0 to 255; there is no level 256. The last voltage level is $U_{\text{ref}} - \Delta U$. The sample A_{in} from the sample and hold unit is simply assigned the integer level number that is closest to the sample value:

$$D_{\text{out}} = \text{round}\left(\frac{A_{\text{in}}}{\Delta U}\right) \tag{11.2}$$

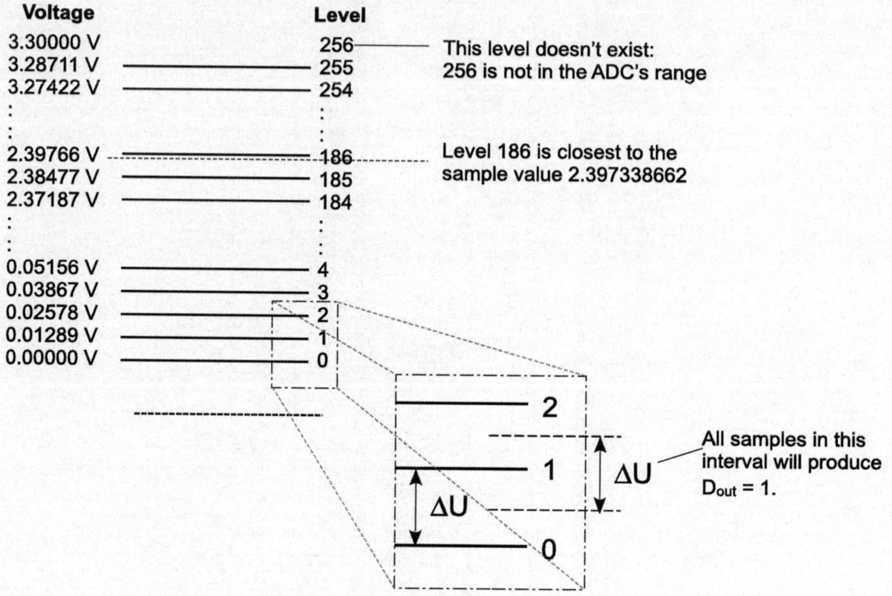

Fig. 11.4 The reference voltage is divided into 2^n levels

Our sample of 2.397338662 in Fig. 11.3 will be converted to

$$D_{\text{out}} = \text{round}\left(\frac{2.397338662}{0.01289}\right) = \text{round}(185.984) = 186$$

Of course, the ADC will produce it in binary format:

$$D_{\text{out}} = 10111010_2 = 186_{10}$$

This is the number that is sampled by the data acquisition computer, and it will be re-converted to a voltage:

$$\widehat{A}_{\text{in}} = D_{\text{out}} \times \Delta U = 186 \times 0.01289 = 2.39765625\ \text{V} \tag{11.3}$$

Notice a few details in Fig. 11.4. The digital output value changes in the middle of the two ΔU levels. A consequence of that is that the width of the first interval is only $\Delta U/2$ and the last interval is $3\Delta U/2$. Figure 11.5 illustrates the in–out characteristics of an 8-bit ADC.

We can see in Eq. (11.3) that there is a small discrepancy between our estimate $\widehat{A}_{\text{in}}$ and the 'true' A_{in} sample, an *uncertainty*, because of the rounding in Eq. (11.2). The 'true' value A_{in} can be anywhere in the interval

$$(D_{\text{out}} - 0.5) \times \Delta U \leq A_{\text{in}} \leq (D_{\text{out}} + 0.5) \times \Delta U \tag{11.4}$$

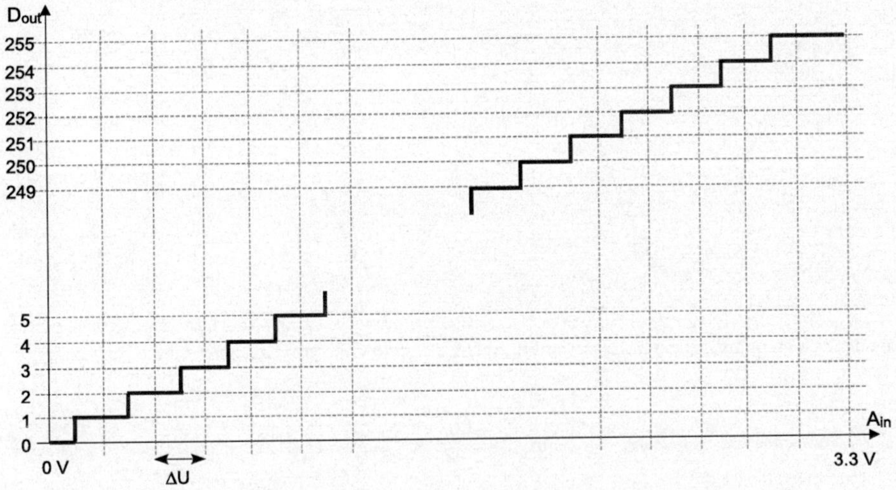

Fig. 11.5 In–out characteristics of 8-bit ADC

Fig. 11.6 The quantization adds noise to the signal

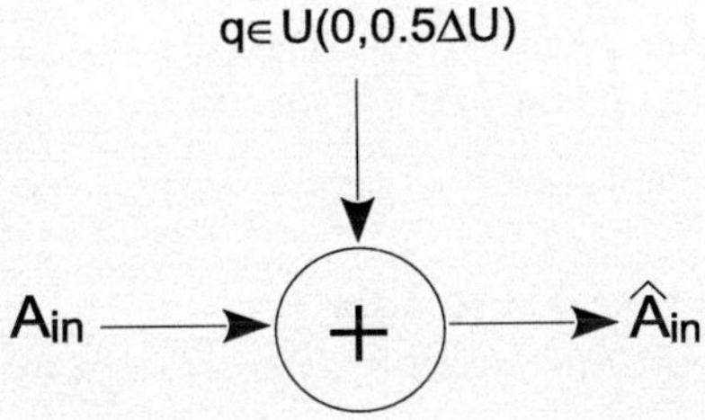

The discrepancy between $\widehat{A}_{\text{in}}$ and A_{in} is called the *quantization noise* (or the 'residual') and is a stochastic variable with a uniform distribution between $\pm\Delta U/2$. We can model this as noise that is added to the sample, see Fig. 11.6.

In Fig. 11.7 we have sampled a sinusoidal signal and plotted the quantization noise in the same graph ($\times 20$); the quantization noise will set a limit to how small signal changes we can detect with the ADC.

We will have more to say about quantization noise later.

11.4 Digital-to-Analog Converters

A Digital-to-Analog converter (DAC) does exactly the opposite of an ADC; the input is an integer (the 'digital') and the output is an analog voltage, see Fig. 11.8. The analog output voltage is

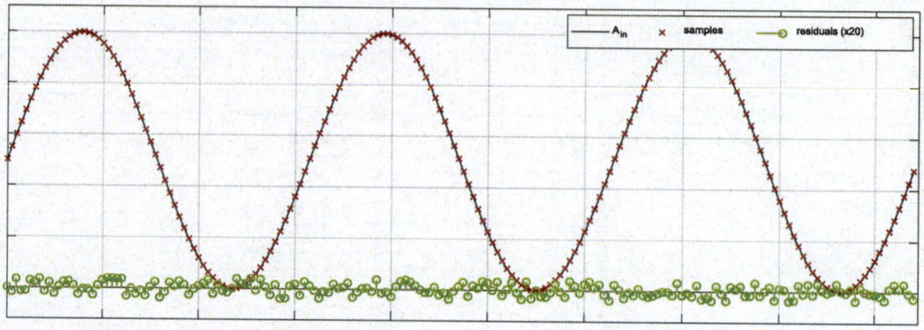

Fig. 11.7 The quantization noise in sampling ($U_{\text{ref}} = 5.0$ V, 8-bit ADC)

$$A_{\text{out}} = \frac{D_{\text{in}}}{2^n} \times U_{\text{ref}} \tag{11.5}$$

We will not go into the details of DACs here. They are *much* easier to implement in hardware than ADCs; the two dominating techniques are (a) an 'R-2R ladder' circuit and (b) lowpass filtering of a PWM signal. (You can easily google that if you are interested.) The only reason we mention them here is because some ADCs depend on a DAC to do an analog-to-digital conversion.

11.5 SAR ADCs

The successive approximation register (SAR) ADC is one of the most common and popular ADCs since it is a good compromise between speed and resolution. Figure 11.9 illustrates the SAR architecture.

The DAC generates an analog voltage that is compared with the analog input sample in a comparator. The comparator output is fed back to the 'SAR logic' which changes the DAC input value until it equals A_{in}. The 'cleverness' in the circuit is the order in which the SAR logic changes the input values to the DAC to minimize the conversion time. In the first input value to the DAC, only the most significant

Fig. 11.8 Digital-to-analog converter

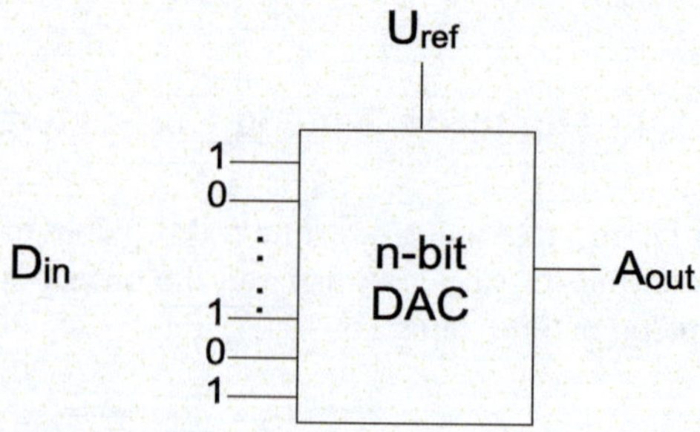

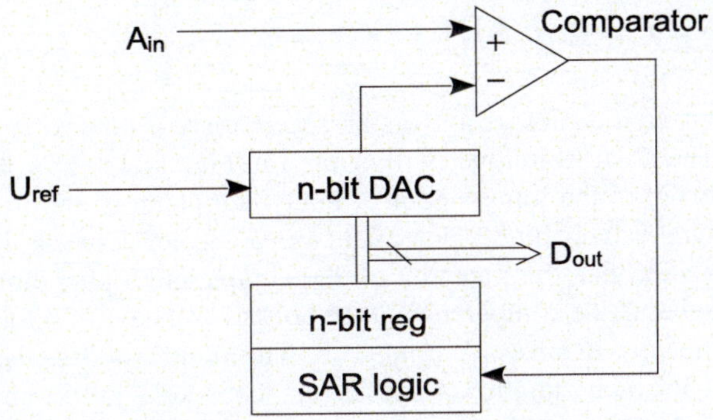

Fig. 11.9 SAR ADC architecture

bit is set: 10000 … 00 (binary). This corresponds to a DAC output value of $U_{ref}/2$. If the comparator output is '1', we know that $A_{in} > U_{ref}/2$, and the only way to get a larger analog output from the DAC is if we keep the most significant bit = 1. If the comparator output is '0', we know that $A_{in} < U_{ref}/2$, and the most significant bit must be = 0; in a single comparison, we have determined the value of the most significant bit. In the next comparison, we set the second most significant bit to '1', and the comparator output determines if we should keep it or not. We continue until all bits have been compared and the digital output value is the output from the *n*-bit register used as the DAC input. An *n*-bit SAR ADC needs *n* comparisons to do an analog-to-digital conversion. The comparison process for an 8-bit SAR is illustrated in Fig. 11.10.

The SAR algorithm is by no means a contemporary invention. It was first suggested by an Italian mathematician, Tartaglia, in 1556. However, he was not concerned with SAR ADCs, he suggested the SAR algorithm to optimize the weighting on balance scales: Start with the heaviest counterweight and keep it if the weight is too small. Next, take the second heaviest counterweight, etc.

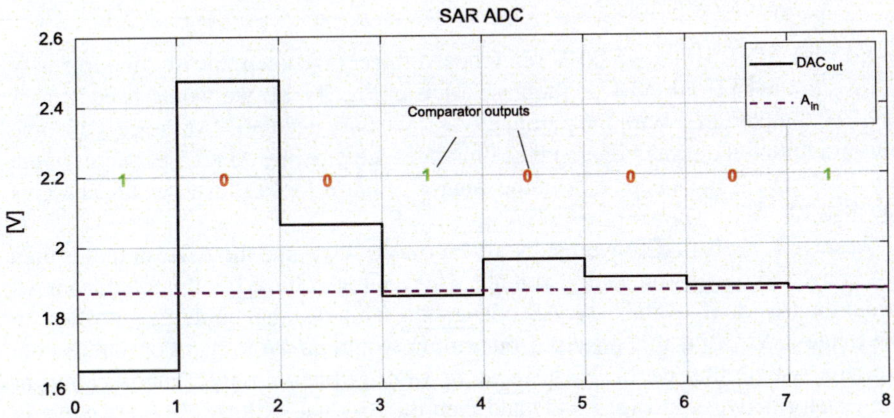

Fig. 11.10 $A_{in} = 1.87$ V $\Rightarrow D_{out} = 1001\ 0001_2 = 145$ ($U_{ref} = 3.3$ V)

11.6 Flash ADCs

When it comes to conversion speed, there is no design that can beat the flash ADC. The flash technique is illustrated in Fig. 11.11. In a flash ADC, the input sample is fed to the minus input of many comparators and the plus inputs are provided a successively higher potential from a resistor network; if the input sample voltage is higher than the potential on the comparator's plus input, the output will be '0'. In general, the comparators at the bottom will have a '0' output, and the comparators at the top will have an output = '1'. The number of 0s at the bottom will be proportional to the input sample voltage and the 'decoder's' job is simply to count the number of 0s and produce this number on the output (a 2^n – to – n decoder). Since all comparisons are performed simultaneously, it is sometimes called a *parallel* ADC.

It is easy to see why this technique is so fast; all comparisons take place at the same time and the only delay is caused by the signal propagation delays in the comparators and the decoder. It is also easy to see the disadvantage; an n-bit flash ADC requires 2^n comparators (minus 1); a 16-bit flash ADC would need 65,535 comparators. This is not possible to implement in silicon and for that reason, you will only find flash ADCs with 'low' resolution (8–10 bits). In the next section we will see how to remedy this.

11.7 Pipeline ADCs

To solve the problem with the large number of comparators in a flash ADC, *pipeline* ADCs are used. We will present the pipeline ADC with an example. First, a sample voltage of $A_{\text{in}} = 2.65$ V would be converted to

$$\text{round}\left(\frac{2.65}{5.00/2^{13}}\right) = 4342 = 10F6_{16} = 1\ 0000\ 1111\ 0110_2 \tag{11.6}$$

in a 13-bit ADC with a +5.00-V reference voltage. (We keep this in mind to verify our result later.) If the ADC is the flash ADC in Fig. 11.11, we would need $2^{13} - 1 = 8191$ comparators. With the pipeline ADC, we only need 36! (At the cost of some *minor* additional delay.) Figure 11.12 illustrates the pipeline ADC. The secret of the pipeline ADC is the electronics in the 'stages'. Figure 11.13 illustrates the contents of stage 1.

In stage 1, A_{in} is first converted by a 3-bit (flash) ADC and the ADC output is then DA converted back to an analog voltage. This analog voltage is subtracted from A_{in} to get the 'residual' of the 3-bit AD conversion. This residual could be anywhere in the range $\pm\Delta U/2$; if it is negative, our circuit would *add* it to A_{in}. For that reason, we will assume that we have a *truncating* ADC (which is only a modification of the resistor network in Fig. 11.11) and then the residual will always be positive in the range 0 to $\Delta U = U_{ref}/2^3$. This residual is multiplied by 4 (2^2) before it is fed

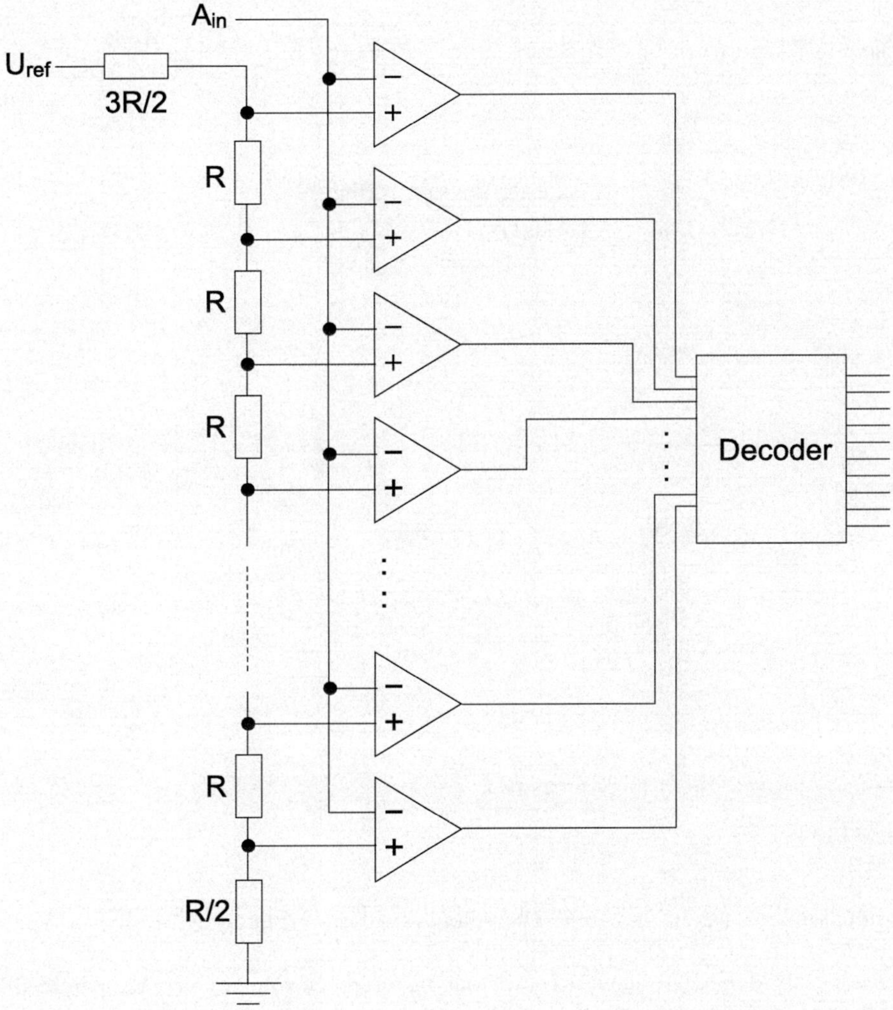

Fig. 11.11 Flash ADC

forward to the next stage. Hence, the maximum voltage fed forward to the next stage is

$$\frac{U_{\mathrm{ref}}}{2^3} \times 2^2 = \frac{U_{\mathrm{ref}}}{2} \tag{11.7}$$

(Which means that the reference voltage of the ADC in the next stage should be half of the reference voltage in the previous stage.) Our 3-bit ADC in Fig. 11.13 has a resolution of $5/2^3 = 0.625$ V. The ADC output is $\lfloor 2.65/0.625 \rfloor = 4 = 100_2$. This is also the input to the DAC and the DAC output will be $4 \times 0.625 = 2.50$ V and the

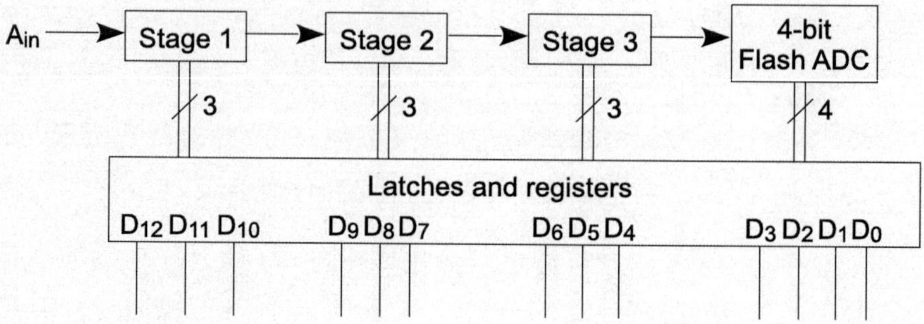

Fig. 11.12 A pipeline ADC

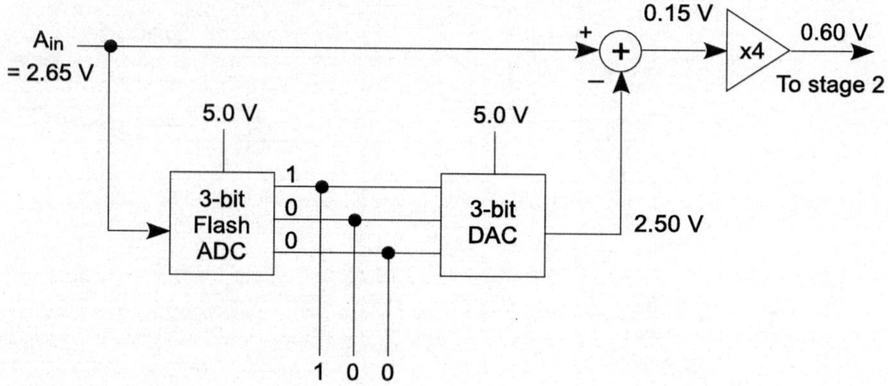

Fig. 11.13 Stage 1

residual is 2.65 – 2.50 = 0.15 V. The voltage fed forward to stage 2 is 4 × 0.15 = 0.60 V.

Stage 2 is identical to stage 1, except that the reference voltages have been divided by 2, see Fig. 11.14.

In stage 2, the ADC resolution is 2.5/8 = 0.3125 and the ADC output is $|0.6/0.3125| = 1 = 001_2$. The DAC output is 1 × 0.3125 = 0.3125 V, and the residual is 0.6000 – 0.3125 = 0.2875 V. The voltage fed forward to stage 3 is 4 × 0.2875 = 1.15 V. Stage 3 is identical to stage 2, except that the reference voltages have again been divided by 2, see Fig. 11.15.

The ADC resolution its now 1.25/8 = 0.15625 V and the ADC output is $|1.15/0.15625| = 7 = 111_2$. The DAC output is 7 × 0.15625 = 1.09375 V, and the residual is 1.15 – 1.09375 = 0.05625 V. The voltage fed forward to the final stage is 4 × 0.05625 = 0.225 V.

Stage 4 is a 4-bit flash ADC (*rounding*, not *truncating*) with a reference voltage that has again been divided by 2, see Fig. 11.16.

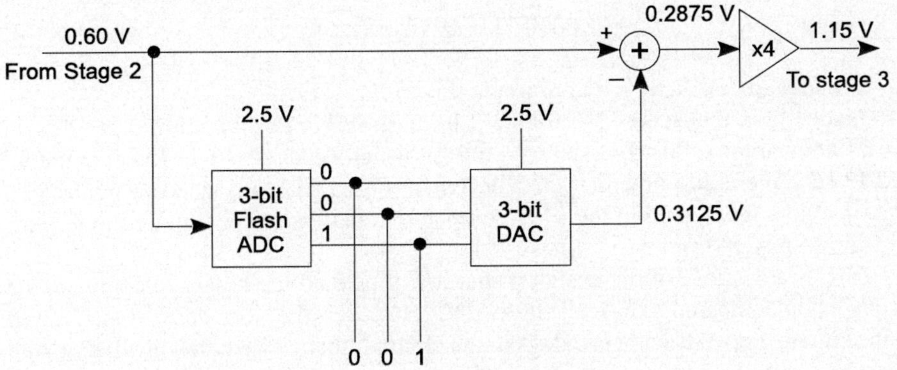

Fig. 11.14 Stage 2; reference voltages have been divided by 2

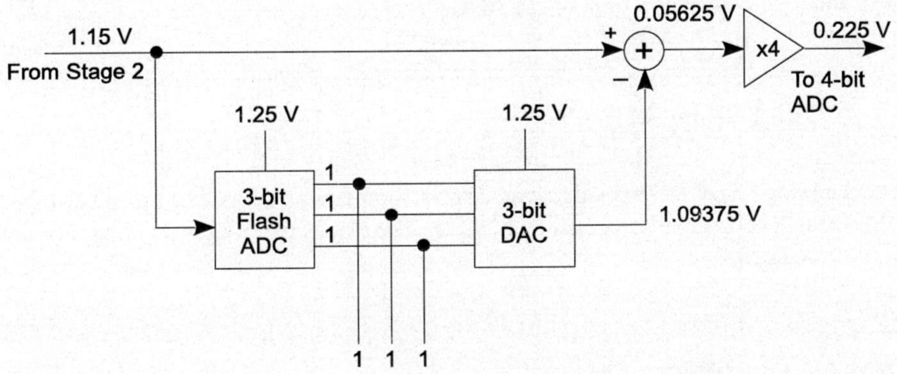

Fig. 11.15 Stage 3

Fig. 11.16 Stage 4

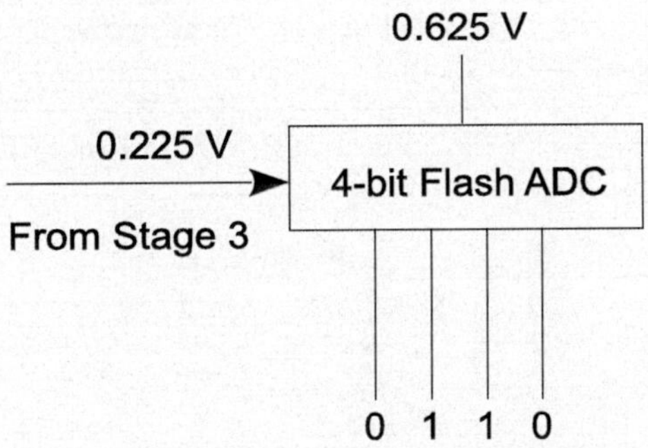

The resolution of this ADC is $0.625/2^4$ and with an input voltage of 0.225, the output is $round(0.225/(0.625/2^4)) = 6 = 0110_2$. Putting all the digital outputs in Figs. 11.13, 11.14, 11.15 and 11.16 together, we see that the 13-bit output is

$$1\,0000\,1111\,0110_2 = 4342 \tag{11.8}$$

Which agrees exactly with our prediction in Eq. (11.6).

Notice first, that stages 1–3 used a 3-bit flash ADC, i.e., a total of $3 \times (2^3 - 1) = 21$ comparators. Stage 4 used a 4-bit flash which needs $2^4 - 1 = 15$ comparators and hence the entire pipeline ADC design in Fig. 11.12 only needs a total of 36 comparators (compared to the 8191 comparators that would be required in a 'real' 13-bit flash ADC).

The disadvantage of the design is that it will take a little longer to complete the conversion compared to a 'real' flash ADC. However, the extra delay is very small since it only depends on gate delays in the circuits (there is no clock involved). And second, we can make up for this delay; if the result from the first stage is 'latched' in a register, then we can start the conversion of the next sample as soon as the first stage is completed (samples are 'pipelined'). Hence, the effective conversion time is ¼ of the total conversion time of the ADC.

11.8 Dual Slope ADCs

The dual slope ADC is the dominating ADC technique used in DMMs. The dual slope ADC is also called the 'integrating' ADC. The reason is that it uses an integrator as part of the design.

11.8.1 The Integrator

Figure 11.17 illustrates an integrator.

The op-amp in Fig. 11.17 has negative feedback, indicating that the inverting input is at virtual ground, and hence the current I must be U_{in}/R. By definition, current is 'charge variation per time unit':

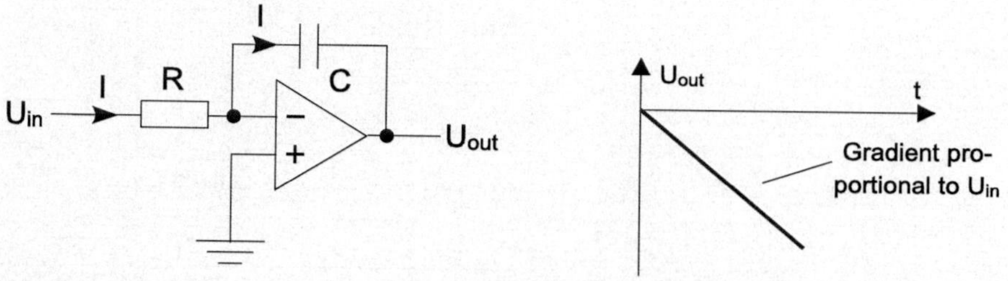

Fig. 11.17 Integrator

$$I = \frac{U_{in}}{R} = \frac{dQ}{dt} \Longrightarrow dQ = Idt = \frac{U_{in}}{R}dt \tag{11.9}$$

$$\Longrightarrow Q = \int Idt = \frac{1}{R}\int U_{in}dt \tag{11.10}$$

The voltage across the capacitor is

$$U_c = \frac{Q}{C} = \frac{1}{RC}\int U_{in}dt \tag{11.11}$$

The op-amp output $U_{out} = -U_c$:

$$U_{out} = -\frac{1}{RC}\int U_{in}dt \tag{11.12}$$

We conclude that the output of the circuit in Fig. 11.17 is the integral of the input signal and if the input signal is constant (positive), the output will be a linearly decreasing signal:

$$U_{out} = -\frac{1}{RC} \cdot U_{in} \cdot t \tag{11.13}$$

That is what we need to explain the dual slope ADC.

11.8.2 The Dual Slope Circuit

The dual slope ADC is illustrated in Fig. 11.18.

We can see an integrator (with output signal U_I), a comparator, and a binary counter. There are two input voltages to the integrator: The input sample A_{in} and the reference voltage U_{ref}. The switches S_1 and S_2 decide which voltage is fed to the integrator. For the design to work, $U_{ref} < 0$, and $|A_{in}| < |U_{ref}|$. A_{in} is assumed to be $\geq$0 V.

At time $t = 0$, the controller closes switch S_1 and opens switch S_2; A_{in} is the input signal to the integrator and hence the integrator output is $-A_{in} \cdot t/RC$ (see Eq. (11.13)); U_I decreases linearly with time at a rate that depends on the input sample A_{in}. Since the integrator output is < 0, the comparator output will be '0'. Also, at $t = 0$, the binary counter is reset, and the clock starts to increase the binary counter value (from 0). The integrator output decreases until the binary counter reaches its maximum count value and 'overflows' (after 2^n clock pulses). This happens after time t_1 and Fig. 11.19 illustrates the integrator's and the comparator's output and the counter value at $t = t_1$.

When the control logic senses the overflow signal from the binary counter, it opens switch S_1 and closes switch S_2; the (negative) reference voltage will now discharge

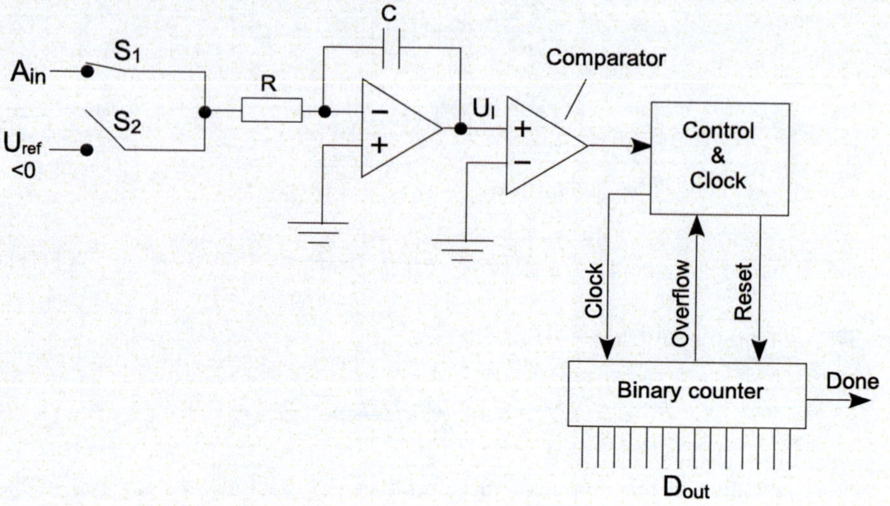

Fig. 11.18 The dual slope ADC

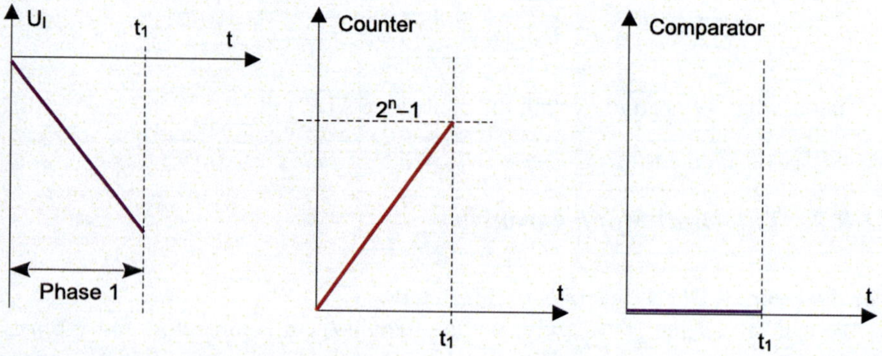

Fig. 11.19 Phase 1: charging

the capacitor, and the voltage on the integrator output will 'turn upwards'. It is still < 0 though, so the comparator output is still '0'. The binary counter just starts over from 0. Figure 11.20 illustrates the signals sometime after t_1.

When the integrator output crosses the 'zero line' (after time t_2) the comparator output goes high and when the control logic senses this, the clock to the binary counter is immediately stopped and the binary counter output stops on D_{out}, see Fig. 11.21.

We will now prove that this an ADC; by definition, it is an ADC if the relationship between the input sample A_{in} and the digital output D_{out} is

$$A_{in} = D_{out} \times \frac{U_{ref}}{2^n} \tag{11.14}$$

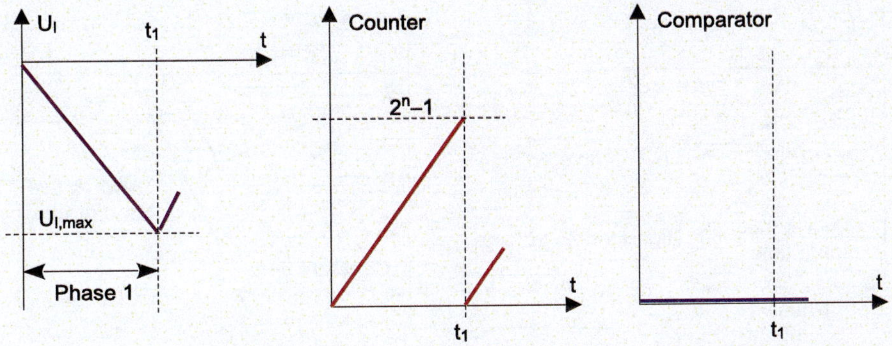

Fig. 11.20 Phase 2: discharging

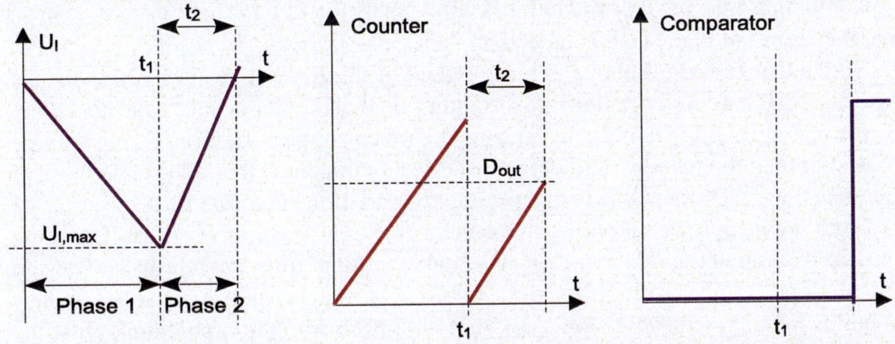

Fig. 11.21 Done!

After time t_1 the integrator output will stop on

$$U_{\mathrm{I,max}} = -\frac{1}{RC} \cdot A_{\mathrm{in}} t_1 \tag{11.15}$$

During time t_2, U_{ref} will discharge the same voltage:

$$U_{\mathrm{I,max}} = -\frac{1}{RC} \cdot U_{\mathrm{ref}} t_2 = -\frac{1}{RC} \cdot A_{\mathrm{in}} t_1 \Rightarrow \frac{t_2}{t_1} = \frac{A_{\mathrm{in}}}{U_{\mathrm{ref}}} \tag{11.16}$$

We also know that the clock frequency is constant during phase 1 and phase 2. The clock frequency f_{c} is

$$f_c = \frac{2^n}{t_1} = \frac{D_{\mathrm{out}}}{t_2} \Rightarrow \frac{t_2}{t_1} = \frac{D_{\mathrm{out}}}{2^n} \tag{11.17}$$

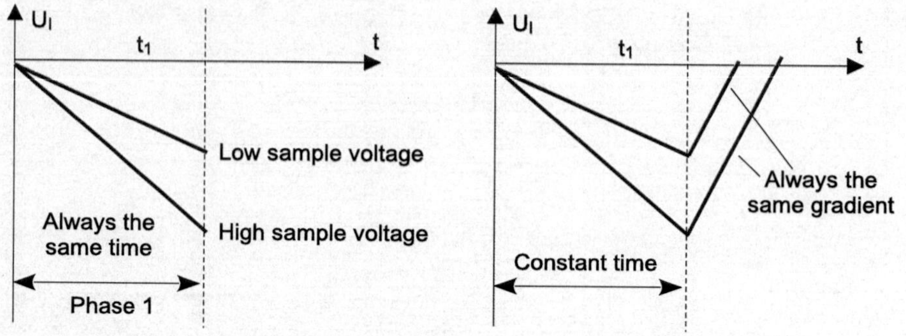

Fig. 11.22 Phase 1: charging *time* is constant. Phase 2: discharging *rate* is constant

Combining Eqs. (11.16) and (11.17) will give us Eq. (11.14) and we have proved that the circuit in Fig. 11.18 is indeed an ADC.

Notice that the result above is independent of the R and C values which means that the design is independent of variations of these components (due to aging, temperature, etc.). We could have designed a much simpler solution, a *single-slope* ADC, but then the result would depend on the component values. The cleverness of the dual slope ADC is that it is independent of variations in R and C.

From the design we can draw two conclusions: (1) It is not very fast. Charging and discharging of the integrator capacitor takes a 'long' time. (2) It is easy to design a dual slope ADC with high resolution; binary counters with many bits are easy to build. Hence, we have a slow ADC but with high-resolution potential. (Just the opposite of flash ADCs.)

Figure 11.22 summarizes the dual slope function. During phase 1 the time is always constant (the time it takes for the binary counter to overflow). During phase 2, the discharging rate is always constant (determined by the reference voltage).

Slow but 'high-resolution' is exactly what we need in a DMM and DMMs are always based on dual slope ADCs.

11.9 Level-Crossing ADCs

Traditional sampling is based on a constant sampling time $T_S = 1/f_S$; samples are taken at regular intervals (see Fig. 11.1). This is sometimes referred to as 'synchronous' sampling. Synchronous ADCs are characterized by a periodicity in time and equidistant quantization levels. Because of the fixed equidistant quantization levels, each sample will have an uncertainty, an *error*, see Fig. 11.23, and the size of the error is determined by the ADC's resolution:

$$\text{Max error} = \pm\frac{1}{2} \times \Delta U = \pm\frac{1}{2} \times \frac{U_{\text{ref}}}{2^N} \tag{11.18}$$

The ‘problem’ with synchronous ADCs is when ‘sparse’ or ‘burst-like’ signals are analyzed. Sparse or burst-like signals are signals with long periods of no, or very low, activity, resulting in a lot of identical samples (that really doesn’t carry any net information). Sparse and burst-like signals are, for example, radar and speech signals and electro cardiograms, see Figs. 11.24 and 11.25. For these situations, *asynchronous* sampling is sometimes used. Asynchronous sampling is also sometimes called *level-crossing* sampling.

The level-crossing ADC (LC-ADC) was first suggested by Inose et al. in 1966 [1] and in an LC-ADC the sampling is triggered by the signal activity rather than

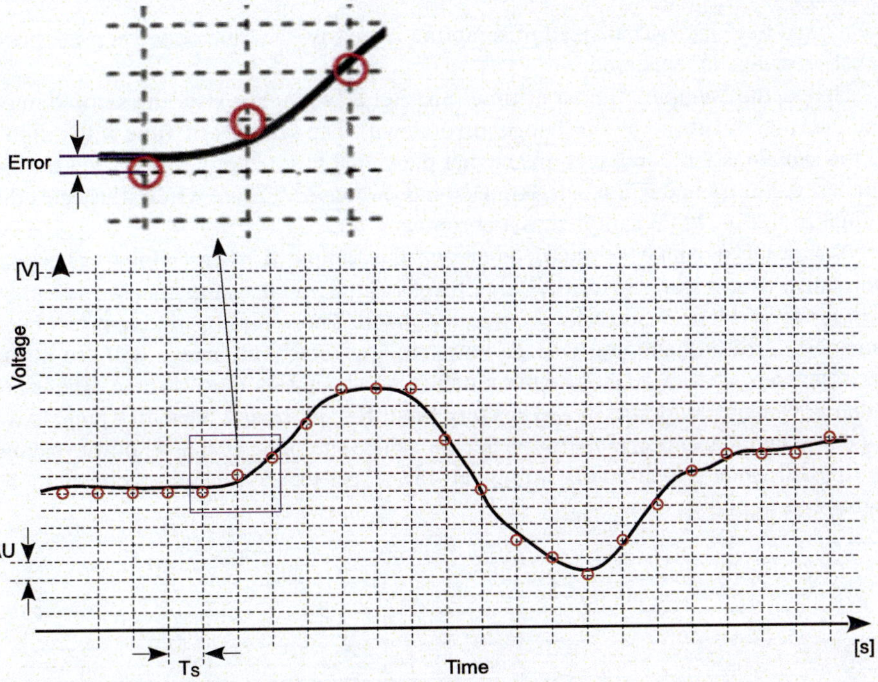

Fig. 11.23 Synchronous sampling: uncertainty in voltage

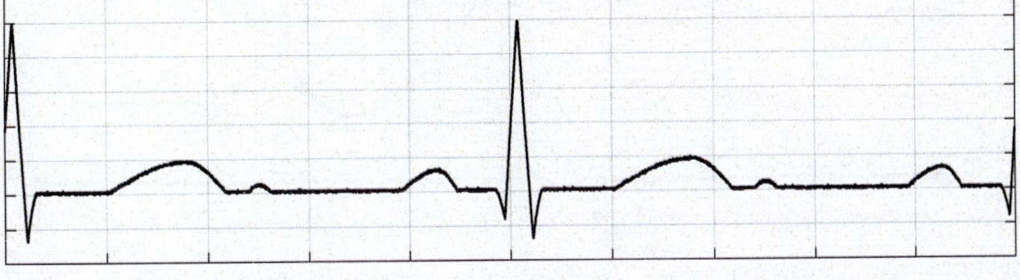

Fig. 11.24 An ECG signal is ‘sparse’

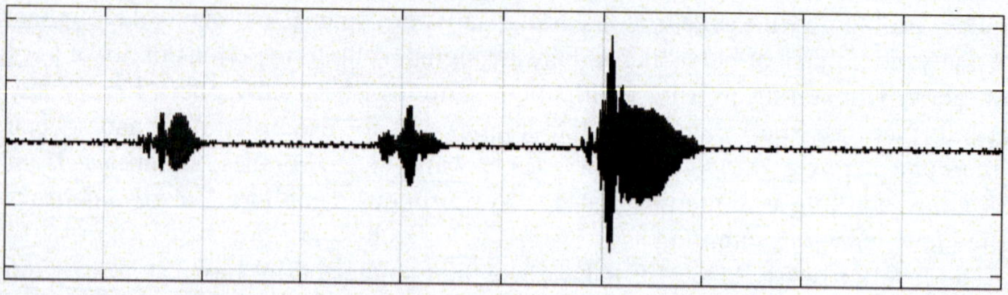

Fig. 11.25 A speech signal is 'burst-like'

by a fixed time interval. Instead of sampling regularly, the time between predefined level-crossings is registered.

Hence, the 'sample' is now a 'time' and not a 'voltage'. Also, the sample must indicate the 'direction' of the change (up or down). The sample is a 'time with a sign'. If the sample is $-0.346\ \mu s$, it means that the signal has *decreased* one level during the last $0.346\ \mu s$ and if it is $+0.346\ \mu s$ it has *increased*. Figure 11.26 illustrates the signal from Fig. 11.23 sampled asynchronously.

Notice the irregular sampling intervals; the sample density follows the signal derivative. But notice most of all, that there is now *no uncertainty in the samples' voltage levels*! The voltage levels are predefined. The sampling problem has been transferred from quantizing *voltage* to quantizing *time*. And, as we will see in the next chapter, we can quantize time much more accurately than voltage. There are other advantages with the LC-ADCs too. First, for sparse and burst-like signals we take less samples and save memory. Second, since sampling is sparse, power-saving is implied since the sampling computer can revert to an 'idle', low-power mode between samples.

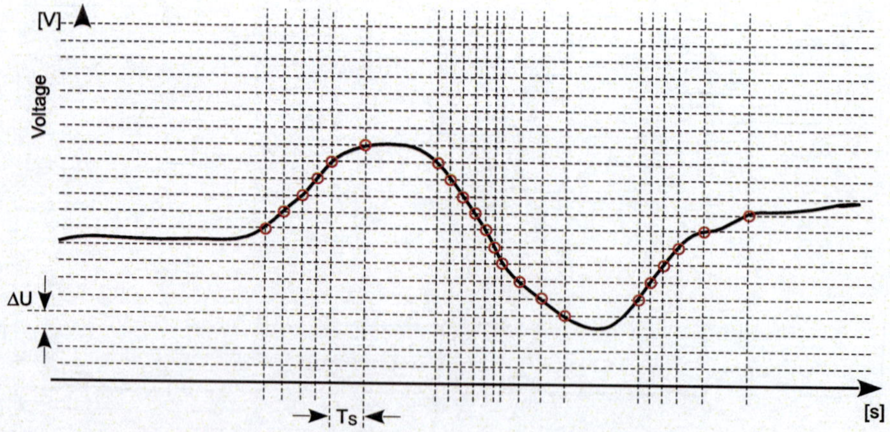

Fig. 11.26 Asynchronous sampling ('level-crossing'): uncertainty in time

11.10 Equivalent Number of Bits

The uncertainty in a synchronous sampling ADC is $\pm\Delta U/2$ (see Eq. (11.18)). We consider this uncertainty as the 'noise' in the ADC output with a 'uniform' probability distribution (see Sect. 13.5). The 'power' of this noise (produced in a 1-Ω resistor) is the variance of the corresponding stochastic variable, and it is (see Eq. (13.27))

$$P_{\text{noise}} = \sigma^2 = \frac{1}{3} \times \left(\frac{\Delta U}{2}\right)^2 = \frac{\Delta U^2}{12} \tag{11.19}$$

The ADC range is $2^n \times \Delta U$; a maximum range sinusoidal would have an amplitude of $2^n \times \Delta U/2$. The power of that signal (produced in a 1-Ω resistor) would be

$$P_{\text{signal}} = \text{RMS}^2 = \left(\frac{A}{\sqrt{2}}\right)^2 = \frac{1}{2}\left(\frac{1}{2}2^n \Delta U\right)^2 = \frac{1}{8}2^{2n}\Delta U^2 \tag{11.20}$$

The signal-to-noise ratio of the ADC output (of the signal in Fig. 11.7) is

$$\begin{aligned} \text{SNR} &= 10 \cdot \log \frac{P_{\text{signal}}}{P_{\text{noise}}} = 10 \cdot \log \frac{2^{2n}\Delta U^2/8}{\Delta U^2/12} = 10 \cdot \log\left(\frac{3}{2} \cdot 2^{2n}\right) = \\ &= 10 \cdot \log \frac{3}{2} + 20n \cdot \log 2 = 1.76 + 6.02n \end{aligned} \tag{11.21}$$

This is the 'raw' signal-to-noise ratio in the ADC's output. Circumstances can make this larger or smaller. External noise can make it smaller and signal processing tricks can make it larger. It is common to express the signal-to-noise ratio as the 'equivalent number of bits' (ENOB); just solve for n in Eq. (11.21):

$$\text{ENOB} = \frac{\text{SNR} - 1.76}{6.02} \tag{11.22}$$

11.11 Oversampling

11.11.1 As a Means to Reduce Noise

According to the sampling theorem, a signal with bandwidth f_b must be sampled at a rate higher than $2f_b$. Most systems sample faster than that and we define the *oversampling rate* (OSR) as how many times faster than the Nyquist limit $2f_b$ we sample:

$$\mathrm{OSR} = \frac{f_\mathrm{S}}{2 f_\mathrm{b}} \tag{11.23}$$

The obvious reason for oversampling is to get a better resolution in the time–space representation of the signal, but there are other advantages too. In Eq. (11.19), we found that the noise in the ADC output, due to the quantization, is $\Delta U^2/12$. Due to aliasing, this noise ends up in the frequency band $0 \ldots f_\mathrm{S}/2$. Hence, the *spectral density* of the quantization noise is

$$p = \frac{\Delta U^2/12}{f_S/2} = \frac{\Delta U^2}{12} \cdot \frac{2}{f_S} \left[\frac{\mathrm{W}}{\mathrm{Hz}}\right] \tag{11.24}$$

The noise power in the frequency range of interest is the noise within the signal's bandwidth f_b:

$$p_b = p \cdot f_b = \frac{\Delta U^2}{12} \cdot \frac{2 f_b}{f_S} = \frac{\Delta U^2}{12} \cdot \frac{1}{\mathrm{OSR}} \tag{11.25}$$

From Eq. (11.25) we can see that the noise power within the signal's bandwidth decreases with the oversampling rate; oversampling improves the signal-to-noise ratio. This is illustrated in Fig. 11.27. In the first case, the signal is sampled at the Nyquist limit (f_S is just above $2f_\mathrm{b}$) and in the second case, the signal is oversampled by a factor of K.

According to Hauser [2], the SNR expression in Eq. (11.21), for a full-scale sinusoidal signal oversampled by a factor of K, is improved to

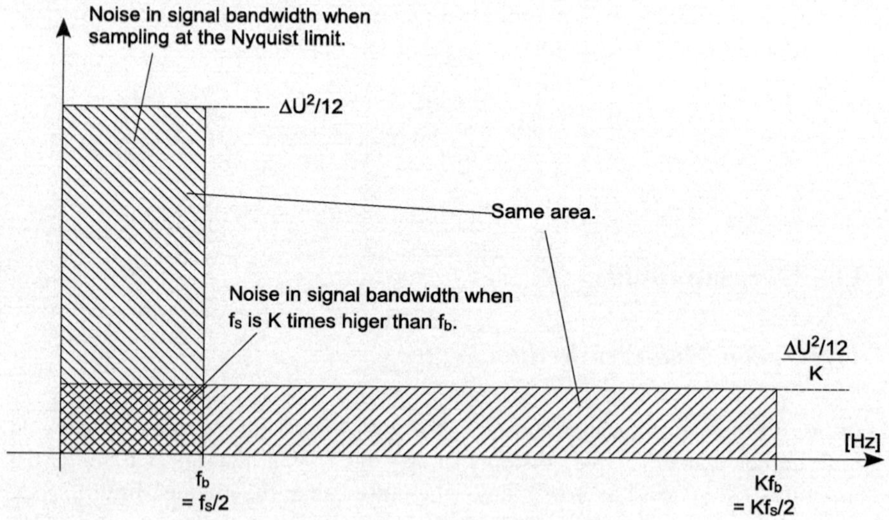

Fig. 11.27 The distribution of quantization noise

$$\text{SNR} = 6.02n + 1.76 + 10\log K \tag{11.26}$$

Or, if we express the oversampling rate in octaves L ($K = 2^L$), then

$$\text{SNR} = 6.02(n + 0.5L) + 1.76\,\text{dB} \tag{11.27}$$

Oversampling an n-bit ADC by a factor of $K = 2^L$ generates the same quantization noise as an $(n + 0.5L)$-bit ADC sampled at the Nyquist rate! For example, an 8-bit ADC oversampling by a factor of 64 ($= 2^6$) will only produce quantization noise (in the signal bandwidth) corresponding to that of an 11-bit ADC sampling at the Nyquist limit.

Another advantage of oversampling is that the anti-aliasing filter requirements are relaxed. If we sample at the Nyquist limit, we need a very 'steep' (high order) anti-aliasing filter, but if we oversample, we might even get away with a first-order filter, see Figs. 11.28 and 11.29.

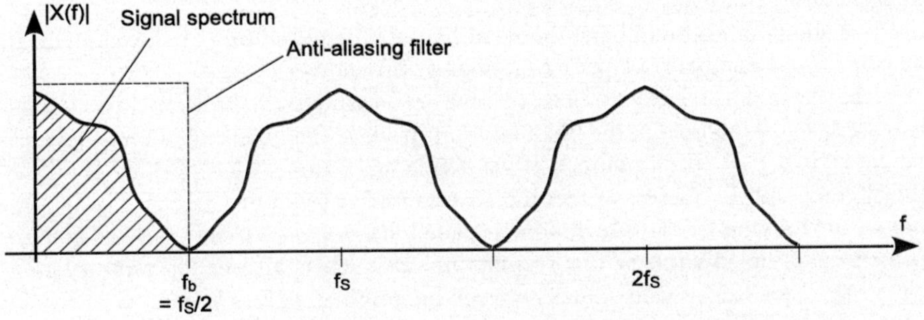

Fig. 11.28 Sampling at the Nyquist limit requires a high-order anti-aliasing filter

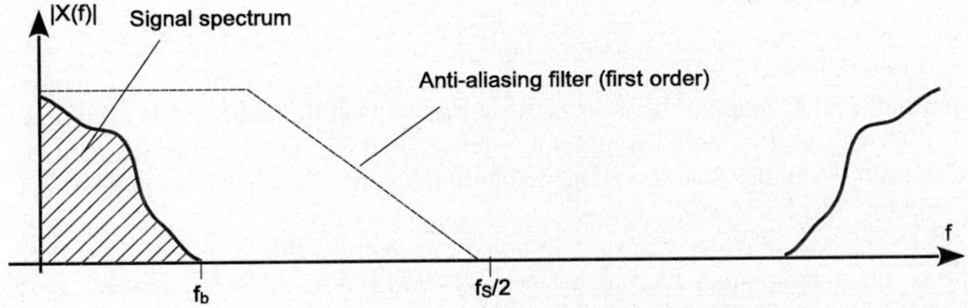

Fig. 11.29 Oversampling relaxes the anti-aliasing filter requirements

11.11.2 As a Means to Improve Resolution

A 10-bit ADC with a reference voltage of +5 V has a resolution of $5/2^{10} = 4.88$ mV. However, suppose our application needs a resolution of 1.0 mV. That would correspond to

$$\frac{5}{2^n} \leq 1.0 \cdot 10^{-3} \Rightarrow n > \log_2 5000 = 12.2877\ldots \text{bits} \tag{11.28}$$

I.e., we would need a 13-bit ADC. According to our results in the previous section, that corresponds to an oversampling rate of

$$10 + 0.5L = 13 \Rightarrow L = 6 \Rightarrow K = 2^6 = 64 \tag{11.29}$$

Hence, if we oversample by a factor of 64, we get the same quantization noise as a 13-bit ADC. However, the ADC itself still produces a 10-bit integer. The 13-bit number must be derived in 'software'.

If we add two n-bit numbers, the result is (in general) an $(n + 1)$-bit number. If we add m n-bit numbers, we get a $(n + \log_2 m)$-bit number.

If we oversample by a factor of 64 (2^6) we get 64 samples in the same time interval as if we take one sample at the Nyquist limit. If we add all these 2^6 n-bit samples,[2] we get a $10 + 6 = 16$-bit number. If we divide this number by 2^8 (which is only a binary right-shift by three), we get the 13-bit number (with the 1.0 mV resolution) that we are looking for. This technique is called *filtering and decimation* and is used to increase the resolution of low-resolution ADCs. (It is also called *interpolation* sometimes because we read values between the original ADC's levels.)

11.12 Dithering

The 'filtering and decimation' trick in Sect. 11.11.2 only works if there is enough noise in the ADC output. If there is no noise, we will get the same output each time, and filtering and decimation would not improve anything. In those cases, where the noise level is smaller than the ADC resolution, we must *add* noise to improve the resolution. This is called *dithering*.

It seems contradictory that *adding* noise can *improve* things, but this has been known for a long time. During the second World War, airplane bombers were controlled by mechanical 'computers', and engineers were puzzled by the fact that the airplanes seemed to perform much better when flying than what was indicated by simulations in the laboratory. They concluded that this was attributed to the vibrations induced (by the engines) into the mechanical control system; the vibrations helped

---

[2] We *average* them, but that includes adding them.

overcome the friction in the mechanical parts. This is the first known example of how noise injection can improve performance.

We can use the same trick to improve the performance of an ADC; the quantization levels correspond to the mechanical system's 'friction'. By inserting noise, we can help the ADC to overcome this 'friction' and read between the quantization levels.

An 8-bit ADC with a reference voltage of +5 V, has a resolution of 5/256 = 19.53 mV. An input signal of 1.12 V will generate *round*(1.12/0.01953) = 57 at the ADC output. The quantization error is $|1.12 - 57 \cdot 0.01953| = 6.72$ mV (an error of 0.6%). If there is no noise in the signal, there is nothing we can do about this; averaging samples won't help, since we would get the same ADC output each time (= 57), see Fig. 11.30.

However, by adding (Gaussian) noise to the signal, we force a variation of the output sample values. By averaging theses samples, we will be able to read 'in between' the quantization levels (interpolating) and get a more accurate estimate of A_{in}, see Figs. 11.31 and 11.32.

For example, adding Gaussian noise with a standard deviation equal to $2 \times \Delta U$ to the 1.12-V signal in Fig. 11.30, and taking 64 samples, see Fig. 11.32, generated the sample distribution illustrated by the histogram in Fig. 11.33. Averaging the samples enables us to interpolate between the quantization levels. The sample average is 57.17 which corresponds to a residual error of only 0.3%.

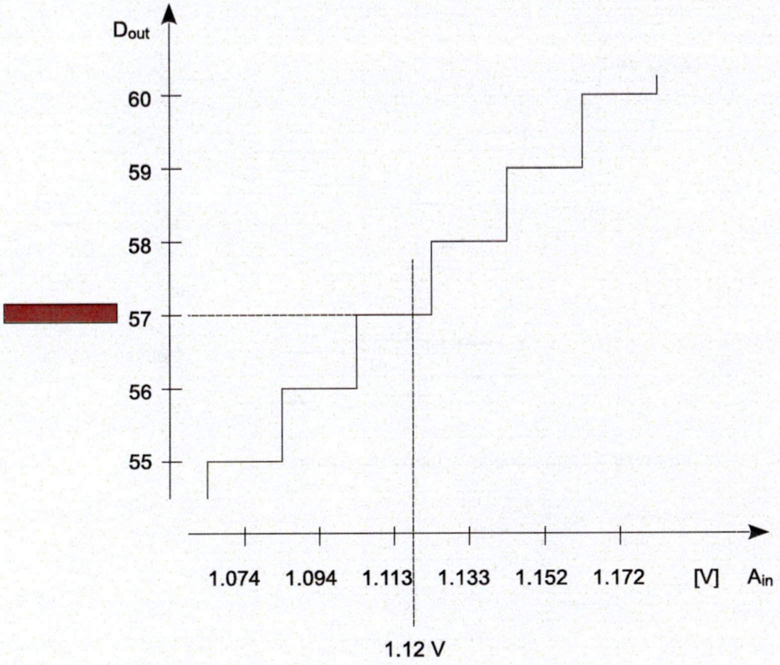

Fig. 11.30 If there is no noise, we get the same sample value every time

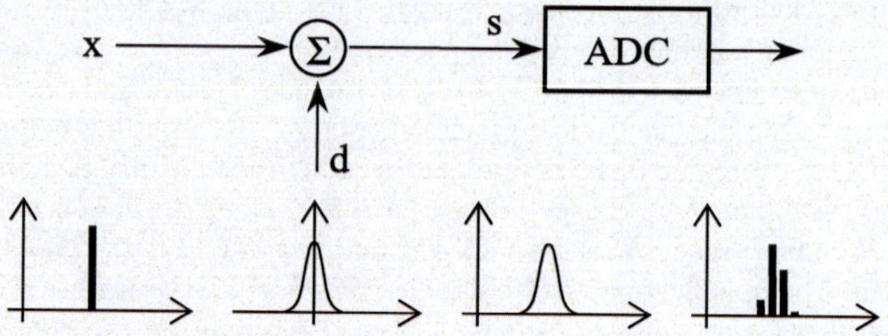

Fig. 11.31 Adding noise (Gaussian)

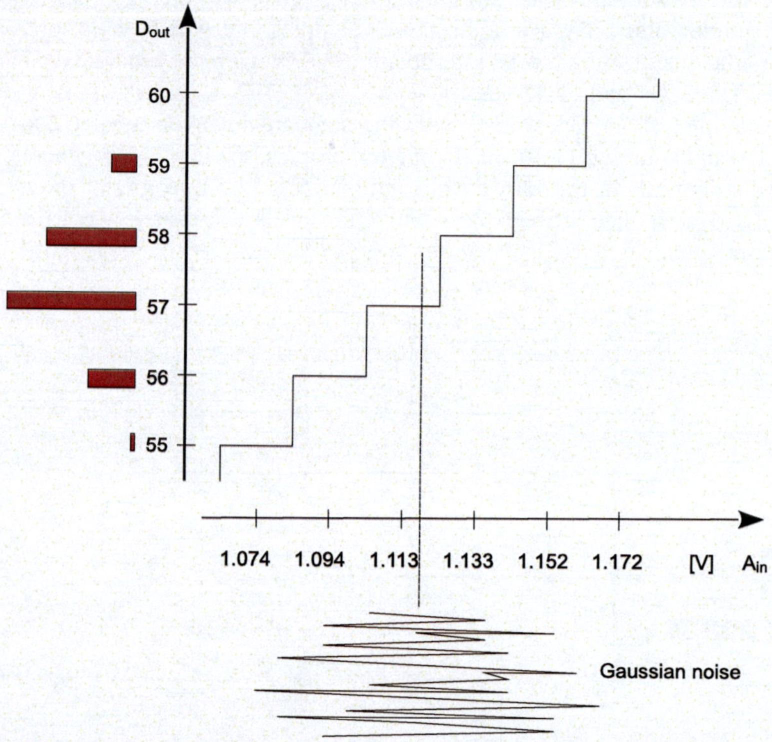

Fig. 11.32 Adding noise will produce a variation in output samples

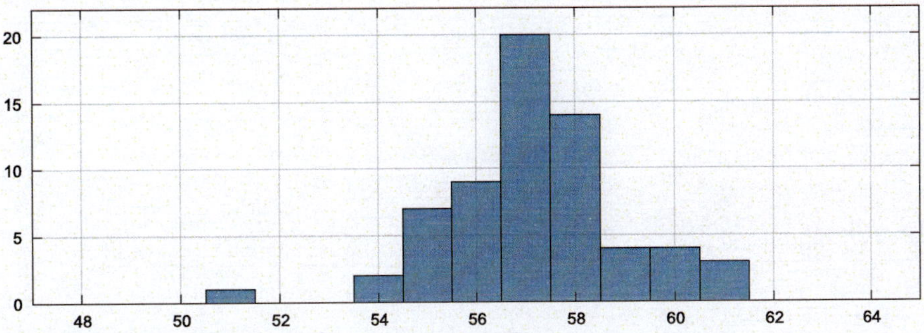

Fig. 11.33 Sample distribution after dithering

11.13 Sigma-Delta ADCs

11.13.1 Background

The Sigma-Delta ADC ($\Sigma\Delta$ ADC) technology emerged from the Δ-modulation technique developed for data transmission. In a Δ-modulator, the actual sample value is not transmitted, but rather the *difference* between successive samples. If the transmitted value is positive, a positive signal change has occurred since the last sample and vice versa; a Δ-modulator tracks the signal's derivative. (And hence, the receiving end must integrate the signal to restore it.) In fact, true Δ-modulators only transmit 1-bit values; 1s or 0s indicating a positive or negative signal change, see Fig. 11.34.

This is implemented by feeding back the quantized signal via an integrator, see Fig. 11.35.

The 'recovered' signal in Fig. 11.34 corresponds to the demodulator signal before the lowpass filter; the lowpass filter 'smooths' the edges of the recovered signal to recover the original signal exactly.

The Δ-modulator was developed for the purpose of improving signal transmission and had nothing to do with ADCs. The modulator was later improved by Inose et al. [3] (still for the purpose of transmitting signals) and they also coined the term '$\Sigma\Delta$ modulation'. Here is how they reasoned:

First, integration is a linear operation; $\int a \cdot x(t)dt = a \int x(t)dt$, which indicates that it doesn't matter if we integrate first and do 'something else' second, or vice versa, see Fig. 11.36. In Fig. 11.35, that means that the integrator at the receiving end can be moved to the front without changing the result, see Fig. 11.37.

Another consequence of the linearity of integration is that it doesn't matter if we integrate first and add second: $\int x(t)dt + \int y(t)dt = \int (x(t) + y(t))dt$. Hence, in Fig. 11.37, we can replace the two integrators with just one if we move it inside the loop, after the summing circuit, see Fig. 11.38.

Inose et al. named it '$\Sigma\Delta$ modulator' because 'sigma' refers to the summing component and 'delta' refers to the differentiator. However, it wasn't until 1969

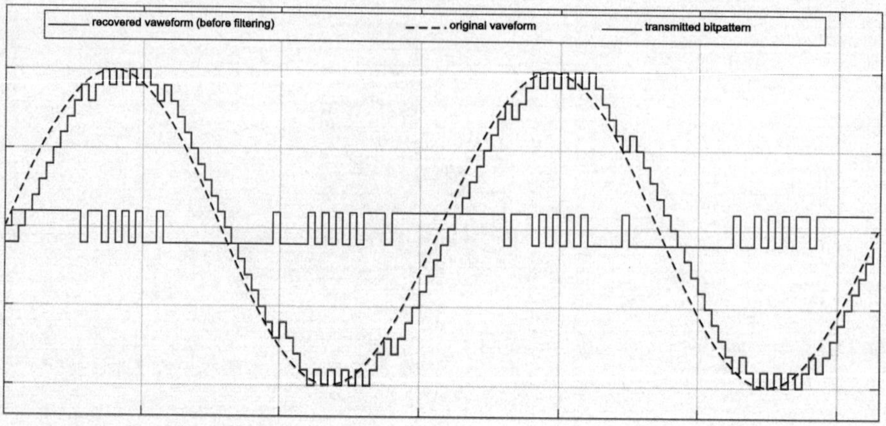

Fig. 11.34 Δ-modulated signal

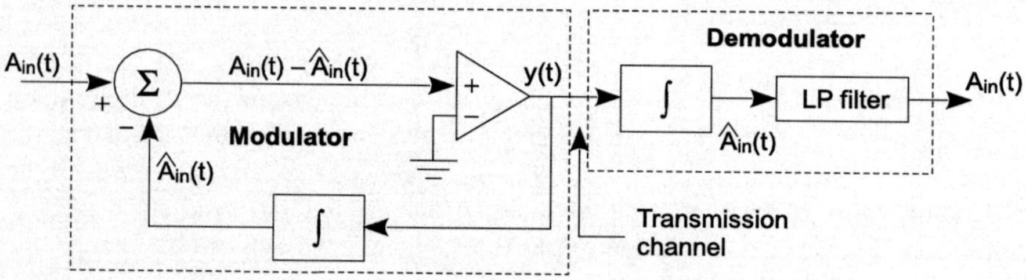

Fig. 11.35 Δ-modulator

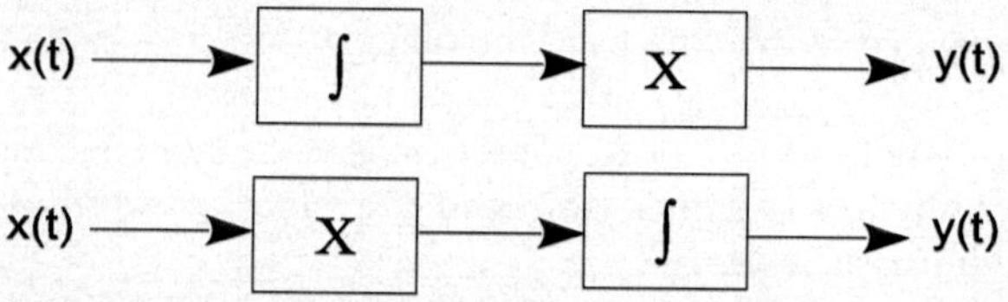

Fig. 11.36 Integration is a linear process; the order doesn't matter

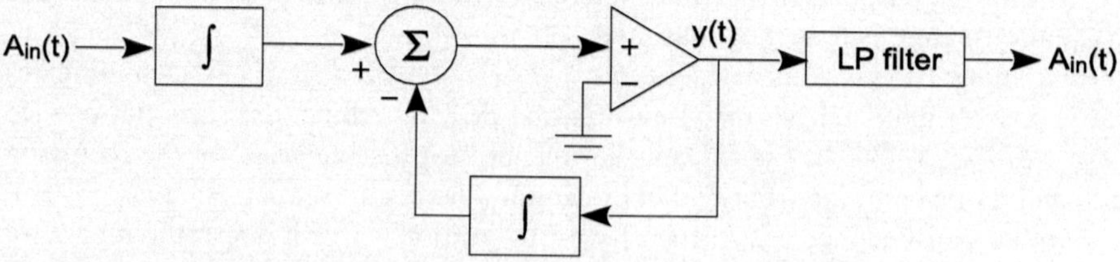

Fig. 11.37 Modified Δ-modulator

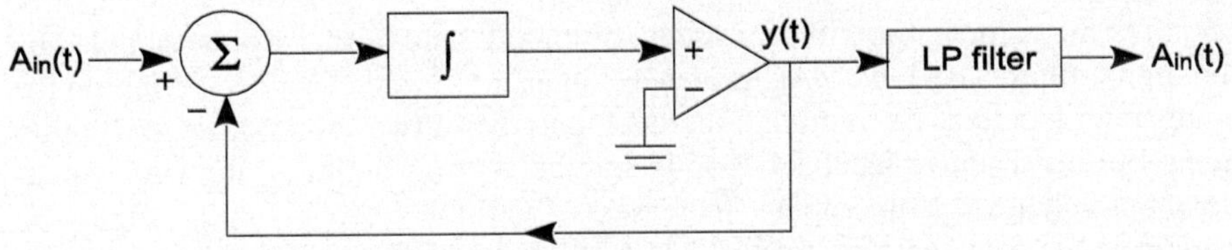

Fig. 11.38 Inose's (et al.) modified Δ-modulator [3]

that it was suggested that this modulator could be used explicitly for the purpose of analog-to-digital conversions [4].

It has been debated in the community whether the correct name is 'ΣΔ ADC' or 'ΔΣ ADC'. In 1990, the editor of Analog Dialog addressed this problem in an editor's note and concluded that the correct name is indeed 'ΣΔ ADC' and urged application engineers in the community to promote that name. That name is now well-established in the community. That was the historical background of the ΣΔ ADC. Let's look at why it has become such a popular ADC technology.

11.13.2 Theory

The ΣΔ ADC differs significantly from the other ADC techniques; it produces (primarily) a bitstream of 1s and 0s and the *density* of 1s in the bitstream is proportional to the sample voltage. A post-processing, digital averaging filter will convert this bitstream to a conventional integer, but the primary output of a ΣΔ ADC is a bitstream whose density of 1s represents the sample voltage. The advantage of ΣΔ ADCs is that it offers extreme resolution (number of bits), but at the expense of speed. (It is a competitor to dual slope ADCs). But, as we will see later, it has another unique property; it will *shape* the quantization noise, boosting the SNR beyond the theoretical limit suggested in Eq. (11.21) and even beyond Eq. (11.27). Here we will only describe the first-order ΣΔ ADC and this description is mostly based on works by Kester [5] and Hauser [2]. Figure 11.39 illustrates a first-order ΣΔ ADC.

If we disregard the digital filter for now, the first-order ΣΔ ADC has four components: An analog summing circuit, an integrator, a comparator, and a 1-bit DAC. The

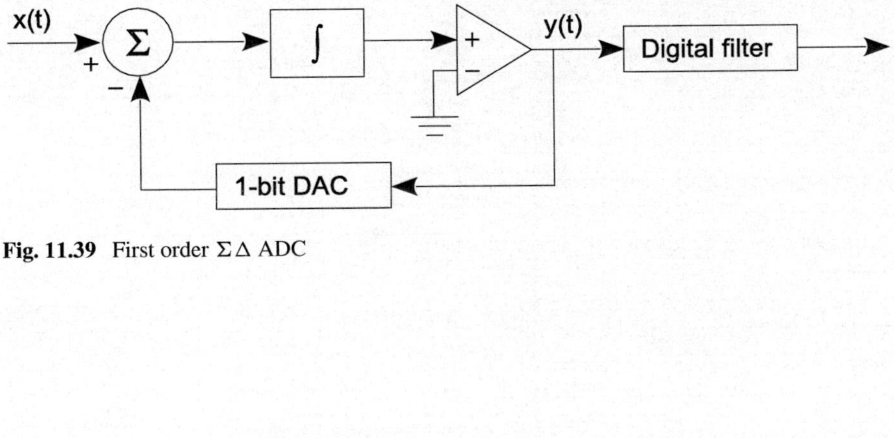

Fig. 11.39 First order ΣΔ ADC

output of the comparator will be a stream of logic 1s and 0s that will be sampled by the digital filter. The 1-bit DAC produces either $+U_{\text{ref}}$ or $-U_{\text{ref}}$ depending on the comparator's output. Even though the DAC only has a two-level output, the consequence of the negative feedback loop is that the *average* output of the DAC equals the input voltage x. If the input x increases, so will the average output of the 1-bit DAC which means that the stream of 1s from the comparator output increases; the density of 1s at the comparator output will be proportional to x.

However, the cleverest feature of the $\Sigma\Delta$ design is its inherent ability to 'shape' the quantization noise, pushing it towards higher frequencies. To understand the noise-shaping, we re-draw Fig. 11.39; we idealize the 1-bit DAC and replace it with a transfer function = 1. The transfer function of the integrator is $1/s$ and the comparator is in fact a 1-bit ADC, which means that the comparator output is a (rough) digitized estimate of the input. Since we have previously modeled an ADC output as the input plus some quantization noise (see Fig. 11.6), we can re-write Fig. 11.39 as in Fig. 11.40.

Figure 11.40 looks simple enough, but this is a *very* clever circuit! To see that we need to figure out what it does both to the signal x and to the noise q. We start with the signal; to see what happens to the signal, we temporarily cancel the noise; $q(t) = 0$. That gives us Fig. 11.41.

The system's transfer function is easily calculated:

$$Y(s) = \frac{1}{s}(X(s) - Y(s)) \Rightarrow sY(s) = X(s) - Y(s)$$

$$(1+s)Y(s) = X(s) \Rightarrow H(s) = \frac{Y(s)}{X(s)} = \frac{1}{s+1} \tag{11.30}$$

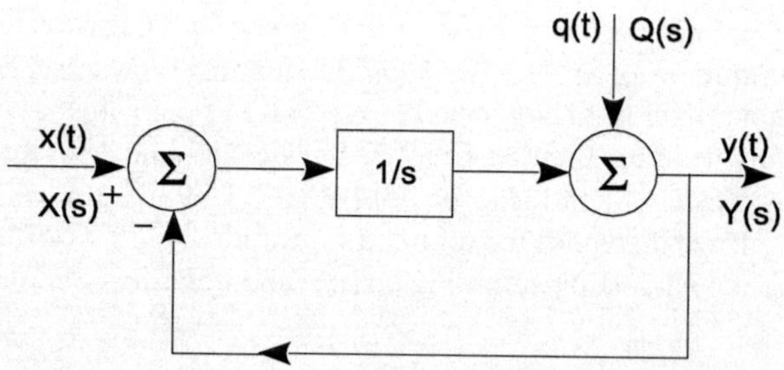

Fig. 11.40 A simplified model of a first-order $\Sigma\Delta$ ADC

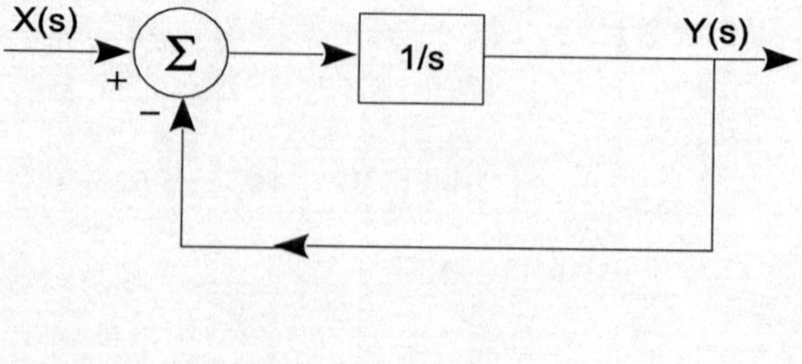

Fig. 11.41 To see what happens to the signal, we cancel the noise

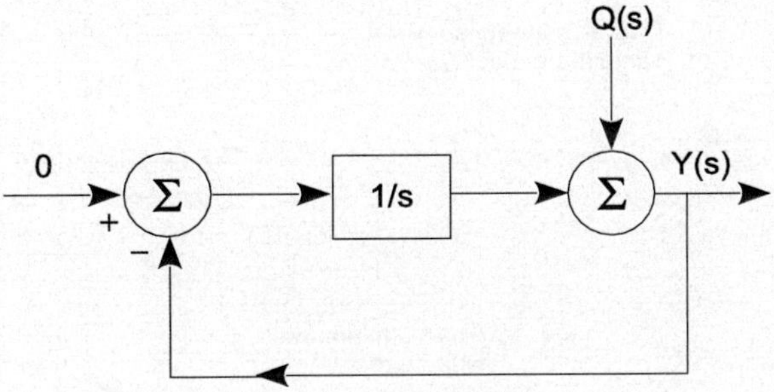

Fig. 11.42 To see what happens to the noise, we cancel the signal

From Eq. (11.30), we can see that as far as the input signal is concerned, the system is a first-order lowpass filter.

To see how the system treats the noise, we cancel the input signal, see Fig. 11.42. The transfer function is now

$$Y(s) = Q(s) - \frac{1}{s}Y(s) \Rightarrow sY(s) = sQ(s) - Y(s)$$
$$sQ(s) = Y(s) \cdot (s+1) \Rightarrow H(s) = \frac{Y(s)}{Q(s)} = \frac{s}{s+1} \tag{11.31}$$

From Eq. (11.31), we can see that the system *highpass filters* the quantization noise! The system lowpass filters the signal and highpass filters the noise. That means that an even larger part of the quantization noise will be attenuated by a lowpass filter and that even less noise ends up within the signal's bandwidth. This is illustrated in Fig. 11.43.

This means that the SNR (due to quantization noise) is increased beyond even Eq. (11.27). According to Hauser (1991) the SNR of a first-order $\Sigma\Delta$ is

$$\text{SNR} = 6.02(n + 1.5L) - 3.41 \text{ dB} \tag{11.32}$$

11.14 Extreme Sampling Rates

There are a few ways to 'boost' the sampling rate to 'extreme' rates without using a flash ADC. The pipeline ADC is one solution, but it still contains too many analog components to be a favorite among ASIC designers of oscilloscope chips. Techniques have been developed that can take 'traditional' ADCs (such as SARs) beyond the sampling rate of what is indicated by the limit of individual ADCs. We will here describe the *interleaved SARs* technique and the *equivalent-time* sampling techniques.

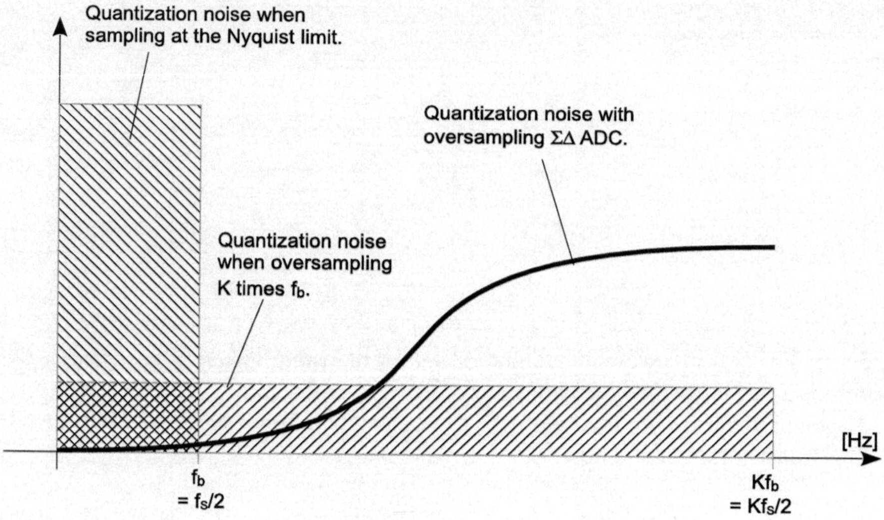

Fig. 11.43 The noise is pushed to high frequencies by the ΣΔ ADC

11.14.1 Interleaved SARs

In an interleaved SAR (sometimes called *time-interleaved*), m SAR ADCs are synchronized to achieve an effective sampling rate that is m times higher than the sampling rate of each individual ADC. Figure 11.44 illustrates an interleaved SAR with $m = 4$.

Each n-bit SAR ADC has a sample and hold circuit, and the 'sample' input signals are phase-shifted 90° relative to each other (360°/m in the general case). The output of each ADC is connected to a multiplexer that interleaves the ADCs' outputs to the common D_{out}. Figure 11.45 illustrates the timing diagram of the clock signals and D_{out}.

In Fig. 11.45 we can see that a new D_{out} is produced at a speed four times higher than the output of each individual ADC. This is the ADC technique used in advanced high-speed digital oscilloscopes (such as Tektronix's 'Mixed Signal Oscilloscopes').

11.14.2 Equivalent-Time Sampling

Another technique used to boost the sampling rate in oscilloscopes is *equivalent-time sampling* (as opposed to *real-time sampling*). With equivalent-time sampling, extreme sampling rates can be achieved, but the restriction is that it only works for periodic signals (which is what we have in most cases anyway). Figure 11.46 illustrates the idea behind equivalent-time sampling.

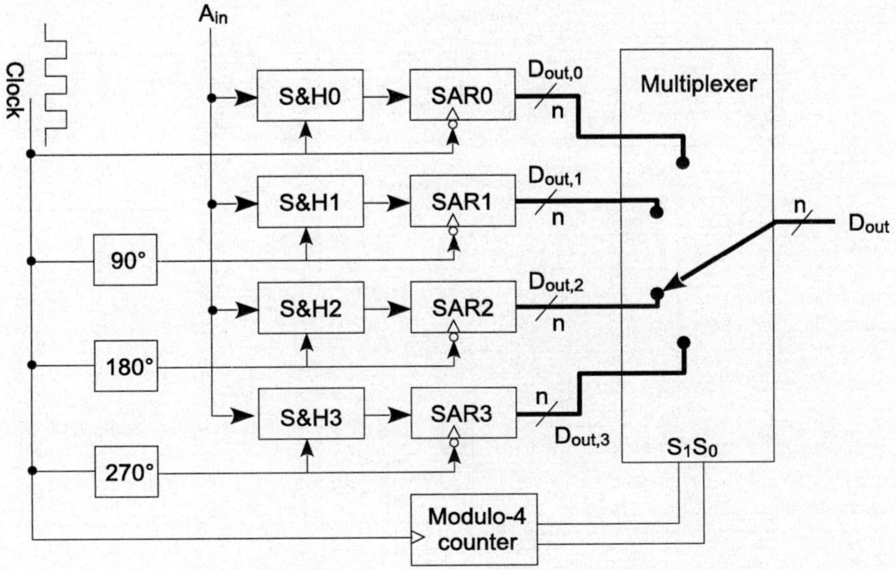

Fig. 11.44 Time-interleaved SARs

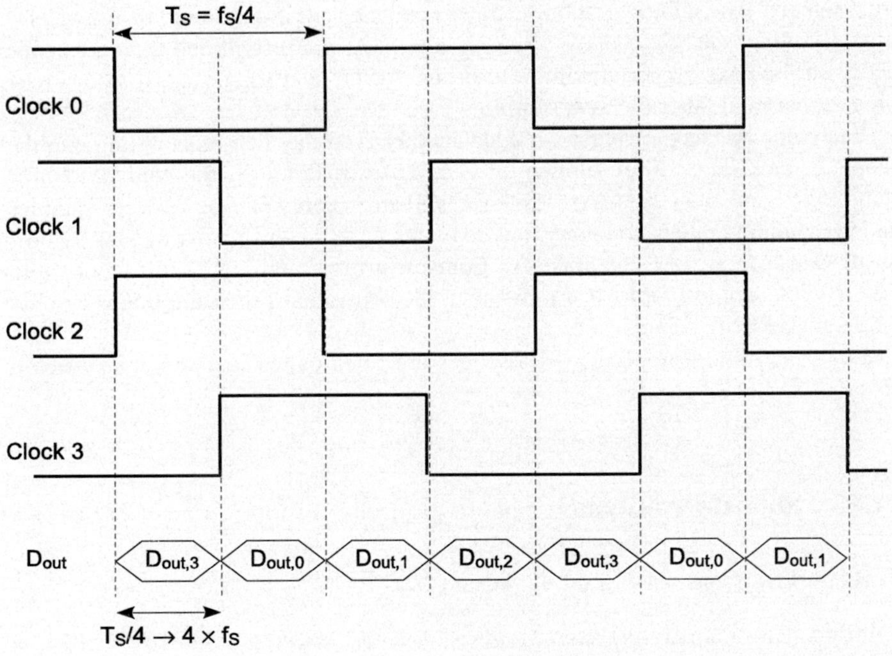

Fig. 11.45 The throughput speed is four times higher

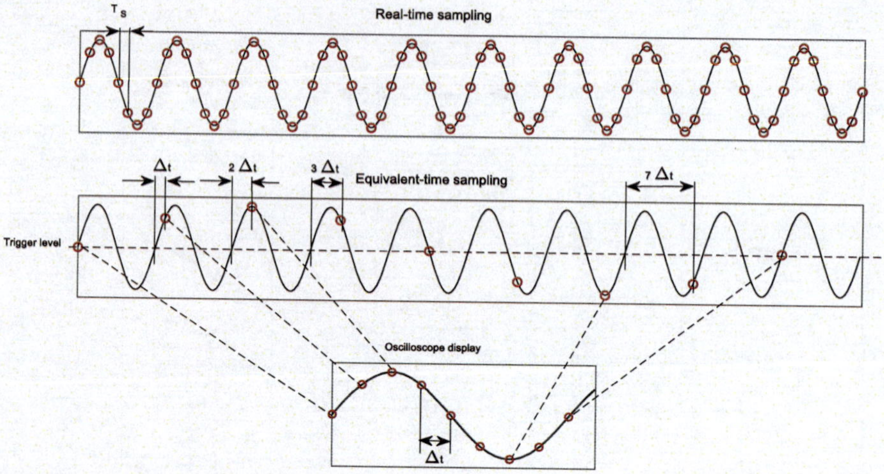

Fig. 11.46 Equivalent-time sampling

The top graph illustrates traditional, real-time sampling where the time between samples is T_S and that is exactly what you see on the oscilloscope screen (you see the 'real signal'). The middle graph illustrates equivalent-time sampling. In equivalent-time sampling, the scope takes a sample, the ADC converts it and then the scope waits for the next trigger condition to occur. At the next triggering, it waits some time Δt before it takes the next sample.

Each time the scope triggers, it adds another Δt delay before sampling. On the scope display, samples are plotted only Δt apart, making Δt the *equivalent-time* sampling period and $f_S = 1/\Delta t$. The ADC will have plenty of time between samples and the sampling rate is no longer limited by the ADC's conversion time, but by how small (and accurate) we can make Δt. Equivalent-time sampling oscilloscopes with $f_S > 10$ GS/s are available. But remember, the equivalent-time sampling trick only works for periodic signals.

Equivalent-time sampling oscilloscopes are often called just *sampling oscilloscopes*.

11.15 Solved Problems

Problem 11.1 Consider Fig. 11.47. What is D_{out}?

Solution The temperature sensor resistance is $100\left(1 + 3.85 \cdot 10^{-3} \cdot 50\right) = 119.25\ \Omega$. $U_- = 1 \cdot 1000/(1000 + 119.25 + 1000) = 0.47187$ volt. $U_+ = 1 \cdot 1119.25/2119.25 = 0.52813 = 0.47187 + 0.05626$ (CM + NM) volt.

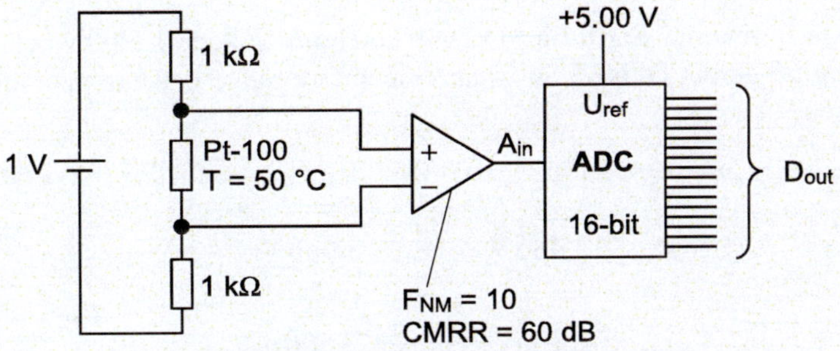

Fig. 11.47 Measuring temperature with an ADC

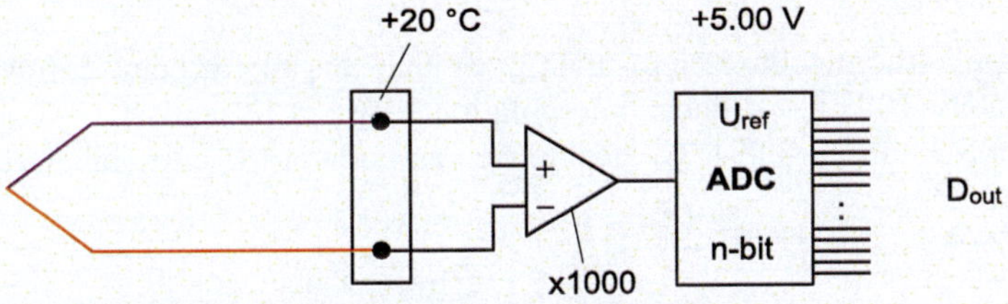

Fig. 11.48 Thermocouple reading with an ADC

The common mode suppression of the instrumentation amplifier is $F_{\mathrm{CM}} = 10/10^{60/20} = 0.01$. The ADC input signal is $A_{\mathrm{in}} = 10 \cdot 0.05626 + 0.01 \cdot 0.47187 = 0.56732$, and hence, the ADC output is $D_{\mathrm{out}} = \mathrm{round}\big(0.56732/(5/2^{16})\big) = \underline{\underline{7436}}$.

Problem 11.2 In Fig. 11.48, an ADC is used to read the temperature from a type T thermocouple.

a What is the temperature range of this system? **b** How many bits resolution does the ADC need, to resolve temperature changes of the order of 0.1 °C?

Solution a The maximum input to the ADC is +5.00 V which means that the maximum thermocouple emf is 5.00 mV. First, we do a cold junction compensation; a google search for a 'thermocouple type T chart' gives first that 20 °C corresponds to an emf of 0.790 mV. Adding 5 mV to that gives us a maximum emf of 5.790 mV, corresponding to 131 °C (see thermocouple table). Hence, the temperature range (for the hot junction) is 20–131 °C.

b The thermocouple chart has a 1 °C resolution only, but the smallest emf change between two adjacent temperatures is 39 μV. This would indicate a 3.9 μV change in the thermo emf for a 0.1 °C change in temperature (assuming a linear interpolation). This would be amplified to 3.9 mV at the ADC input; the ADC needs to be able to resolve input changes of 3.9 mV:

$$\frac{5\,\mathrm{V}}{2^n} \leq 3.9\,\mathrm{mV} \Rightarrow \underline{\underline{n = 11}}$$

Problem 11.3 If we have an 8-bit ADC with reference voltage +5.00 V in Fig. 11.48, and the digital output is 0xB3, in what temperature range is the temperature at the hot junction?

Solution $\Delta U = 5/256 = 0.01953$ V. 0xB3 = 179 $\Rightarrow A_{\text{in}} = (179 \pm 0.5) \times 0.01953$ V.

$$3.486\,\text{V} \leq A_{\text{in}} \leq 3.506\,\text{V}$$

$$3.486\,\text{mV} \leq \text{emf} \leq 3.506\,\text{mV}$$

$$\text{CJC (add 0.790 mV): } 4.276\,\text{mV} \leq \text{emf} \leq 4.296\,\text{mV}$$

The type T thermocouple emf table gives that this thermo emf corresponds to approximately 100 °C. A linear interpolation between 99 and 101 °C gives that $T = 21.505 \times \text{emf} + 7.991\,°\text{C}$. Hence, we can convert the emf values above to a temperature range:

$$\underline{\underline{99.9\,°\text{C} \leq T_{\text{hot}} \leq 100.4\,°\text{C}}}$$

Problem 11.4 In Fig. 11.49, an ADC is used to measure time. **a** Prove that Δt is proportional to D_{out}. **b** If $D_{\text{out}} = $ 0x3AC, what is Δt?

Solution First, 0x3AC = 940. The charge on the capacitor is $Q = I \cdot \Delta t = U_C \cdot C = A_{\text{in}} \cdot C$. Hence,

$$\Delta t = \frac{C}{I} \times A_{\text{in}} = \underline{\underline{\frac{C}{I} \times D_{\text{out}} \times \frac{U_{ref}}{2^n}}} = \frac{100 \cdot 10^{-9}}{1 \cdot 10^{-3}} \cdot 940 \cdot \frac{5}{2^{12}} = \underline{\underline{114.7\,\mu\text{s}}}$$

Problem 11.5 The signal $x(t) = 2(\cos(100t - 0.875) + 1)$ is sampled at a rate of 150 S/s.

a Is the sampling theorem met?

b If the sampling starts at $t = 0$, what are the exact values of the first three samples (*after* the sample and hold unit, but *before* the ADC)?

c What are the values of these three samples *after* the ADC, if we use a 12-bit ADC with a reference voltage of +5 V?

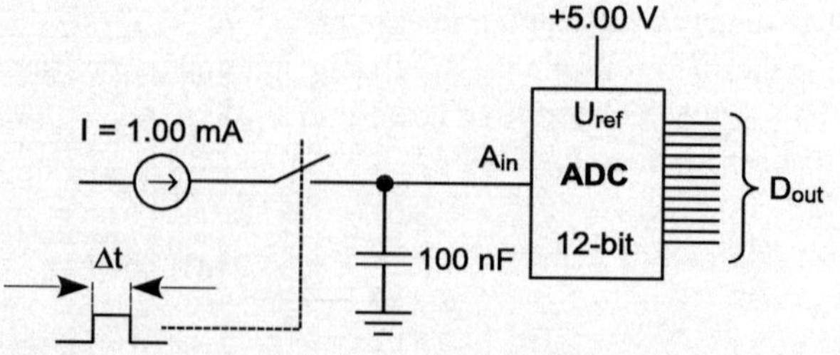

Fig. 11.49 Measuring time with an ADC

d If the ADC produces the integer 2075, in what range is then the input sample voltage?

Solution a $f = 100/2\pi = 15.9$ Hz $< 150/2 = 75$ Hz. Yes, the sampling complies with the sampling theorem.

b $x(0) = 2(\cos(0 - 0.875) + 1) = \underline{3.282\text{ V}}$ $x(1) = 2(\cos(100/150 - 0.875) + 1) = \underline{3.957\text{ V}}$. $x(2) = 2(\cos(100 \cdot 2/150 - 0.875) + 1) = \underline{3.794\text{ V}}$.

c $\Delta U = 5/2^{12} = 1.22$ mV $\Rightarrow$ $3.282/0.00122 = \underline{2689}$ $3.957/0.00122 = \underline{3242}$ $3.794/0.00122 = \underline{3108}$.

d $x = \Delta U \cdot (D_{out} \pm 0.5) = 0.00122 \cdot (2075 \pm 0.5) \Rightarrow \underline{2.5323 < x < 2.5336\text{ V}}$.

Problem 11.6 A scientist uses instruments in a 'NIM' rack and one of the instruments is an ADC module. To use it, he/she needs to know what the sampling rate of this ADC is. However, the module is old, and nobody knows where the manual is. The scientist decides to figure out the sampling rate by performing a simple experiment.

He/she first connects a 10 kHz sinusoidal signal from a waveform generator to the ADC input. The computer, reading the samples from the NIM module, correctly recreates the 10 kHz sine signal.

Next, the scientist slowly increases the frequency of the sine signal and when the sine wave frequency reaches 190 kS/s, the computer again displays a 10 kHz sine signal.

a What is the sampling rate of the ADC module?

b If the frequency is increased even more, what would be the *next* sine frequency that the computer would interpret as 10 kHz?

Solution a Since the frequency was increased *slowly*, 190 kHz, was the *first* frequency that produced 10 kHz as an aliasing signal. For a signal with frequency f_0, sampled at f_S, aliasing frequencies appear at $nf_S \pm f_0$. Hence, the *first* one is $1 \cdot f_S - 10 = 190 \Rightarrow \underline{f_S = 200\text{ kS/s}}$.

b The second one is $1 \cdot f_S + 10 = 200 + 10 = \underline{210\text{ kHz}}$.

Problem 11.7a In an experiment, a 10-bit AD converter was used to measure a voltage X, see Fig. 11.50. In this measurement, D_{out} was consistently = 852 (dec). Find X and the quantization uncertainty: In what range is X?

Solution $X = D_{out} \cdot \frac{U_{ref}}{2^n} \pm \frac{1}{2} \cdot \frac{U_{ref}}{2^n}$ $X = 852 \cdot \frac{5}{2^{10}} \pm \frac{1}{2} \cdot \frac{5}{2^{10}}$ $\underline{X = 4.1602 \pm 0.0024\text{ V}}$

Problem 11.7b It was decided that the uncertainty was too high. Unfortunately, they did not have an ADC with higher resolution. To overcome the problem, they added some noise (zero mean, Gaussian) to X, see Fig. 11.51. By doing that, D_{out} varied from sample to sample. Instead of taking just one sample, they took 16 samples and added them. The 16 samples were: 855, 850, 847, 855, 851, 851, 850, 850, 852, 854, 854, 853, 851, 854, 851, 851.

Make a new estimate of X from these samples.

Solution Adding 16 10-bit numbers gives us a $10 + \log_2 16 = 14$-bit number. The 14-bit sample is (= the sum of the samples) $855 + 850 + 847 + \ldots 851 = 13{,}629$. Hence:

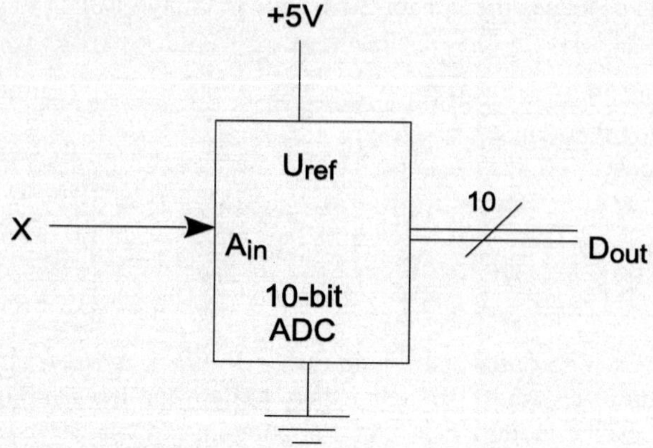

Fig. 11.50 AD conversion of sample

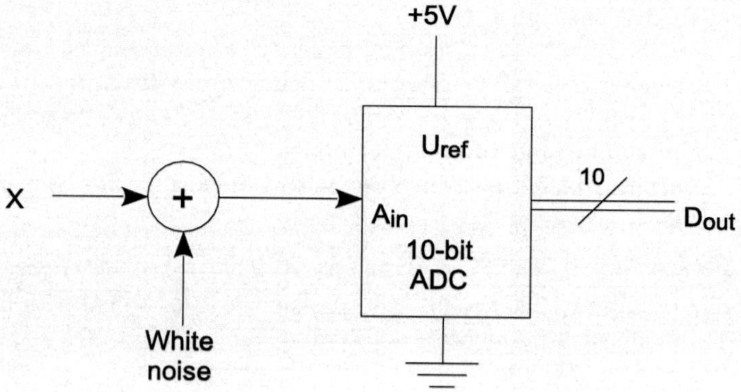

Fig. 11.51 AD conversion with dithering

$$X = 13629 \cdot \frac{5}{2^{14}} \pm \frac{1}{2} \cdot \frac{5}{2^{14}} = \underline{4.15924 \pm 0.00015\ \mathrm{V}}$$

References

1. Inose, H., T. Aoki, and K. Watanabe. 1966. Asynchronous delta-modulation system. *Electronics Letters* 3 (2): 95–96.
2. Hauser, M.W. 1991. Principles of oversampling A/D conversion. *Journal of the Audio Engineering Society* 39 (1/2): 3–26.
3. Inose, H., Y. Yasuda, and J. Murakami. 1962. A telemetering system by code modulation-δ-σmodulation. *IRE Transactions on Space Electronics and Telemetry* 3: 204–209.
4. Goodman, D.J. 1969. The application of Delta modulation to analog-to-PCM encoding. *Bell System Technical Journal* 48 (2): 321–343.
5. Kester, W. 2008. ADC architectures III: Sigma-delta ADC basics. *Analog Devices*, MT022.

Chapter 12
Time-to-Digital Converters

Abstract Accurate time measurements are critical in many disciplines such as laser, atomic, and nuclear physics and we need a way to convert time to a digital number with extreme resolution and extreme accuracy. That is what a TDC does (Time-to-Digital converter). This chapter presents the two dominating TDC techniques: the Vernier principle and time stretching.

12.1 Introduction

A lot of experiments in a physics laboratory depend on accurate time measurements (decay times, time-of-flight mass spectroscopy, reaction times, etc.). And, just as in the 'voltage problem' addressed in the previous chapter, we prefer to measure time in digital units. Since time is by nature an analog quantity, we will need a 'Time-to-Digital Converter', a *TDC*. There are basically two situations we encounter; either we need to measure the time between a start and a stop signal or, we need to measure the duration of a pulse. Some TDCs are designed for the first case and others are designed for the latter case, but that is not important; one case can easily be translated to the other case with some simple digital electronics. For example, a 'start' and a 'stop' signal pair can be translated to a pulse with just an xor gate, see Fig. 12.1.

Similarly, a pulse signal can easily be translated to a start/stop pair with two D flip-flops, see Fig. 12.2.

(Notice that in both Figs. 12.1 and 12.2, both signals suffer from the same gate delay.) Even if we usually prefer *digital* TDCs, some are analog, or at least 'semi-analog'. A common semi-analog TDC technique is to integrate the pulse and then use an ADC to digitize the time, see Fig. 12.3. (See Fig. 11.7 for an integrator circuit.)

The main disadvantage of the analog TDC is that it contains analog circuitry that doesn't scale very well; all-digital circuitry scales much better in VLSI designs than mixed-signal circuitry. For that reason, TDCs are almost always *counter* based. That means that they in principle, simply count the pulses from an oscillator during the start and stop interval (or during the pulse duration).

L. Bengtsson, *Electrical Measurement Techniques*,
https://doi.org/10.1007/978-981-99-8187-8_12

Fig. 12.1 From start/stop to pulse (an SR latch would work too)

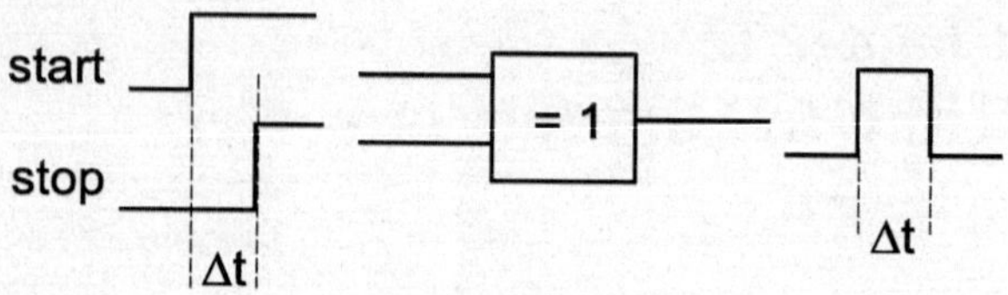

Fig. 12.2 Pulse to start/stop pair

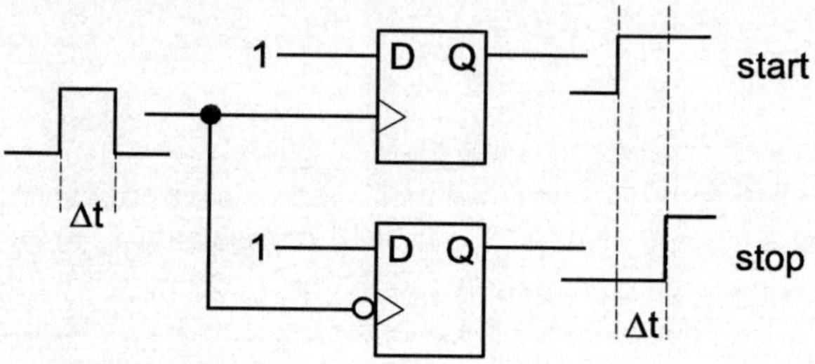

Fig. 12.3 Analog TDC

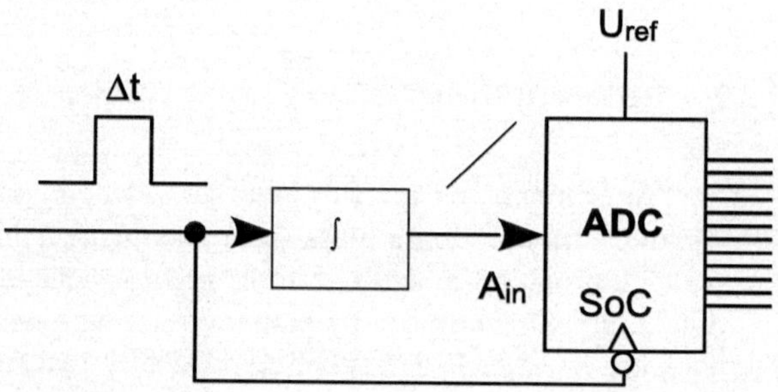

Figure 12.4 looks simple enough but notice that the start and stop signals are asynchronous (to each other and to the reference clock). Figure 12.5 illustrates a typical timing diagram.

If we assume that we have a positive edge-triggered counter, we can see from Fig. 12.5 that the start-stop interval T is

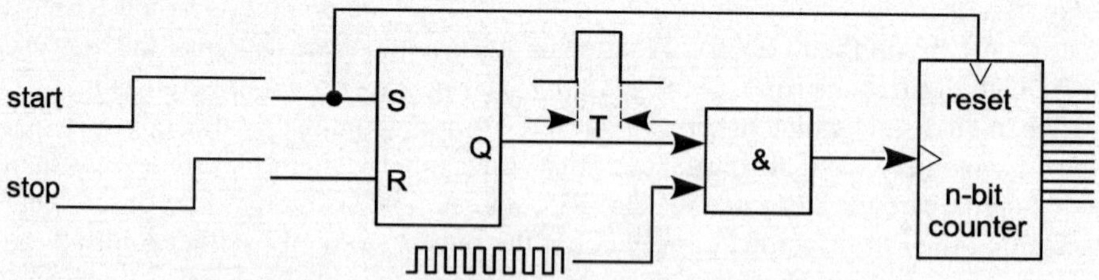

Fig. 12.4 Digital TDC

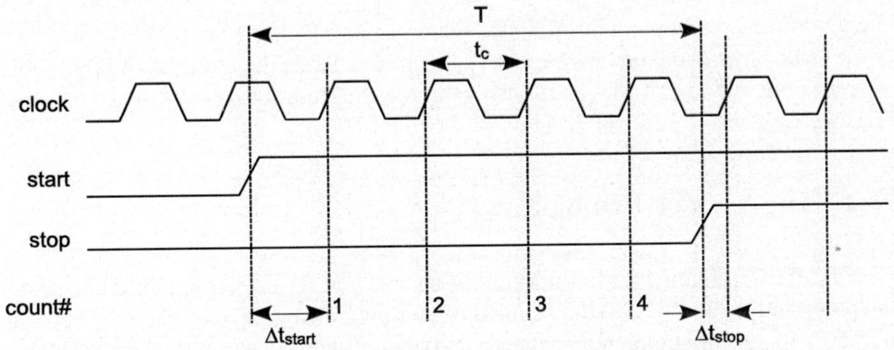

Fig. 12.5 Timing diagram of asynchronous TDC

$$T = N \cdot t_c + \Delta t_{start} - \Delta t_{stop} \tag{12.1}$$

(Where $N = 4$ in Fig. 12.5). Since both Δt_{start} and $\Delta t_{stop} \in [0, t_c]$, the inherent quantization error of counting TDCs is $\pm\, t_c$. Hence, the quantization error scales with the reference clock's period. However, increasing the clock frequency raises two other issues; first, the power consumption increases. Second, there is a limit to the maximum oscillator frequency that can be implemented in CMOS technology. Other tricks must be implemented to overcome the inherent quantization uncertainty. For example, if the signal is repetitive, we could average several measurements; if we average n measurements the uncertainty will decrease to $t_c/\sqrt{n}$ (see Eq. (13.18)). For non-repetitive transients, more advanced tricks are needed.

Most of the tricks improve the resolution by interpolating between the clock cycle pulses (without increasing the clock frequency). These techniques are referred to as *Vernier* time measurements. The name refers to the inventor of the metric caliper, Pierre Vernier (1580–1637), which can indeed perform a mechanical interpolation between the millimeter markers of a ruler; it has a 'Nonie' scale (Fig. 12.6).

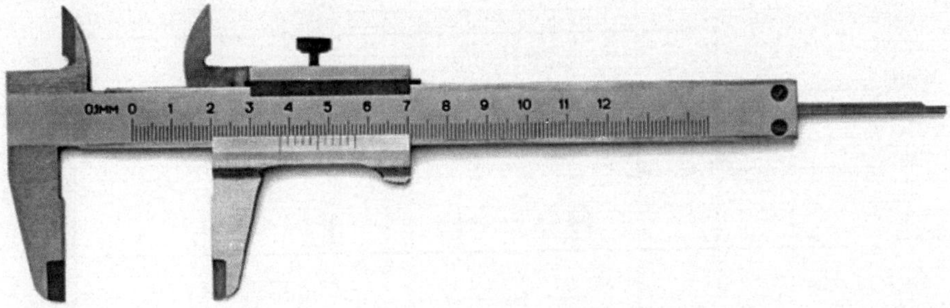

Fig. 12.6 A Vernier caliper

So how can we implement a Nonie scale in our TDC in Fig. 12.4? In fact, the Nonie scale is what characterizes, or even *defines* a TDC; the counter gives you the 'coarse' time only, but a TDC will also give you the 'fine structure' (interpolation).

12.2 The Vernier Principle

Even if all interpolation techniques could be referred to as 'Vernier' methods, the one presented here is the one that is most often implied when we refer to the 'Vernier method'. In this method, interpolation between clock cycles is implemented by engaging two oscillators with slightly different frequencies, $f_1 = 1/T_1$ and $f_2 = 1/T_2$, respectively, where $f_2 > f_1$ (*slightly* larger). There are two different implementations of the Vernier TDC principle: With or without a reference clock.

12.2.1 Vernier TDC with no Reference Clock

Figure 12.7 illustrates the timing diagram of the first method (not using a reference clock) [1].

Oscillator 1, with frequency $f_1 < f_2$, starts on the positive edge of the start signal. The second oscillator with frequency, f_2, is triggered by the positive edge of the stop signal. Since $f_2 > f_1$, the pulses from the f_2 oscillator will eventually 'catch up' with the pulses from the f_1 oscillator. When this happens, both oscillators are stopped ('moment of coincidence') and at this point both oscillators have generated the same number of pulses, i.e., $N_1 = N_2 = N$. From Fig. 12.7, we can see that

$$\Delta t = N_1 T_1 - N_2 T_2 = N(T_1 - T_2) = N \cdot \Delta T \tag{12.2}$$

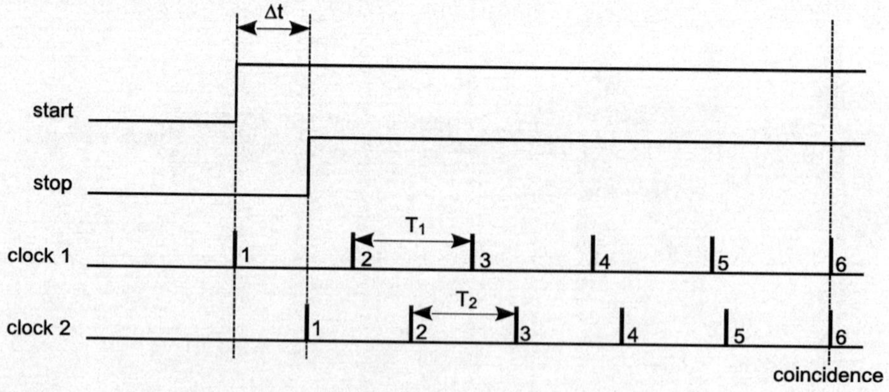

Fig. 12.7 The Vernier TDC (no reference clock)

From Eq. (12.2) we can see that the time resolution depends on the difference ΔT in the clocks' cycle periods; we can read values in between the clock pulses of the individual clocks.

12.2.2 Vernier TDC with a Reference Clock

The alternative approach is to use a reference clock that runs asynchronously to the Vernier clocks, see Fig. 12.8. The reference clock's cycle period is T_{ref} and we make the Vernier clocks' cycle period slightly longer; $T_{vern} = T_{ref}(1 + 1/N)$, where N is an integer that determines the overall time resolution. The 'start' Vernier clock starts on the positive edge of the pulse and the 'stop' Vernier clock starts on the negative edge. t_{start} is the time it takes for the start clock's edges to align with the reference clock's edges and t_{stop} is the time it takes for the stop clock's edges to align with the reference clock's edges.

From Fig. 12.8, we can see that

$$\Delta t + t_{stop} = t_{start} + t_{diff}$$
$$\Delta t = t_{start} + t_{diff} - t_{stop} = n_0 T_{vern} + n_1 T_{ref} - n_2 T_{vern} \tag{12.3}$$

where n_0 is the number of T_{vern}-pulses counted during t_{start}, n_1 is the number of T_{ref}-pulses counted during t_{diff}, and n_2 is the number of T_{vern}-pulses counted during

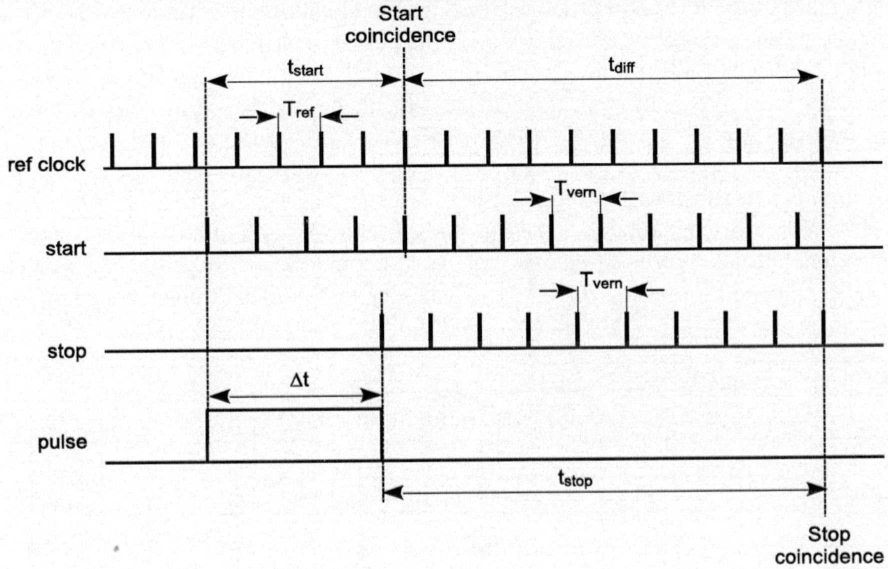

Fig. 12.8 The Vernier TDC (with reference clock)

t_{stop}. If we insert $T_{vern} = T_{ref}(1 + 1/N)$, we get that

$$\begin{aligned}\Delta t &= T_{ref}\left(n_0\left(1+\frac{1}{N}\right)+n_1-n_2\left(1+\frac{1}{N}\right)\right)=\\ &= T_{ref}\left(n_1+(n_0-n_2)\left(1+\frac{1}{N}\right)\right)=\\ &= T_{ref}(n_1+n_0-n_2)+\frac{T_{ref}}{N}(n_0-n_2)\end{aligned} \tag{12.4}$$

and we can see from (12.4) that this design offers a time resolution of T_{ref}/N. We can easily translate a specified time resolution into a difference in clock cycle periods:

$$T_{vern} = T_{ref}\left(1+\frac{1}{N}\right) \Longrightarrow \Delta T = T_{vern} - T_{ref} = T_{ref} \times \frac{1}{N} \tag{12.5}$$

12.3 Delaylines

If you design a TDC in CMOS technology (VLSI designers) there are a few alternative implementations to the Vernier techniques in Sect. 12.2. Figure 12.9 illustrates the basic idea.

The start signal is connected to the first of an array of cascaded buffers. The start signal's high level will propagate through the chain of buffers at a speed corresponding to each buffer's gate delay τ_{delay}. The output of each buffer is the data input to an edge-triggered flip-flop. The flip-flops are latched by the stop signal arriving sometime later. When the stop signal arrives and latches the flip-flops, some of them will have a high ('1') data input, and some will have a low ('0') data input, depending on how far the start edge has propagated through the buffer chain when the stop edge appears. This is illustrated in Fig. 12.10.

When the stop signal latches the flip-flops, the flip-flops' output will be a 'thermometer' bit code representing how far the start signal's positive edge has propagated. The resolution of this TDC is τ_{delay}. Notice that it doesn't involve a counter/

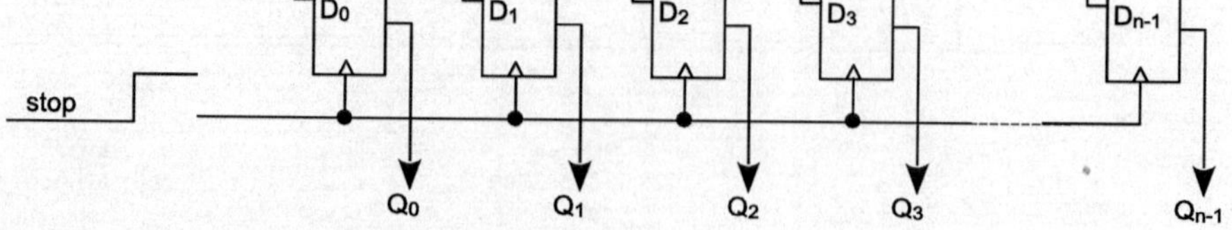

Fig. 12.9 Delayline for clock cycle subdivision

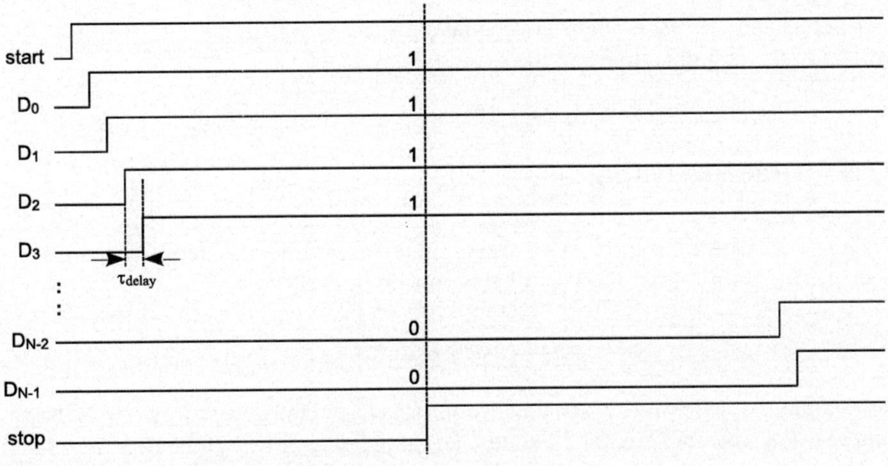

Fig. 12.10 The stop signal latches the flip-flops

oscillator. This means that the range is limited to the number of buffers (=N); the range is $N \cdot \tau_{\text{delay}}$. However, it is all-digital and therefore scalable. Since time-to-digital conversion in this case is immediate, it is sometimes referred to as the 'flash' TDC.

Figure 12.9 represents the 'basic' tapped delayline; the resolution depends on the buffers' delay τ_{delay}. The 'next-generation' TDCs take this technique one step further; the resolution depends on the *difference* in buffers' delay. This is illustrated in Fig. 12.11.

The buffers in the top delayline have a delay of τ_1 that is slightly longer than the delay τ_2 of the buffers in the bottom delayline; $\tau_1 > \tau_2$. Hence, when the start and stop signals arrive, the stop signal will propagate faster through the delayline than the start signal does. If the start signal is leading, 1s will be latched into each flip-flop when the stop signal (= latch signal) arrives. At some point though, the stop signal will catch up and pass the start signal (since it propagates faster) and from that point on, 0s will be latched into each flip-flop. The 'temperature' code formed by the flip-flops' outputs represents the time difference between the arrivals of the start and

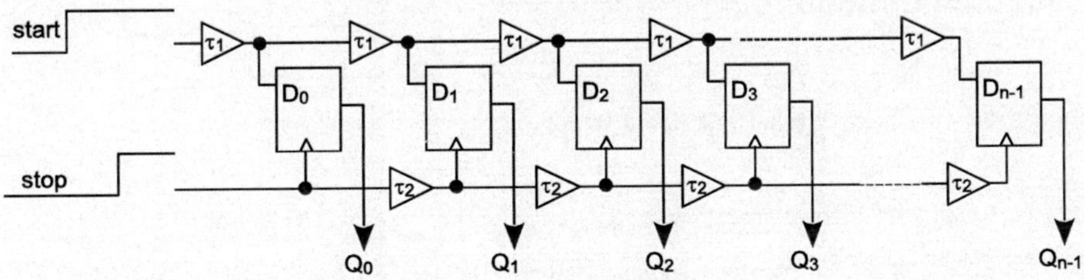

Fig. 12.11 A 'Vernier' delayline [2]

stop signal and the resolution is now equal to the *difference* in the delay between the buffers in the first delayline and the second delayline.

12.4 Time Stretching

Figure 12.12 illustrates the 'basic' digital time measurement system. We learned in Sect. 12.1 that this system has an inherent uncertainty of $\pm 1\ t_c$, i.e.,

$$\widehat{\Delta t} = (N_0 \pm 1) \times t_c \tag{12.6}$$

if we count N_0 pulses during Δt; the resolution is t_c and the uncertainty is $\pm\, t_c$. Next, suppose that we could *stretch* the time interval by a factor of k (Fig. 12.13).

If we measure the stretched time interval ΔT with the same instrument as in Fig. 12.12, we get

$$\widehat{\Delta T} = (N_1 \pm 1) \times t_c = N_1 t_c \pm t_c \tag{12.7}$$

(if we count N_1 pulses during ΔT). But, since $\Delta t = \Delta T / k$, then

$$\widehat{\Delta t} = \frac{1}{k}\widehat{\Delta T} = N_1 \frac{t_c}{k} \pm \frac{t_c}{k} \tag{12.8}$$

Hence, if the time interval is stretched by a factor of k, both the resolution and the uncertainty improve by a factor of k. Figure 12.14 illustrates how a pulse can be stretched.

In Fig. 12.14, $i_1 >> i_2$. The time stretching is a two-step process. In step one, the switch is closed during Δt and the capacitor is charged by a constant current i_1

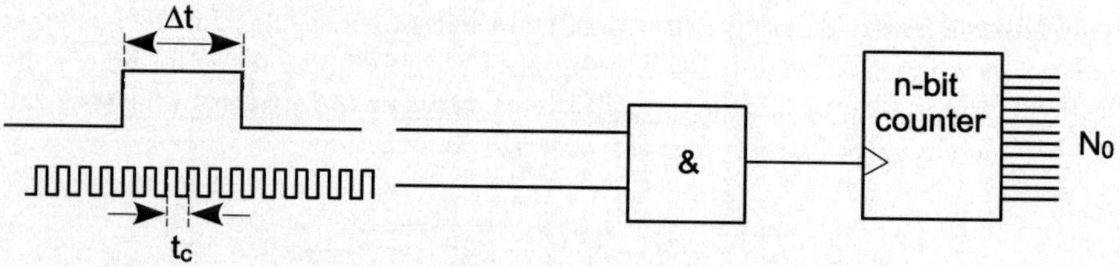

Fig. 12.12 A basic digital time measurement system

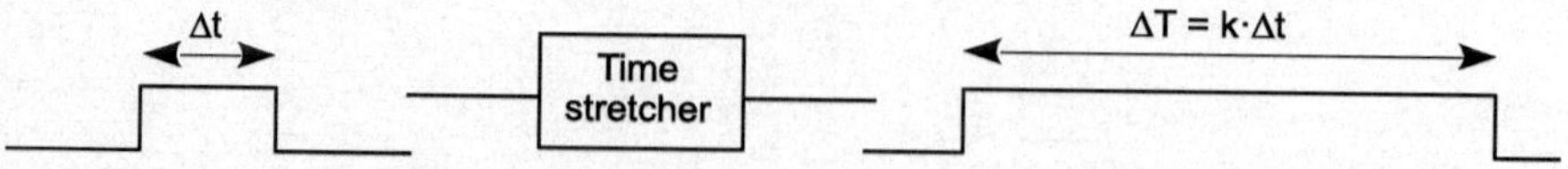

Fig. 12.13 A pulse stretcher

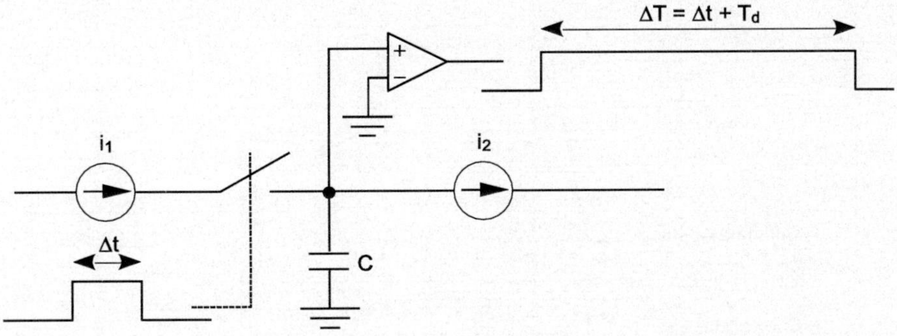

Fig. 12.14 Time stretching [3]

– i_2. The comparator's output goes high immediately after the switch is closed. The switch opens when the time interval Δt expires and at that time the voltage across the capacitor is

$$U_C = \frac{Q}{C} = \frac{1}{C}\int (i_1 - i_2)dt = \frac{1}{C}(i_1 - i_2) \cdot \Delta t \tag{12.9}$$

When the switch opens, the capacitor is discharged by the constant current i_2. The capacitor will be discharged after some time T_d:

$$\frac{1}{C} i_2 T_d = \frac{1}{C}(i_1 - i_2) \cdot \Delta t \Longrightarrow T_d = \frac{i_1 - i_2}{i_2} \cdot \Delta t = \left(\frac{i_1}{i_2} - 1\right) \cdot \Delta t \tag{12.10}$$

The comparator's output will be high for a time

$$\Delta t + T_d = \Delta t + \left(\frac{i_1}{i_2} - 1\right) \cdot \Delta t = \frac{i_1}{i_2} \Delta t \tag{12.11}$$

Hence, if we compare the comparator's output with the input pulse, we can see that the time interval has been stretched by a factor of $k = i_1/i_2$, see Fig. 12.15.

Time resolutions of <10 ps have been reported [4] with the time stretching technique and is used in, for example, pulsed time-of-flight laser radars with a 4.5 mm precision over a range from 1.5 to 370 m [5].

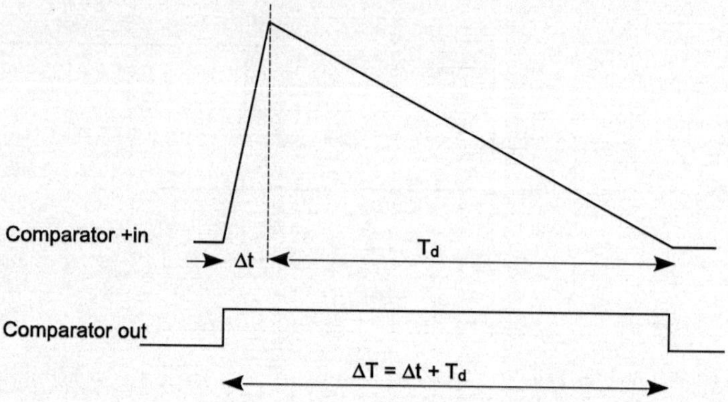

Fig. 12.15 Pulse stretching

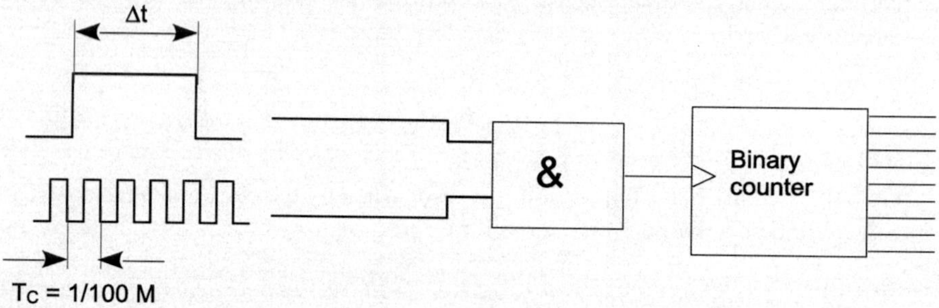

Fig. 12.16 'Basic' TDC

12.5 Solved Problems

Problem 12.1 A 'basic' TDC has a reference clock of 100 MHz, see Fig. 12.16.

a If you want to detect/measure a pulse of a few hundred ns with a resolution of 50 ps, how much would you have to stretch it if you are going to use the TDC in Fig. 12.16?

b Suggest a Vernier solution for this problem, assuming the reference clock above is one of the Vernier clocks.

Solution **a** The resolution of the 'basic' TDC in Fig. 12.16 is $1/100 \cdot 10^6 = 10$ ns. After stretching: 10 ns/$k = 0.050$ ns $\Rightarrow$ $\underline{k = 200 \text{ times}}$.

b $\Delta t = T_1 - T_2 \Rightarrow 0.05 = 10 \text{ ns} - T_2 \Rightarrow T_2 = 9.95 \text{ ns} \Rightarrow \underline{f_2 = 100.5 \text{ MHz}}$.

Problem 12.2 In a physics lab, scientists are ionizing a graphite sample and they want to know the distribution of carbon-12 and carbon-13 in the sample. They use a 1-m-long time-of-flight mass spectrometer where the ions are accelerated by 1 kV, see Fig. 12.17. They have a TDC that measures the time between the ionization pulse (= 'start') and the detector pulse ('stop').

In this experiment, they used the basic TDC in Fig. 12.16 to measure the flight times. After how many 'counts' will the carbon-12 and carbon-13 ions show up on the mass spectrum?

Solution The ions are accelerated by a voltage $U = 1$ kV, which means that they enter the field-free flight area with a kinetic energy of qU. Hence,

$$qU = \frac{1}{2}mv^2 \Rightarrow v = \sqrt{\frac{2qU}{m}}$$

The flight time is

$$t = \frac{s}{v} = \frac{1 \text{ m}}{v} = v^{-1} = \sqrt{\frac{m}{2qU}}$$

The flight times for the two carbon isotopes are

$$t_{13} = \sqrt{\frac{13 \cdot 1.66 \cdot 10^{-27}}{2 \cdot 1.602 \cdot 10^{-19} \cdot 10^3}} = 8.207 \text{ µs}$$

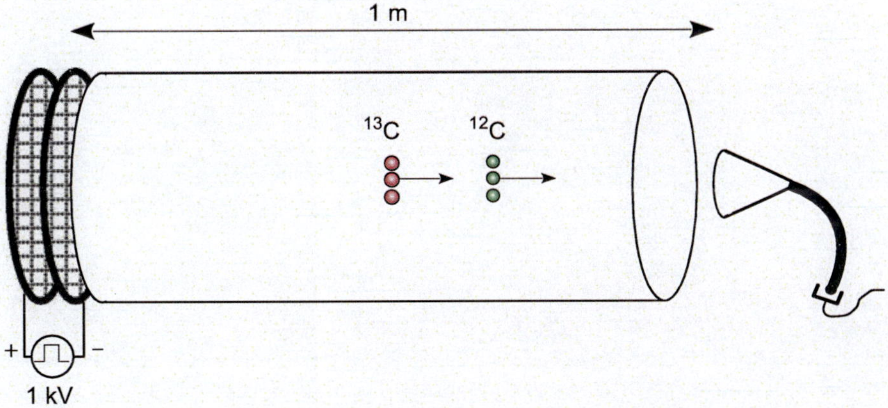

Fig. 12.17 Time-of-flight mass spectrometer

$$t_{12} = \sqrt{\frac{12 \cdot 1.66 \cdot 10^{-27}}{2 \cdot 1.602 \cdot 10^{-19} \cdot 10^{3}}} = 7.885\ \mu\text{s}$$

The basic TDC has a resolution of 10 ns, so the flight times correspond to 8.207/0.01 = 820 counts and 7.885/0.01 = 788 counts, respectively.

References

1. Porat, D.I. 1973. Review of sub-nanosecond time-interval measurements. *IEEE Transactions on Nuclear Science* 20 (5): 36–51.
2. Henzler, S. 2010. Time-to-digital converters with sub-gatedelay resolution–the third generation. In *Time-to-digital converters*, 69–102.
3. Nutt, R. 1968. Digital time intervalometer. *Review of scientific instruments* 39 (9): 1342–1345.
4. Kalisz, J., M. Pawlowski, and R. Pelka. 1985. A method for autocalibration of the interpolation time interval digitiser with picosecond resolution. *Journal of Physics E: Scientific Instruments* 18 (5): 444.
5. Raisanen-Ruotsalainen, E., T. Rahkonen, and J. Kostamovaara. 2000. An integrated time-to-digital converter with 30-ps single-shot precision. *IEEE Journal of Solid-State Circuits* 35 (10): 1507–1510.

Chapter 13
Statistics

Abstract This chapter summarizes the basic concepts of statistics with the only purpose of laying the ground for the next chapter (about measurement uncertainty). Basic statistical concepts are defined, such as stochastic variables and the most common probability distribution functions (normal, uniform). The expectation value, the variance, and the standard deviation of a stochastic variable are defined, and the difference between the *population* variance and the *sample* variance is stressed. This leads to interval estimations, the Student-*t* distribution, and the central limit theorem.

13.1 Introduction

In any measurement, noise is omnipresent (see Chap. 2); it is only a matter of what level of accuracy you are considering. Noise will cause a 'flickering' on the display of a voltage meter; if there is no flickering, it only means that the noise is less than the voltage represented by the least significant digit on the display. Figure 13.1 illustrates our signal model of a DMM measurement.

Noise adds to the system in either normal or common mode; the DMM's sample value is in general:

$$U_m = U_0 + U_{NM}(+F_{CM} \cdot U_{CM}) \tag{13.1}$$

However, in the following, we will disregard the CM residual in the output and only focus on the normal mode noise. If we assume that the noise (U_{NM}) is 'white Gaussian' (variance σ^2), then the voltage measured by the DMM is

$$U_{\mathrm{NM}} \in \mathrm{N}(0, \sigma) \Rightarrow U_m \in \mathrm{N}(U_0, \sigma) \tag{13.2}$$

Figure 13.2 illustrates the density function of the measured voltage.

Hence, the conclusion is that the voltage we measure with the DMM is a *stochastic variable*. The density function expression is

L. Bengtsson, *Electrical Measurement Techniques*,
https://doi.org/10.1007/978-981-99-8187-8_13

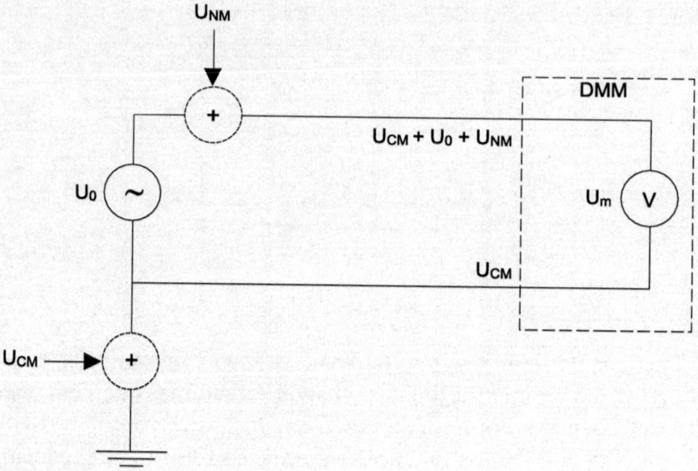

Fig. 13.1 Signal model

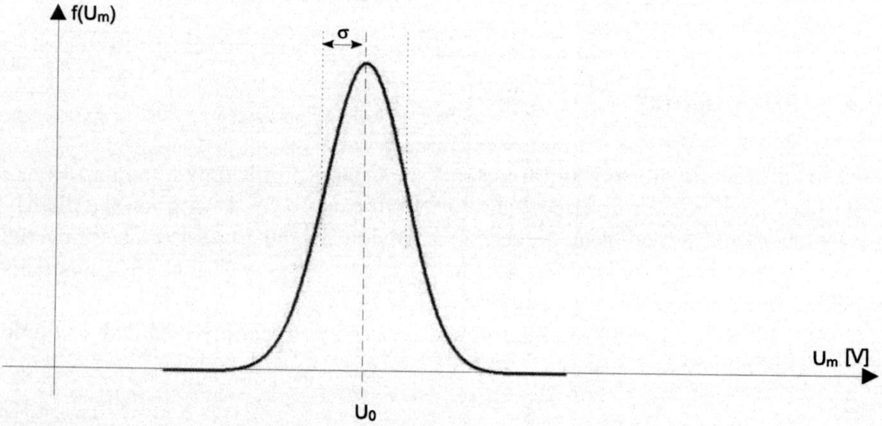

Fig. 13.2 The sample is a stochastic variable

$$f(U_m) = \frac{1}{\sigma\sqrt{2\pi}} \cdot \mathrm{e}^{-\frac{(U_m - U_0)^2}{2\sigma^2}} \tag{13.3}$$

13.2 Expectation and Variance

Usually, we don't need the density function expression; instead, we use three parameters that characterize the stochastic variable; the *expectation value*, the *variance*, and the *standard deviation*:

The expectation value of a stochastic variable X is defined as

$$\mathrm{E}(X) = \int_{-\infty}^{\infty} xf(x)dx = \mu \tag{13.4}$$

We will refer to μ as the *mean* value. In our example in Fig. 13.2, $\mathrm{E}(U_\mathrm{m}) = U_0$. The variance of the stochastic variable X is defined as

$$\mathrm{V}(X) = \mathrm{E}\{(x-\mu)^2\} = \int_{-\infty}^{\infty} (x-\mu)^2 f(x)dx = \sigma^2 \tag{13.5}$$

The variance is a number that tells us something about the 'spread' of the values around the expectation value. However, the variance unit is 'voltage squared' [V^2] and when we talk about 'spread', it makes more sense to express spread in the same unit as the stochastic variable. For that reason, we have the *standard deviation* as the square root of the variance:

$$\sigma = \sqrt{\mathrm{V}(X)} \tag{13.6}$$

We also need to find expressions for the expectation and variance values of some functions of X. First, we multiply X by a constant a:

$$Y = aX \tag{13.7}$$

The expectation value of Y is $\mathrm{E}(aX)$, but since the expectation is an integral, which is a 'linear' operation, $\mathrm{E}(aX) = a\mathrm{E}(X)$ and hence:

$$\mathrm{E}(Y) = a\mathrm{E}(X) = a\mu \tag{13.8}$$

We also need the variance of Y:

$$\mathrm{V}(Y) = \mathrm{E}\{(Y-\mathrm{E}(Y))^2\} = \mathrm{E}\{(aX-a\mu)^2\} = \mathrm{E}\{a^2(X-\mu)^2\} =$$

$$= a^2\mathrm{E}\{(X-\mu)^2\} = a^2\mathrm{V}(X) = a^2\sigma^2 \tag{13.9}$$

The next function of stochastic variables that we need to analyze is the sum of two variables:

$$Y = X_1 + X_2 \tag{13.10}$$

It is important for the following that the stochastic variables X_1 and X_2 in Eq. (13.10) are *iid*, i.e., *independent, and identically distributed.* We justify that assumption by remembering that X_1 and X_2 are two DMM samples and they are iid because we assume that the first sample has no influence on the second sample, which seems reasonable. Under what circumstances would they not be iid? Well, if we look at the sample and hold circuit in Fig. 11.2, they could be dependent if the 'holding' capacitor is not allowed enough time to charge/discharge between samples, i.e., if we sample too fast, but that would be a design flaw in the measurement system. Since the samples come from the same 'population' in a measurement, they are always identically distributed.

The expectation of Y is

$$\mathrm{E}(Y) = \mathrm{E}(X_1 + X_2) = \mathrm{E}(X_1) + \mathrm{E}(X_2) = \mu + \mu = 2\mu \tag{13.11}$$

We can easily see how this result can be generalized for a sum of N variables:

$$Y = \sum_{i=1}^{N} X_i \Rightarrow \mathrm{E}(Y) = N\mu \tag{13.12}$$

The variance of Y is

$$\begin{aligned}
\mathrm{V}(Y) &= \mathrm{V}(X_1 + X_2) = \mathrm{E}\{(X_1 + X_2 - 2\mu)^2\} = \mathrm{E}\{((X_1 - \mu) + (X_2 - \mu))^2\} = \\
&= \mathrm{E}\{(X_1 - \mu)^2 + (X_2 - \mu)^2 + 2 \cdot (X_1 - \mu) \cdot (X_2 - \mu)\} = \\
&= \mathrm{E}\{(X_1 - \mu)^2\} + \mathrm{E}\{(X_2 - \mu)^2\} + 2 \cdot \mathrm{E}\{(X_1 - \mu)(X_2 - \mu)\} = \\
&= \mathrm{V}(X_1) + \mathrm{V}(X_2) + 2\,\mathrm{cov}(X_1, X_2) = \sigma^2 + \sigma^2 + 0 = 2\sigma^2
\end{aligned} \tag{13.13}$$

where the last equal sign comes from the assumption that our samples are iid; the covariance of iid variables is zero. Again, we can easily generalize this result:

$$Y = \sum_{i=1}^{N} X_i \Rightarrow \mathrm{V}(Y) = N\sigma^2 \tag{13.14}$$

13.3 Unbiased Estimators

In a typical measurement we try to estimate the value of some unknown parameter and our estimation is usually based on sampled data (not always). The 'estimator' is 'unbiased' if its expectation value equals the mean. For example, the best unbiased estimator of the mean is the average:

$$\text{Average: } \overline{X} = \frac{1}{N} \cdot \sum\nolimits_{i=1}^{N} X_i \tag{13.15}$$

This is an unbiased estimator of the mean because:

$$\begin{aligned} \mathrm{E}\{\overline{X}\} &= \mathrm{E}\left\{\frac{1}{N}\sum_{i=1}^{N} X_i\right\} = \frac{1}{N}\mathrm{E}\left\{\sum_{i=1}^{N} X_i\right\} = \frac{1}{N}\sum_{i=1}^{N}\mathrm{E}\{X_i\} = \\ &= \frac{1}{N}\sum\nolimits_{i=1}^{N}\mu = \frac{1}{N}N\mu = \mu \end{aligned} \tag{13.16}$$

(We can change places between the 'sum' and the 'expectation' operators since integration is a linear operation). We also need the variance of the average:

$$\mathrm{V}(\overline{X}) = \mathrm{V}\left(\frac{1}{N}\sum\nolimits_{i=1}^{N} X_i\right) = \frac{1}{N^2}\mathrm{V}\left(\sum\nolimits_{i=1}^{N} X_i\right) = \frac{1}{N^2}N\sigma^2 = \frac{\sigma^2}{N} \tag{13.17}$$

where we have used the results from Eqs. (13.9) and (13.14). We now define the *standard error* as the standard deviation of the mean:

$$\sigma_{\overline{X}} = \sqrt{\mathrm{V}(\overline{X})} = \frac{\sigma}{\sqrt{N}} \tag{13.18}$$

Hence, if our 'raw' samples have the distribution $\mathrm{N}(\mu, \sigma)$, the average of N samples has the distribution

$$\overline{X} \in \mathrm{N}(\mu, {}^{\sigma}/_{\sqrt{N}}) \tag{13.19}$$

Notice that the mean value μ is 'what we are looking for', and the parameter value that we try to estimate (by our samples); μ is the 'signal level'. σ represents the 'noise' in our samples. Hence, we can define the signal-to-noise ratio (for normal mode coupled noise) as

$$\mathrm{SNR}(X) = \frac{\mu}{\sigma} \tag{13.20}$$

The signal-to-noise ratio for the averaged value is then:

$$\mathrm{SNR}(\overline{X}) = \frac{\mu}{\sigma/\sqrt{N}} = \sqrt{N}\frac{\mu}{\sigma} = \sqrt{N} \times \mathrm{SNR}(X) \tag{13.21}$$

From Eq. (13.21) we conclude that averaging improves the SNR by a factor of $\sqrt{N}$. Figure 13.3 illustrates a sinusoidal signal superimposed with white Gaussian noise and what it looks like after 4 and 64 averages, respectively.

Equation (13.15) gives us the unbiased estimator for the mean. We also need an unbiased estimator for the variance; in most cases we will not know the variance and we need to estimate it. We refer to σ^2 as the *population variance* (the 'true'

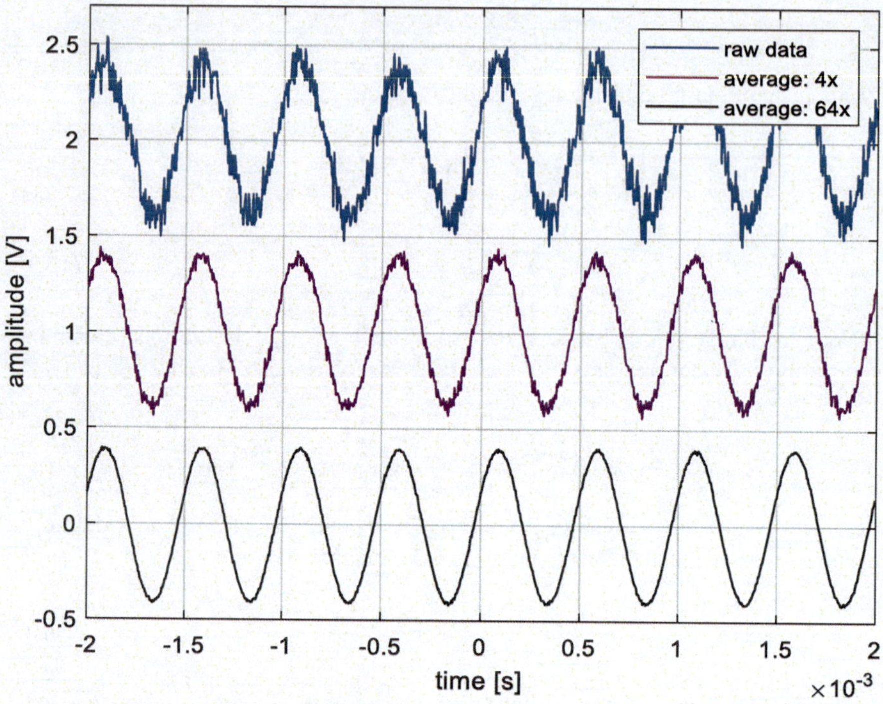

Fig. 13.3 Averaging improves the SNR

variance), and we estimate it with the *sample variance,* s^2:

$$s^2 = \frac{1}{N-1}\sum_{i=1}^{N}\left(X_i - \overline{X}\right)^2 \tag{13.22}$$

(The reason we are using $N-1$ and not N in the denominator in (13.22) is because we lose one degree of freedom when we use $\overline{X}$ instead of μ).

13.4 Interval Estimations

The estimator in Eq. (13.15) is 'only' a 'point estimator'; it has some 'uncertainty'. In a typical measurement, you don't report only the point estimation, you report an 'interval estimation'. You report an interval $x_0 = \hat{x} \pm U$, but remember that we are dealing with stochastic variables here that in most cases are normally distributed, or close to normally distributed. That implies that to be 100% sure of that x_0 is in the interval $x_0 \pm U$, U would have to be infinitely large, and that information would be useless (since x_0 is obviously in the interval $\pm\infty$). For that reason, we will have to settle with a U value that does not guarantee 100% inclusion of x_0. This is expressed

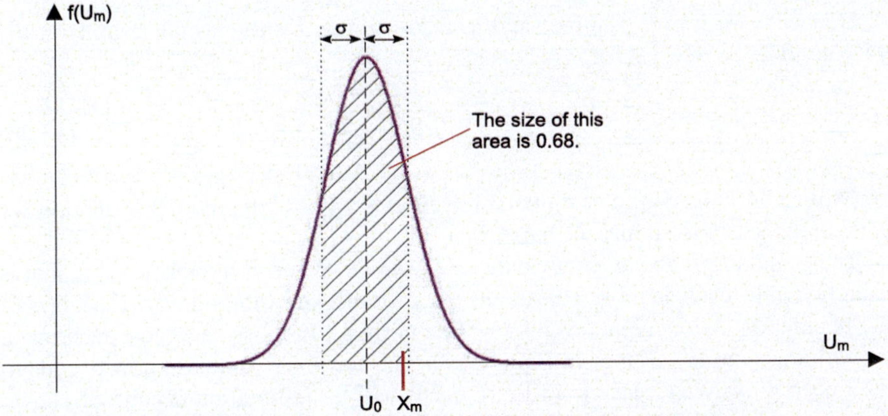

Fig. 13.4 The 68% confidence interval

as a *confidence level*: 'I am X % sure that x_0 is in the interval $x_0 \pm U$'. The larger U is the more 'confident' we are that x_0 is within the interval (the 'confidence interval'). For example, if we take one sample from a normal distribution, we can be 68% confident that the sample will be in the interval $X_m = U_0 \pm \sigma = \mu \pm \sigma$, see Fig. 13.4.

However, we don't really want to know the probability that our *sample* is in a certain interval; our objective with the measurement is to estimate the *unknown parameter* μ; we want to find an interval for μ , not for X_m. Well, we can remedy that with some simple probability juggling. The fact that X_m is in the interval $\mu \pm \sigma$ with a probability of 68%, can be expressed as

$$P(\mu - \sigma < X_m < \mu + \sigma) = 0.68 \tag{13.23}$$

(where 'P' is the 'Probability'). Next, we subtract, X_m and μ from each value:

$$P(-X_m - \sigma < -\mu < -X_m + \sigma) = 0.68 \tag{13.24}$$

We can change the ' < ' to ' > ' if we also change all the signs:

$$P(X_m + \sigma > \mu > X_m - \sigma) = 0.68 \tag{13.25}$$

From Eq. (13.24) we conclude that if we take a sample X_m, we can say that μ (the parameter that we are trying to estimate) is in the interval $X_m \pm \sigma$ with a probability of 68%; we are 68% *confident* that μ is in this interval.

The interval $\pm\sigma$ corresponds to 68% probability. This is usually considered to be 'too uncertain'; there is a 32% probability that μ is *not* in this interval! For that reason, it is generally recommended that you use the interval $\pm 2\sigma$ which represents a 95% probability of finding μ within the interval. Actually, the probability is 95.4%; you need to multiply by 1.96 if you want 95% *exactly*.

Hence, if we take N samples and instead use the *average* as our estimator, the 95% confidence interval for μ becomes

$$\mu = \overline{x} \pm 1.96\frac{\sigma}{\sqrt{N}} \tag{13.26}$$

(Notice how the confidence interval decreases with N; the more information we collect, the less uncertainty we have).

The calculations above assumed that σ is *known*. When you think about it, that is usually *not* the case; in most measurements, we don't know the standard deviation of the population (We may not even know it is normal). In those cases, we must make do with the sample variance in Eq. (13.22). The fact that we don't know σ adds to the uncertainty; we replace σ in Eq. (13.26) with an estimation (s) and of course that makes our estimation a little more precarious. The consequence is that we will probably have to multiply by a larger number (a larger *coverage factor*) than 1.96 to get a confidence level of 95%.

Even if that is true in general, we may keep the 1.96 if the circumstances are 'right'. Here, we lean on the *central limit theorem* (CLT) which says that if you average enough samples, the distribution is still normal even if the original samples are not normal. 'Enough' samples are generally considered to be '30 or more'; if $N \geq 30$, we will still use Eq. (13.26) and just replace σ with s. Our problems appear when $N < 30$; in these cases, we need to find a new value for the coverage factor (a *larger* value) because the average value's distribution is in general no longer an exact normal distribution; it has a *Student-t* distribution.

The Student-t distribution looks like the normal distribution, but it is 'wider' (reflecting a larger spread of the samples). As a matter of fact, the Student-t distribution is not just *one* distribution, it is a series of distributions: one for each degree of freedom. (The degree of freedom $v = N-1$). When $N \to \infty$, the t-distribution becomes the normal distribution. Figure 13.5 illustrates the t density function for different degrees of freedom.

If our average value has a t-distribution, then the 95% confidence interval is

$$\mu = \overline{x} \pm t_{v,\alpha} \times \frac{s}{\sqrt{N}} \tag{13.27}$$

where $\alpha = 1-\text{confidence} = 1-0.95 = 0.05$. For example, if $N = 10$ and we want a 95% confidence level, we must find the $t_{9,0.05}$ value. A quick googling for 'two-tailed t-table' immediately gives us something like Table 13.1; the value we are looking for is $t_{9,0.05} = 2.262$.

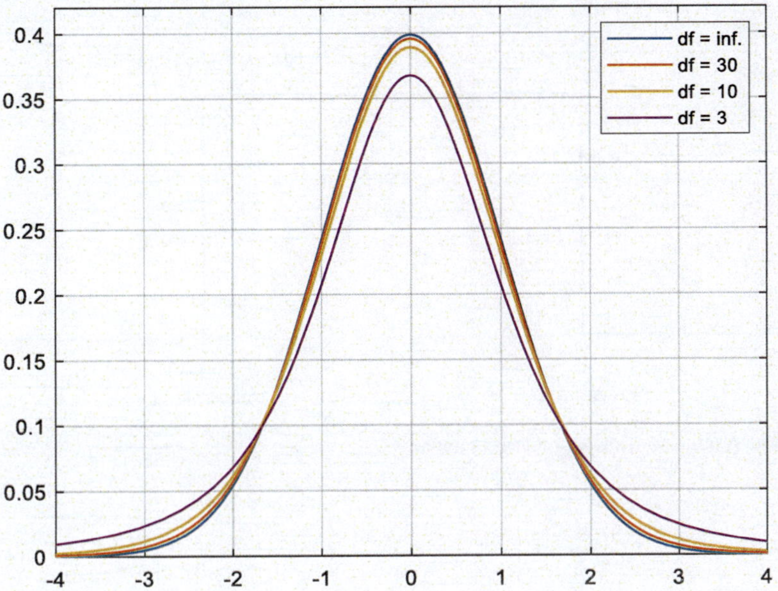

Fig. 13.5 The *t*-distribution (plotted in MATLAB using the *tpdf* command)

Table 13.1 The two-tailed *t*-table

	Significance level (α)				
Degrees of freedom (*df*)	0.2	0.15	0.1	0.05	0.025
:					
8	1.397	1.592	1.860	2.306	2.752
9	1.383	1.574	1.833	**2.262**	2.685
10	1.372	1.559	1.812	2.228	2.634
:					

13.5 The Uniform Distribution

Finally, we will need the variance and standard deviation of a uniform probability distribution, see Fig. 13.6. A uniformly distributed stochastic variable can take any value between the upper and lower limits with equal probability ('uniform' probability). The variance of a (symmetric) uniform distribution as the one in Fig. 13.6 is straightforward:

$$\sigma^2 = \mathrm{V}(X) = \mathrm{E}\{(X-\mu)^2\} = \mathrm{E}\{X^2\} = \int_{-\infty}^{+\infty} x^2 f(x)dx = \int_{-c}^{+c} x^2 \frac{1}{2c} dx =$$

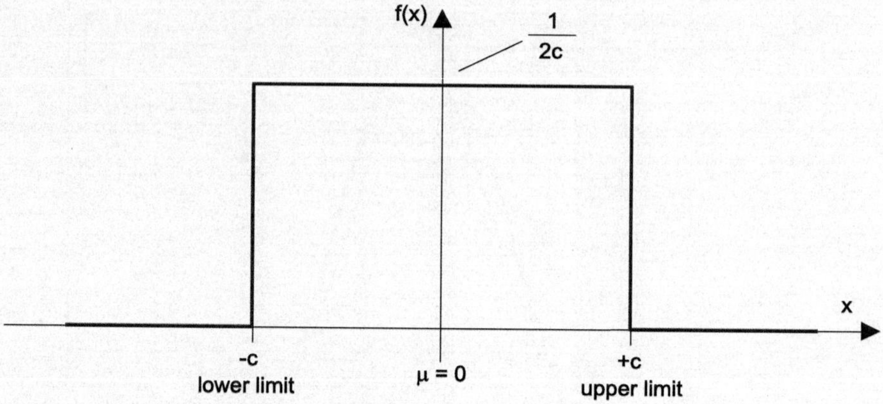

Fig. 13.6 A uniform distribution ('rectangular')

$$= \frac{1}{2c} \cdot \frac{1}{3} \left[x^3 \right]_{-c}^{+c} = \frac{1}{2c} \cdot \frac{1}{3} \left(c^3 + c^3 \right) = \frac{1}{3} c^2 \tag{13.28}$$

Hence, the standard deviation of a uniformly distributed variable is

$$\sigma = \frac{1}{\sqrt{3}} \cdot c \tag{13.29}$$

This is all we need to set up an 'uncertainty budget' to find the uncertainty of a measurement.

13.6 Solved Problems

Problem 13.1 Using a 6 ½ DMM, N samples were taken of a DC voltage. The average of the samples was 4.17698 V, and the sample standard deviation was $s =$ 0.76 mV. Find the 95% confidence interval of this measurement if **a** $N = 100$, **b** $N = 12$.

Solution a The 'standard error' is $s/\sqrt{N} = 0.76/\sqrt{100} = 0.076$ mV, and 1.96·0.076 = 0.149 mV ≈ 0.15 mV. Hence, $U_m = 4.17698 \pm 0.00015$ V (95%).

b $0.76/\sqrt{12} = 0.219$ mV. Degrees of freedom = 12 – 1 = 11 ⇒ cover factor 2.201. 2.201·0.219 = 0.483 mV ≈ 0.49 mV. Hence, $U_m = 4.17698 \pm 0.00049$ V (95%).

Problem 13.2 The quantization uncertainty of an ADC is $\pm\frac{1}{2}\Delta U$ (see Eq. (11.4)). What is the standard deviation of the output from a 12-bit ADC with + 5 V reference voltage?

Solution Eq. (13.29) gives that

$$\sigma = \frac{1}{\sqrt{3}} \times \frac{1}{2} \times \frac{U_{\text{ref}}}{2^n} = \frac{1}{\sqrt{3}} \times \frac{1}{2} \times \frac{5}{2^{12}} = \underline{0.352\ \text{mV}}$$

Problem 13.3 A noisy DC voltage is normally distributed with $\mu = 3.30$ V and $\sigma =$ 60 mV. If we measure this voltage once, what is the probability that we will measure a voltage **a** < 3.35 V and **b** > 3.20 V?

Solution a First we 'normalize' the distribution by calculating the z value:

$$z = \frac{x - \mu}{\sigma} = \frac{3.35 - 3.30}{0.06} = 0.83$$

Looking up this number in a normal z-table gives that this corresponds to a probability

$$p(U < 3.35) = p(Z < 0.83) = \underline{0.7967}$$

b

$$z = \frac{3.2 - 3.3}{0.06} = -1.67 \Rightarrow p(Z < -1.67) = 0.0475 = p(U < 3.20)$$

$$\Rightarrow p(U > 3.20) = 1 - p(U < 3.20) = 1 - 0.0475 = \underline{0.9525}$$

Problem 13.4 If we measure the voltage in problem 13.3, 15 times, what is the probability that the *average* of these samples is > 3.33 V?

Solution Since the individual samples are normally distributed, the average will also be normally distributed with a standard deviation of $\sigma/\sqrt{N} = 0.06/\sqrt{15} =$ 0.0155 volts.

Hence,

$$z = \frac{3.33 - 3.30}{0.0155} = 1.94$$

$$p(U > 3.33) = 1 - p(U < 3.33) = 1 - p(z < 1.94) = 1 - 0.9738 = \underline{0.0262}$$

Problem 13.5 In problem 13.4 we *knew* that the samples were normally distributed, and we knew the mean and the standard deviation. That is usually *not* the case. What do we do if we don't know the distribution, the mean, and the standard deviation?

Solution Then we must use the t-table instead of the z-table (for N–1 degrees of freedom).

Chapter 14
Uncertainty Budgets

Abstract All measurement numbers have finite accuracy, and it is good practice to always state the uncertainty in a measurement. The uncertainty is typically expressed as an uncertainty interval $\pm U$ around the measurement number and it is extremely important to understand how to find U and what it represents. This should always be done by setting up an *uncertainty budget*. This chapter introduces the uncertainty budget and defines concepts like the coverage factor, the standard uncertainty, and the effective degrees of freedom.

14.1 Introduction

All measurements should be reported as a confidence interval; a point estimation is in general not enough. It is the responsibility (and privilege) of *The International Bureau of Weights and Measures* (the BIPM[1]) in Paris to provide guidelines for the community of exactly how to report uncertainties. These guidelines have been published in a document that the community refers to as the 'GUM' document [1]. However, this document is very extensive and could be overwhelming for the average engineer. For that reason, local organizations have published 'light versions' of the original GUM document with step-by-step instructions on how to conduct a proper uncertainty analysis. In this chapter, we will follow the guidelines presented in the European Accreditation's publication *Evaluation of the Uncertainty of Measurement in calibration* [2].

14.2 Signal Models

First, we need to update our signal model in Fig. 13.1; instead of separating the signal and the (external) noise, we combine them and attribute the noise to the signal source itself (which may very well be the case anyway). Hence, if the noise has a

[1] BIPM: Bureau International des Poids et Mesures.

L. Bengtsson, *Electrical Measurement Techniques*,
https://doi.org/10.1007/978-981-99-8187-8_14

normal distribution, we model the measurand as a stochastic variable with a normal distribution, an expectation value of X_0, and a standard deviation of σ, see Fig. 14.1.

Next, we also need a signal model for the DMM; the DMM is not perfect, it too has some inherent 'noise'. Figure 14.2 illustrates an excerpt from a DMM datasheet.

Notice in Fig. 14.2 how the instrument's accuracy is specified as ' $\pm$ (% of reading + % of range)'. This represents the uncertainty of the instrument display value, and this implies that even if there is *no* noise in the input signal, there is still an uncertainty in the measurement because of the limitations of the instrument itself. We will take the instrument's uncertainty into account by modeling that as an internal noise source, see Fig. 14.3.

The $\pm$ accuracy number in Fig. 14.2 should be interpreted as the upper and lower limits, $\pm\ c$, in a uniform distribution, and according to Eq. (13.29), the standard deviation of such a distribution is $c/\sqrt{3}$. From Fig. 14.3, we have that

$$X_m = \underbrace{X_1}_{N(X_0,\sigma/\sqrt{N})} + \underbrace{X_2}_{U(0,c/\sqrt{3})} \tag{14.1}$$

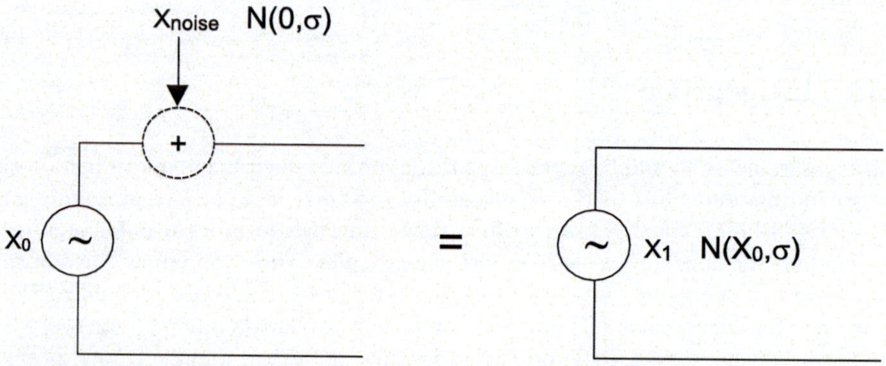

Fig. 14.1 Signal model

Specifications 34460A

- 34460A accuracy specifications: ± (% of reading + % of range) [1].
- These specifications are compliant to ISO/IEC 17025 for K = 2.

Range [2]/frequency	24 hours [3] $T_{CAL} \pm 1$ °C	90 days $T_{CAL} \pm 5$ °C	1 year $T_{CAL} \pm 5$ °C	2 years $T_{CAL} \pm 5$ °C	Temperature coefficient/°C [4]
DC voltage					
100 mV	0.0040 + 0.0060	0.0070 + 0.0065	0.0090 + 0.0065	0.0115 + 0.0065	0.0005 + 0.0005
1 V	0.0030 + 0.0009	0.0060 + 0.0010	0.0080 + 0.0010	0.0105 + 0.0010	0.0005 + 0.0001
10 V	0.0025 + 0.0004	0.0050 + 0.0005	0.0075 + 0.0005	0.0100 + 0.0005	0.0005 + 0.0001
100 V	0.0030 + 0.0006	0.0065 + 0.0006	0.0085 + 0.0006	0.0110 + 0.0006	0.0005 + 0.0001

Fig. 14.2 Typical instrument specifications [3] (Published courtesy of Keysight Technologies, Inc.)

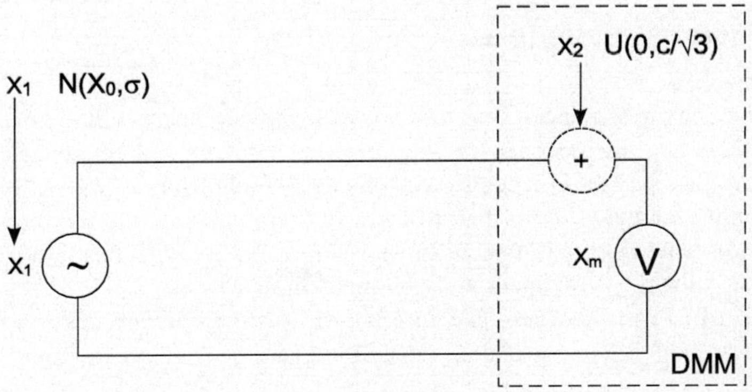

Fig. 14.3 Instrument model

where we have assumed that X_1 is the average of a sample from a population with known variance σ^2. (If variance is unknown, we use s^2.)

Our objective here is to present the measurement result as a 95% confidence interval. In the publication reference EA4-02 M, this is expressed as follows:

$$X_0 = \hat{X} \pm U = \hat{X} \pm k \cdot u\left(\hat{X}\right) \tag{14.2}$$

where U is the *expanded uncertainty* of the measurement (and should represent a 95% confidence interval), k is the *coverage factor* (and should 'almost always be = 2', see Sect. 14.3.) and $u\left(\hat{X}\right)$ is the *standard uncertainty of the estimate output*. Hence, we need to find $u\left(\hat{X}\right)$. If all the contributions are uncorrelated (which we will always assume), we can add the variances (according to Eq. (13.14)):

$$u^2\left(\hat{X}\right) = \sum_{\text{all } i} u^2(X_i) \tag{14.3}$$

Finally, if $y = f(x)$, then

$$u(y) = c(x)u(x) \text{ where } c(x) = \left.\frac{df}{dx}\right|_{x=\hat{x}} \tag{14.4}$$

where $c(x)$ is the *sensitivity coefficient* (a number that represents the uncertainty *propagation*). The GUM document recommends that an uncertainty analysis is performed by using an *uncertainty budget*.

14.3 Uncertainty Budgets

When you calculate the uncertainty of an output estimate there will inevitably be a lot of numbers and just keeping track of all these numbers is a challenge. The use of an uncertainty budget is a suggested remedy for this. But, as we will see later, the uncertainty budget is more than just a way to organize all the numbers; it will provide important information about the measurement. We will illustrate that later.

Table 14.1 illustrates an uncertainty budget template.

In Table 14.1, $\hat{y}$ is the *output estimate* and $u(\hat{y})$ is the standard uncertainty of $\hat{y}$. If all contributions are uncorrelated, we add the variances to get the total standard uncertainty:

$$u(\hat{y}) = \sqrt{\sum_{i=1}^{n} (c(x_i)u(x_i))^2} \tag{14.5}$$

The expanded uncertainty U (representing the 95% confidence interval) is the coverage factor k times $u(\hat{y})$. When the conditions of the central limit theorem can be assumed to be sufficiently fulfilled ('enough data'), the coverage factor $k = 2$, should be used. When the conditions of the central limit theorem are not met, we must first find the *effective degrees of freedom*, ν_{eff}, and then find the proper k value. The effective degrees of freedom are given by the *Welch–Satterthwaite* formula:

$$\nu_{eff} = \frac{u^4(\hat{y})}{\sum_{i=1}^{n} \frac{(c(x_i)u(x_i))^4}{\upsilon_i}} \tag{14.6}$$

In Eq. (14.6), ν_i is the degrees of freedom for each individual contribution. For contributions from *type A* uncertainties (uncertainties based on data samples), the degrees of freedom are $N-1$. If the uncertainty is not based on data samples (= *type B* uncertainties), we must estimate the degrees of freedom in each case. However, according to the GUM document, when the uncertainty comes from a contribution where it has been estimated with the upper and lower limits of a uniform distribution, the degrees of freedom can be assumed to be *infinite*. Also, the effective degrees of

Table 14.1 An uncertainty budget

Quantity	Estimate	$u(x_i)$	$c(x_i)$	$c(x_i)\cdot u(x_i)$
x_1	$\hat{x}_1$	$u(x_1)$	$c(x_1)$	$c(x_1)\cdot u(x_1)$
x_2	$\hat{x}_2$	$u(x_2)$	$c(x_2)$	$c(x_2)\cdot u(x_2)$
$\vdots$	$\vdots$	$\vdots$	$\vdots$	$\vdots$
x_n	$\hat{x}_n$	$u(x_n)$	$c(x_n)$	$c(x_n)\cdot u(x_n)$
$y = f(x_1,..,x_n)$	$\hat{y}$			$u(\hat{y})$

Table 14.2 Coverage factors k for 95% confidence (95.45%)

veff	1	2	3	4	5	6	7	8	9	10
k	13.97	4.53	3.31	2.87	2.65	2.52	2.43	2.37	2.32	2.28
veff	11	12	13	14	15	16	17	18	19	20
k	2.25	2.23	2.21	2.20	2.18	2.17	2.16	2.15	2.14	2.13
veff	25	30	35	40	45	50	∞			
k	2.11	2.09	2.07	2.06	2.06	2.05	2.00			

freedom calculated from Eq. (14.6) will in general not be an integer, and the number should then be truncated to the nearest *lower* integer.

Once we know the effective degrees of freedom, the coverage factor is given by Table 14.2 (which is really the two-tailed t-table for a 95.45% confidence interval).

The use of an uncertainty budget is best illustrated by examples.

14.3.1 Examples

Example 14.1 Figure 14.4 illustrates a DC voltage measurement where the DMM range is 10 V and according to the DMM's manual, the instrument's uncertainty on this range is ± (0.04% of reading + 0.03% of range).

We took ten samples:

$$\left.\begin{aligned} x(1) &= 9.0125 \text{ V} \\ x(2) &= 8.9763 \text{ V} \\ x(3) &= \ldots \\ &\vdots \\ x(10) &= \ldots \end{aligned}\right\} \quad \begin{aligned} \overline{x} &= 9.0068 \text{ V} \\ s &= 0.01752 \text{ V} \\ s/\sqrt{10} &= 0.00554 \text{ V} \end{aligned}$$

According to Eq. (14.1) we have that

$$\hat{y}(= X_m) = X_1 + X_2 = \overline{x} + 0 = 9.0068 \text{ V}$$

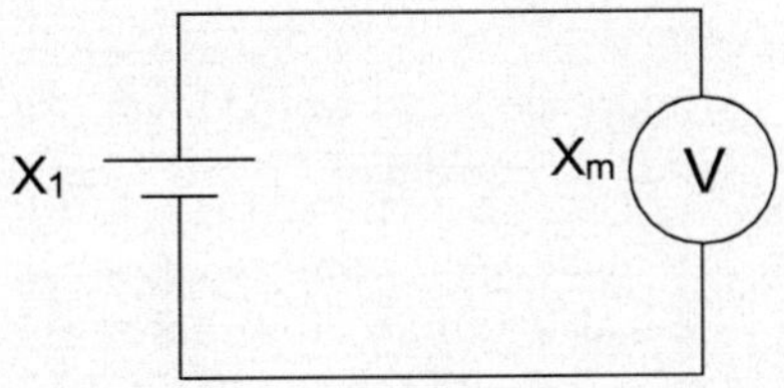

Fig. 14.4 DCV measurement

Table 14.3 An uncertainty budget

Quantity	Estimate	$u(x_i)$	$c(x_i)$	$c(x_i) \cdot u(x_i)$
x_1	9.0068 V	5.54 mV	1	5.54 mV
x_2	0 V	3.81 mV	1	3.81 mV
y	9.0068 V			**6.72 mV**

The uncertainty of X_1 is $s/\sqrt{N} = 0.00554$ V and the uncertainty of X_2 is (see Eq. (14.1)):

$$\frac{1}{\sqrt{3}}\left(0.04\frac{1}{100} \cdot 9.0068 + 0.03\frac{1}{100} \cdot 10\right) = 3.81 \text{ mV}$$

and since all $df/dx_i = 1$ in this case, we have the uncertainty budget in Table 14.3.

Equation (14.5) gives us the total uncertainty for the output estimate.

$$u(\hat{y}) = \sqrt{5.54^2 + 3.81^2} = 6.72 \text{ mV}$$

Since we only took ten samples in this case, we cannot assume that the output estimate has a normal distribution, hence we need to calculate the effective degrees of freedom to find the proper coverage factor.

$$\upsilon_{eff} = \frac{6.72^4}{\frac{5.54^4}{9} + \frac{3.81^4}{\infty}} = 19.5 \rightarrow \upsilon_{eff} = 19 \rightarrow k = 2.14$$

Hence, the expanded uncertainty U is

$$U = 2.14 \times 6.72 \text{ mV} = 14.38 \text{ mV} = 0.015 \text{ V}$$

Since the uncertainty is of the order of 15 mV, it really doesn't make sense to report the output estimate with four decimals. This is how we would present the result of our measurement in a 'scientific' report:

> *In order to estimate the accuracy in the measurement, an uncertainty analysis was conducted according to the guidelines in reference* [1, 2]. *The uncertainty budget produced a standard uncertainty of 6.72 mV, and a coverage factor of 2.14 was used to get the 95% confidence interval:*
>
> $y = 9.007 \pm 0.015$ V

(Reference [1, 2] would be the GUM and the EA-4/02 documents.)

Notice a few details in this example:

- We made no 't compensation' in the uncertainty budget for $u(x_1)$ even though we had less than 30 samples; the 't compensation' was only introduced at the last stage with the use of the Welch–Satterthwaite formula to find the expanded uncertainty. This makes sense; since the total uncertainty has more contributions (just one more in this case) the 't problem' is mitigated; according to the central limit theorem, the more things we add, the closer to a normal distribution we get. For that reason, we should 't compensate' the final uncertainty value only.

- The number of significant digits in the uncertainty is *two*; this is what is generally recommended.
- The estimate and the uncertainty have the same number of decimals; the uncertainty determines how many significant digits are meaningful.
- The estimate and the uncertainty have the same unit, if the estimate is in [V], the uncertainty should be in [V] (not [mV]!).

Example 14.2 Figure 14.5 illustrates a current measurement; we use a DMM to measure the voltage across a resistor to get the current I. The uncertainty of the DMM is $\pm$ (0.04% of reading + 0.02% of range). We took 40 samples, and the average was 6.62953 V (range: 10 V) and the sample standard deviation was 6.83 mV. The resistor has color code marking that can be interpreted as '1800 Ω, ±1%'. What is the 95% confidence interval of this current measurement?

Solution $I = \frac{U}{R} = \frac{X_1+X_2}{R} = f(X_1, X_2, R)$

$$I = \frac{4.62953 + 0}{1800} = 2.5719611 \text{ mA} \quad u(X_1) = 6.83/\sqrt{40} = 1.08 \text{ mV}$$

$$u(X_2) = \frac{1}{\sqrt{3}}(0.0004 \cdot 4.62953 + 0.0002 \cdot 10) = 2.22 \text{ mV}$$

The uncertainty of the resistor is specified as ' ± 1%'. Since we have no other information about this value, we must assume that it represents the upper and lower limits in a uniform distribution:

$$u(R) = \frac{1}{\sqrt{3}} \cdot 1800 \cdot \frac{1}{100} = 10.4 \ \Omega$$

Before we design the uncertainty budget, we calculate the sensitivity coefficients:

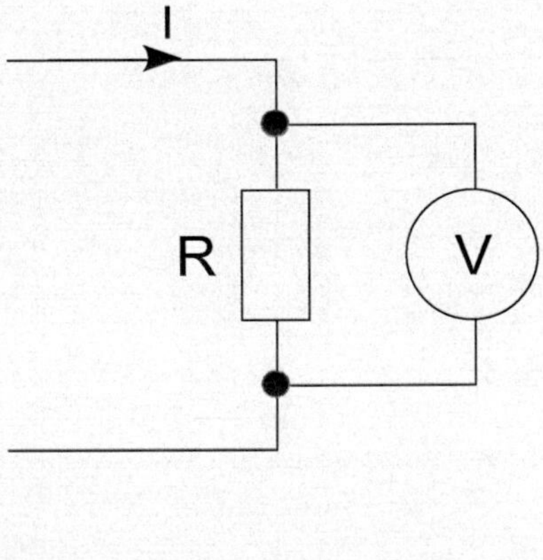

Fig. 14.5 Current measurement

$$c(X_1) = \frac{df}{dX_1} = \frac{1}{R} = \frac{1}{1800} = 5.56 \cdot 10^{-4}\ \Omega^{-1} = c(X_2)$$

$$c(R) = \frac{df}{dR} = (-)\frac{X_1 + X_2}{R^2} = \frac{4.62953}{1800^2} = 1.43 \cdot 10^{-6}\ \mathrm{V}\Omega^{-2}$$

Now we have what we need to setup the uncertainty budget (Table 14.4):
We sum the squares of the uncertainties to get the uncertainty of the estimate:

$$u(I) = \sqrt{0.601^2 + 1.24^2 + 14.9^2} = 14.96\ \mu A$$

In this case, we have enough samples not worry about any '*t* compensation'; we use the coverage factor $k = 2$ to get the expanded uncertainty:

$$U = 2 \times u(I) = 29.93\ \mu A = 0.030\ \mathrm{mA}$$

Hence, only three decimals make sense when we report the measurement:

$$I = 2.572 \pm 0.030\ \mathrm{mA}\ (95\ \%\ )$$

From this example we can learn something more about the use of uncertainty budgets:

- Use at least three significant digits in the budget; only round (upwards) to *two* digits in the last stage when you calculate the expanded uncertainty.
- We get important information from the budget. From the budget it is obvious that it is the lack of information about the resistor that is hurting our accuracy. To improve the accuracy in this example we should try to get a more accurate value for R (by *measuring* it!); buying a new (more expensive) DMM would not help and taking more samples wouldn't help either!

Table 14.4 The uncertainty budget

Quantity	Estimate	$u(x_i)$	$c(x_i)$	$c(x_i) \cdot u(x_i)$ (μA)
X_1	4.62953 V	1.08 mV	$5.56 \cdot 10^{-4}\ \Omega^{-1}$	0.601
X_2	0 V	2.22 mV	$5.56 \cdot 10^{-4}\ \Omega^{-1}$	1.24
R	1800 Ω	10.4 Ω	$1.43 \cdot 10^{-6}\ \mathrm{V}\ \Omega^{-2}$	14.9
I	2.5719611 mA			**14.96**

14.4 'Guesstimating'

Sometimes we don't have any information about the uncertainty of a quantity. For example, suppose we use a DMM to measure the resistance of a Pt-100 temperature sensor. We use the 4-wire method, a 7½ digit DMM and we take a thousand samples to really minimize the noise (the 'type A' uncertainty); the X_1 and X_2 uncertainties are so small that it implies a ppm accuracy (part per million). However, when we translate the resistance to temperature, we use the following formula:

$$R = R_0(1 + \gamma T) \Rightarrow T = \frac{1}{\gamma}\left(\frac{R}{R_0} - 1\right) = f(\gamma, R, R_0) \tag{14.7}$$

(In Eq. (14.7), $R = R_1 + R_2$ if we use a DMM, see Fig. 14.3). From Eq. (14.7), it is clear that the uncertainty of T doesn't depend only on the uncertainty of the measured quantity R, it also depends on the accuracy of γ and R_0. Suppose we use $R_0 = 100\ \Omega$ and $\gamma = 3.85 \cdot 10^{-3}\ ^\circ\text{C}^{-1}$, to calculate the temperature. What will the uncertainty of T be?

Well, we don't have any information about the uncertainties of R_0 and γ, so we will have to estimate it by guessing ('guesstimating'). We only know that $\gamma = 3.85 \cdot 10^{-3}\ ^\circ\text{C}^{-1}$, and we will have to assume that this number has been 'correctly rounded'. That implies that the 'true' value of γ is somewhere in the range

$$3.845 \cdot 10^{-3} < \gamma < 3.855 \cdot 10^{-3}$$

Any γ value in this range would be rounded to $3.85 \cdot 10^{-3}\ ^\circ\text{C}^{-1}$ if you only use three significant digits. Hence, if the only information we have is that $\gamma = 3.85 \cdot 10^{-3}\ ^\circ\text{C}^{-1}$, then it is a reasonable assumption that $\gamma = (3.850 \pm 0.005) \cdot 10^{-3}\ ^\circ\text{C}^{-1}$. And since we base that assumption on a 'rounding', the correct value can be anywhere in that range; it follows that we must assume a uniform distribution function. The standard uncertainty of γ, that we would use in the uncertainty budget, would be (see Eq. (13.29))

$$u(\gamma) = \frac{1}{\sqrt{3}} \cdot 0.005 \cdot 10^{-3} = 0.00289 \cdot 10^{-3}\ ^\circ\text{C}^{-1}$$

With the same reasoning, the reasonable range of R_0 would be $100.0 \pm 0.5\ \Omega$, with a standard uncertainty of $0.289\ \Omega$. Both the uncertainties of γ and R_0 are of the order of ‰. We would have to find the sensitivity coefficients to really understand the impact they have on the uncertainty of T, but considering the relative uncertainties of γ and R_0, a ppm accuracy in the measured R-value is likely to be redundant. We will illustrate this with another example.

Example 14.2 A BPW21 photo diode and a resistor are used to measure light flux. A 4½ digits DMM measures the voltage over the resistor, see Fig. 14.6. The resistor has a nominal resistance of 10 kΩ and has a 2% precision. The DMM uncertainty

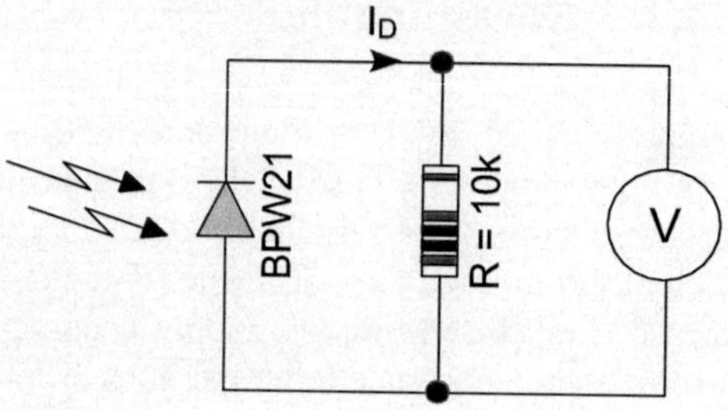

Fig. 14.6 Measuring light flux

is ± (0.08% of rdg + 2 digits) and the nominal sensor constant for the photodiode is 9.2 nA/lx. In a previous measurement, the sample standard deviation was $s =$ 4 mV. One sample reads 1.2843 V. If you were asked to determine the light flux in this experiment, how many samples would you take (how many samples would you average)?

('x digits' is sometimes used instead of 'range'. It means 'x' units of the last display digit's weight.) Using E for the light flux, we get

$$U_m = kER \Rightarrow E = \frac{U_m}{kR} = \frac{X_1 + X_2}{kR} = f(X_1, X_2, k, R)$$

$$u(X_2) = \frac{1}{\sqrt{3}}\left(0.08\frac{1}{100} \cdot 1.2843 + 0.002\right) = 0.7087 \text{ mV}$$

$$k = 9.20 \pm 0.05 \text{ nA/lx} \Rightarrow$$

$$u(k) = \frac{1}{\sqrt{3}}0.05 = 0.0289 \text{ nA lx}^{-1} \quad u(R) = \frac{1}{\sqrt{3}}10000 \cdot \frac{2}{100} = 115.5\ \Omega$$

$$c(X_1) = \frac{df}{dX_1} = \frac{1}{kR} = \frac{1}{9.2 \cdot 10^{-9} \cdot 10000} = 1.087 \cdot 10^4 \text{ lx V}^{-1} = c(X_2)$$

$$c(k) = \frac{df}{dk} = \frac{X_1 + X_2}{k^2 R} = \ldots = 1.517 \cdot 10^{12} \text{ A}^{-1}\text{lx}^2$$

$$c(R) = \frac{df}{dR} = \frac{X_1 + x_2}{kR^2} = \ldots = 1.396 \text{ lx } \Omega^{-1}$$

This gives us the uncertainty budget in Table 14.5. From this budget we can see that it is the uncertainty of R that dominates the contributions to the estimate's uncertainty; it is about 20 times larger than the smallest contribution (from X_2). We can conclude that an uncertainty of, say 10 lx, from X_1 wouldn't have any significant impact on the total uncertainty. That means that

$$u(X_1) \cdot c(X_1) = \frac{s}{\sqrt{N}} \cdot c(X_1) \le 10 \Rightarrow N \ge \left(\frac{s \cdot c(X_1)}{10}\right)^2$$

Table 14.5 The uncertainty budget

Quantity	Estimate	$u(x_i)$	$c(x_i)$	$c(x_i)\cdot u(x_i)$
X_1	1.2843 V	$s/\sqrt{N}$ V	$1.087\cdot 10^4$ lx V^{-1}	$u(X_1)\cdot c(X_1)$
X_2	0 V	0.709 mV	$1.087\cdot 10^4$ lx V^{-1}	7.70 lx
k	$9.2\cdot 10^{-9}$ nA lx^{-1}	0.0289 nA lx^{-1}	$1.5171.087\cdot 10^{12}$ A^{-1} lx^2	43.9 lx
R	10,000 Ω	115.5 Ω	1.396 lx Ω^{-1}	161.2 lx
E	13,959.5 lx			***u*(*E*)**

$$N \geq \left(\frac{4 \cdot 10^{-3} \cdot 1.087 \cdot 10^4}{10}\right)^2 = 18.9$$

Conclusion: We don't need to take more than 20 samples; after that, the uncertainty of the other quantities hurts us more than the uncertainty from the sample variation and the DMM uncertainty.

14.5 Summary

Being able to determine the uncertainty in a measurement is extremely important and should be considered as a 'fundamental' skill for any measurement personal. The math is not 'advanced', but there are a lot of numbers to handle, and the uncertainty budget is a good way to organize them. As we have seen in this chapter, the budget does not only produce the standard uncertainty of the output estimate, but it also provides important information about which quantity is hurting our overall accuracy most; we know what to do first if we need to improve the accuracy.

In Fig. 14.2, we demonstrated that the uncertainty of a digital DMM is typically stated as ± (% of reading + % of range). Students often ask where these uncertainties come from. That information is not so easy to find in the literature (or DMM vendors' manuals), but we can draw some conclusions from what we have learned so far. A digital DMM is of course based on an ADC, and we learned in Chap. 11 that ADCs have an inherent uncertainty of ± 0.5 LSB, which accounts for the '% of range' uncertainty.

The other contribution, '% of reading', is a little harder to motivate; this is an uncertainty that increases with the input sample's voltage level! To explain that we need to consider the hardware design of the dual slope. During the charging phase (see Fig. 11.18) a capacitor is charged, and this charging must be linear for the design to work. It is reasonable to assume that it is not linear and that it becomes more non-linear the more charge we have on the capacitor. That would explain why the uncertainty increases with the 'reading'.

14.6 Solved problems

Problem 14.1 In Fig. 14.7 we measure the electric power generated in a resistor. The amp meter is a 3½ digits handheld DMM with uncertainty ± (0.2% of rdg + 1 digit) and the display reads 0.467 A (stable, no flickering). The voltage meter is a 6½ digits desktop DMM with uncertainty ± (0.02% of rdg + 0.04% of range). We took 18 samples with an average of 0.7594175 V and a sample standard deviation of $s = 367\ \mu V$. What is the 95% confidence interval of the power in this measurement?

Solution: The power is $P = UI = (U_1 + U_2) \cdot (I_1 + I_2) =$
$= (0.7594175 + 0) \cdot (0.467 + 0) = \underline{354.648\ \text{mW}} = f(U_1, U_2, I_1.I_2)$
$u(U_1)$ is the type A uncertainty of the voltage measurement: $u(U_1) = \frac{367}{\sqrt{18}} = \underline{86.50\ \mu V}$. $u(U_2)$ is the type B uncertainty of the voltage measurement. Since the reading (average) is 0.7594175 V, we conclude that the range used was '1 V':

$$u(U_2) = \frac{1}{\sqrt{3}}\left(0.02\frac{1}{100} \cdot 0.759417 + 0.04\frac{1}{100} \cdot 1\right) = \underline{318.6\ \mu V}$$

The type A uncertainty of the current measurement, $u(I_1)$, is $= 0$ (since there was 'no flickering' on the DMM display). The type B uncertainty is

$$u(I_2) = \frac{1}{\sqrt{3}}\left(0.2\frac{1}{100} \cdot 0.467 + 0.001\right) = \underline{1.117\ \text{mA}}$$

Sensitivity coefficients are

$$c(U_1) = \frac{df}{dU_1} = I_1 + I_2 = \underline{0.467\ \text{A}} = c(U_2)$$

$$c(I_1) = \frac{df}{dI_1} = U_1 + U_2 = \underline{0.7594175\ \text{V}} = c(I_2)$$

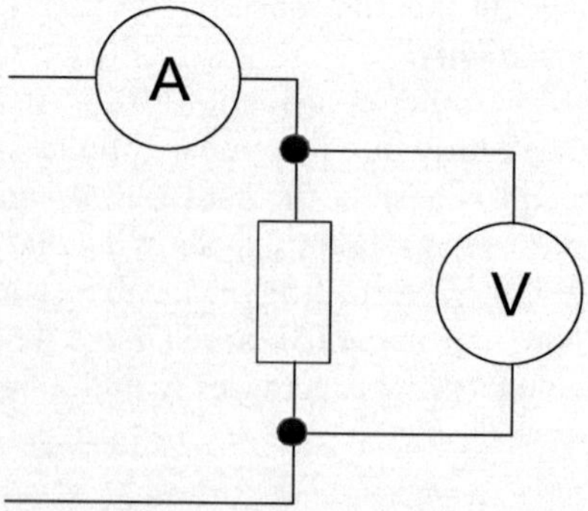

Fig. 14.7 Power measurement

This gives us the uncertainty budget in Table 14.6. The standard uncertainty of the power is $u\left(\hat{P}\right) = \sqrt{40.40^2 + 148.8^2 + 848.3^2} = \underline{862.2}\ \mu\text{W}$. Since the number of samples are less < 30, we need to find the effective degrees of freedom:

$$\upsilon_{eff} = \frac{862.2^4}{\frac{40.40^4}{17} + \frac{148.8^4}{\infty} + \frac{848.3^4}{\infty}} \gg 30 \rightarrow k = 2$$

$$U = 2 \times 862.2\ \mu\text{W} = 1.72\ \text{mW} \Rightarrow 1.8\ \text{mW} \Rightarrow \underline{\underline{P = 354.6 \pm 1.8\ \text{mW}\ (95\ \%\ )}}$$

Problem 14.2 Figure 14.8 illustrates a flow measurement in a water pipe. The output voltage of the sensor depends on the volume flow q as $U_m = \alpha \times \sqrt{q}$, and according to the datasheet, the sensor constant is 5.0 mV$\cdot$(l/min)$^{-0.5}$. The differential amplifier has an amplification of 175 and a 5½ digits DMM was used. The DMM uncertainty was ± (0.08% of rdg + 0.05% of range). We took eight readings as reported in Table 14.7. What is the 95% confidence interval of the volume flow q in this case?

Solution: From Table 14.7 we conclude that the DMM range used was 10 V.

$$u_m = \alpha\sqrt{q} \cdot F \Rightarrow q = \frac{u_m^2}{\alpha^2 F^2} = \frac{(X_1 + X_2)^2}{\alpha^2 F^2} = f(X_1, X_2, \alpha, F)$$

$$X_1 = \overline{u}_m = \frac{1}{8}(2.6946 + \ldots. + 2.8276) = \underline{2.7414375\ \text{V}}$$

$$\hat{q} = \frac{(2.7414375 + 0)^2}{0.005^2 \cdot 175^2} = \underline{9.81614\quad 1 \cdot \text{min}^{-1}}$$

$$s = \sqrt{\frac{1}{8-1}\left((2.6946 - 2.7414375)^2 + \ldots + (2.8276 - 2.7414375)^2\right)} = 0.07802\ \text{V}$$

$$u(X_1) = s/\sqrt{8} = 0.07802/\sqrt{8} = \underline{0.0276\ \text{V}}$$

$$u(X_2) = \frac{1}{\sqrt{3}}\left(0.08\frac{1}{100} \cdot 2.7414375 + 0.05\frac{1}{100} \cdot 10\right) = \underline{0.00415\ \text{V}}$$

Table 14.6 The uncertainty budget

Quantity	Estimate	$u(x_i)$	$c(x_i)$	$c(x_i)\cdot u(x_i)$
U_1	0.7594175 V	86.50 µV	0.467 A	40.40 µW
U_2	0 V	318.6 µV	0.467 A	148.8 µW
I_1	0.467 A	0 A	0.7594175 V	0 µW
I_2	0 A	1.117 mA	0.7594175 V	848.3 µW
P	354.648 mW			**862.2 µW**

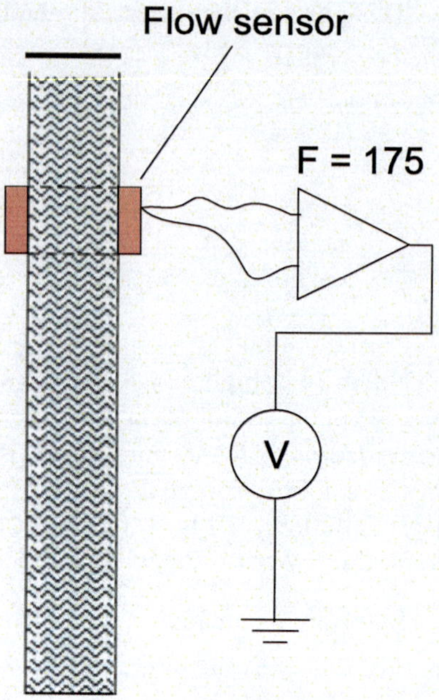

Fig. 14.8 Flow measurement

Table 14.7 We took eight samples

U_m (V)	02.6946	02.6650	02.7918	02.7186	02.7885	02.8256	02.6198	02.8276

We know nothing about the uncertainties of α and k, so we will have to 'guestimate':

$$\alpha = 5.00 \pm 0.05 \text{ mV} \cdot \text{l}^{-0.5}\text{min}^{0.5} \Rightarrow u(\alpha) = \frac{1}{\sqrt{3}} \cdot 0.05 = \underline{0.0289 \text{ mV} \cdot \text{l}^{-0.5}\text{min}^{0.5}}$$

$$F = 175.0 \pm 0.5 \Rightarrow u(F) = \frac{1}{\sqrt{3}} \cdot 0.5 = \underline{0.289}$$

$$c(X_1) = \frac{df}{dX_1} = \frac{2 \cdot (X_1 + X_2)}{\alpha^2 F^2} = \ldots = \underline{7.161 \text{ V}^{-1} \cdot \text{l}^{-1} \cdot \text{min}} = c(X_2)$$

$$c(\alpha) = \frac{df}{d\alpha} = (-)2 \times \frac{(X_1 + X_2)^2}{\alpha^3 F^2} = \ldots = \underline{3926.5 \text{ V}^{-1} \cdot \text{l}^{3/2} \cdot \text{min}^{-3/2}}$$

$$c(F) = \frac{df}{dF} = (-)2 \times \frac{(X_1 + X_2)^2}{\alpha^2 F^3} = \ldots. = \underline{0.1122 \ \text{l} \cdot \text{min}^{-1}}$$

$$u(\hat{q}) = \sqrt{0.198^2 + 0.0297^2 + 0.114^2 + 0.0325^2} = 0.233 \ \text{l} \cdot \text{min}$$

Table 14.8 The uncertainty budget

Quantity	Estimate	$u(x_i)$	$c(x_i)$	$c(x_i)\cdot u(x_i)$
X_1	2.7414375	0.0276	7.161	0.198 l/min
X_2	0	0.00415	7.161	0.0297 l/min
α	5.0	0.0289	3926.5	0.114 l/min
F	175	0.289	0.1122	0.0325 l/min
q	9.81614 l/min			**0.233 l/min**

$$\nu_{eff} = \frac{0.233^4}{\frac{0.198^4}{7} + \frac{0.0297^4}{\infty} + \frac{0.0297^4}{\infty} + \frac{0.0325^4}{\infty}} = 13.4 = 13 \rightarrow k = 2.21$$

$$U = 2.21 \times 0.233 = 0.52 \quad \underline{\underline{q = 9.82 \pm 0.52\ \text{l/min}\ (95\%)}}$$

That gives us the uncertainty budget in Table 14.8.

From the uncertainty table, it is obvious that we should take more samples.

References

1. BIPM. (2008) Evaluation of measurement data—guide to the expression of uncertainty in measurement, JCGM 100: 2008 GUM 1995 with minor corrections. Jt Comm Guid Metrol 98
2. Expression of the Uncertainty of Measurement in Calibration (1999) European co-operation for accrediation
3. Keysight Technologies I (2022) Digital multimeters 34460A, 34461A, 34465A (6 1/2 digit), 34470A (7 1/2 digit)

Chapter 15
The Lock-In Amplifier

Abstract The lock-in amplifier (LIA) is a common instrument in physics laboratories that can detect signals with extremely low signal-to-noise ratios. However, to really take advantage of its potential, a basic understanding of its operating principle is necessary. This chapter first introduces the phase sensitive detector (PSD) and evolves it into a lock-in amplifier. At the end of this chapter, the dual-phase lock-in amplifier is introduced and the I and Q signals are defined.

15.1 Introduction

The fundamental problem that we address in this book is how to find a 'sinusoidal signal in noise'. We have provided several solutions for that already (and will provide a few more) and the solution depends on the circumstances: What kind of noise do we have? (NM/CM? Random/periodic?) How large is the noise? (SNR?). Do we know the period of the sine or not? etc. Each problem has its 'best' solution. In this chapter we will address the problem where the sinusoidal frequency is *known*; in fact, we *control* the frequency. This is not unusual, think of the sinusoidal as the 'excitation frequency' of the experiment. For example, in a physics lab we might use a pulsed laser (or a laser beam 'chopper') and then the laser pulse frequency is the 'excitation frequency' (which we control) of the experiment. So, we know/control the frequency, and we want to determine the amplitude. The 'catch' is that the sine (or any periodic 'response signal') is 'buried' in noise; here we will assume an *extremely* low SNR (<< 1).

Here is a common argument (misunderstanding): 'If we *know* the frequency, why don't we just design a resonance filter (as described in Chap. 9) with a resonance frequency that matches our sinusoidal signal?'.

That won't work for several reasons. First, the SNR is so low, that we would need an *extremely* narrow resonance filter (Q > 10,000) and such narrow resonance filters are impossible (?) to design with standard analog components. Second, even if we could design such a narrow filter it would need to have a stability on a 'ppm level' or the signal frequency would 'slip off' the filter resonance, and the signal would

L. Bengtsson, *Electrical Measurement Techniques*,
https://doi.org/10.1007/978-981-99-8187-8_15

be 'killed' with the noise. All it would take is that the signal frequency, or the filter parameters, change just a little bit (due to temperature, humidity....) and we would fail to detect our sinusoidal signal.

We *will* build a 'resonance filter' with an extremely high-Q value ($> 10^6$), but for that to work, we can't use ordinary components as we did in Chap. 9. For such a high-Q filter to work, the signal frequency must be 'locked' to the filter's resonance frequency; if one changes, the other must follow. Such a narrow 'resonance filter' is called a *lock-in amplifier* (LIA).

15.2 Phase Sensitive Detector

15.2.1 PSDs

Before we get into the lock-in amplifiers, there are some electronics that we need to introduce. First, we need a *phase sensitive detector* (a 'PSD'). Imagine that we have the following signal:

$$x(t) = 0.01 \sin 2\pi 100t + 10 \sin 2\pi 120t + \text{'lots of white noise'} \tag{15.1}$$

We assume that the 100 Hz signal is the 'good' signal that we want to detect, and the other signal parts are 'noise'. By 'detect' we mean that we want to find its *amplitude* (we *know* the frequency, remember?). This signal is illustrated in Fig. 15.1.

The signal-to-noise ratio here is so bad that there is not even the slightest hint of it in Fig. 15.1. Earlier, when we learned about Fourier transforms, we demonstrated that we could just do an FFT of the signal to 'detect' a small signal in a lot of noise.

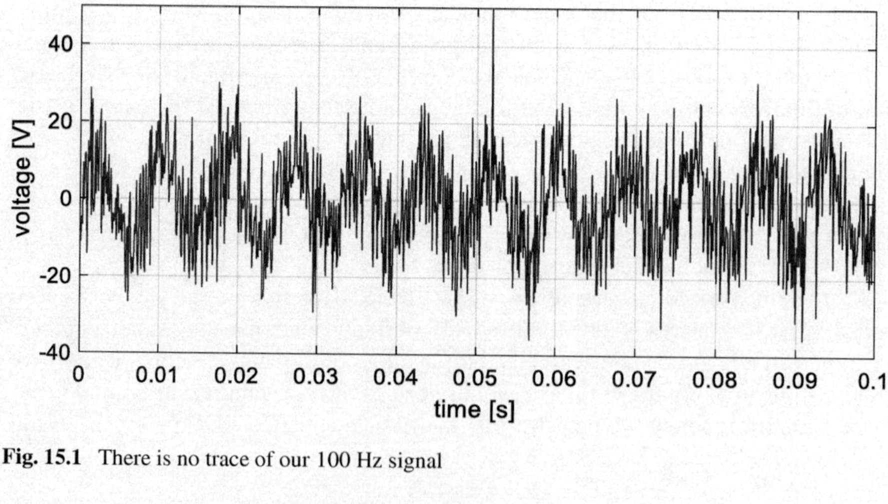

Fig. 15.1 There is no trace of our 100 Hz signal

That won't work here. First, we *know* what frequency we are looking for, we don't need to do a spectral analysis to find the frequency. Second, the SNR is so poor that we wouldn't see it anyway. (If you don't believe it, Fig. 15.2 illustrates the FFT spectrum of the signal in Fig. 15.1.) Nevertheless, we need to detect it. That seems like 'mission impossible', but since we *know* what frequency we are looking for, we will find it; all it takes is some clever sampling and some time (or lots of time, depending on the SNR). Here is what we are going to do: The frequency is 100 Hz, i.e., the period of the signal is 10 ms. We are going to sample the signal with a sampling frequency of 5 ms (200 S/s), taking *exactly two* samples per period, see Fig. 15.3.

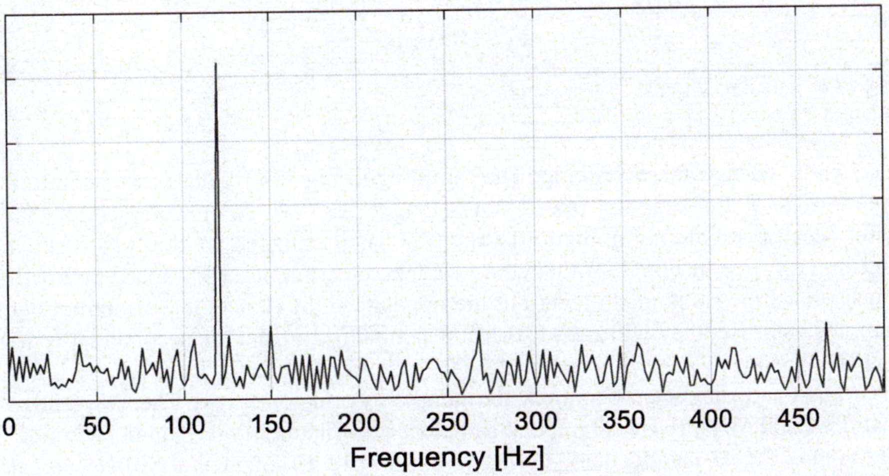

Fig. 15.2 There is nothing at 100 Hz

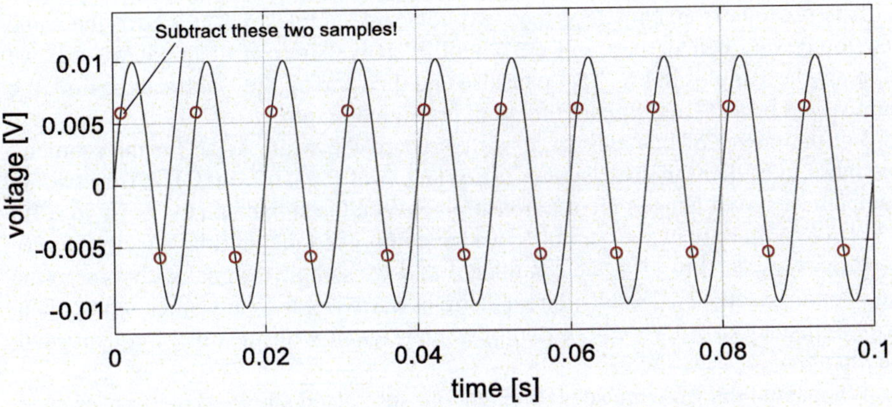

Fig. 15.3 PSD in software: take samples 180° apart and subtract [1]

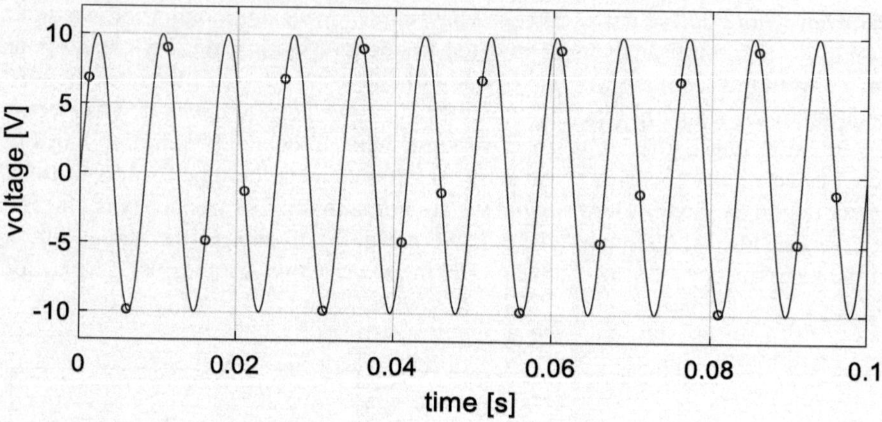

Fig. 15.4 Sampling a 120 Hz signal at 200 S/s

(I know what you are thinking: The sampling in Fig. 15.3 violates the sampling theorem since the sampling rate is not > $2f_{signal}$. But we don't worry about that here, because we are not going to do any FFT or filtering, so we don't care about aliasing!) Notice in Fig. 15.3 that since we take our samples 180° apart, they will always have the same magnitude, but the opposite sign! Hence, if we subtract our samples pairwise, they will *add*; if the first sample is $x(t)$ and the second one is $x(t + T/2)$, then $x(t + T/2) = -x(t)$, and $x(t) - x(t + T/2) = 2x(t)$. If we take samples 180° apart and subtract, we amplify the sample by a factor of two. Then we realize that if we add N 'pairwise subtracted' samples, we will amplify the signal $2N$ times! The sum of the 10 'sample pairs' in Fig. 15.3 is ≈ 0.12. Compare that with the signal amplitude which is 0.01.

Let's see what happens to the other signal components in Eq. (15.1). In Fig. 15.4 we sample the 120 Hz signal with the same sampling rate.

It is obvious from Fig. 15.4 that two adjacent samples will not have the same magnitude and hence, they will not 'amplify' if we subtract them pairwise. If we subtract the samples in Fig. 15.4 pairwise and then add all the 'subtracted pairs', the sum would be ≈ -1! Compare that with the signal amplitude of 10.

Let the power of this sink in; if we disregard the white noise for the moment, we have an SNR in the raw signal in Eq. (15.1) of $0.01/10 = 0.001$. After twenty samples we have changed the relationship between the signal and the noise to $0.12/1 = 0.12$. With only twenty samples, we have improved the SNR by a factor of 120!

From Figs. 15.3 to 15.4, we conclude that if we sample a signal with a sampling rate that is exactly $T_0/2$, and subtract samples pairwise, then any signal component with frequency $f_0 = 1/T_0$ will be *amplified* and signals with any other frequency will be *attenuated*!

What about the white noise? If we treat our subtracted pairs as a random variable, $Y = X_1 - X_2$, where the X variables are normally distributed with a standard deviation of σ (same as the noise), then it is easy to prove that Y has a standard deviation of

$\sqrt{2}\sigma$. Hence, if we take the average of N subtracted pairs (just accumulating will not work, we will overflow), then we know from Chap. 13 that the standard deviation of the average will be $\sigma\sqrt{2/N}$; the random noise will also be reduced as we take more samples.

Hence, if we just keep sampling the signal in Eq. (15.1) (at 200 S/s) and average 'subtracted pairs', the signal with frequency f_0 will gradually materialize from the noise since this sampling technique gradually amplifies the f_0 signal and attenuates anything else. By 'materialize', we mean that we will be able to detect its presence despite the extremely poor SNR in the raw signal.

The conclusion is that if the signal we are looking for is periodic, with a *known* period, we can always find it (at least in theory); it is only a matter of time (take enough samples).

We can also learn something else from Fig. 15.3. We take the first sample at a phase angle of approximately 45° and the sample value is approximately 0.006. We realize that we could have done better! If we instead take the first sample at a phase angle of 90°, the first sample would be 0.01 and the amplification by accumulating pairwise subtractions would have been even larger.

On the other hand, we could have done worse too. If we take the first sample at a phase angle of 0°, all samples would be 0 and there would be no amplification at all. So, this sampling strategy is a little precarious; the result depends on the phase angle of the first sample! In fact, what we have here is a *phase sensitive detector*, a PSD; the outcome depends on the phase angle.

15.2.2 Analog PSDs

The sampling technique illustrated in Fig. 15.2 is the 'digital' version of a PSD. It is 'digital' because the PSD is implemented in software. Figure 15.5 illustrates the classic 'analog' PSD that can be implemented in hardware.

Here, the reference signal is a square signal that controls a bipolar switch; when the reference signal is 'on' the signal $x(t)$ is routed through the switch, and when the reference signal is 'off', the signal $-x(t)$ is routed through. If there is a sinusoidal in $x(t)$ with a frequency that is exactly $f_0 = 1/T_0$, and in phase with $r(t)$ ($\varphi = 0°$),

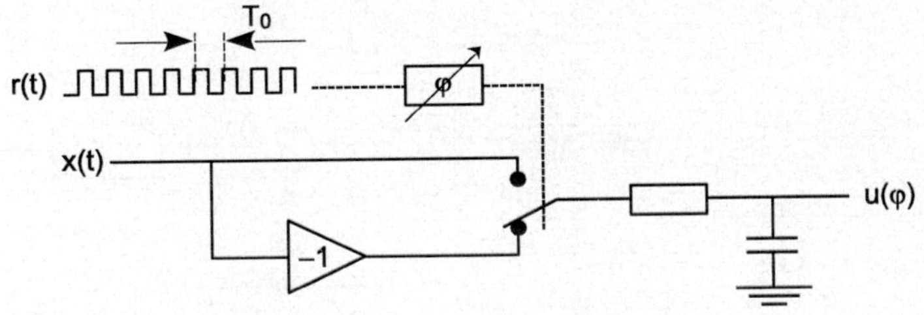

Fig. 15.5 The classic analog PSD

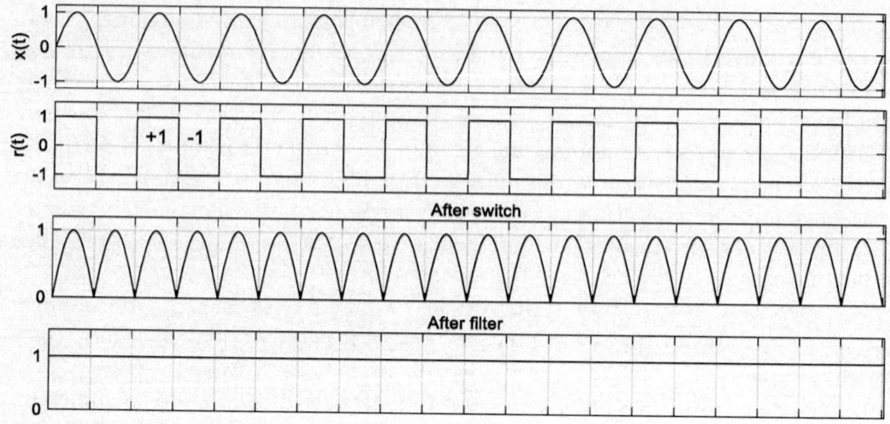

Fig. 15.6 We have a rectifier

then the first part of the electronics in Fig. 15.5 is a *rectifier*. This is illustrated in Fig. 15.6.

The post-processing lowpass filter will 'smooth' the signal to a DC voltage. Notice the phase angle regulator in Fig. 15.5; with this 'knob' we can adjust the phase angle arbitrarily to match the phase of $x(t)$. In this case, the phase should be exactly 0° for a maximum output. A gradual phase shift from 0° to 180° would generate a change in $u(\varphi)$ from $+1$ to -1; we have a phase sensitive detector.

15.2.3 *Multiplying PSDs*

The PSD in Fig. 15.5 has a disadvantage, the switch. Even if this switch is a semiconductor relay, it limits the frequency range we can use. Lock-in amplifiers, depend on PSDs, but they use a different technique, like the heterodyne technique we used in Chap. 8; we multiply the signal with the reference signal, see Fig. 15.7.

If we multiply them, the multiplier output $y(t)$ is (see Chap. 8)

$$y(t) = A\{\cos((\omega_x + \omega_0)t + \varphi_x + \varphi_0) + \cos((\omega_x - \omega_0)t + \varphi_x - \varphi_0)\} \quad (15.2)$$

Fig. 15.7 A multiplying PSD

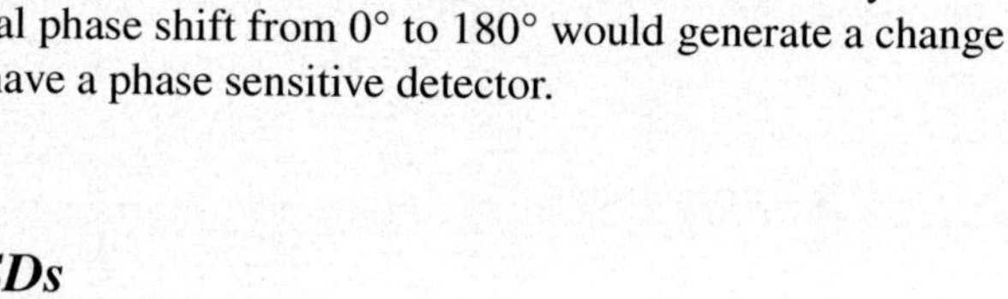

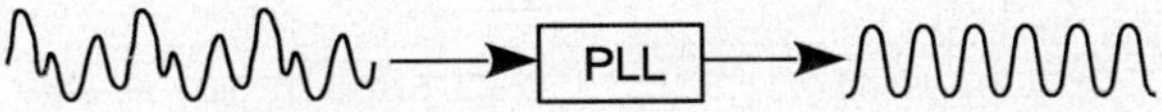

Fig. 15.8 A PLL shapes the reference signal

The lowpass filter will stop the 'sum frequency' signal. We assume that the cutoff frequency of the lowpass filter is so low, that it blocks all frequencies $\neq 0$; anything but DC is blocked by the lowpass filter. Hence, the only case where *anything* comes out of the lowpass filter is if $\omega_x = \omega_0$. In that case

$$u(t) = A\cos(\varphi_x - \varphi_0) \tag{15.3}$$

From Eq. (15.3) it is obvious that the multiplier + filter in Fig. 15.7 is a phase detector. For a 'perfect' detection, we would need a phase angle shifter (as in Fig. 15.5) to match the phases of the signal and the reference signal. A *lock-in amplifier* does not need one though, at least not the more expensive ones. We explain that in Sect. 15.4.

15.3 Phase-Locked Loops

Before we present the lock-in amplifier, there is one more component we need to introduce. From Fig. 15.7 and Eq. (15.2) it is obvious that the detection depends on a reference signal that is a 'good' cosine. Keep in mind that the reference signal is the signal that we use to 'excite' the experiment. That could be a cosine, but more often it is not. In laser experiments, we either use a 'pulsed' laser or a 'chopper' to periodize a photo detector signal. In both cases the reference signal will be a square signal. It may also be 'noisy'. For that reason, the reference signal in a lock-in amplifier, is first passed through a *phase-locked loop* (PLL). We will not go into the details of PLLs here. It is a versatile, standard component that can divide or multiply a signal's frequency, but for our purposes, we will only use it as a 'signal formatting' component; we input a 'noisy' signal with some period, and the PLL will output a 'good' cosine with the same frequency (Fig. 15.8).

15.4 LIAs

A lock-in amplifier (LIA) is an 'advanced PSD'. The 'simple' LIAs have indeed a phase control knob that you have to adjust (manually) to 'align' the reference signal with the detector signal. (Which can be quite a challenge!). The 'advanced' LIAs produce a signal that is independent of the phase angle (but they still produce and display the phase angle). They achieve that by adding another multiplying PSD where the reference signal is phase-shifted 90° (converting a 'cosine' to a 'sine'). To see how this works, we need a few basic trigonometric expressions. In Chap. 8 we multiplied two cosines. If we multiply a sine and cosine, we get Eq. (15.4).

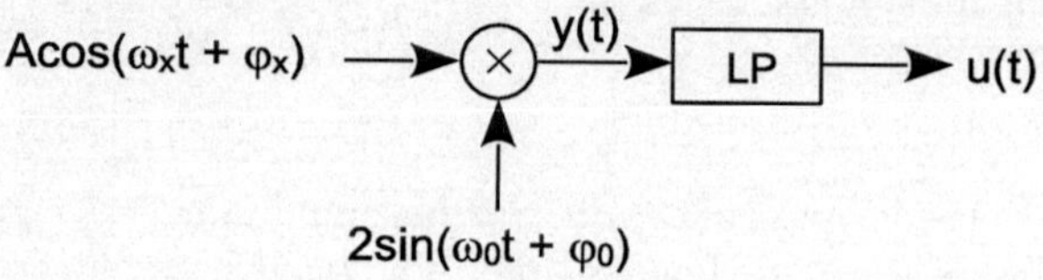

Fig. 15.9 Reference signal is phase-shifted 90°

$$\sin\alpha \times \cos\beta = \frac{1}{2}(\sin(\alpha+\beta) + \sin(\alpha-\beta)) \tag{15.4}$$

In Fig. 15.9 we have a multiplying PSD where the reference signal is phase-shifted 90° versus the signal.

If we apply Eq. (15.4), and the same reasoning that gave us Eq. (15.3), we see that the filter output in Fig. 15.9 will be

$$u(t) = A\sin(\varphi_x - \varphi_0) \tag{15.5}$$

Next, we apply the Pythagorean identity expression $\sin^2\alpha + \cos^2\alpha = 1$ to Eqs. (15.3) and (15.5):

$$\sqrt{A^2\cos^2(\varphi_x - \varphi_0) + A^2\sin^2(\varphi_x - \varphi_0)} = A \tag{15.6}$$

From Eq. (15.6), we see that if we have 'dual phase' detectors, with a 90° phase shift, we can get an output that is independent of the phase shift between the signal and the reference (which saves you a lot of time and frustration in the lab, believe me!). Figure 15.10 illustrates a 'dual phase' lock-in amplifier.

The 'output box' in Fig. 15.10 (the 'squaring' and 'root squaring') can be implemented either in hardware or software; here we will just treat it as a 'black box'. In a commercial LIA you will also find pre- and post-amplifiers, pre- and post-filters, several input signal options (current inputs, differential-ended inputs), etc.

When you use a lock-in amplifier, *you* provide the reference signal (from the 'excitation device' in your experiment) and the general rule is that you select a frequency

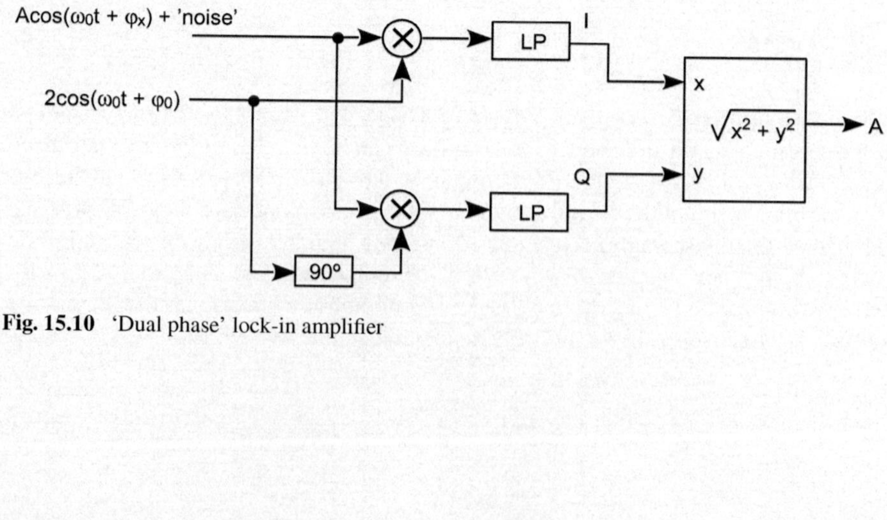

Fig. 15.10 'Dual phase' lock-in amplifier

'far away' from the local power line frequency (50/60 Hz) and its harmonics. The reason is that in many situations the noise comes from the power line (see Chap. 2) and so you should use some 'odd frequency number'. If the power line frequency is 50 Hz, you don't excite your system with 50 or 100 Hz, you use something 'odd' (like 217 Hz, for example).

Finally, a few comments about LIAs. First, notice in Fig. 15.10 that we have named the phase detectors' outputs 'I' and 'Q', respectively. 'I' stands for 'In-phase' (with the measurement signal) and 'Q' stands for 'phase shifted a Quarter of a period'. That is 'standard terminology'. Second, the LIA in Fig. 15.10 produces the signal magnitude only, but since we have access to both the I and Q parts, we can easily also produce the phase angle:

$$\varphi = \tan^{-1} Q/I \tag{15.7}$$

Some LIAs indeed do that and since they produce both the magnitude and the phase angle, they are sometimes referred to as 'vector voltage meters'.

Third, we can implement the 'I' and 'Q' parts in the digital (software) phase detector too. In 2011, Li et al. [2] presented an algorithm that accomplishes that. They took samples only π/2 apart (twice as fast as in Fig. 15.3) and showed that the I-part corresponds to $(x(1) - x(3))/4$ and the Q-part corresponds to $(x(0) - x(2))/4$.

15.5 Solved Problems

Problem 15.1 In a bioscience laboratory, researchers want to measure the resistance in a very thin cell tissue. The resistance is very small (mΩ) and since the properties of the cell sample are very temperature dependent, it is paramount that the sample is not heated during the measurement; the power must be minimized, i.e., the experiment conditions don't allow you to just crank up the current through your sample. It has been estimated that the current must not exceed 1 μA. How would you solve this problem?

Solution We measure resistance by sending current through the sample and measure the voltage across it. We will excite the sample with a cosine from a waveform generator (amplitude 1 V, frequency 217 Hz), and by adding a 1-MΩ series resistor, we make sure we don't violate the 1 μA current restriction. This circuit is connected to a lock-in amplifier as illustrated in Fig. 15.11.

In Fig. 15.11, the sample resistance is <<1 MΩ, so we may assume that the current in the circuit is 1 μA. The output voltage has been amplified 200,000 times; hence, the voltage across the sample is 0.134 V/200,000 = 0.67 μV. The sample resistance is 0.67 μV/1 μA = 0.67 Ω.

Problem 15.2 In an atomic physics experiment, atoms in a vacuum chamber are excited by a continuous wave (CW) laser, and the relaxation light is detected by a

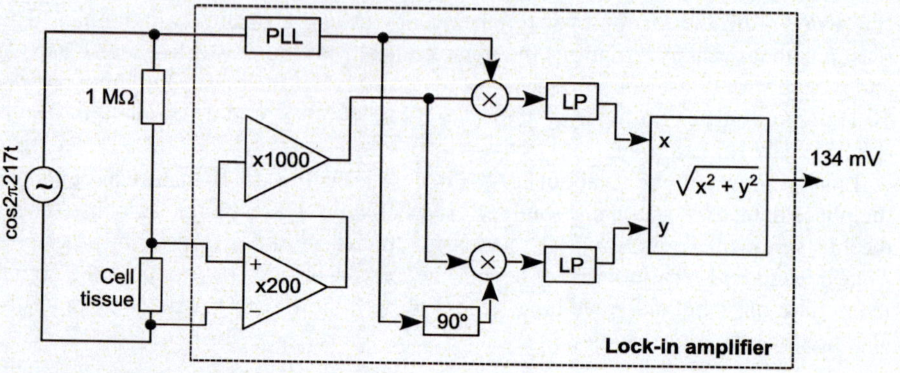

Fig. 15.11 Cell tissue experiment

photo sensor, see Fig. 15.12. However, due to low signal levels and ambient light, they are unable to detect the scattered relaxation light.

How would you suggest they solve this problem?

Solution The experiment must be 'excited', i.e., we must make sure that the light signal we are trying to detect only appears at the photo sensor with a certain frequency; a frequency that we control, or at least can measure confidently. To achieve that, we 'chopper' the laser beam. This is 'standard' laser accessory; a round plate with holes in, rotated by a stepper or DC motor, see Fig. 15.13.

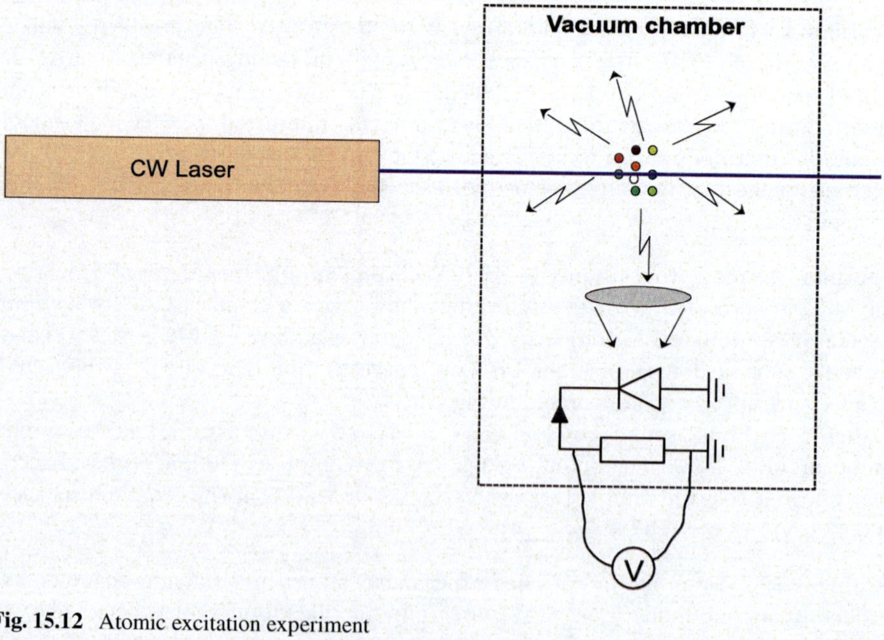

Fig. 15.12 Atomic excitation experiment

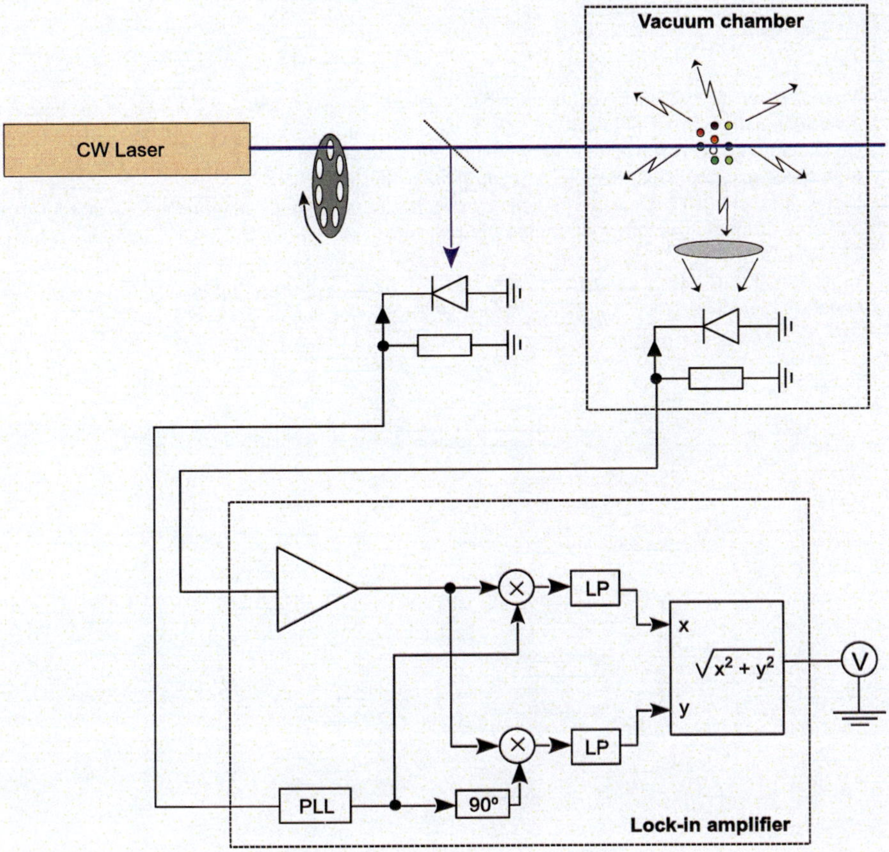

Fig. 15.13 'Excite' the experiment

That will 'excite' the system. The chopper system may provide a 'sync' signal, but if it doesn't, we use a glass plate to deflect a small fraction of the laser beam to another photo sensor; that photo sensor provides the reference signal to our LIA.

The entire solution is illustrated in Fig. 15.13. Even if this excites the detection signal to a unique frequency, it is usually a good idea to reduce the ambient light in the room anyway. Single photon detections have been reported with this technique [3].

For an introductory laboratory exercise, I recommend a paper by Libbrecht et al. [4].

References

1. Momo, F., et al. 1981. Microcomputer based phase sensitive detector. *Journal of Physics E: Scientific Instruments* 14 (11): 1253.
2. Li, G., et al. 2011. A novel algorithm combining oversampling and digital lock-in amplifier of high speed and precision. *Review of Scientific Instruments* 82 (9).
3. Wang, X., et al. 2012. Detection efficiency enhancement of single-photon detector at 1.55-μm by using of single photons lock-in and optimal threshold. *Optics & Laser Technology* 44 (6): 1773–1775.
4. Libbrecht, K., E. Black, and C. Hirata. 2003. A basic lock-in amplifier experiment for the undergraduate laboratory. *American Journal of Physics* 71 (11): 1208–1213.

Chapter 16
Correlation

Abstract By 'correlation', this book primarily refers to temporal correlation. This chapter defines the auto- and cross-correlation functions and presents their applications in signal processing. The cross-correlation function is compared to the convolution operator and from this a *matched* filter can be designed. Both analog and discrete-time correlation is treated, and computer implementations of correlation algorithms (such as 'circular' correlation) are discussed.

16.1 Introduction

In this context, 'correlation' refers to 'time-correlation' (or 'temporal' correlation). The (time) correlation function of two, time functions $x(t)$ and $y(t)$ is

$$R_{xy}(t) = \int_{-\infty}^{+\infty} x(\tau)y(t+\tau)d\tau \tag{16.1}$$

Does this expression look familiar? If we compare Eq. (16.1) with the convolution expression Eq. (9.27), we can see that they are identical, except that we don't time-reverse the second function. Hence, we can treat correlation just like convolution, except for the time-reversion (which for most students is easier to comprehend, conceptually).

There are two cases of Eq. (16.1); when $x(t) = y(t)$ and when $x(t) \neq y(t)$. The former case is called 'auto-correlation' ('correlate with yourself') and the second case is called cross-correlation ('correlate with someone else'). Both are important in physics, and they are used in quite different applications. Auto-correlation is primarily used to find a periodic signal in stochastic noise and cross-correlation is used to find a 'known' signal in a lot of noise (such as a radar echo, for example).

The close relationship with the convolution integral is also very important, and we will discuss this in detail in Sect. 16.2.1 ('matched filters'). We will also present the discrete-time correlation functions and some practical computational problems when it comes to correlation.

L. Bengtsson, *Electrical Measurement Techniques*,
https://doi.org/10.1007/978-981-99-8187-8_16

But first, we need some correlation theory, and we will start with cross-correlation.

16.2 Cross-Correlation

To illustrate the cross-correlation process (Eq. (16.1)), we will use the same signals that we used when we illustrated convolution in Sect. 9.8.1. First, we look at the signal $y(\tau) = 2\tau - 1$. This signal is illustrated in Fig. 16.1, and Fig. 16.2 illustrates $y(t + \tau)$ for different times t.

Comparing with Figs. 9.38 and 9.39, we can see that $y(t + \tau)$ is first, not time-reversed and second, when t goes from $-\infty$ to $+\infty$, $y(t + \tau)$ slides *from right to left* on the τ axis (the opposite direction compared to convolution).

Example 16.1 Figure 16.3 illustrates $x(t)$ and Fig. 16.4 illustrates $y(t)$. Find the cross-correlation between $x(t)$ and $y(t)$. (Compare the result with Example 9.5).

Solution Figure 16.5 illustrates how $y(t + \tau)$ moves along the τ axis relative $x(\tau)$. We can see that there is no overlap between the two signals for times $t < 0$ and $t > 3$. Figure 16.6 illustrates the signals for $0 < t < 1$. For this time interval, Eq. (16.1) is

$$R_{xy}(t) = \int_{1-t}^{1} (\tau - 1) \cdot 1 d\tau = \left[\frac{1}{2}\tau^2 - \tau\right]_{1-t}^{1}$$

$$= \frac{1}{2} - 1 - \frac{1}{2}(1 - \tau)^2 + (1 - \tau) = \cdots = \underline{-\frac{1}{2}\tau^2}$$

Figure 16.17 illustrates the two signals for $1 < t < 2$. Equation (16.1) is now:

Fig. 16.1 $y(\tau) = 2\tau - 1$

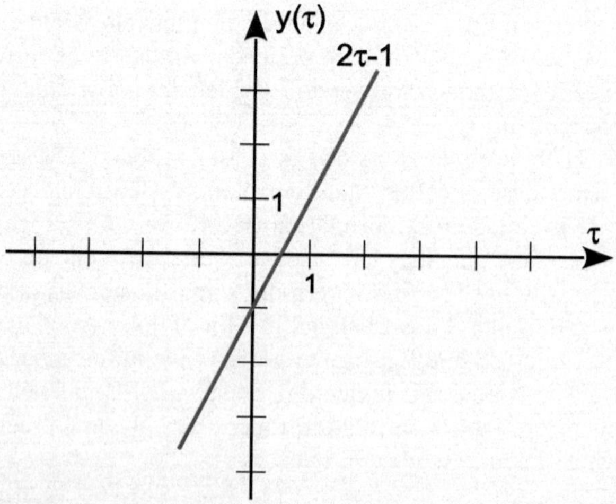

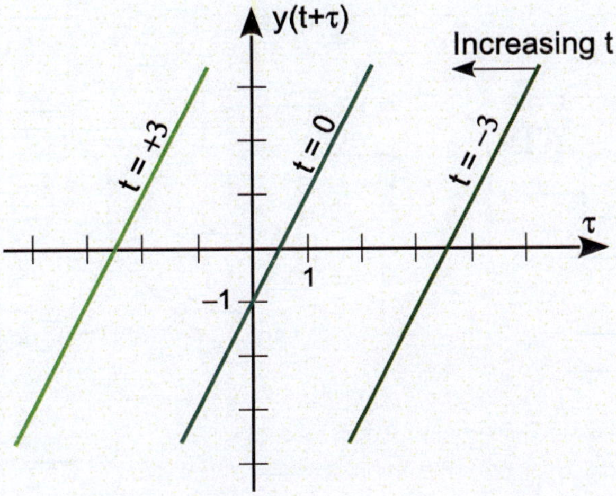

Fig. 16.2 $y(t + \tau)$ for $t = +3$, 0 and -3

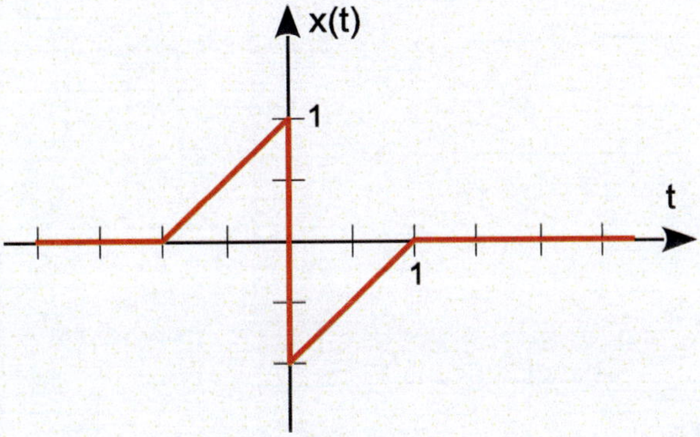

Fig. 16.3 $x(t)$

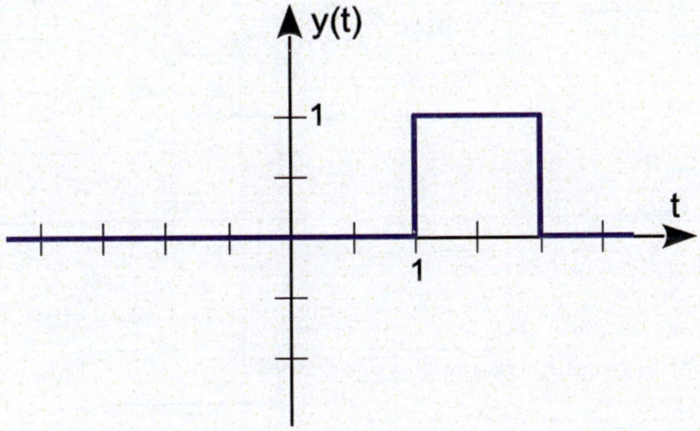

Fig. 16.4 $y(t)$

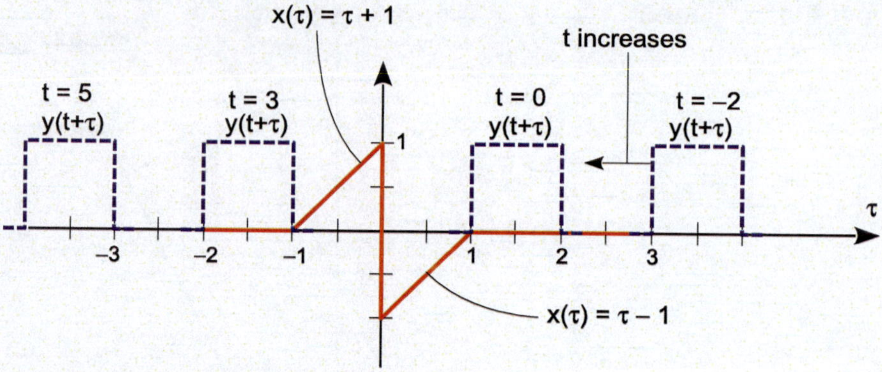

Fig. 16.5 $x(t)$ and $y(t+\tau)$ for some t values

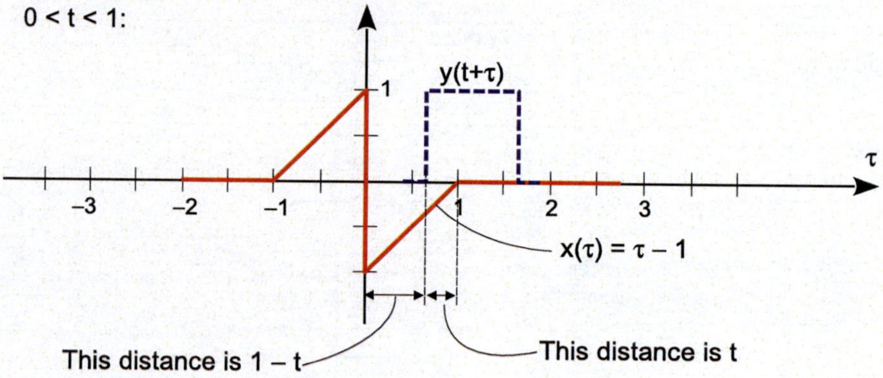

Fig. 16.6 $x(t)$ and $y(t+\tau)$ for some $0 < t < 1$.

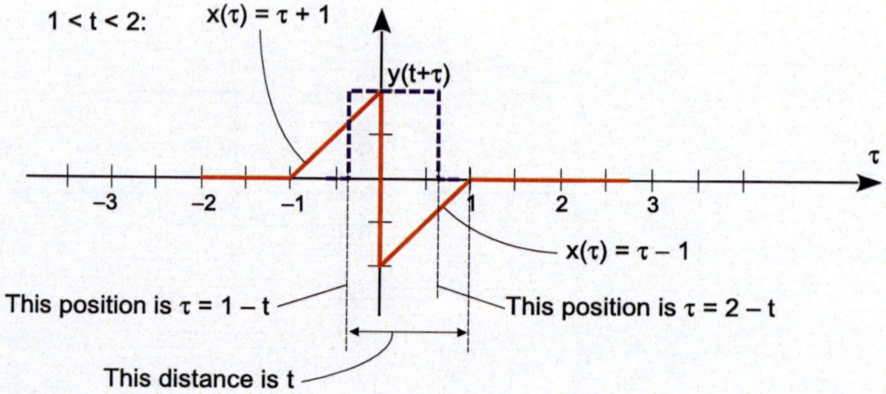

Fig. 16.7 $x(t)$ and $y(t+\tau)$ for some $1 < t < 2$

$$\int_{1-t}^{0} (\tau + 1) \cdot 1 d\tau + \int_{0}^{2-t} (\tau - 1) \cdot 1 d\tau = \left[\frac{1}{2}\tau^2 + \tau\right]_{1-t}^{0} + \left[\frac{1}{2}\tau^2 - \tau\right]_{0}^{2-t}$$

$$= 0 - \left(\frac{1}{2}(1-t)^2 + 1 - t\right) + \frac{1}{2}(2-t)^2 - 2 + t = \cdots = \underline{t - 1.5}$$

Finally, Fig. 16.8 illustrates the signals for $2 < t < 3$. The cross-correlation is

$$\int_{-1}^{2-t} (\tau + 1) d\tau = \left[\frac{1}{2}\tau^2 + \tau\right]_{-1}^{2-t} = \frac{1}{2}(2-t)^2 + 2 - t - \left(\frac{1}{2} - 1\right)$$

$$= \cdots = \underline{\frac{1}{2}t^2 - 3t + 4.5}$$

The resulting cross-correlation function is plotted in Fig. 16.9.

Cross-correlation has the following properties: First, it is commutative, i.e., $R_{xy}(t) = R_{yx}(t)$. Second, if x and y correlate to R_{xy} and x and z correlate to R_{xz}, then x and

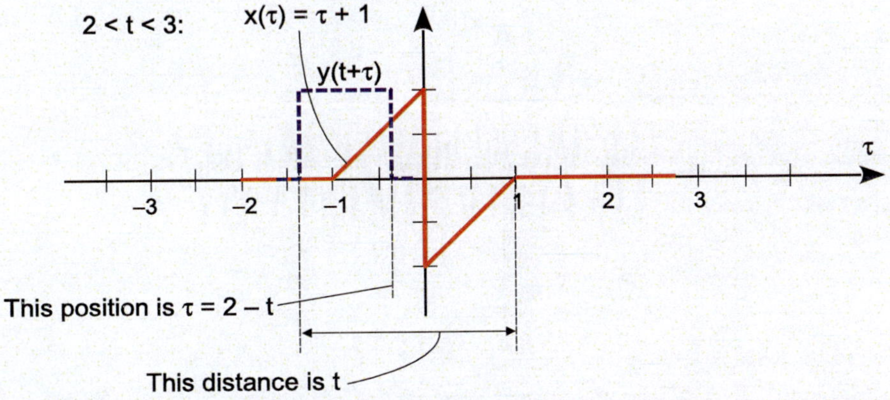

Fig. 16.8 $x(t)$ and $y(t + \tau)$ for some $2 < t < 3$

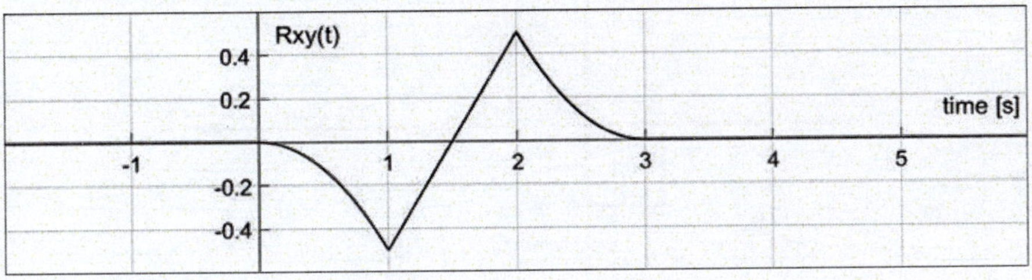

Fig. 16.9 The cross-correlation function

$(y + z)$ correlate to $R_{xy} + R_{xz}$. Also, if we compare Fig. 16.9 with Fig. 9.49, we can confirm that cross-correlation corresponds to convolution with a time-reversed impulse response.

So, what are the applications of cross-correlation? To see that, let's first cross-correlate white noise ($x(t)$, Fig. 16.10) with the symmetric, bipolar square signal in Fig. 16.11.

In Fig. 16.12, we have plotted $x(t)$ and $y(t + \tau)$ for some different times t. If we multiply $x(t)$ and $y(t + \tau)$, the resulting 'area' will be ≈ 0 for any time t; the cross-correlation of $x(t)$ and $y(t + \tau)$ is ≈ 0 everywhere.

In Fig. 16.13, $y(t)$ is a 'small', delayed copy of $x(t)$. If we cross-correlate $x(t)$ and $y(t)$ (let $y(t)$ 'slide' left), $R_{xy}(t)$ will be zero until they start to overlap (for $t = t_0 - 2$). $R_{xy}(t)$ will have a maximum for $t = t_0$ and then it will decrease to 0 again. Figure 16.14 illustrates $R_{xy}(t)$.

Hence, the 'peak' in the cross-correlation function indicates the position of the 'small copy' of $x(t)$. This is how radar and sonar systems work. They emit a known 'signature' chirp ($x(t)$), and they cross-correlate it with the echo detector signal and the 'peak' will correspond to the time it took for the chirp to travel back and forth to the 'target'.

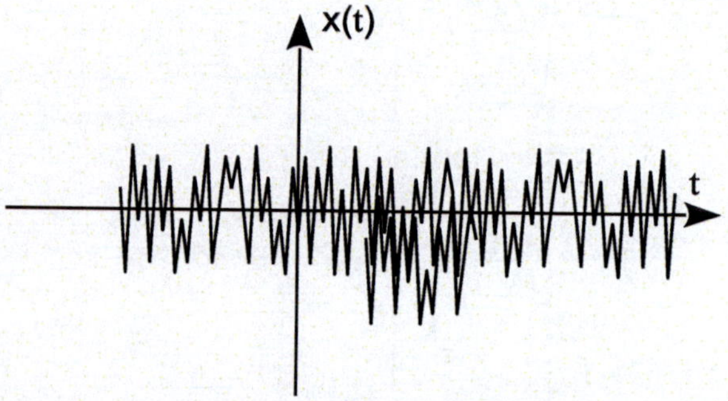

Fig. 16.10 White noise

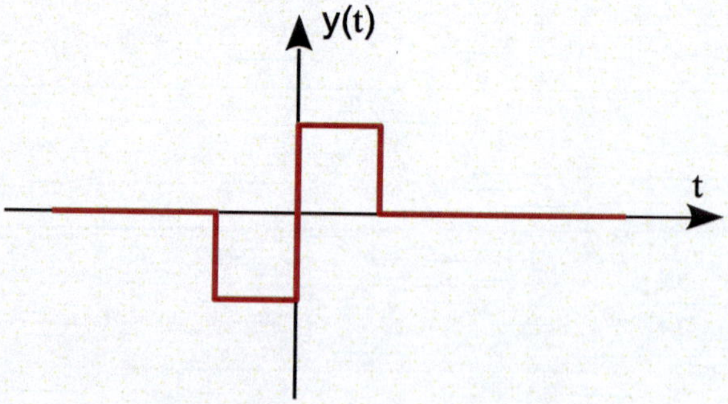

Fig. 16.11 Symmetric and bipolar

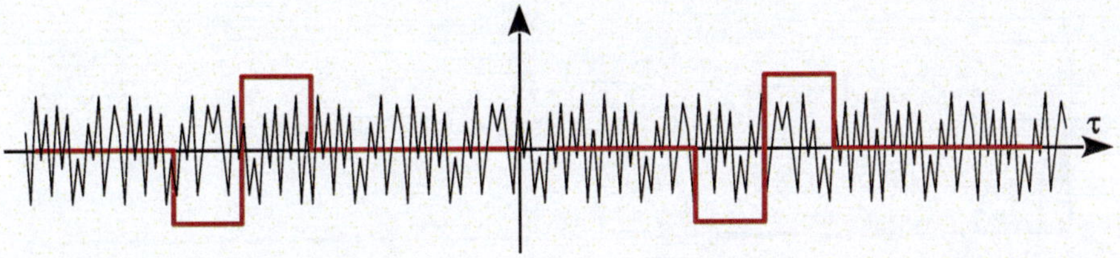

Fig. 16.12 $R_{xy}(t) \approx 0$ for all t

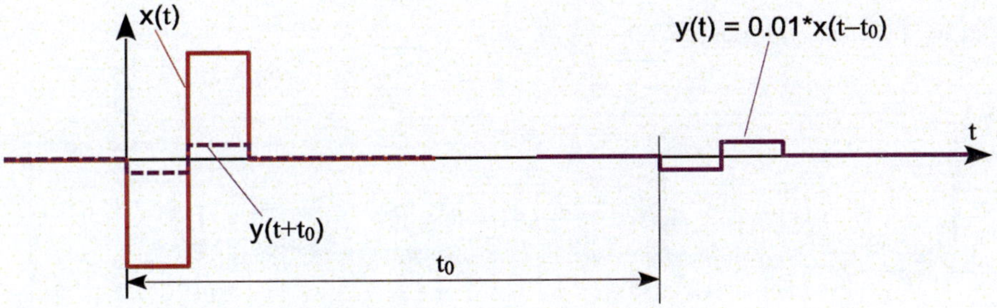

Fig. 16.13 $y(t)$ is a small delayed 'copy' of $x(t)$

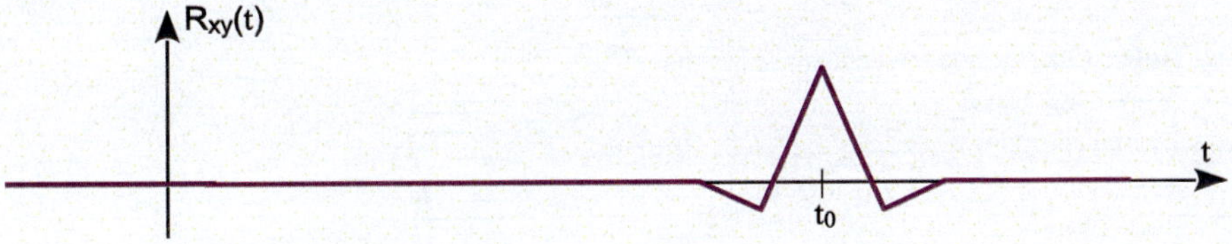

Fig. 16.14 The cross-correlation function

Figure 16.15 illustrates a radar chirp and an 'echo' and in Fig. 16.16 you can see the detector signal. The echo signal is not visible in the detector signal, but we have plotted it separately to indicate where it is. In Fig. 16.17, we have plotted the cross-correlation between the detector signal in Fig. 16.16 and the chirp signal in Fig. 16.15. From the R_{xy} peak in Fig. 16.17, we can easily determine where the echo is in Fig. 16.16.

We conclude that cross-correlation is used to find 'known' signals buried in random noise. NB! The cross-correlation technique is extremely selective and in radar applications, the echo signal in Fig. 16.16 may be frequency-shifted due to the Doppler effect and that has a severe impact on the detectability. For that reason, radar detection is complemented with statistical hypothesis testing (called 'Neyman-Pearson' detection).

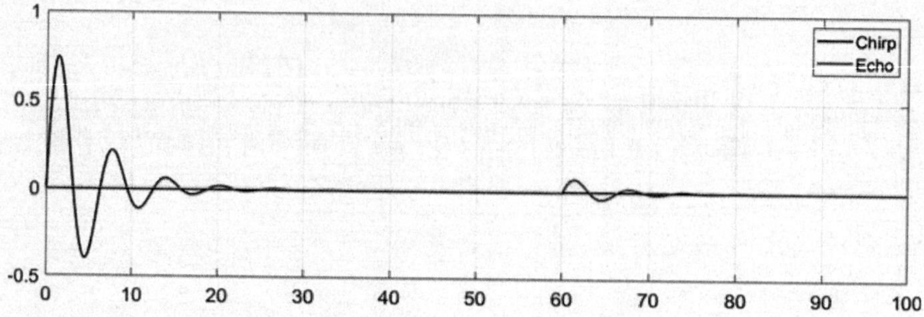

Fig. 16.15 A radar chirp and an 'echo'

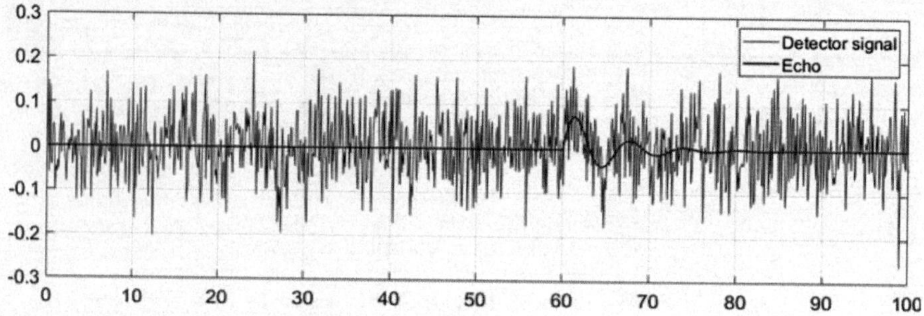

Fig. 16.16 Can you see the echo signal in the noise?

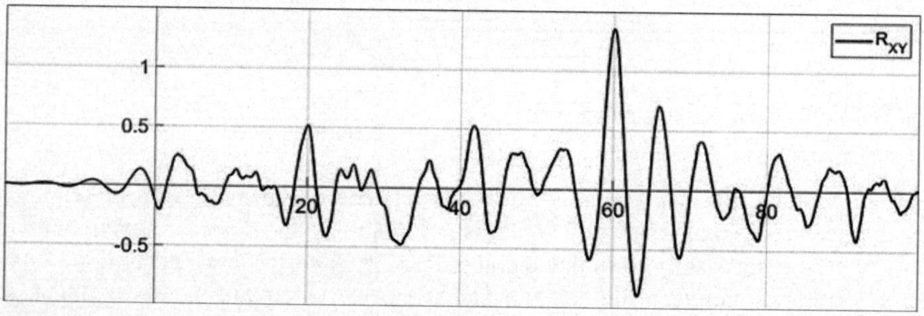

Fig. 16.17 The cross-correlation function

16.2.1 Implementation: Matched Filters

Consider the signal $x(t)$. If we cross-correlate it with $y(t)$ we get

$$\text{Cross-correlation:} \int_{-\infty}^{\infty} x(\tau)y(t+\tau)d\tau = \int_{-\infty}^{\infty} y(\tau)x(t+\tau)d\tau \tag{16.2}$$

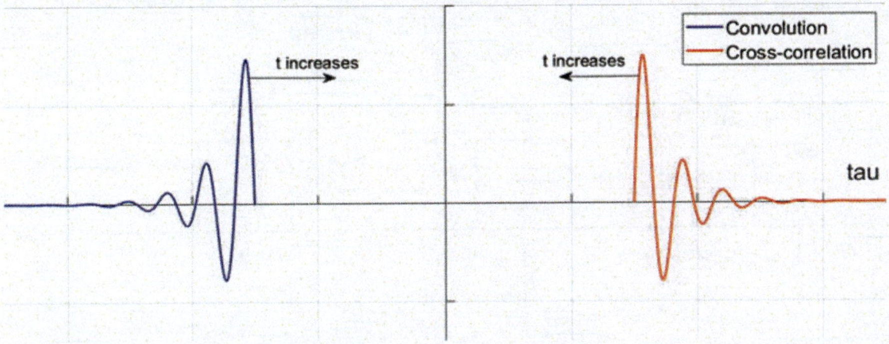

Fig. 16.18 Convolution versus cross-correlation

Fig. 16.19 A cross-correlator

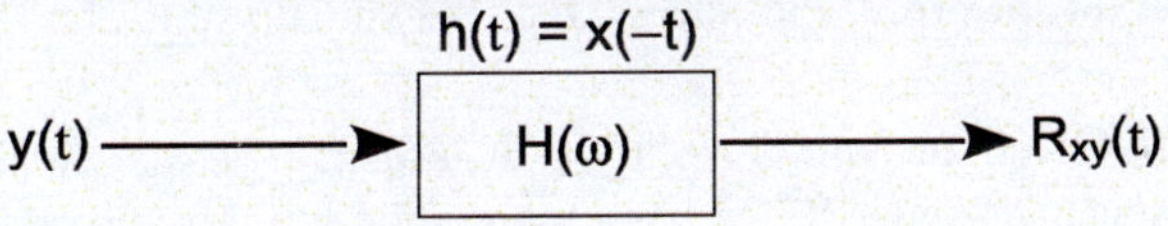

If we instead filter the signal $y(t)$ in a filter with impulse response $x(t)$ we get

$$\text{Convolution: } \int_{-\infty}^{\infty} x(\tau)y(t-\tau)d\tau = \int_{-\infty}^{\infty} y(\tau)x(t-\tau)d\tau \tag{16.3}$$

In the cross-correlation case, $x(t + \tau)$ slides from right to left when t increases, and in the convolution case, the time-reversed signal slides from left to right, see Fig. 16.18.

That means that we can implement a cross-correlator in a filter by designing the filter such that its impulse response $h(t) = x(-t)$. This is called a 'matched' filter, see Fig. 16.19.

16.3 Auto-Correlation

The auto-correlation function (ACF) is

$$R_{xx}(t) = \int_{-\infty}^{\infty} x(\tau)x(t+\tau)d\tau \tag{16.4}$$

In auto-correlation, a time-shifted copy of the signal 'slides over itself', see Fig. 16.20. An inherent property of the ACF is that it is always symmetric around $t = 0$, i.e., $R_{xx}(t) = R_{xx}(-t)$.

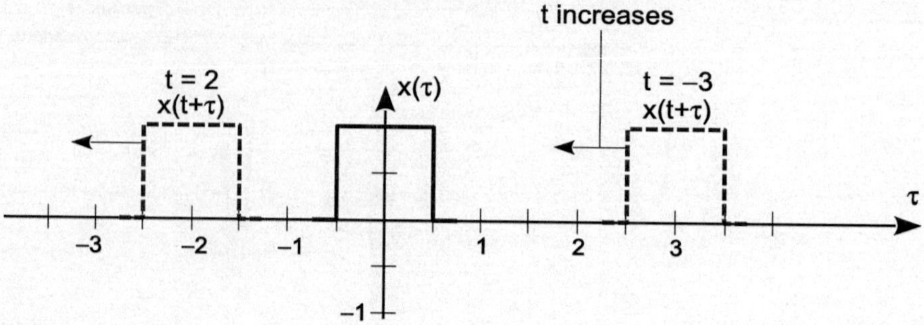

Fig. 16.20 A copy of the signal 'slides over itself'

Example 16.2 Plot the ACF of the signal in Fig. 16.21.

Solution In auto-correlation, it is usually easier to find the integration limits by starting from $x(0 + \tau)$ and then 'slide left' (in the positive t direction). Because of the inherent symmetric property of the ACF, we only need to find $R_{xx}(t)$ for positive times t. In Fig. 16.22, we can see that for $0 < t < 2$, we must integrate between $\tau = -1$ and $1 - t$.

$$R_{xx}(t) = \int_{-1}^{1-t} 1 \cdot 1 d\tau = [\tau]_{-1}^{1-t} = 1 - t + 1 = \underline{2 - t}$$

The ACF for the time interval $-2 < t < 0$ is then $2 + t$. Figure 16.23 illustrates the ACF.

Fig. 16.21 Auto-correlate a square pulse

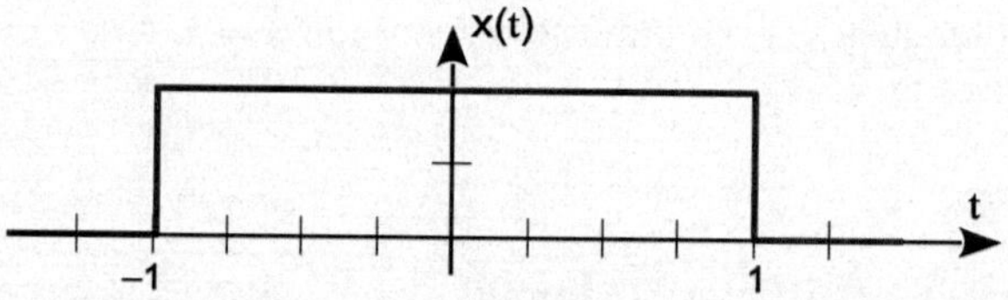

Fig. 16.22 We only need to find the ACF for positive times t

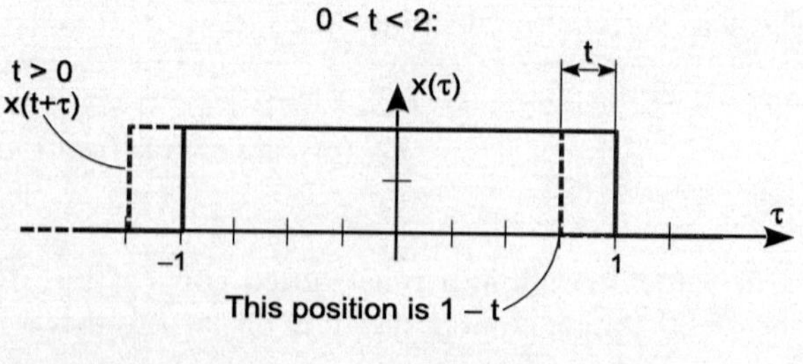

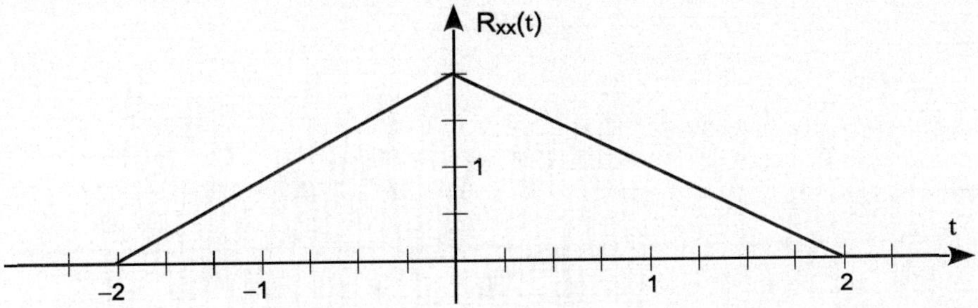

Fig. 16.23 The ACF

From Example 16.2, we can draw another conclusion about the ACF; it will always have a maximum for $t = 0$.

Example 16.3 Find the ACF of a sine function, $x(t) = \sin t$.

Solution For a periodic signal, we only need to integrate over one period, and the auto-correlation expression should also be divided by the period T:

$$R_{xx}(t) = \frac{1}{T}\int_0^T \sin\tau \times \sin(t+\tau)d\tau$$

$$= \frac{1}{2T}\int_0^T \left(\underbrace{\cos(-t)}_{\substack{=\cos t \\ \text{Independent of } \tau}} + \underbrace{\cos(2\tau + t)}_{\substack{=0 \\ \text{Integrated over two periods}}} \right) d\tau$$

$$R_{xx}(t) = \frac{1}{2T}\cos t \int_0^T d\tau = \underline{\frac{1}{2}\cos t} \tag{16.5}$$

From Example 16.3, we can draw some more conclusions about the ACF. First, if $x(t)$ is periodic, then $R_{xx}(t)$ has the same period as $x(t)$; *the frequency information is* ***preserved*** *in the auto-correlation process.* Second, we auto-correlated a sine and got a cosine. As a matter of fact, we would always get a cosine for any $\sin(\omega t + \varphi)$ function, independent of the phase angle φ; *the phase information is* ***lost*** *in the auto-correlation process.*

So, what are the applications of auto-correlation? To see that, we first auto-correlate white noise. Figure 16.24 illustrates white noise and a time-shifted copy of it. Imagine that we multiply the noise signal and the time-shifted copy at each time instant. If the time-shift is zero, then we just square the noise signal, and the product signal would be all-positive and if we integrated it, we would get a number > 0; this number is equal to the *variance* σ^2 of the noise (the noise power).

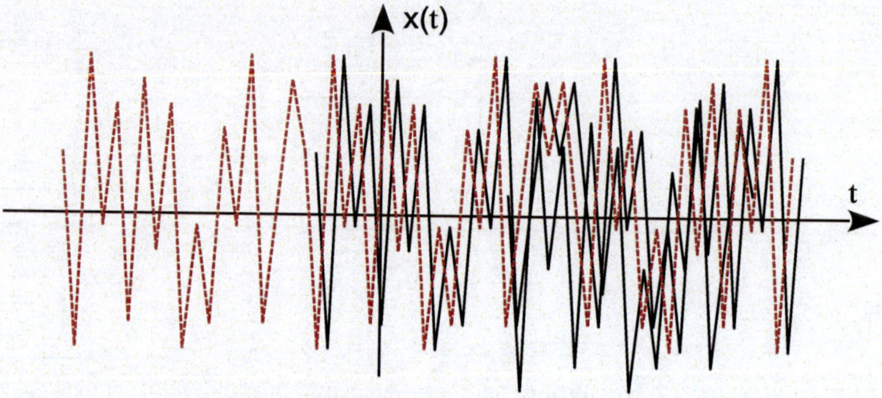

Fig. 16.24 White noise (bandwidth-limited)

Fig. 16.25 Auto-correlation of white noise

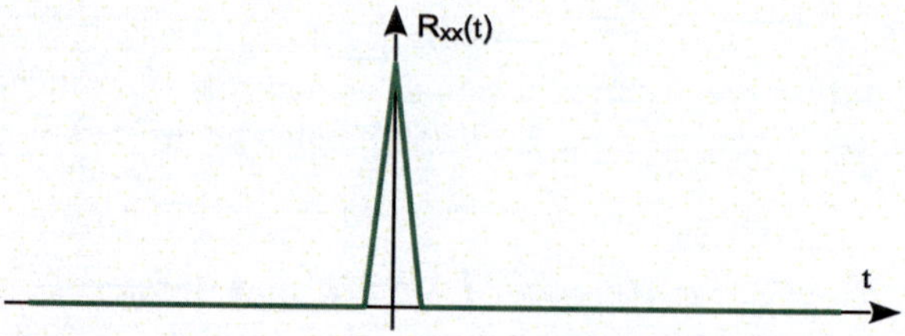

However, if the time shift is > 0, then soon the product function between the noise and the time-shifted copy will be a random noise signal and integrating it over some time interval would just be ≈ 0 everywhere. This is illustrated in Fig. 16.25.

How fast R_{xx} goes to zero from $t = 0$ depends on the noise bandwidth; if the noise is not bandlimited, then the ACF is actually a delta function:

$$R_{yy}(t) = \begin{cases} \sigma^2 & t = 0 \\ 0 & t \neq 0 \end{cases} \tag{16.6}$$

i.e., $R_{xx}(t) = \sigma^2 \delta(t)$.

Next, suppose that the signal $x(t)$ has the ACF $R_{xx}(t)$, and that $y(t)$ has the ACF $R_{yy}(t)$. Then the signal $x(t) + y(t)$ has the ACF $R_{xx}(t) + R_{yy}(t)$ (provided that the two functions are statistically independent). The proof of this relies on statistics and is presented in Problem 16.6.

Now, let's suppose that we have a periodic signal (sin$t$) with white noise; $x(t) =$ (sin$t$ + *white noise*), see Fig. 16.26. The ACF of this signal is the sum of the ACF of the sine (= a cosine) and the ACF of the noise (a delta function). The ACF is illustrated in Fig. 16.27. Notice in Fig. 16.27 that (a) the noise is concentrated to $t =$ 0 and (b) the huge improvement of the signal-to-noise ratio.

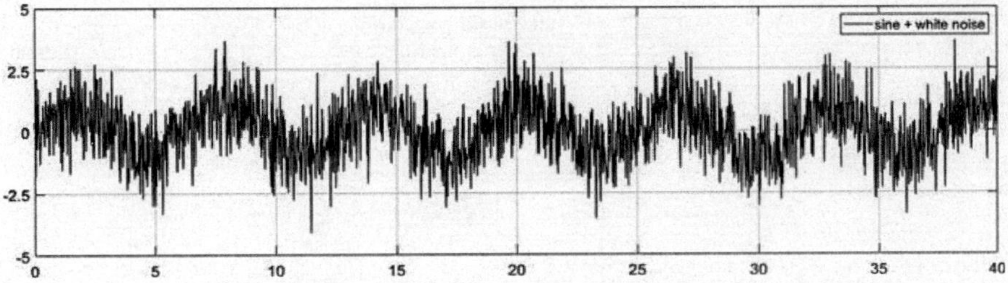

Fig. 16.26 Sine + white noise

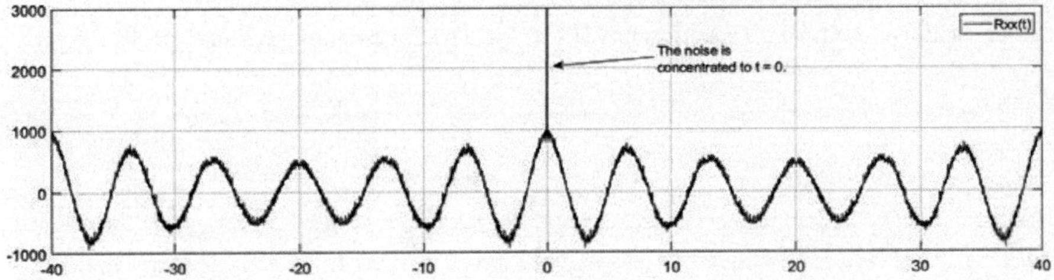

Fig. 16.27 The ACF

Improving the signal-to-noise ratio in periodic signals with white noise is the obvious application of autocorrelation, but it doesn't stop there. Auto-correlation is used in a huge range of applications, and we will present some of them in the next section.

16.3.1 Auto-Correlation Applications

Figure 16.28 illustrates a *Photon Correlation Spectroscopy* experiment (PCS). Here, auto-correlation is used to determine the diffusion coefficient and the size of colloidal particles in a solvent.

If the size of the particles is smaller than the laser wavelength, then Rayleigh scattering will occur. If the distance $d \times \sin\theta$ between two adjacent light-scattering particles equals a multiple of the laser wavelength ($m\lambda$), then there will be constructive interference in the photodetector resulting in a high photodetector current $I(t)$. On the other hand, if the distance equals a multiple of $\lambda/2$, then there will be destructive interference resulting in a low photodetector current. Due to Brownian motion of the particles, the photodetector signal will vary randomly as the phase shift of the light from two adjacent particles will change gradually when the particles move. Figure 16.29 illustrates the photodetector signal.

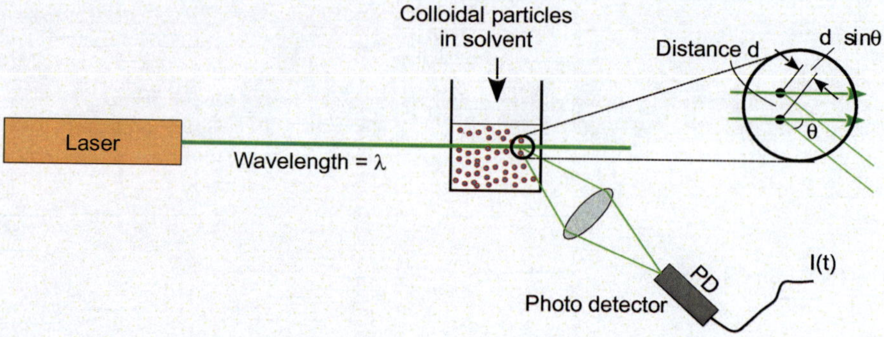

Fig. 16.28 Photo correlation spectroscopy (PCS) (or 'Dynamic light scattering', DLS)

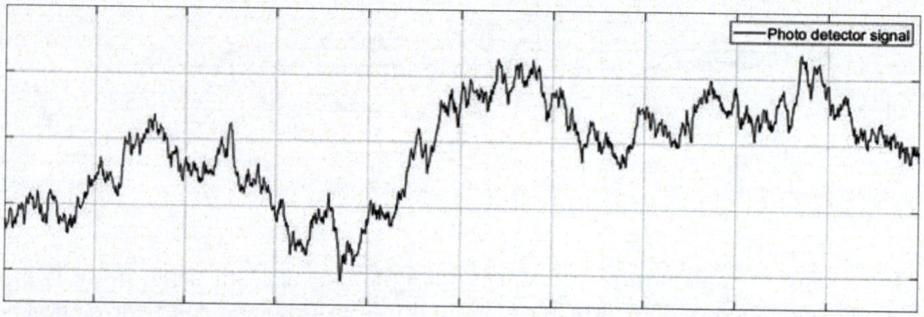

Fig. 16.29 The photodetector signal

The photodetector signal is random, but not 'white' random; there is some correlation between adjacent samples. This is reflected in the ACF, which will be an exponentially decaying signal, see Fig. 16.30; the correlation between samples decreases with time. From the ACF parameters (time constant, baseline, max value), both the diffusion coefficient and the particle size can be derived [1]. For example, this technique has been used to study the homogeneity of proteins [2].

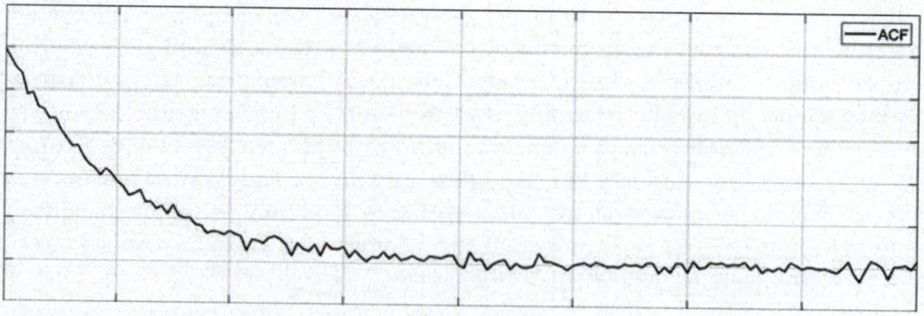

Fig. 16.30 The ACF

It has also been used to measure particle-flow velocity [3]. When the particles pass through a laser beam, the moving particles 'encode' an intensity fluctuation in the backscattered light and the slope of the ACF is proportional to the particle velocity.

Spatial autocorrelation has its own applications. For example, it is used in ecology to study the synchronous fluctuation of ecological variables over wide geographical areas (such as birds, butterflies, trees, hares...) [4]. It is also used to study the fluctuation of socio-economic variables over regional areas [5].

In the physics lab though, we mostly use it to detect periodic signals in random noise.

16.4 Discrete-Time Correlation

16.4.1 Cross-Correlation

The discrete-time expression for cross-correlation is

$$R_{xy}(n) = \sum_{i=-\infty}^{\infty} x_i y_{n+i} \tag{16.17}$$

If we compare this expression with Eq. (10.3) (discrete-time convolution), we can see that it is in principle the same thing, except we don't time-reverse y_n. Writing it out explicitly we get (assuming all signals = 0 for $n < 0$)

$$R_{xy}(n) = x_0 y_n + x_1 y_{n+1} + x_2 y_{n+2} + \ldots.$$

Table 16.1 illustrates the case where $x = [x_0, x_1, x_2]$ and $y = [y_0, y_1, y_2, y_3]$. If we compare this to Table 10.1, we can see that data is now 'sliding' in from the right side (and is not time-reversed).

Example 16.4 Cross-correlate the signals in Figs. 16.31 and 16.32.

Table 16.1 Cross-correlation in discrete time

$R_{xy}(n)$	x_0	x_1	x_2	y
$R_{xy}(-2) = x_2 y_0$			y_0	$y_1\ y_2\ y_3$
$R_{xy}(-1) = x_1 y_0 + x_2 y_1$		y_0	y_1	$y_2\ y_3$
$R_{xy}(0) = x_0 y_0 + x_1 y_1 + x_2 y_2$	y_0	y_1	y_2	y_3
$R_{xy}(1) = x_0 y_1 + x_1 y_2 + x_2 y_3$	y_1	y_2	y_3	
$R_{xy}(2) = x_0 y_2 + x_1 y_3$	y_2	y_3		
$R_{xy}(3) = x_0 y_3$	y_3			

Fig. 16.31 x_n

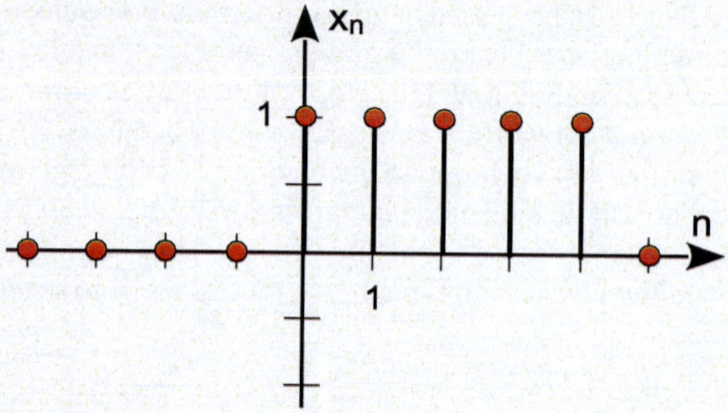

Fig. 16.32 y_n

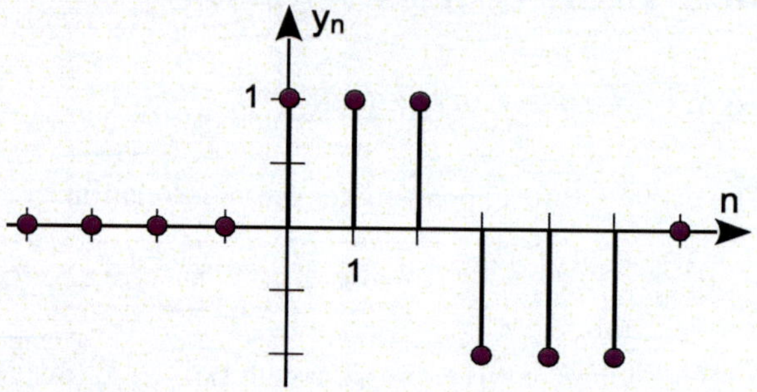

Solution We start by delaying the y_i signal by five samples ($n = -5$). From Fig. 16.33, we can see that there is no overlap until $n = -4$.

$$R_{xy}(-4) = x_4 y_0 = 1 \cdot 1 = 1$$

$$R_{xy}(-3) = x_3 y_0 + x_4 y_1 = 1 + 1 = 2$$

$$R_{xy}(-2) = x_2 y_0 + x_3 y_1 + x_4 y_2 = 1 + 1 + 1 = 3$$

$$R_{xy}(-1) = x_1 y_0 + x_2 y_1 + x_3 y_2 + x_4 y_3 = 1 + 1 + 1 - 1 = 2$$

$$R_{xy}(0) = x_0 y_0 + x_1 y_1 + x_2 y_2 + x_3 y_3 + x_4 y_4 = 1 + 1 + 1 - 1 - 1 = 1$$

$$R_{xy}(1) = x_0 y_1 + x_1 y_2 + x_2 y_3 + x_3 y_4 + x_4 y_5 = 1 + 1 - 1 - 1 - 1 = -1$$

$$R_{xy}(2) = x_0 y_2 + x_1 y_3 + x_2 y_4 + x_3 y_5 = 1 - 1 - 1 - 1 = -2$$

$$R_{xy}(3) = x_0 y_3 + x_1 y_4 + x_2 y_5 = -1 - 1 - 1 = -3$$

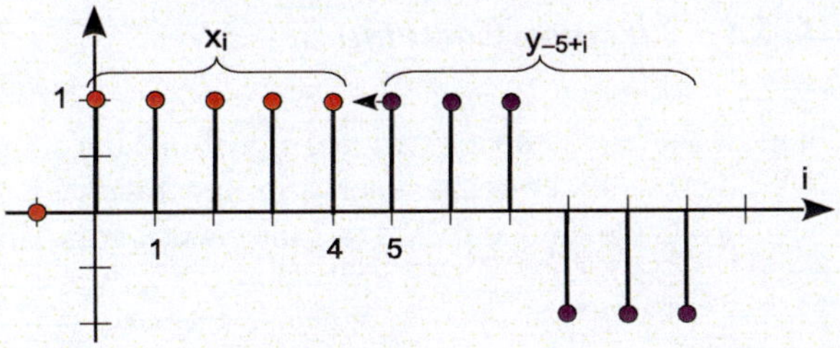

Fig. 16.33 No overlap until $n = -4$

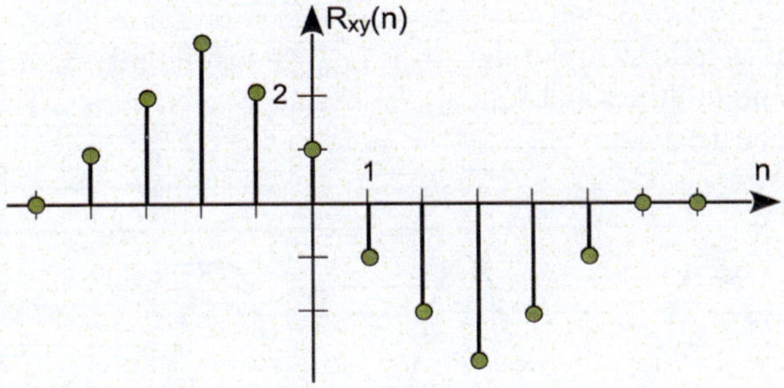

Fig. 16.34 The cross-correlation function

$$R_{xy}(4) = x_0 y_4 + x_1 y_5 = -1 - 1 = -2$$

$$R_{xy}(5) = x_0 y_5 = -1$$

The cross-correlation function is illustrated in Fig. 16.34.

16.4.2 Auto-Correlation

We get the discrete-time auto-correlation expression by just substituting y_{n+i} for x_{n+i} in expression (16.17):

$$R_{xx}(n) = \sum_{i=-\infty}^{\infty} x_i x_{n+i} \tag{16.18}$$

We don't give any examples of discrete-time autocorrelation. It works just like the cross-correlation in Eq. (16.17), and it has the same properties as the continuous-time ACF; frequency is preserved, but not the phase and it has a maximum for $n = 0$.

Instead, we will investigate a 'computational' problem concerned with discrete-time correlation.

16.4.3 Circular Correlation

Suppose that we sample a periodic signal like the sinusoidal in Fig. 16.35.

When we auto-correlate a periodic signal with a period of N samples, we should only correlate over one period and the correct correlation expression is

$$R_{xx}(n) = \frac{1}{N}\sum_{i=0}^{N-1} x_i x_{n+i} \tag{16.19}$$

However, we could execute this expression in two different ways. We could do it as indicated in Table 16.1, i.e., we just let the 'copy slide' by the 'original' signal and multiply and sum all the overlapping samples. This is illustrated in Figs. 16.36 and 16.37.

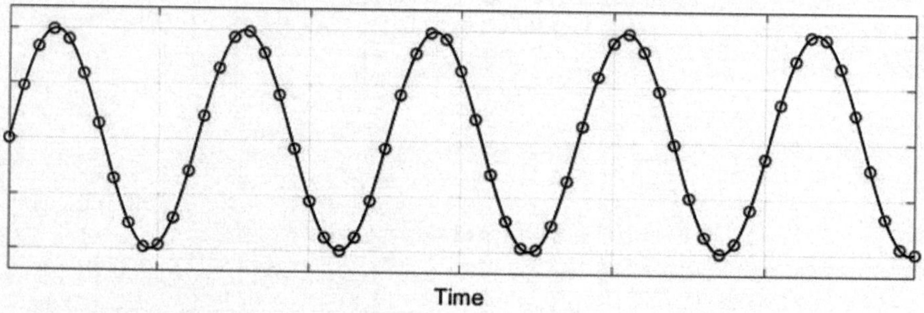

Fig. 16.35 Sampling a sine

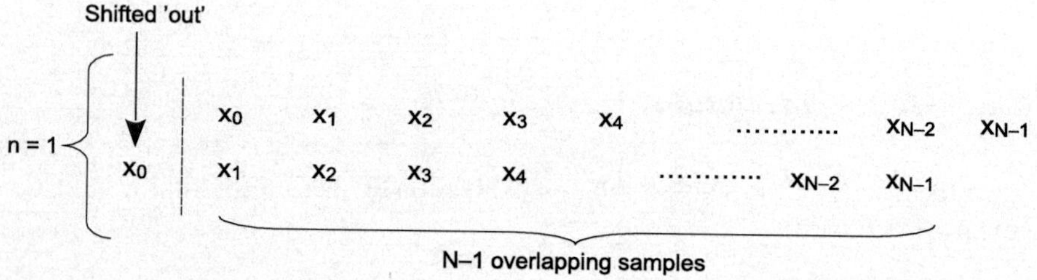

Fig. 16.36 'Common' correlation

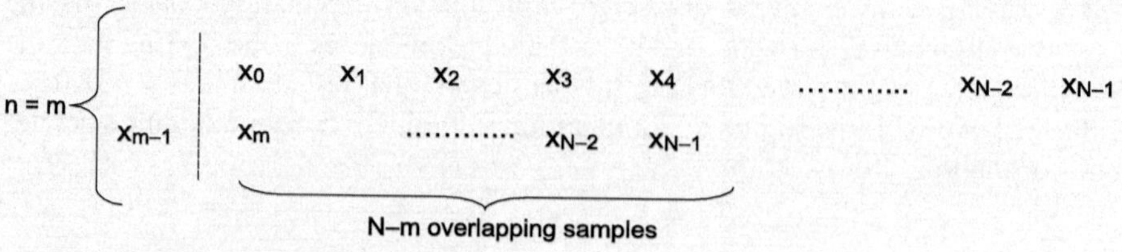

Fig. 16.37 'Common' correlation

With the correlation technique in Fig. 16.36, samples that are 'shifted' out are just discarded and the number of overlapping samples decreases gradually and that will 'distort' the ACF. We can make up for this in two different ways. One simple solution would be to 're-define' N in Eq. (16.19) from representing the signal period to representing the number of overlapping samples. Then N would decrease with the shift to compensate for the decreasing number of overlapping samples. The downside of this solution is that there will be very few overlapping samples at the 'ends' and that gives us 'poor statistics'.

A better way to make up for it is to use 'circular' correlation. In circular correlation, the sample that is shifted out is not discarded but is instead 'rotated' back to the beginning of the sample array, see Figs. 16.38 and 16.39.

Using this technique, we will always have N overlapping samples. Circular correlation was used to produce the ACF in Fig. 16.27. Figure 16.40 illustrates what it would have looked like if we had used 'common' correlation.

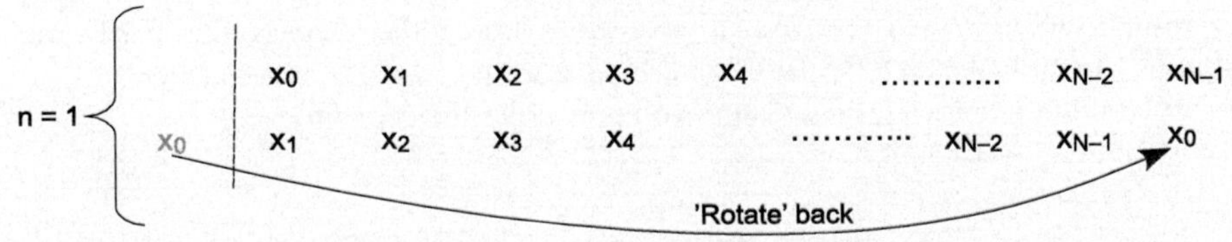

Fig. 16.38 'Circular' correlation

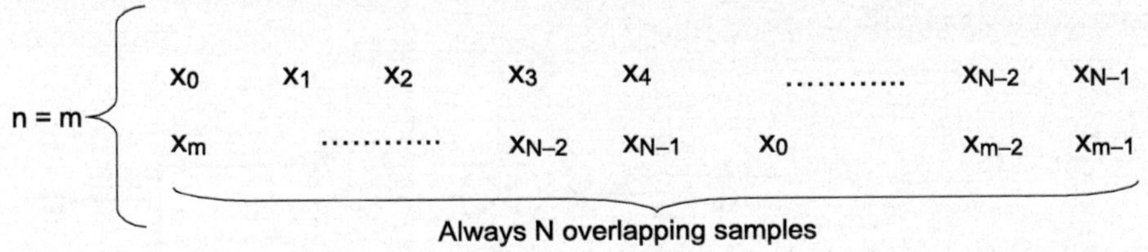

Fig. 16.39 'Circular' correlation

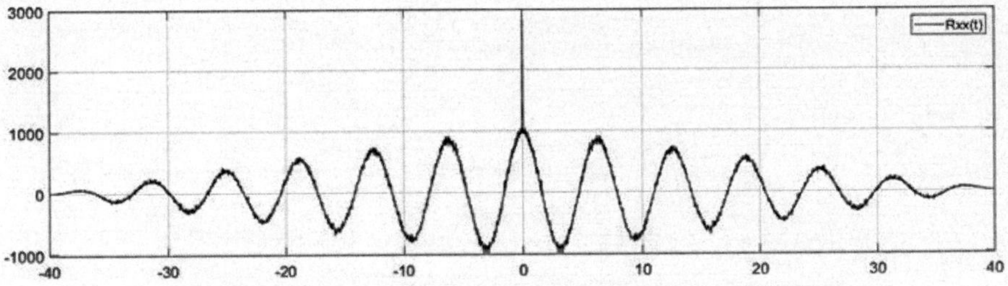

Fig. 16.40 Auto-correlation without 'circular' correlation (compare with Fig. 16.27)

Notice in Fig. 16.40 how the amplitude of the sine is decreasing gradually because of the decreasing number of samples that we 'multiply-and-add'.

In MATLAB, you use the *xcorr* command for 'common' correlation and the *cxcorr* command for circular correlation. (However, you need the Signal Processing Toolbox to get access to the *cxcorr* command.)

16.5 Solved Problems

Problem 16.1 Prove that the lock-in amplifier we introduced in Chap. 15 is just a special case of cross-correlation; the lock-in amplifier output $= R_{xy}(0)$.

Solution Fig. 16.41 illustrates the lock-in amplifier system. The multiplier obviously produces the signal product. All we need to do is to prove that the lowpass filter integrates the product. To prove that we need to look at the lowpass filter hardware. Figure 16.42 illustrates a first-order RC lowpass filter.

The output voltage $b(t)$ equals the voltage across the capacitor:

$$b(t) = U_c = \frac{Q}{C} = \frac{1}{C}\int_{-\infty}^{t} i(\tau)d\tau$$

Hence, $b(t)$ is the integral of the *current*. If $R >> 1/\omega C$, then almost all of $a(t)$ falls over R and then $i \approx a(t)/R$. Hence

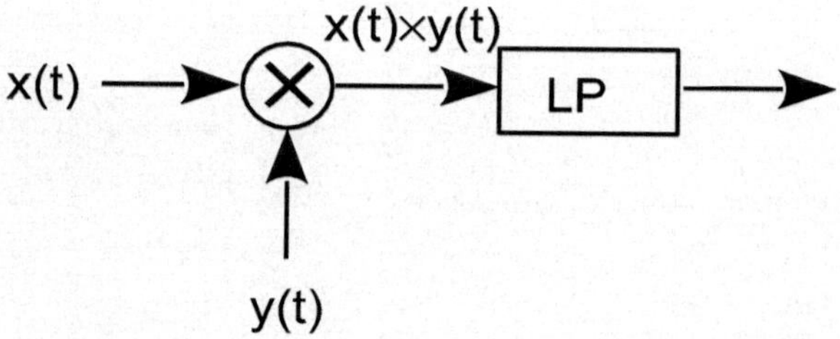

Fig. 16.41 Lock-in amplifier

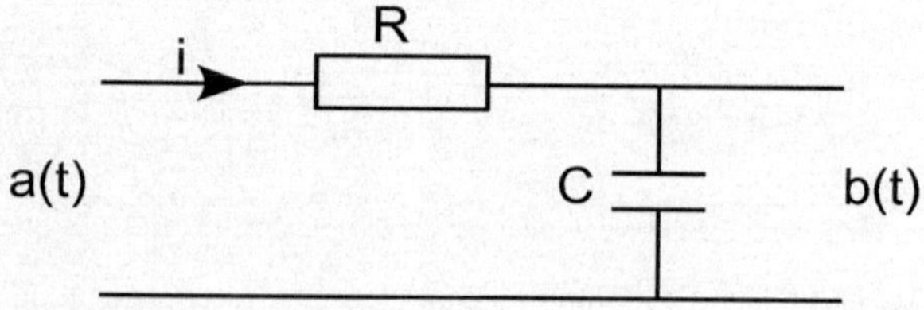

Fig. 16.42 First-order lowpass filter

$$b(t) = \frac{1}{C}\int_{-\infty}^{t} \frac{a(\tau)}{R} d\tau = \frac{1}{RC}\int_{-\infty}^{t} a(\tau) d\tau$$

We can conclude that if $R >> 1/\omega C$, i.e., if $RC >> 1/\omega, = T/2\pi$ (where T is the signal period), then the lowpass filter is indeed integrating the input signal and the lock-in amplifier output is

$$\frac{1}{RC}\int_{-\infty}^{t} x(\tau)y(\tau)d\tau = \frac{1}{RC}R_{xy}(0)$$

Problem 16.2 Find the auto-correlation function of a square signal (duty cycle 50%).

Solution Figure 16.43 illustrates a square wave $x(\tau)$ and a shifted copy $x(t+\tau)$ $(t > 0)$.

We only need to find one period of the ACF; we derive the ACF for $-1 < t < 1$. Figure 16.43 represents the two signals when $0 < t < 1$. In this interval, the ACF is

$$R_{xx}(t) = \frac{1}{2}\int_{0}^{1-t} 1 \cdot 1 d\tau = \underline{\frac{1}{2}(1-\tau)}$$

Figure 16.44 represents the signals when $-1 < t < 0$.
The ACF is

$$R_{xx}(t) = \frac{1}{2}\int_{-t}^{1} 1 \cdot 1 d\tau = \underline{\frac{1}{2}(1+t)}$$

The ACF is plotted in Fig. 16.45.

Problem 16.3 In Example 16.3, we concluded that the frequency information is *preserved* in the auto-correlation and that the phase information is *lost*. What about the amplitude information. Lost or preserved?

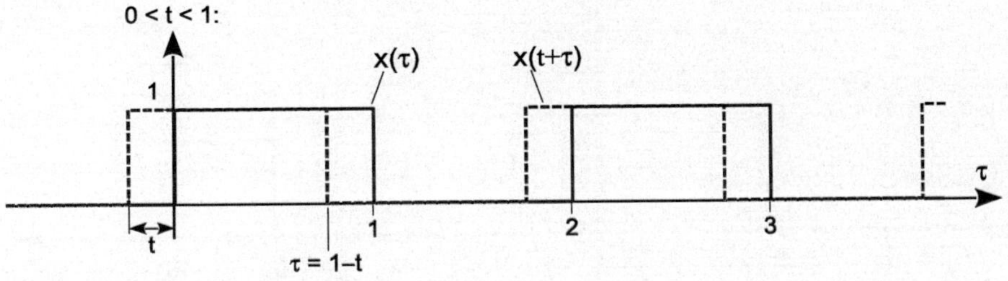

Fig. 16.43 The two signals $x(\tau)$ and $x(t+\tau)$

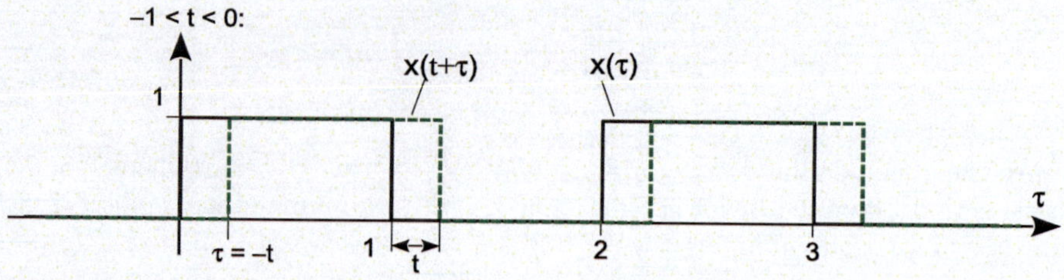

Fig. 16.44 The two signals $x(\tau)$ and $x(t+\tau)$ for $-1 < t < 0$

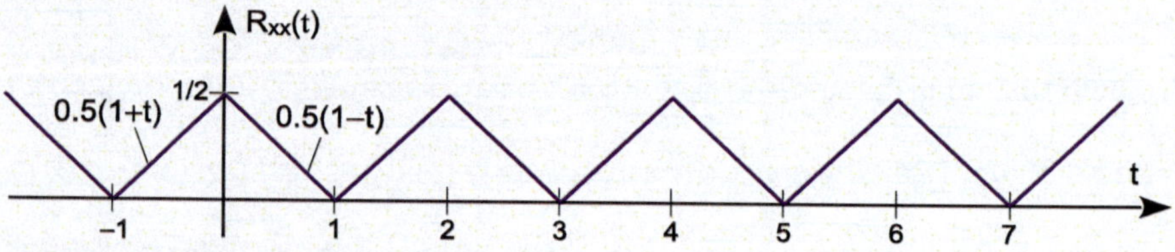

Fig. 16.45 The ACF

Solution It is lost (or at least 'scrambled'). We have seen two examples of this. In Example 16.3, the ACF of a sine (amplitude = 1) became a cosine with amplitude = 0.5. Also, in Problem 16.2 above, the ACF of a square wave became a sawtooth signal. We know from Fourier transform theory that the square and the sawtooth have the same frequencies, but their amplitude spectrums are different.

Problem 16.4 What is the auto-correlation function of the square pulse signal in Fig. 16.46?

Solution This is the same problem as in Example 16.2; the auto-correlation function is independent of time delays. Hence, the auto-correlation function of the signal in Fig. 16.46 is illustrated in Fig. 16.23.

Problem 16.5 In Sect. 16.3, we asserted that for the auto-correlation of white noise, $R_{xx}(0) = \sigma^2$. Prove that this is true.

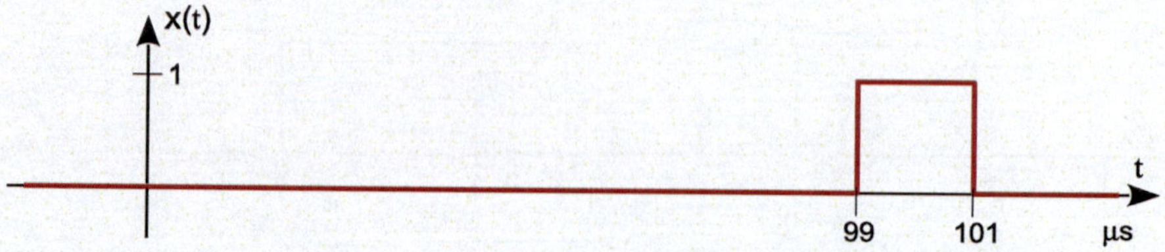

Fig. 16.46 Squared pulse

Solution To prove that we need some statistical theory. First, the noise signal is random, and we need to treat it as a *stochastic process*, $X(t)$. The auto-correlation function of a stochastic process is the expectation value of $X(\tau)\cdot X(t+\tau)$. If $t=0$, then

$$R_{xx}(0) = \mathrm{E}\{X(t)\cdot X(t)\} = \mathrm{E}\left(X^2\right) = \sigma^2$$

from the definition of variance and from the fact that the expectation value of white noise is $= 0$. (Also, assuming that we have a *stationary* process.)

Problem 16.6 Prove that if the signal $x(t)$ has the auto-correlation function $R_{xx}(t)$, and if $y(t)$ has the auto-correlation function $R_{yy}(t)$, then signal $x(t) + y(t)$ has the auto-correlation function $R_{xx}(t) + R_{yy}(t)$.

Solution We can prove it if we treat the two signals as two stochastic processes. The auto-correlation function is in general the expectation value of $X(\tau)$ and $X(t+\tau)$. In our case, we have a sum of functions:

$$\begin{aligned}
&E((X(\tau)+Y(\tau))\cdot(X(t+\tau)+Y(t+\tau)))\\
&= E\{X(\tau)X(t+\tau)+X(\tau)Y(t+\tau)\\
&\quad +Y(\tau)X(t+\tau)+Y(\tau)Y(t+\tau)\}\\
&= \underbrace{E\{X(\tau)X(t+\tau)\}}_{=R_{xx}(t)} + \underbrace{E\{X(\tau)Y(t+\tau)\}}_{=0\text{ (Independent)}}\\
&\quad + \underbrace{E\{Y(\tau)X(t+\tau)\}}_{=0\text{ (Independent)}} + \underbrace{E\{Y(\tau)Y(t+\tau)\}}_{=R_{yy}(t)}\\
&= R_{xx}(t) + R_{yy}(t)
\end{aligned}$$

(Because the correlation between independent processes is 0.)

Problem 16.7 The ACF $R_{xx}(t)$ and the power spectrum $|X(\omega)|^2$ is a 'Fourier transform pair', i.e.,

$$\int_{-\infty}^{\infty} R_{xx}(t)\cdot \mathrm{e}^{-j\omega t}\,dt = |X(\omega)|^2$$

'Prove' that this is true by calculating first the Fourier transform of the pulse in Fig. 16.21 and then calculating the Fourier transform of its auto-correlation function.

Solution We already found in example 7.2 that the Fourier transform of a square pulse is

$$|H(\omega)| = \frac{2}{\omega}\sin\frac{\omega}{2} \Rightarrow |H(\omega)|^2 = \frac{4}{\omega^2}\sin^2\omega$$

We also derived the ACF of the square pulse in Problem 16.2. The Fourier transform is:

$$\int_{-\infty}^{\infty} R_{xx}(t)\mathrm{e}^{-\mathrm{j}\omega t}dt = \int_{-2}^{0}(2+t)\cdot\mathrm{e}^{-\mathrm{j}\omega t}dt + \int_{0}^{2}(2-t)\cdot\mathrm{e}^{-\mathrm{j}\omega t}dt$$

$$= 2\int_{-2}^{2}\mathrm{e}^{-\mathrm{j}\omega t}dt + \int_{-2}^{0}t\cdot\mathrm{e}^{-\mathrm{j}\omega t}dt - \int_{0}^{2}t\cdot\mathrm{e}^{-\mathrm{j}\omega t}dt$$

$$= -\frac{2}{\mathrm{j}\omega}\left[\mathrm{e}^{-\mathrm{j}\omega t}\right]_{-2}^{2} + \left[-\frac{t}{\mathrm{j}\omega}\mathrm{e}^{-\mathrm{j}\omega t}\right]_{-2}^{0} + \frac{1}{\mathrm{j}\omega}\int_{-2}^{0}\mathrm{e}^{-\mathrm{j}\omega t}dt - \left[-\frac{t}{\mathrm{j}\omega}\mathrm{e}^{-\mathrm{j}\omega t}\right]_{0}^{2} - \frac{1}{\mathrm{j}\omega}\int_{0}^{2}\mathrm{e}^{-\mathrm{j}\omega t}dt =$$

$$= -\frac{2}{\mathrm{j}\omega}\mathrm{e}^{-\mathrm{j}2\omega t} + \frac{2}{\mathrm{j}\omega}\mathrm{e}^{\mathrm{j}2\omega t} - \frac{2}{\mathrm{j}\omega}\mathrm{e}^{\mathrm{j}2\omega t} + \frac{1}{\omega^2}\left[\mathrm{e}^{-\mathrm{j}\omega t}\right]_{-2}^{0} + \frac{2}{\mathrm{j}\omega}\mathrm{e}^{-\mathrm{j}2\omega t} - \frac{1}{\omega^2}\left[\mathrm{e}^{-\mathrm{j}\omega t}\right]_{0}^{2}$$

$$= \frac{1}{\omega^2}\left(1 - \mathrm{e}^{\mathrm{j}2\omega t} - \mathrm{e}^{-\mathrm{j}2\omega t} + 1\right) = \frac{2}{\omega^2}(1-\cos 2\omega) = \frac{4}{\omega^2}\sin^2\omega$$

And hence, $\int_{-\infty}^{\infty} R_{xx}(t)\mathrm{e}^{-\mathrm{j}\omega t}dt = |H(\omega)|^2$.

Problem 16.8 Find the ACF of $x(t) = \mathrm{e}^{-t}$ $(t > 0)$, $x(t) = 0$ if $t < 0$.

Solution Figure 16.47 illustrates $x(\tau)$ and $x(t+\tau)$. The ACF is

$$R_{xx}(t) = \int_{0}^{\infty}\mathrm{e}^{-\tau}\mathrm{e}^{-(t+\tau)}d\tau = \mathrm{e}^{-t}\int_{0}^{\infty}\mathrm{e}^{-2\tau}d\tau = -\frac{1}{2}\mathrm{e}^{-t}\left[\mathrm{e}^{-2\tau}\right]_{0}^{\infty} = \underline{\frac{1}{2}\mathrm{e}^{-t}}$$

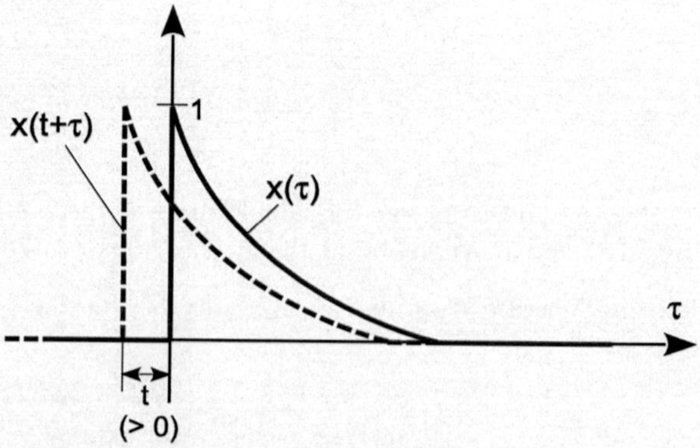

Fig. 16.47 Auto-correlating an exponential function

Table 16.2 Sampling the exponential

n	0	1	2	3	4	5	6	7	8	9	10
x_n	1.000	0.819	0.670	0.549	0.449	0.368	0.301	0.247	0.202	0.165	0.135

Problem 16.9 Suggest a digital FIR filter that you could use to 'detect' the exponential function in problem 16.8 in a noise signal.

Solution First we need to decide a sampling rate; $f_S = 5$ S/s. That means that the samples are taken at times $n/5 = n{\cdot}0.2$ s. To 'detect' the exponential signal in noise, we need to cross-correlate it with an identical exponential function and we implement that by designing a filter with an impulse response that is just the time-reversed copy of the signal. In a FIR filter, the filter coefficients are also the impulse response coefficients.

First, we sample the exponential: $x_n = \mathrm{e}^{-n\cdot 0.2}$ (Table 16.2).

Then we just time-reverse the sample order (and shift to make it causal) to get an 11-tap FIR filter that cross-correlates:

$$\underline{y_n = 0.135x_n + 0.165x_{n-1} + 0.202x_{n-2} + 0.247x_{n-3} + \cdots + 0.819x_{n-9} + x_{n-10}}$$

Problem 16.10 Figure 16.48 illustrates a seismograph that determines the direction to the epicentrum of earthquakes. It cross-correlates the signals from three vibration sensors placed as indicated in Fig. 16.48. Chock waves propagate through the earth's crust at a speed of 4000 m/s, and they are assumed to be 'far away', i.e., the shock waves impact as plane waves on the seismograph.

At an earthquake, the vibration signals from sensors y_1 and y_2 were cross-correlated with the x signal, and the cross-correlation functions are illustrated in Figs. 16.49 and 16.50. Determine the direction to the epicentrum.

Solution In $R_{xy1}(t)$, y_1 is shifted left and we get a 'hit' for $t > 0$. That means that the shock wave hit sensor x before it hit sensor y_1; from $R_{xy1}(t)$ we conclude that the shock wave came either from the South-West or from the North-West. When x was correlated with y_2, we got a hit for $t < 0$; the shock wave hit sensor y_2 before it hit sensor x, and that happens if the chock wave comes from either South-West or South-East. The conclusion is that the shock wave came from the South-West. Once we know the approximate direction, we can calculate the exact direction.

Figure 16.51 illustrates how plane waves hit the sensors (from South-West). We need to find the angle θ.

According to $R_{xy1}(t)$, the time difference in the impact between sensor x and sensor y_1 is 100 ms; hence the distance $L = 4000 \times 0.1 = 400$ m. We can then find the angle θ:

$$\sin\theta = \frac{400}{500} \quad \Rightarrow \theta = \sin^{-1}0.8 = \underline{53^\circ}$$

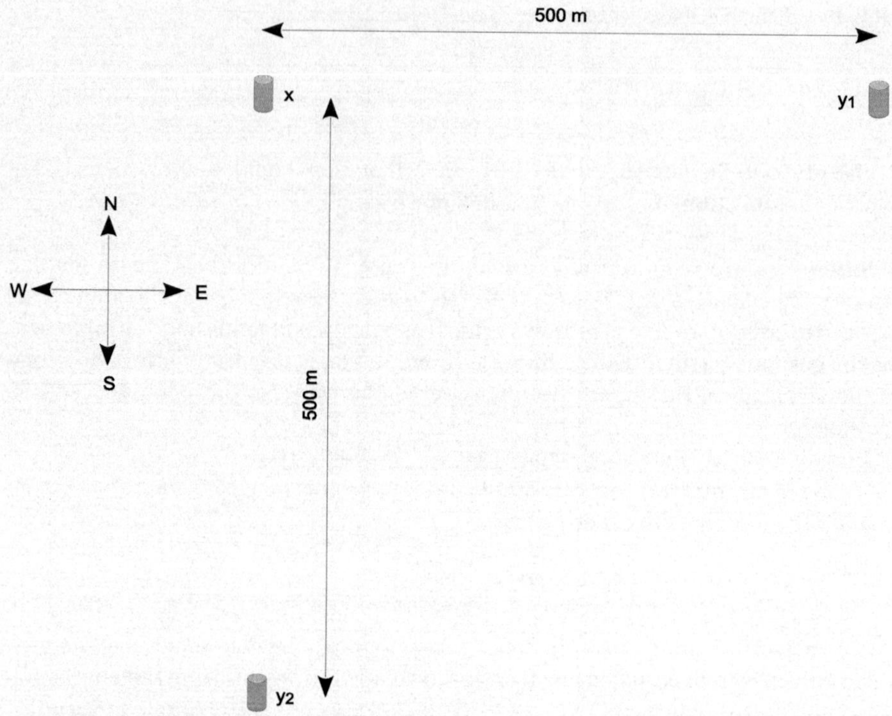

Fig. 16.48 Epicentrum detector

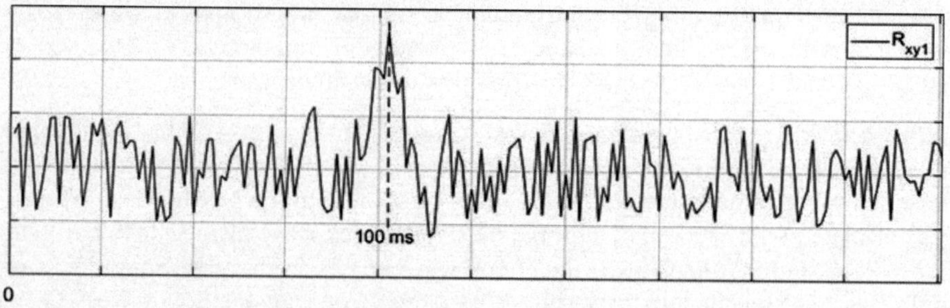

Fig. 16.49 $R_{xy1}(t)$

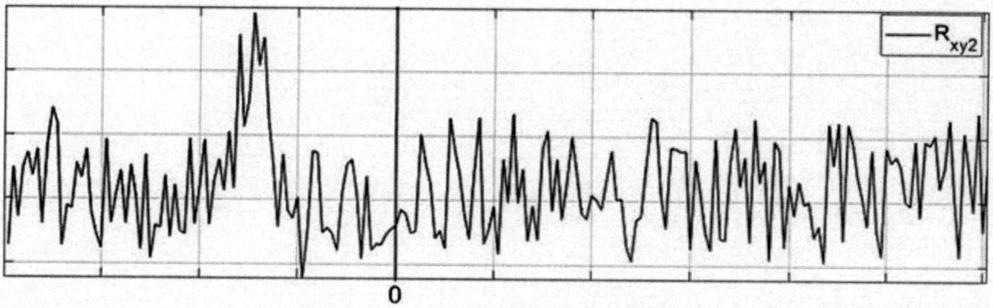

Fig. 16.50 $R_{xy2}(t)$

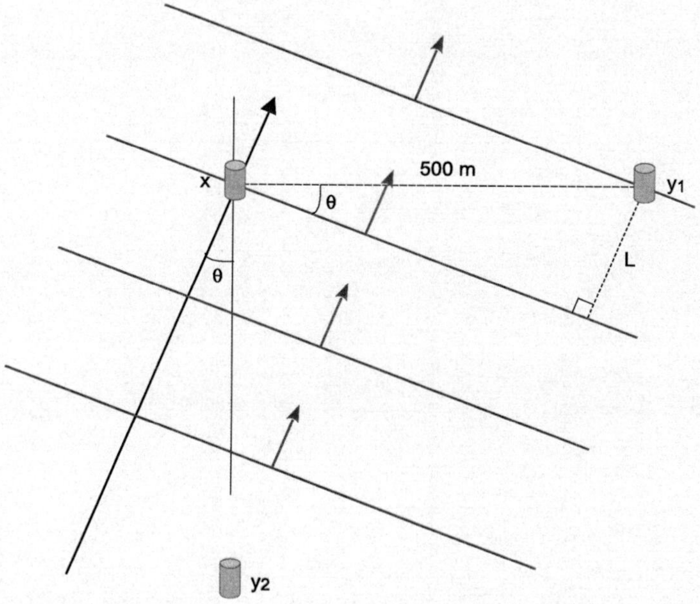

Fig. 16.51 Finding the impact angle

References

1. Tscharnuter, W. 2000. Photon correlation spectroscopy in particle sizing. *Encyclopedia of analytical chemistry*, 5469–5485.
2. Stetefeld, J., S.A. McKenna, and T.R. Patel. 2016. Dynamic light scattering: A practical guide and applications in biomedical sciences. *Biophysical reviews* 8: 409–427.
3. Wang, Y., and R. Wang. 2010. Autocorrelation optical coherence tomography for mapping transverse particle-flow velocity. *Optics letters* 35 (21): 3538–3540.
4. Koenig, W.D. 1999. Spatial autocorrelation of ecological phenomena. *Trends in Ecology & Evolution* 14 (1): 22–26.
5. Getis, A. 2007. Reflections on spatial autocorrelation. *Regional Science and Urban Economics* 37 (4): 491–496.

Chapter 17
Curve Fitting

Abstract One of the most common 'post-measurement' data processing operations is *curve fitting*, i.e., fitting samples to a predicted expression. This is usually derived by least-square calculations, but this chapter will use the orthogonality principle to derive the curve fitting expressions. The advantage of this is that it offers a graphical argument for its legitimacy. The pseudo inverse of a non-square matrix is defined and some common pitfalls due to error propagation in matrix operations are highlighted. Once curve fitting is understood, it can be used to understand how sampling instruments, such as digital oscilloscopes, retrieve the original signal from only a few samples; in Sect. 17.6 the sampling theorem is revisited, and this section explains the difference between linear interpolation and sin$x$/x interpolation.

17.1 Introduction

In a typical measurement, we observe (measure) some quantity y as some other quantity x varies; the objective is to figure out how y depends on x ($y = f(x)$). In a typical case, we know the 'general' dependence of y on x, i.e., we know they are related by a first-order polynomial or an exponential function, but we don't know the function coefficients. So, we take data, but since data are noise-infected, they will not follow the expected function graph exactly.

In Fig. 17.1, we denote the measured data y^m, and the 'theoretical' value just y. In the general case, if we take r data points, the result of the measurement would be an $r \times 2$ table (an $r \times 2$ *matrix*), see Table 17.1. The deviation of the measured data from the theoretical value is called the *error,* ε:

$$\varepsilon_n = y_n - y_n^m \tag{17.1}$$

Our objective is to find the theoretical function $y = f(x)$ from the data. For this introduction, we will assume a straight line:

$$f(x) = c_0 + c_1 x \tag{17.2}$$

L. Bengtsson, *Electrical Measurement Techniques*,
https://doi.org/10.1007/978-981-99-8187-8_17

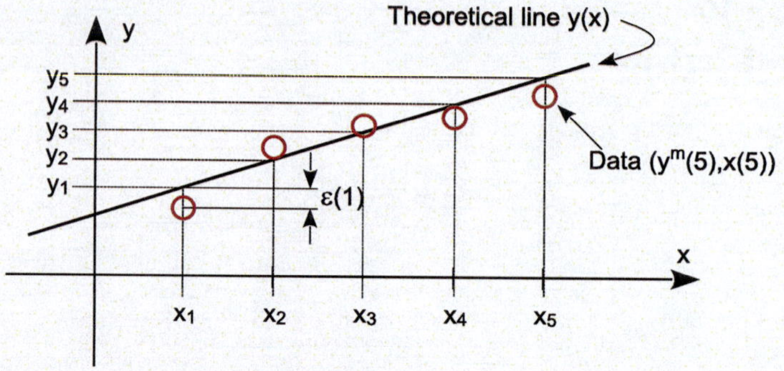

Fig. 17.1 Data won't fit exactly to the expected line

Table 17.1 The data table

x	y_m
x_1	y_1^m
x_2	y_2^m
:	:
x_r	y_r^m

If there was no noise in the data, we could choose any two data pairs in Table 17.1, insert them into Eq. (17.2), and we would have a system of two equations with two unknowns, and we could solve for the unknown coefficients c_0 and c_1. And we would be done. But that won't work because of the noise; we would get a different result every time depending on which two pairs we select.

If we insert *all* of them into Eq. (17.1), we get the following system of equations:

$$\begin{cases} c_0 + c_1 x_1 = y_1^m \\ c_0 + c_1 x_2 = y_2^m \\ \quad\vdots \\ c_0 + c_1 x_r = y_r^m \end{cases} \Rightarrow \begin{pmatrix} 1 & x_1 \\ 1 & x_2 \\ \vdots & \vdots \\ 1 & x_r \end{pmatrix} \cdot \begin{pmatrix} c_0 \\ c_1 \end{pmatrix} = \begin{pmatrix} y_1^m \\ y_2^m \\ y_3^m \\ y_4^m \\ y_5^m \end{pmatrix}$$
$$\Rightarrow A \cdot C = M \tag{17.3}$$

First, this is an *overdetermined* system of equations (because we have more equations than we need to solve it). Second, it doesn't have an *exact* solution (because of the noise). The only thing we can do is to try to find the *best* solution. We'll get back to what we mean by 'best' in a minute. In Eq. (17.3) we also wrote the equation system in matrix form and A is the *observation* matrix, C is the *coefficient* matrix and M is the *measurement data* matrix. Our objective is to find the C matrix.

By the 'best' solution, we mean the straight line in Fig. 17.1 that will minimize the sum of all the squared errors. That makes sense; we must square them before we add them since the errors have different signs. So, we square and add, and then it becomes a classical 'minimum' problem (take the derivative and set it equal to zero). That is how it is derived in undergraduate classes, and it is called 'linear regression'. However, here we will try to go a little 'deeper' (to promote profound understanding).

We concluded above that our overdetermined system of equations doesn't have an exact solution because of the noise. Well, let's assume that we don't understand that (or don't care) and try to solve it anyway. Maybe we are lucky... If A had been quadratic ($r \times r$), then the solution would have been easy; just invert A and multiply both sides from (the left) with A^{-1} and we get $C = A^{-1} M$. But A isn't quadratic, it is an $r \times 2$ matrix and A^{-1} does not exist. Well, we can fix that; multiply both sides (from the left) by A^{T} (the 'transpose' of A, which is a $2 \times r$ matrix):

$$A^{\mathrm{T}} A C = A^{\mathrm{T}} M \tag{17.4}$$

$A^{\mathrm{T}}A$ *is* a quadratic, 2×2 matrix and *does have* an inverse matrix (if it only has 'full rank', which all 'normal' measurements produce). So, if we multiply, from the left, with the inverse of $A^{\mathrm{T}}A$ we get:

$$\left(A^{\mathrm{T}} A\right)^{-1} A^{\mathrm{T}} A C = \left(A^{\mathrm{T}} A\right)^{-1} A^{\mathrm{T}} M \Rightarrow C = \left(A^{\mathrm{T}} A\right)^{-1} A^{\mathrm{T}} M = A^{\#} M \tag{17.5}$$

($A^{\#} = (A^{\mathrm{T}}A)^{-1}A^{\mathrm{T}}$ is the 'pseudo inverse'.) We solved it! We found a solution to the system of equations with no solution! Well, we said that it doesn't have an *exact* solution. So, what does the solution in Eq. (17.5) represent? It is the same solution we would get if we solved the minimum square problem with the sum of errors. Here, will prove it to you in a different way. Equation (17.5) is the 'best' solution and we will prove it using geometry.

17.2 The Orthogonality Principle

The starting point is that we consider C to be a *vector*; in this case, it is a vector in $\mathbf{R}^2$, but in the general case it would be a vector in $\mathbf{R}^n$ (when we try to fit data to an n – 1 order polynomial). Hence, the C matrix is a *column vector* (c_0,c_1) (temporarily lying down here). Similarly, A also consists of two (column) vectors (1,1,...,1) and (x_1,x_2,...x_r):

$$AC = c_0 \begin{pmatrix} 1 \\ 1 \\ \vdots \\ 1 \end{pmatrix} + c_1 \begin{pmatrix} x_1 \\ x_2 \\ \vdots \\ x_r \end{pmatrix} = \begin{pmatrix} c_0 + c_1 x_1 \\ c_0 + c_1 x_2 \\ \vdots \\ c_0 + c_1 x_r \end{pmatrix} \tag{17.6}$$

In Eq. (17.6), we multiply the column vector (1,1,...1) by c_0 and the vector (x_1,x_2,...x_r) is multiplied by c_1. This creates a new vector, see Fig. 17.2.

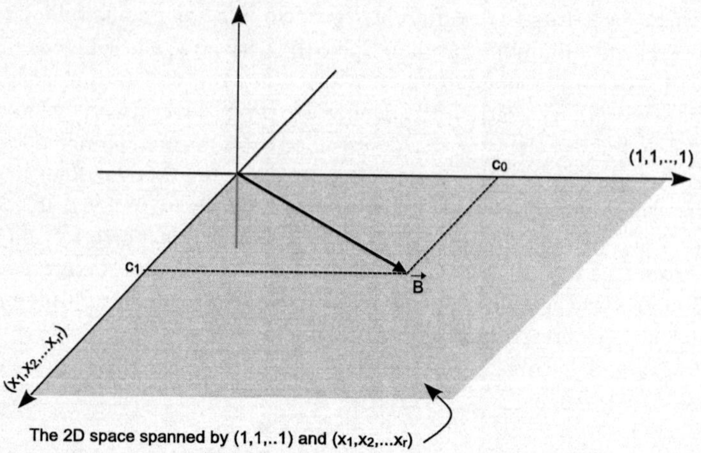

Fig. 17.2 Vectors (1,1,…1) and $(x_1,x_2,\ldots x_r)$ define a space

From Fig. 17.2, we can see first that the two column vectors (1,1,…1) and $(x_1,x_2,\ldots x_r)$ define a 2D space (they *span* a 2D space) and second that the matrix product AC is a vector (that we call $\overrightarrow{B}$ in Fig. 17.2). So, if the matrix product AC defines a vector in the space defined by the vectors (1,1,…1) and $(x_1,x_2,\ldots x_r)$, the matrix equation $AC = M$, can only have a solution if the vector M *is in this space*! But because of the noise in the measurement, the measurement data matrix M *is not in this space*, see Fig. 17.3.

Since we are limited to the space spanned by vectors (1,1,…1) and $(x_1,x_2,\ldots x_r)$, we cannot find an exact solution to the $AC = M$ equation. It that case, we instead find the best solution. The 'best' solution is the vector in the space [(1,1,…1),

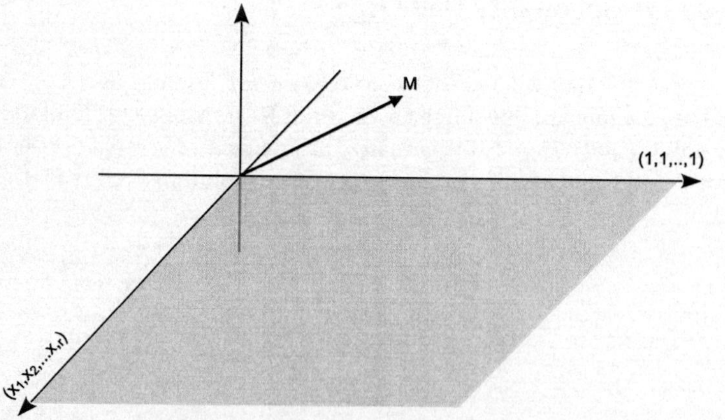

Fig. 17.3 Our measurement matrix M is not in the 2D space spanned by (1,1,…1) and $(x_1,x_2,\ldots x_r)$

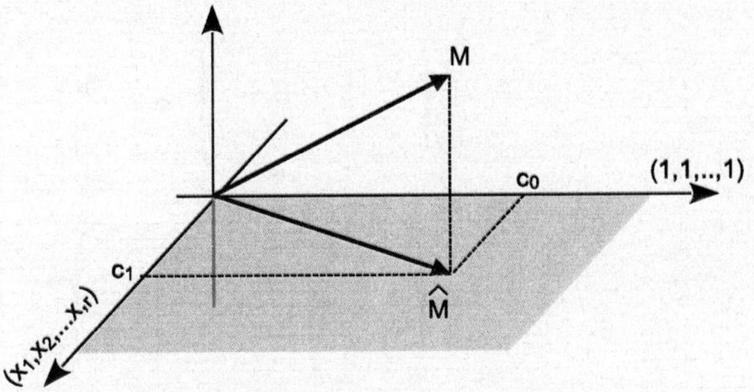

Fig. 17.4 The projection of M is closest to M

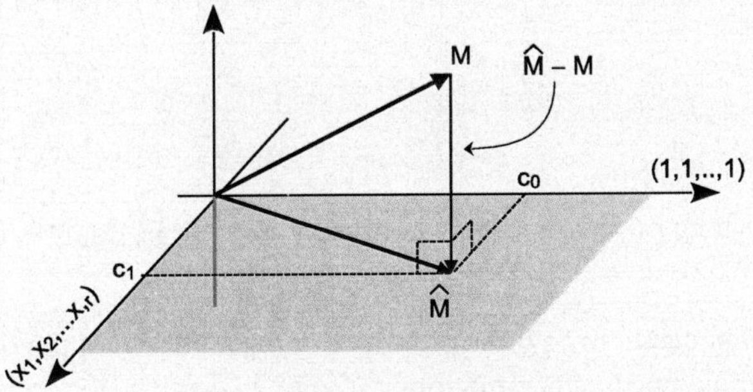

Fig. 17.5 $\widehat{M} - M$ is perpendicular to the plane

$(x_1,x_2,\ldots x_r)$], that is closest to M; the vector closest to M is the projection of M onto the [(1,1,...1),$(x_1,x_2,\ldots x_r)$] space, see Fig. 17.4.

So how do we find the coefficients c_0 and c_1 that define the closest vector $\widehat{M}$ ('M hat')? That is easy; that's when the vector $\widehat{M} - M$ is perpendicular to the plane spanned by [(1,1,..1), $(x_1,x_2,\ldots x_r)$],[1] see Fig. 17.5.

To be perpendicular to the plane spanned by [(1,1,...1), $(x_1,x_2,\ldots x_r)$], it must be perpendicular to both vectors [(1,1,...1) and $(x_1,x_2,\ldots x_r)$], i.e., the scalar products between $\widehat{M} - M$ and both vectors [(1,1,...1) and $(x_1,x_2,\ldots x_r)$] must be zero:

$$\begin{cases} (1\ \ 1\ldots 1)^{\mathrm{T}}\left(\widehat{M} - M\right) = 0 \\ (x_1\ \ x_2\ldots x_r)^{\mathrm{T}}\left(\widehat{M} - M\right) = 0 \end{cases} \Rightarrow \begin{pmatrix} 1 & 1\ldots 1 \\ x_1 & x_2\ldots x_r \end{pmatrix}^{\mathrm{T}}\left(\widehat{M} - M\right) = 0 \tag{17.7}$$

$$A^T(\widehat{M} - M) = 0 \tag{17.8}$$

But,

[1] This is the'orthogonality principle'.

$$\widehat{M} - M = \begin{pmatrix} y_1 \\ y_2 \\ \vdots \\ y_r \end{pmatrix} - \begin{pmatrix} y_1^m \\ y_2^m \\ \vdots \\ y_r^m \end{pmatrix} = \begin{pmatrix} c_0 + c_1 x_1 \\ c_0 + c_1 x_2 \\ \vdots \\ c_0 + c_1 x_r \end{pmatrix} - \begin{pmatrix} y_1^m \\ y_2^m \\ \vdots \\ y_r^m \end{pmatrix} =$$

$$= \begin{pmatrix} 1 & x_1 \\ 1 & x_2 \\ \vdots \\ 1 & x_r \end{pmatrix} \begin{pmatrix} c_0 \\ c_1 \end{pmatrix} - \begin{pmatrix} y_1^m \\ y_2^m \\ \vdots \\ y_r^m \end{pmatrix} = AC - M$$

Substituting $AC–M$ for $\widehat{M} - M$ in Eq. (17.8) gives us

$$\begin{aligned} &A^T(\widehat{M} - M) = A^T(AC - M) = 0 \\ &\Rightarrow A^T AC = A^T M \Rightarrow C = (A^T A)^{-1} A^T M = A^{\#} M \end{aligned} \tag{17.9}$$

which proves that the solution we derived in Eq. (17.5) is indeed the best solution. We summarize this in the following theorem.

Theorem (The orthogonality principle.) We find the best solution to the overdetermined system of equations $AC = M$, by setting the error vector $\widehat{M} - M$ perpendicular to the columns in A.

The orthogonality principle holds true for any polynomial order; if we want to fit data to an nth order polynomial, the C matrix is an $(n + 1) \times 1$ matrix and A is $r \times (n + 1)$.

It is also common to state the total error, i.e., the sum or all r errors in Eq. (17.1). However, since this number increases with the data size, we divide it by the number of samples:

$$e^2 = \frac{1}{r}\sum_i \varepsilon_i^2 = \frac{1}{r}\sum_{i=0}^{r-1}\left(f(x) - y_i^m\right)^2 \tag{17.10}$$

Equation (17.10) is not just a number that quantifies the quality of the fitting; it is the power of the noise that interfered with our samples. (Assuming that we are fitting to the right polynomial, see discussion in Problem 17.1.)

Example 17.1 Table 17.2 represents data samples from a calibration of a temperature sensor. Use this data to find a first-order calibration expression for the sensor. What was the noise level in the measurement?

Solution We will fit data to a first-order polynomial: $U = c_0 + c_1 T$:

Table 17.2 Temperature calibration data

T [°C]	−10	0	10	20	40
U^m [V]	−2.50	−0.09	2.21	4.82	9.31

$$\begin{cases} c_0 - c_1 10 = -2.50 \\ c_0 + c_1 0 = -0.09 \\ c_0 + c_1 10 = 2.21 \\ c_0 + c_1 20 = 4.82 \\ c_0 + c_1 40 = 9.31 \end{cases} \Rightarrow \begin{pmatrix} 1 & -10 \\ 1 & 0 \\ 1 & 10 \\ 1 & 20 \\ 1 & 40 \end{pmatrix} \begin{pmatrix} c_0 \\ c_1 \end{pmatrix} = \begin{pmatrix} -2.50 \\ -0.09 \\ 2.21 \\ 4.82 \\ 9.31 \end{pmatrix} \Rightarrow AC = M$$

$$A^{\mathrm{T}}A = \begin{pmatrix} 1 & 1 & 1 & 1 & 1 \\ -10 & 0 & 10 & 20 & 40 \end{pmatrix} \begin{pmatrix} 1 & -10 \\ 1 & 0 \\ 1 & 10 \\ 1 & 20 \\ 1 & 40 \end{pmatrix} = \begin{pmatrix} 5 & 60 \\ 60 & 2200 \end{pmatrix}$$

$$\Rightarrow \left(A^{\mathrm{T}}A\right)^{-1} = \begin{pmatrix} 0.2973 & -0.081 \\ -0.081 & 0.0007 \end{pmatrix}$$

$$\left(A^{\mathrm{T}}A\right)^{-1}A^{\mathrm{T}} = \begin{pmatrix} 0.2973 & -0.081 \\ -0.081 & 0.0007 \end{pmatrix} \begin{pmatrix} 1 & 1 & 1 & 1 & 1 \\ -10 & 0 & 10 & 20 & 40 \end{pmatrix} =$$

$$= \begin{pmatrix} 0.0374 & 0.2973 & 0.2162 & 0.1351 & -0.0270 \\ -0.0149 & 0.0081 & -0.0014 & 0.0054 & 0.0189 \end{pmatrix}$$

$$C = \left(A^{\mathrm{T}}A\right)^{-1}A^{\mathrm{T}}M = \begin{pmatrix} -0.0951 \\ 0.2371 \end{pmatrix} \Rightarrow \underline{U = -0.0951 + 0.2371 \times T}$$

This line is plotted in Fig. 17.6 together with the samples. In Table 17.3, we have included the fitted data and the errors.

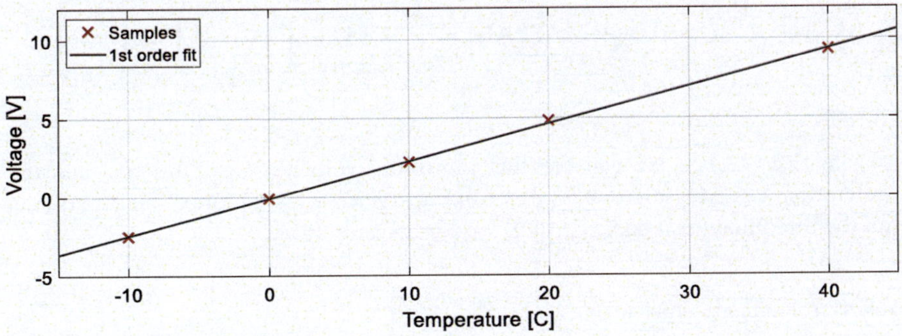

Fig. 17.6 The fit and the samples

Table 17.3 Temperature calibration data

T [°C]	−10	0	10	20	40
U^m [V]	−2.50	−0.09	2.21	4.82	9.31
U [V]	−2.466	−0.095	2.276	4.647	9.389
ε [V]	0.0339	−0.005	0.0658	−0.173	0.079

The average squared error is

$$e^2 = \frac{1}{r}\sum_{i=0}^{4}\varepsilon_i^2 = 0.0083\,\mathrm{V}^2 \Rightarrow u_{\mathrm{rms}}^{\mathrm{noise}} = \sqrt{0.0083} = 0.0913 = \underline{91\,\mathrm{mV}}$$

Example 17.2 A projectile's position was registered at some regular time intervals, see Table 17.4. Use this data to determine the projectile's initial velocity and its acceleration.

Solution A projectile's position as a function of initial position s_0, initial speed, v_0 and its acceleration a is given by

$$s = s_0 + v_0 t + \frac{1}{2}at^2 = c_0 + c_1 t + c_2 t^2 \qquad (17.11)$$

Inserting our measurement data gives us:

$$\begin{cases} c_0 + c_1 0 + c_2 0^2 = 0.02 \\ c_0 + c_1 5 + c_2 5^2 = 18.21 \\ c_0 + c_1 10 + c_2 10^2 = 48.95 \\ c_0 + c_1 15 + c_2 15^2 = 99.31 \\ c_0 + c_1 20 + c_2 20^2 = 158.46 \end{cases} \Rightarrow \underbrace{\begin{pmatrix} 1 & 0 & 0 \\ 1 & 5 & 25 \\ 1 & 10 & 100 \\ 1 & 15 & 225 \\ 1 & 20 & 400 \end{pmatrix}}_{A} \underbrace{\begin{pmatrix} c_0 \\ c_1 \\ c_2 \end{pmatrix}}_{C} = \underbrace{\begin{pmatrix} 0.02 \\ 18.21 \\ 48.95 \\ 99.31 \\ 158.46 \end{pmatrix}}_{M}$$

$$\left(A^{\mathrm{T}}A\right)^{-1}A^{\mathrm{T}}M = \{\text{Using MATLAB}\} = \begin{pmatrix} -0.1003 \\ 2.1573 \\ 0.2901 \end{pmatrix} = \begin{pmatrix} c_0 \\ c_1 \\ c_2 \end{pmatrix} = \begin{pmatrix} s_0 \\ v_0 \\ \frac{1}{2}a \end{pmatrix} \qquad (17.12)$$

From Eq. (17.12), we can see that the initial velocity was 2.16 m/s, and the acceleration was $2{\cdot}0.2901 = 0.58$ m/s^2. In Fig. 17.7, we have plotted the data points with the fitted line (as a 'sanity test').

Table 17.4 Position at different times

t [s]	0.00	5.00	10.00	15.00	20.00
s^m [m]	0.02	18.21	48.95	99.31	158.46

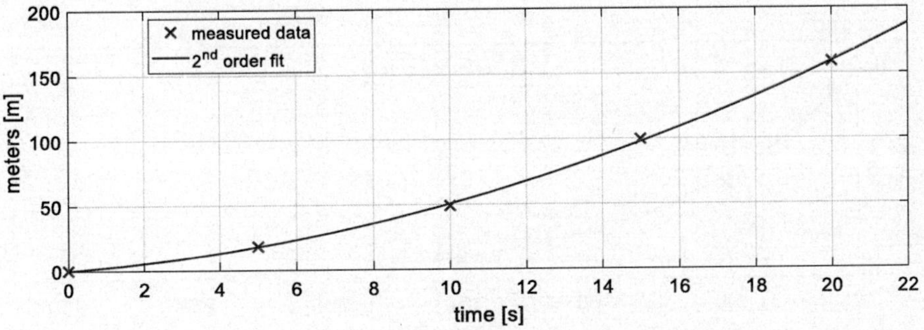

Fig. 17.7 The fit and the data

17.3 Curve Fitting to Exponential Functions

Our ‘pseudo inverse’ formula above only works when data are fitted to a polynomial, i.e., when data can be fitted to a linear combination of the coefficients. That is not always the case. For example, there are lots of examples of exponential relationships in science (nuclear decay, cooling, population growth, etc.). Suppose we have an expected relationship as in Eq. (17.13):

$$y(x) = c_0 \cdot \mathrm{e}^{c_1 x} \tag{17.13}$$

Equation (17.9) cannot be applied here, because we don’t have a linear dependence on all the coefficients. However, we can turn it into a linear combination of coefficients by taking the logarithm of both sides:

$$\ln y = \ln c_0 + c_1 x = \acute{c}_0 + c_1 x \tag{17.14}$$

Hence, we just proceed exactly as above, but when we are done, we transform $\acute{c}_0$ back to c_0. We illustrate this with an example.

Example 17.3 In a nuclear experiment, the radioactivity of a sample was measured at some times, see Table 17.5. What was the decay constant and the half-life time of the sample?

Solution The radioactivity decays exponentially, so we need to take the logarithm of both sides:

Table 17.5 Radioactivity from a sample

t [s]	10	20	50	100	150	300
A^m [Bq]	90,345	71,491	44,609	25,873	10,077	1329
$\ln A^m$	11.41	11.18	10.71	10.16	9.218	7.192

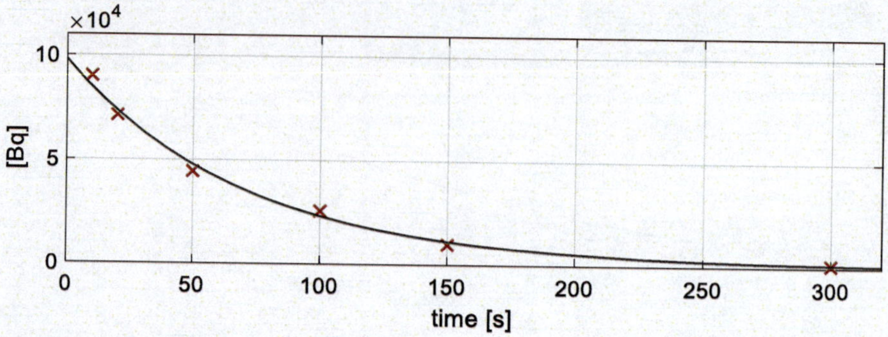

Fig. 17.8 The fit and the data points

$$A = A_0 e^{-\lambda t} \Rightarrow \ln A = \ln A_0 - \lambda t = c_0 - c_1 t \qquad (17.15)$$

In Table 17.5, we have already calculated the logarithms:

$$\begin{cases} c_0 - c_1 \cdot 10 = 11.41 \\ c_0 - c_1 \cdot 20 = 11.18 \\ c_0 - c_1 \cdot 50 = 10.71 \\ c_0 - c_1 \cdot 100 = 10.16 \\ c_0 - c_1 \cdot 150 = 9.218 \\ c_0 - c_1 \cdot 300 = 7.192 \end{cases} \Rightarrow \underbrace{\begin{pmatrix} 1 & 10 \\ 1 & 20 \\ 1 & 50 \\ 1 & 100 \\ 1 & 150 \\ 1 & 300 \end{pmatrix}}_{A} \underbrace{\begin{pmatrix} c_0 \\ -c_1 \end{pmatrix}}_{C} = \underbrace{\begin{pmatrix} 11.41 \\ 11.18 \\ 10.71 \\ 10.16 \\ 9.218 \\ 7.192 \end{pmatrix}}_{M} \qquad (17.16)$$

$$C = \left(A^{\mathrm{T}} A\right)^{-1} A^{\mathrm{T}} M = \begin{pmatrix} 11.4941 \\ -0.0144 \end{pmatrix} = \begin{pmatrix} \ln A_0 \\ -\lambda \end{pmatrix} \Rightarrow \begin{matrix} A_0 = 98135 \text{ Bq} \\ \lambda = 0.0144 \text{ s}^{-1} \end{matrix}$$

$$\Rightarrow A = 98135 \cdot e^{-0.0144t}$$

In Fig. 17.8 we have plotted the data points and the fit.

The half-life time is: $0.5 = e^{-0.0144t} \Rightarrow t_{1/2} = -\ln 2 / -0.0144 =$ <u>48 seconds</u>. The decay constant is $\lambda = -0.0144 \text{ s}^{-1}$.

17.4 MATLAB Tips

If you have access to MATLAB, you don't need to use Eq. (17.9); the '\' operator in MATLAB solves the $AC = M$ equation immediately: $C = A \backslash M$.

Example 17.4 Solve the problem in Example 7.1 using the backslash operator in MATLAB.

Table 17.6 Temperature calibration data

T [°C]	−10	0	10	20	40
U^m [V]	−2.50	−0.09	2.21	4.82	9.31
$2 \times u$	0.25	0.27	0.23	0.24	0.22

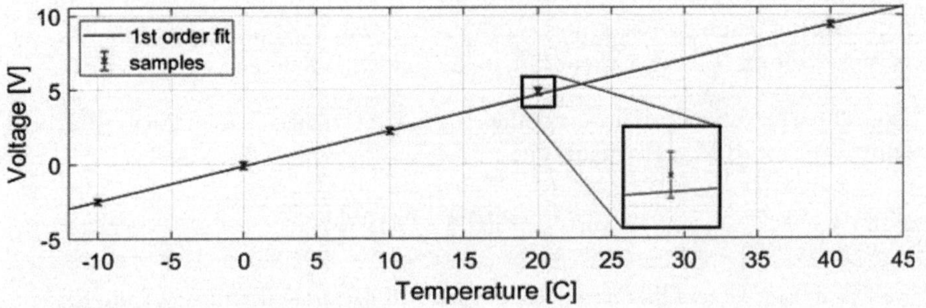

Fig. 17.9 The plot with error bars (95% confidence)

Solution

>> A = [1−10;1 0;1 10;1 20;1 40];

>> M = [−2.50; −0.09;2.21;4.82;9.31];

>> C = A\M.

C = −0.0951

0.2371

In Chap. 14, we learned to calculate the 95% confidence interval for a measurement (the 'uncertainty') and when we fit data to a polynomial, we should indicate each measured value's (y^m) uncertainty as an *error bar*. MATLAB can handle that for you if you just plot the graphs with the *errorbar(x,y,e)* command.

Example 17.5 In Table 17.6, we have added the uncertainties for each sample (see Sect. 14.2). Plot the fitted line and samples with error bars in the same diagram.

Solution Using *errorbar(T,U,e)* in MATLAB, we get the plot in Fig. 17.9.

17.5 Matrix Uncertainties and Pitfalls

17.5.1 Error Propagation in Matrices

In the matrix equation $AC = M$, M represents *measured* data, and in Chap. 14, we learned that all measured data has an uncertainty. The question to ask now is of course how the uncertainty in the measured data propagates to an uncertainty in the coefficients in the C matrix? Our data are supposed to follow some polynomial: $y = c_0 + c_1x + c_2x^2 + \cdots + c_px^p$ but because of noise there is a stochastic contribution to y^m, and hence,

$$y^m(n) = c_0 + c_1x(n) + c_2x^2(n) + \ldots c_px^p(n) + b(n) \tag{17.17}$$

where $b(n)$ is white gaussian noise: $b(n) \in \mathrm{N}(0, \sigma)$. This noise propagates to the coefficients in the C matrix according to the following theorem:

Theorem The variance of the c_i coefficients in the C matrix is found in the diagonal elements of the $\sigma^2(A^{\mathrm{T}}A)^{-1}$ matrix, i.e.,

$$\mathrm{var}(c_i) = \left\{\sigma^2\left(A^{\mathrm{T}}A\right)^{-1}\right\}_{ii} \tag{17.18}$$

We don't prove that theorem here, but we will illustrate it with an example.

Example 17.6 What is the uncertainty of the coefficients in Example 17.1?

Solution The variance of the noise level was 0.0083 V^2, so

$$\sigma^2\left(A^{\mathrm{T}}A\right)^{-1} = 0.0083 \cdot \begin{pmatrix} 0.2973 & -0.081 \\ -0.081 & 0.0007 \end{pmatrix} = \begin{pmatrix} 2.47 & -0.672 \\ -0.672 & 0.00581 \end{pmatrix} \times 10^{-3}$$

$$\Rightarrow \begin{cases} u(c_0) = \sqrt{2.47 \cdot 10^{-3}} = 0.050 \\ u(c_1) = \sqrt{5.81 \cdot 10^{-6}} = 0.0024 \end{cases}$$

With a coverage factor of 2, we get the following coefficients:

$$\begin{cases} c_0 = -0.095 \pm 0.100 \\ c_1 = 0.2371 \pm 0.0048 \end{cases} (95\% \text{ confidence})$$

Notice in the example above that the uncertainty was much smaller for c_1 than for c_0; this is true in general. Higher order coefficients are more sensitive to variations in data and can therefore be determined with higher precision.

17.5.2 Ill-Conditioned Matrices

In some situations, the A matrix can be 'ill-conditioned' and that can have severe consequences for the precision. We will illustrate that with an example.

Example 17.6 Table 17.7 illustrates data from some measurement. Fit this data to a first-order polynomial, $y = c_0 + c_1x$.

Table 17.7 Some measurement

x	1000	1001	1003
y	4	6	9

Solution A and M are; $A = \begin{pmatrix} 1 & 1000 \\ 1 & 1001 \\ 1 & 1003 \end{pmatrix}$ and $M = \begin{pmatrix} 4 \\ 6 \\ 9 \end{pmatrix}$. Hence,

$$\left(A^{\mathrm{T}}A\right)^{-1} = \begin{pmatrix} 3 & 3004 \\ 3004 & 3008010 \end{pmatrix}^{-1} = \frac{1}{14}\begin{pmatrix} 3008010 & -3004 \\ -3004 & 3 \end{pmatrix} \tag{17.19}$$

$$A^T M = \begin{pmatrix} 1 & 1 & 1 \\ 1000 & 1001 & 1003 \end{pmatrix} \cdot \begin{pmatrix} 4 \\ 6 \\ 9 \end{pmatrix} = \begin{pmatrix} 19 \\ 19033 \end{pmatrix} \tag{17.20}$$

$$\Rightarrow C = \frac{1}{14}\begin{pmatrix} 3008010 & -3004 \\ -3004 & 3 \end{pmatrix}\begin{pmatrix} 19 \\ 19033 \end{pmatrix} = \begin{pmatrix} -1638.7 \\ 1.6429 \end{pmatrix} = \begin{pmatrix} c_0 \\ c_1 \end{pmatrix} \tag{17.21}$$

Inserting these coefficients into our fit gives us $y(1000) = 4.20$, $y(1001) = 5.84$ and $y(1003) = 9.13$.

The result in Example 17.6 seems reasonable. However, these calculations are usually carried out by computers and all computers have limited accuracy. To illustrate the problem, we will 'amplify' it here, by assuming that our computer has a precision of only four digits. (A real computer has much higher precision, but the problem is the same, it is just on a smaller scale.). That means that 3,008,010 will be rounded to 3,008,000 and 19,033 is rounded to 19,030. In the first case, it is an error of 3.3 ppm, and in the second case, it is an error of 158 ppm. You wouldn't expect such small rounding errors to have any significant impact on the result, would you? Let's see:

$$\frac{1}{14}\begin{pmatrix} 3008000 & -3004 \\ -3004 & 3 \end{pmatrix}\begin{pmatrix} 19 \\ 19030 \end{pmatrix} = \begin{pmatrix} -1009 \\ 1.000 \end{pmatrix} = \begin{pmatrix} c_0 \\ c_1 \end{pmatrix} \tag{17.22}$$

Comparing Eq. (17.22) with Eq. (17.21), we can see that the ppm level rounding has catastrophic consequences on the calculations! The problem is that the $A^{\mathrm{T}}A$ matrix is 'ill-conditioned'. (This is implied by the size of the highest eigenvalue of the $A^{\mathrm{T}}A$ matrix; in this case, it is 1734 and rounding troubles are expected.) There is an easier way to predict the problems in this case. If we take a closer look at the column vectors in A, we can see that they are almost parallel; the vector (1,1,1) is almost parallel to (1000, 1001, 1003). The angle between these two vectors is only 0.07° and that makes it hard for the two vectors to span the space properly (the $A^{\mathrm{T}}A$ matrix is very close to being singular). Ideally, we want the angle between the 'spanning' vectors to be 90°. We can fix this problem by fitting to a savvier polynomial.

Example 17.7 In Example 17.6, fit data to the polynomial $y = c_0 + c_1(x - 1000)$ instead.

Solution That gives us the following system of equations:

$$\begin{cases} c_0 + c_1 0 = 4 \\ c_0 + c_1 1 = 6 \\ c_0 + c_1 3 = 9 \end{cases} \Rightarrow \begin{pmatrix} 1 & 0 \\ 1 & 1 \\ 1 & 3 \end{pmatrix}\begin{pmatrix} c_0 \\ c_1 \end{pmatrix} = \begin{pmatrix} 4 \\ 6 \\ 9 \end{pmatrix} \Rightarrow A^{\mathrm{T}}A$$

$$= \begin{pmatrix} 1 & 1 & 1 \\ 0 & 1 & 3 \end{pmatrix}\begin{pmatrix} 1 & 0 \\ 1 & 1 \\ 1 & 3 \end{pmatrix} = \begin{pmatrix} 3 & 4 \\ 4 & 10 \end{pmatrix}$$

(The four-digit restriction doesn't have any influence on the numbers anymore.)

$$A^{\mathrm{T}}M = \begin{pmatrix} 1 & 1 & 1 \\ 0 & 1 & 3 \end{pmatrix}\begin{pmatrix} 4 \\ 6 \\ 9 \end{pmatrix} = \begin{pmatrix} 19 \\ 33 \end{pmatrix} \Rightarrow \left(A^T A\right)^{-1} A^{\mathrm{T}}M$$

$$= \frac{1}{14}\begin{pmatrix} 10 & -4 \\ -4 & 3 \end{pmatrix}\begin{pmatrix} 19 \\ 33 \end{pmatrix} = \begin{pmatrix} 4.143 \\ 1.643 \end{pmatrix}$$

$$\Rightarrow y = 4.143 + 1.643 \cdot (x - 1000) = -1639 - 1.643x$$

In Example 17.7, we have the column vectors (1,1,1) and (0,1,3). The angle between them is:

$$X \cdot Y = |X| \cdot |Y| \cos\alpha \Rightarrow \alpha = \cos^{-1}\frac{X \cdot Y}{|X| \cdot |Y|} = \cos^{-1}\frac{4}{\sqrt{3} \cdot \sqrt{10}} = 43°$$

which indicates a more stable 'spanning' of the space.

(To what polynomial should you fit the data to get perpendicular column vectors?).

Advanced calculation programs, like MATLAB, avoid this problem by first factorizing the matrices. For example, the backslash ('\') operator in MATLAB uses QR factorization to avoid rounding errors in matrices with 'almost parallel' column vectors.

17.6 The Sampling Theorem Revisited

In this section, we will investigate a problem that is not exactly curve fitting, but closely related and we will answer a question that is often asked by students.

According to the sampling theorem, the sampling rate f_S must exceed $2f_{\max}$, i.e., $f_S > 2{\cdot}f_{\max}$. Hence, $f_S = 3{\cdot}f_{\max}$ should be enough. That sampling rate indicates that

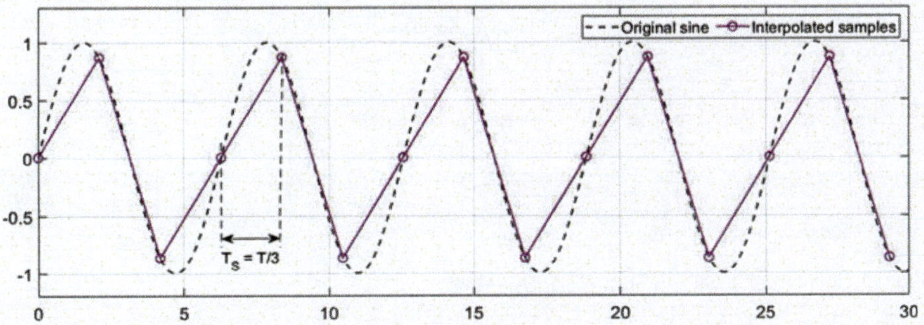

Fig. 17.10 Sampling a sine; $f_S = 3f$

we would only take three samples of each period of a sine. This is illustrated in Fig. 17.10.

The question often asked by students is: 'How can you recover the original sine from only three samples of a period'?

First, *it is possible* to recover the original signal shape from only a few samples (as long as you don't violate the sampling theorem). Second, *linear interpolation* is obviously not going to do it, see Fig. 17.10. We need a more cunning plan than that.

To understand how to recover the original sine from only a few samples, we first need to look at the Fourier transforms of the original sine, let's call it $x_a(t)$ ('a' for 'analog'), and the sampled sine, $x_d(t)$ ('d' for 'discrete-time'). Figure 17.11 illustrates the Fourier spectrum of the analog sine; there is just one pair of peaks at the positive and negative sine frequency.

We know from chapter 7 that when we sample $x(t)$, the Fourier transform becomes periodic, with a period equal to the sampling frequency. The Fourier spectrum of the sampled signal is illustrated in Fig. 17.12.

After the sampling, we only have the samples and the question is, how we can retrieve the original $x_a(t)$ signal? Linear interpolation of $x_d(t)$ doesn't work and if we take the inverse Fourier transform of $X_d(\omega)$, we will only get our samples back, not the analog signal $x_a(t)$.

But the key to the retrieval of $x_a(t)$ is still in the Fourier transforms. To retrieve $x_a(t)$, we first need to recreate $X_a(\omega)$. We can do that by multiplying $X_d(\omega)$ with a 'square' frequency function covering only frequencies between $\pm\, \omega_S/2$, see Fig. 17.13.

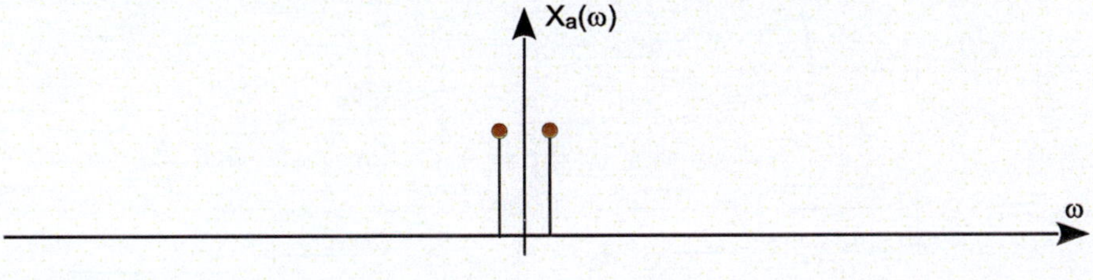

Fig. 17.11 Fourier spectrum of analog sine

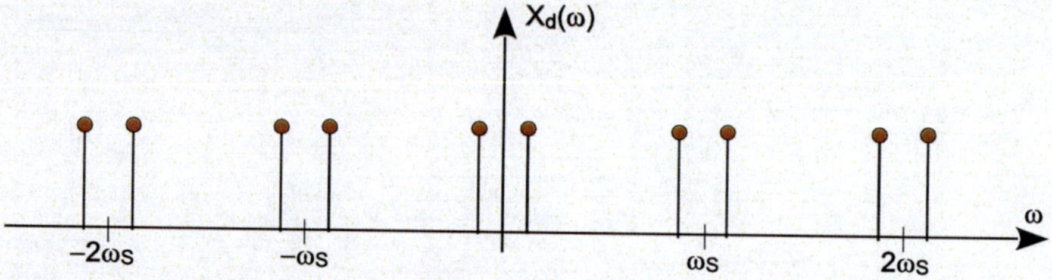

Fig. 17.12 Fourier spectrum of discrete-time sine

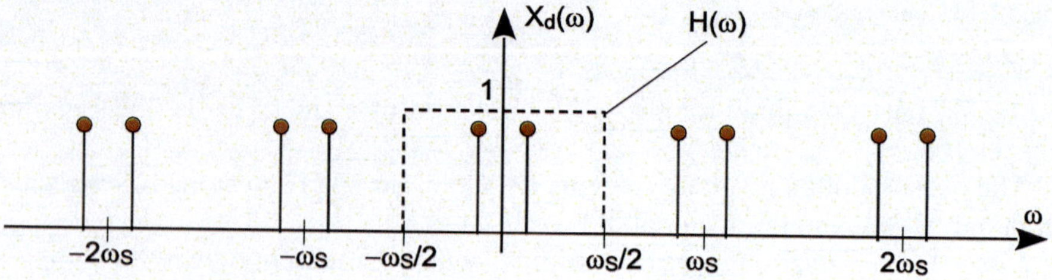

Fig. 17.13 Multiplying by a 'square' filter

The resulting product will be $X_{\mathrm{a}}(\omega)$:

$$X_{\mathrm{a}}(\omega) = X_{\mathrm{d}}(\omega) \cdot H(\omega) \tag{17.23}$$

So, to retrieve the original signal in frequency space, we *multiply* $X_{\mathrm{d}}(\omega)$ by $H(\omega)$. We know from Sect. 9.8 that multiplication in frequency space corresponds to *convolution* in time space. Hence, to retrieve the original signal $x_{\mathrm{a}}(t)$, we must convolve $x_{\mathrm{d}}(t)$ with $h(t)$ $(= H^{-1}(\omega))$:

$$x_{\mathrm{a}}(t) = x_{\mathrm{d}}(t) \otimes h(t) = \int_{-\infty}^{t} x_{\mathrm{d}}(\tau) h(t-\tau) d\tau \tag{17.24}$$

Before we can evaluate Eq. (17.24) we need to find $h(t)$:

$$\begin{aligned} h(t) &= \frac{1}{\omega_s} \int_{-\omega_s/2}^{\omega_s/2} 1 \cdot \mathrm{e}^{\mathrm{j}\omega t} d\omega = \frac{1}{\omega_s} \cdot \frac{1}{\mathrm{j}t} \left[\mathrm{e}^{\mathrm{j}\omega t}\right]_{-\omega_s/2}^{\omega_s/2} \\ &= \frac{1}{\omega_s t} \cdot \frac{2}{2\mathrm{j}} \left(\mathrm{e}^{\mathrm{j}\omega_S t/2} - \mathrm{e}^{-\mathrm{j}\omega_S t/2}\right) = \frac{1}{\omega_s t/2} \sin \frac{\omega_S}{2} t = \operatorname{sinc} \frac{\omega_S}{2} t \\ &= \operatorname{sinc} \frac{2\pi}{2T_S} t \end{aligned}$$

(In the inverse Fourier transform, we divide by ω_S to 'normalize', to make the area under $|H(\omega)| = 1$.)

$h(t)$ is a sinc function with period $2T_S$. Next, we insert that into Eq. (17.24), and remember that $x_d(t)$ is a discrete-time function, $\neq 0$ only if $t = nT_S$:

$$
\begin{aligned}
x_a(t) &= \int_{-\infty}^{t} x_d(\tau)\,\mathrm{sinc}\frac{2\pi}{2T_S}(t-\tau)\mathrm{d}\tau \\
&= \sum_{\substack{i \\ \text{over all samples}}} x_d(i)\,\mathrm{sinc}\,\frac{2\pi}{2T_S}(t - i \cdot T_S) \\
&= x_d(0)\cdot \mathrm{sinc}\frac{2\pi}{2T_S}t + x_d(1)\cdot \mathrm{sinc}\frac{2\pi}{2T_S}(t - T_S) \\
&\quad + x_d(2)\cdot \mathrm{sinc}\frac{2\pi}{2T_S}(t - 2T_S) + \ldots
\end{aligned}
$$

Hence, to recover the original analog signal $x_a(t)$, we multiply each sample by a sinc function that has period $2T_S$ and is centered around the sample position.

This is illustrated in Fig. 17.14 and in Fig. 17.15, we have plotted the sum of them and the original $x_a(t)$ signal.

This is called 'sin$x$/x interpolation' and this is what digital oscilloscopes use to recreate the signal on the screen when there are not enough samples to do regular linear interpolation.

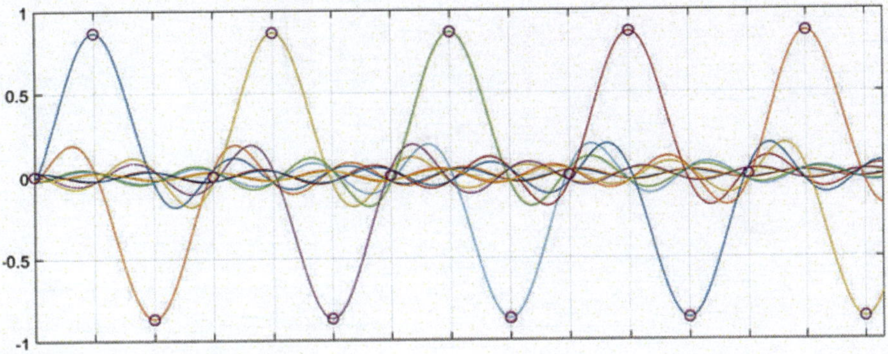

Fig. 17.14 A sin$x$/x interpolation

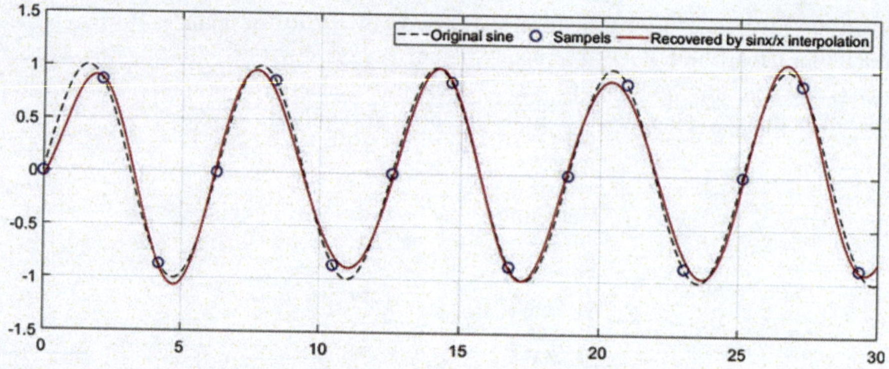

Fig. 17.15 The recreated analog signal

17.7 Solved Problems

Problem 17.1 Fit the data in Table 17.8 to (a) a constant, (b) a first-order polynomial, (c) a second-order polynomial, and (d) an exponential function. Plot them all in the same graph and find the total error in each fit. Also, discuss the differences in the total errors and what conclusions to draw from it.

Solution $y = c_0$:

$$\begin{cases} c_0 = 7 \\ c_0 = 5 \\ c_0 = 2 \\ c_0 = 1 \end{cases} \Rightarrow \begin{pmatrix} 1 \\ 1 \\ 1 \\ 1 \end{pmatrix} c_0 = \begin{pmatrix} 7 \\ 5 \\ 2 \\ 1 \end{pmatrix} \Rightarrow A^{\mathrm{T}}A = (1 \quad 1 \quad 1 \quad 1)\begin{pmatrix} 1 \\ 1 \\ 1 \\ 1 \end{pmatrix}$$

$$= 4 \Rightarrow \left(A^{\mathrm{T}}A\right)^{-1} = 0.25$$

$$\left(A^{\mathrm{T}}A\right)^{-1}A^{\mathrm{T}}M = 0.25 \cdot (1 \quad 1 \quad 1 \quad 1) \cdot \begin{pmatrix} 7 \\ 5 \\ 2 \\ 1 \end{pmatrix} = \underline{3.75 = c_0}$$

First-order fit: $y = c_0 + c_1 x$

Table 17.8 Data

x	-2	0	1	3
y^m	7	5	2	1

$$\begin{cases} c_0 - c_1^2 = 7 \\ c_0 - c_1^0 = 5 \\ c_0 - c_1^1 = 2 \\ c_0 - c_1^3 = 1 \end{cases} \Rightarrow \underbrace{\begin{pmatrix} 1 & -2 \\ 1 & 0 \\ 1 & 1 \\ 1 & 3 \end{pmatrix}}_{A} \underbrace{\begin{pmatrix} c_0 \\ c_1 \end{pmatrix}}_{C} = \underbrace{\begin{pmatrix} 7 \\ 5 \\ 2 \\ 1 \end{pmatrix}}_{M}$$

$$\underline{C = A \backslash M = \begin{pmatrix} 4.3846 \\ -1.2692 \end{pmatrix} = \begin{pmatrix} c_0 \\ c_1 \end{pmatrix}}$$

Second-order fit: $y = c_0 + c_1 x + c_2 x^2$

$$\begin{cases} c_0 - c_1 2 + c_2 2^2 = 7 \\ c_0 + c_1 0 + c_2 0^2 = 5 \\ c_0 + c_1 1 + c_2 1^2 = 2 \\ c_0 + c_1 3 + c_2 3^2 = 1 \end{cases} \Rightarrow \underbrace{\begin{pmatrix} 1 & -2 & 4 \\ 1 & 0 & 0 \\ 1 & 1 & 1 \\ 1 & 3 & 9 \end{pmatrix}}_{A} \underbrace{\begin{pmatrix} c_0 \\ c_1 \\ c_2 \end{pmatrix}}_{C} = \underbrace{\begin{pmatrix} 7 \\ 5 \\ 2 \\ 1 \end{pmatrix}}_{M}$$

$$\underline{C = A \backslash M = \begin{pmatrix} 4.1346 \\ -1.3526 \\ 0.0833 \end{pmatrix} = \begin{pmatrix} c_0 \\ c_1 \\ c_2 \end{pmatrix}}$$

Exponential fit: $y = c_0 \mathrm{e}^{c_1 x} \Rightarrow \ln y = \ln c_0 + c_1 x$

$$\begin{cases} \ln c_0 - c_1 2 = \ln 7 \\ \ln c_0 + c_1 0 = \ln 5 \\ \ln c_0 + c_1 1 = \ln 2 \\ \ln c_0 + c_1 3 = \ln 1 \end{cases} \Rightarrow \underbrace{\begin{pmatrix} 1 & -2 \\ 1 & 0 \\ 1 & 1 \\ 1 & 3 \end{pmatrix}}_{A} \begin{pmatrix} \ln c_0 \\ c_1 \end{pmatrix} = \underbrace{\begin{pmatrix} \ln 7 \\ \ln 7 \\ \ln 7 \\ \ln 7 \end{pmatrix}}_{M}$$

$$\Rightarrow \begin{pmatrix} \ln c_0 \\ c_1 \end{pmatrix} = A \backslash M = \begin{pmatrix} 1.2669 \\ -0.4095 \end{pmatrix} \Rightarrow \begin{pmatrix} c_0 \\ c_1 \end{pmatrix} = \begin{pmatrix} 3.5498 \\ -0.4095 \end{pmatrix}$$

All four fits are plotted in Fig. 17.16.

The total error of each fit is plotted in Table 17.9. From the table, we can see that the fit to a second-order polynomial has the least error. However, it is deceiving to only focus on the total error to find a relationship. Table 17.9 cannot be used to deduce that the relationship between y and x is a second-order polynomial; the total error will decrease with the polynomial order. For higher order polynomials, we end up fitting data to the noise. We *need* a priori knowledge of the relationship before we do the fitting.

Problem 17.2 The resistance of an unknown temperature sensor was measured for some temperatures, see Table 17.10. What kind of temperature sensor was used?

Solution Assuming a first-order fit: $R = R_0 + \alpha T$:

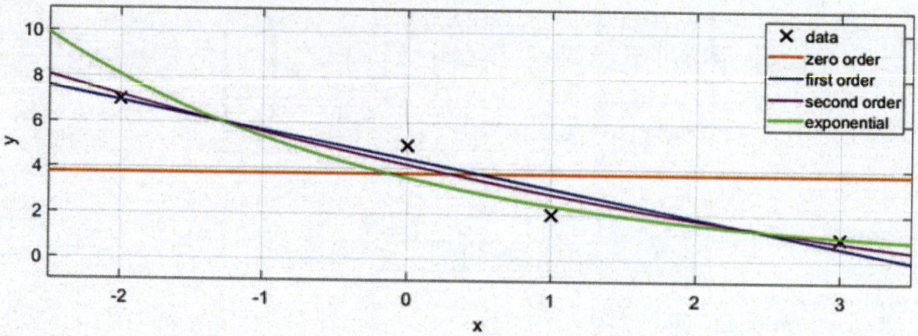

Fig. 17.16 Four different fits

Table 17.9 Error table

x	y^m	y_0-y^m	y_1-y^m	y_2-y^m	$y_{exp}-y^m$
−2	7	−3.25	−0.0770	0.1735	1.0518
0	5	−1.25	−0.6154	−0.8654	−1.4502
1	2	1.75	1.1154	−0.8653	0.3570
3	1	2.75	−0.4230	−0.1735	0.0391
$\sum \varepsilon^2/4$		1.192	0.336	0.312	0.457

Table 17.10 Temperature sensor data

T [°C]	20	60	80	120	150
R [Ω]	1080	1209	1312	1526	1641

$$\begin{cases} R_0 + \alpha 20 = 1080 \\ R_0 + \alpha 60 = 1209 \\ R_0 + \alpha 80 = 1312 \\ R_0 + \alpha 120 = 1526 \\ R_0 + \alpha 150 = 1641 \end{cases} \Rightarrow \begin{pmatrix} 1 & 20 \\ 1 & 60 \\ 1 & 80 \\ 1 & 120 \\ 1 & 150 \end{pmatrix} \begin{pmatrix} R_0 \\ \alpha \end{pmatrix} = \begin{pmatrix} 1080 \\ 1209 \\ 1312 \\ 1526 \\ 1641 \end{pmatrix}$$

$$\Rightarrow \begin{pmatrix} R_0 \\ \alpha \end{pmatrix} = A \backslash M = \begin{pmatrix} 967.6 \\ 4.489 \end{pmatrix}$$

We can write our linear expression as:

$$R = 976.6 + 4.489T = 977(1 + 4.49 \cdot 10^{-3} T)$$

Most likely the temperature sensor was a Cu-1000 sensor (where $R = 1000(1 + 4.33 \cdot 10^{-3} T)$).

Problem 17.3 Eight samples of the signal $y = a_0 + a_1 \sin 2\pi 250 t$ (+ noise) are taken, see Table 17.11. Find the DC offset a_0 and the amplitude a_1.

Solution We get the following system of equations:

$$\begin{cases} a_0 + a_1 \sin\ 2\pi 250 * 0 = 1.196 \\ a_0 + a_1 \sin\ 2\pi 250 * 0.5 * 10^{-3} = 4.242 \\ a_0 + a_1 \sin\ 2\pi 250 * 1.0 * 10^{-3} = 5.291 \\ a_0 + a_1 \sin\ 2\pi 250 * 1.5 * 10^{-3} = 3.707 \\ a_0 + a_1 \sin\ 2\pi 250 * 2.0 * 10^{-3} = 1.117 \\ a_0 + a_1 \sin\ 2\pi 250 * 2.5 * 10^{-3} = -2.143 \\ a_0 + a_1 \sin\ 2\pi 250 * 3.0 * 10^{-3} = -2.645 \\ a_0 + a_1 \sin\ 2\pi 250 * 3.5 * 10^{-3} = -2.287 \end{cases}$$

$$= \begin{cases} a_0 + a_0 0 = 1.196 \\ a_0 + a_1 0.707 = 4.242 \\ a_0 + a_1 1 = 5.291 \\ a_0 + a_1 0.707 = 3.707 \\ a_0 + a_1 0 = 1.117 \\ a_0 - a_1 0.707 = -2.143 \\ a_0 - a_1 1 = -2.645 \\ a_0 - a_1 0.707 = -2.287 \end{cases}$$

$$\underbrace{\begin{pmatrix} 1 & 0 \\ 1 & 0.707 \\ 1 & 1 \\ 1 & 0.707 \\ 1 & 0 \\ 1 & -0.707 \\ 1 & -1 \\ 1 & -0.707 \end{pmatrix}}_{A} \begin{pmatrix} a_0 \\ a_1 \end{pmatrix} = \underbrace{\begin{pmatrix} 1.196 \\ 4.242 \\ 5.291 \\ 3.707 \\ 1.117 \\ -2.143 \\ -2.645 \\ -2.287 \end{pmatrix}}_{M}$$

$$\Rightarrow \begin{pmatrix} a_0 \\ a_1 \end{pmatrix} = A \backslash M = \begin{pmatrix} 1.0597 \\ 4.1726 \end{pmatrix}$$

Hence, the DC offset is 1.0597 V, and the amplitude is 4.1726. In Fig. 17.17, we have plotted $y = 1.0597 + 4.1726 \times \sin 2\pi 250 t$ and the sampled data.

Table 17.11 Samples of a sine signal (with noise)

t [ms]	0.000	0.500	1.000	1.500	2.000	2.500	3.000	3.500
y [V]	1.196	4.242	5.291	3.707	1.117	−2.143	−2.645	−2.287

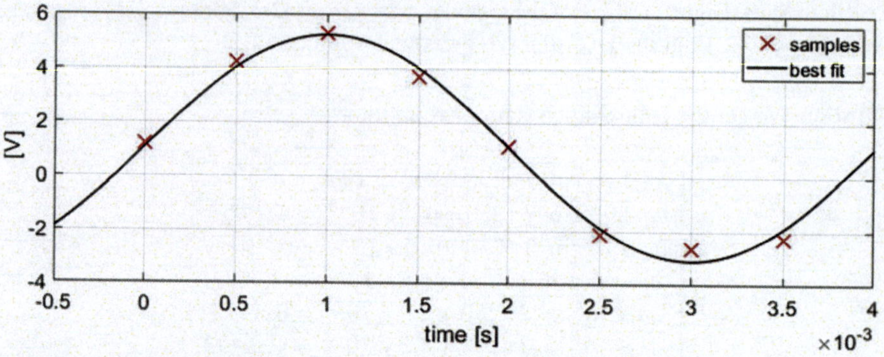

Fig. 17.17 Data and the best fit

Chapter 18
Introduction to Control Theory

Abstract Control theory may or may not be part of the electrical measurement curriculum, but it is such a common instrument in a physics laboratory that a fundamental understanding of its operation is necessary. This chapter focuses on the PID controller, feedback models and stability criteria, the need for integration and differentiation of the error signal and how to identify an unknown system. It is explained how control parameters are derived using rules of thumb (Ziegler–Nichol's) or phase/gain margins. Finally, this chapter illustrates how a control algorithm can be implemented in a computer system using either Euler transformation or bilinear transformation.

18.1 Control Systems

Figure 18.1 illustrates a control system (Fig. 18.1).

$G(s)$ represents the 'plant' that we want to control, for example, a furnace whose temperature we want to control, $y(t)$ is the 'process value' (the actual temperature) and $x(t)$ is the 'set value' (the temperature we would like the oven to have). The process value is fed back via a (temperature) sensor and subtracted from the set value to produce the 'error signal' $e(t)$. $C(s)$ is the 'controller' whose job it is to produce a voltage $u(t)$ depending on $e(t)$ such that the process value is always equal to $x(t)$.

There are several different kinds of controllers, but here we will only describe the *PID controller* (since it is the most common type of controller in a physics lab). Our objective here is to find the $C(s)$ control function so that $y(t) = x(t)$, and we need to do that so that the system is first of all *stable*; all feedback systems have a potential risk of instability. After a change in the set value (or some other disturbance), there will be some transient events on all signals, but after some time ('settling time'), we should again have $y(t) = x(t)$, i.e., $e(t) = 0$; our system should not have a *steady state error*.

Other properties of interest are the system's reaction to step changes in the set value (the 'step response'), see Fig. 18.2.

L. Bengtsson, *Electrical Measurement Techniques*,
https://doi.org/10.1007/978-981-99-8187-8_18

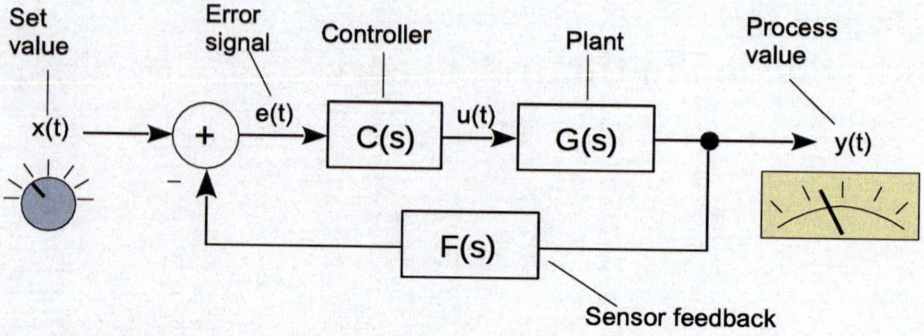

Fig. 18.1 A control system

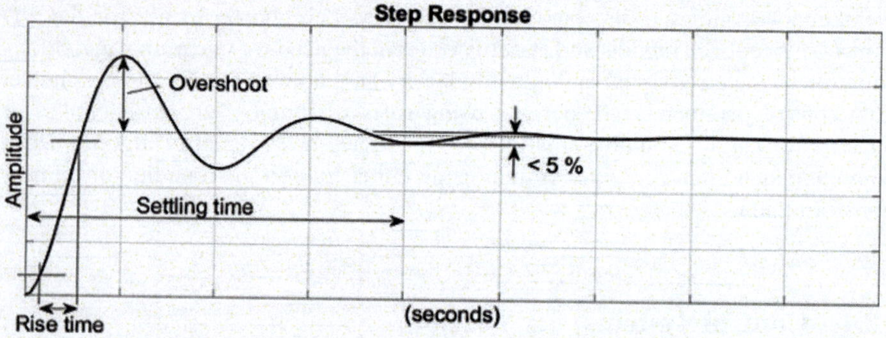

Fig. 18.2 Step response

Fig. 18.3 Signal model

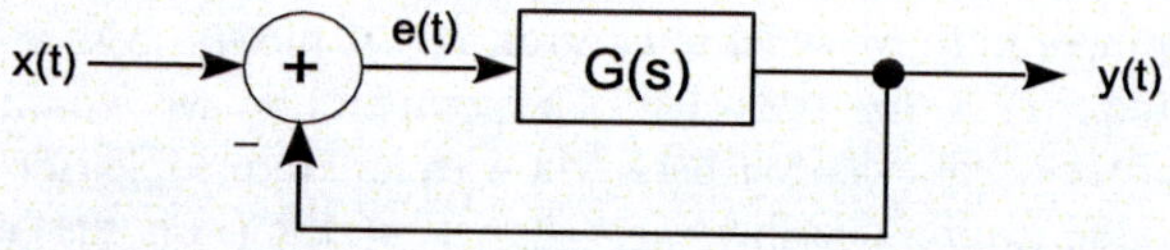

The *risetime* is the time it takes to go from 10 to 90% of the final value and the *settling time* is the time it takes for the output to stabilize within 5% of the final value. The *overshoot* is the maximum voltage above the final voltage level. All these parameters are affected when we change the parameters of the controller, and different applications have different priorities (overshoot, rise time or settling time). In this context, we will not worry too much about them; our main priority here is to find the conditions where the system is stable and has no steady state error.

In most systems, it is also assumed that the sensor feedback system has a transfer function $F(s) = 1$ and that is what we will assume here. To make sure that we don't get too complicated equations, initially, we will limit our plant systems to first-order systems; $G(s) = 1/(s + a)$. That will allow us to focus more on the understanding of the control function's influence on the system's behavior. We will look at second-order systems later.

18.2 Feedback Systems

In our first analysis of the control system, we will cancel the controller and just look at the plant with feedback and analyze its behavior, see Fig. 18.3.

The signal $e(t) = x(t) - y(t)$ is the difference between the input signal and the output signal. Let's find the transfer function of this system:

$$Y(s) = G(s) \cdot E(s) = G(s)(X(s) - Y(s)) = G(s)X(s) - G(s)Y(s)$$
$$Y(s)(1 + G(s)) = X(s)G(s) \Rightarrow H(s) = \frac{Y(s)}{X(s)} = \frac{G(s)}{1 + G(s)} \quad (18.1)$$

From Eq. (18.1) we can see that this system will be unstable if $1 + G(s) = 0$, i.e., if

$$G(s) = -1 = 1 \cdot \mathrm{e}^{-\mathrm{j}180^\circ} = |G(s)| \cdot \mathrm{e}^{\varphi(\omega)} \quad (18.2)$$

Hence, if the system has a gain of $+1$ (or greater) for the frequency where the phase shift is $-180°$, the system will be unstable. So, we can see immediately from the Bode plot if the system is stable or not; just check the phase shift at amplification $= 1$ (0 dB) and the amplification at $\varphi = -180°$, see Fig. 18.4. The distance from the phase diagram to $-180°$ at 0 dB is the *phase margin*, which tells you how far the phase diagram is from instability. The distance from the gain diagram to 0 dB at phase angle $-180°$ is the *gain margin*, which tells you how far the gain diagram is from instability. (We want some margin because there are some uncertainties in the system that could push it over the 'edge' and become unstable if we are too close to the margins.)

From Fig. 18.4, we can see that there are two things that could make our system unstable; either if we 'lift' (amplify) the gain diagram or if we 'lower' the phase diagram. Any action we take that either lifts the gain or lowers the phase may render the system unstable.

Something else we can see in Eq. (18.1) is that if $G(s) >> 1$, then $H(s) \approx 1$, which would mean that $y(t) \approx x(t)$, (which is what we are looking for in a *control system*), and for that reason it is tempting to amplify $G(s)$ with some factor K_P, see Fig. 18.5.

However, by doing that we push the gain diagram in the Bode plot upwards, see Fig. 18.6, which means that the gain margin decreases. In Fig. 18.6, we can see that we amplified the signal too much; at the phase shift angle $-180°$, we have an amplification >0 dB, and the system is unstable (it will oscillate).

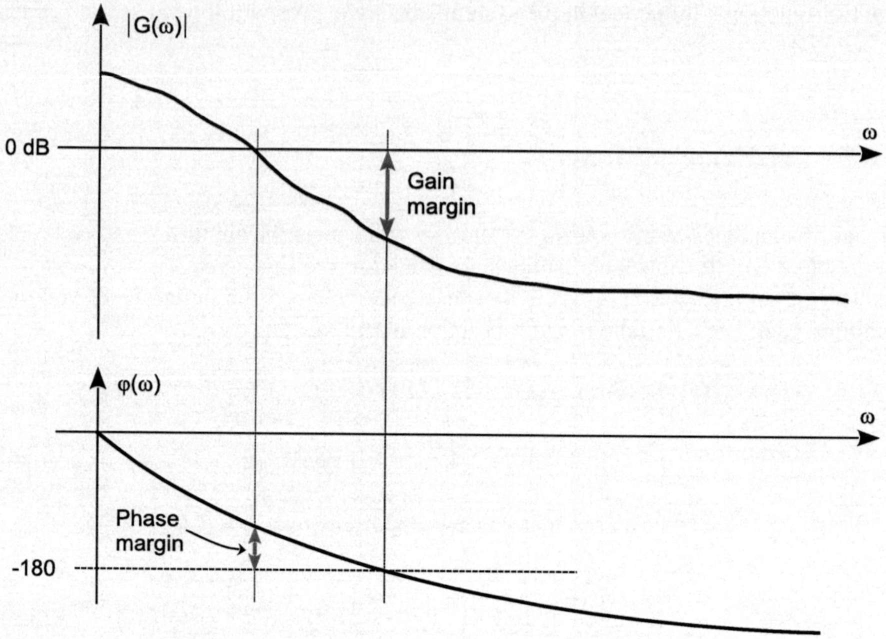

Fig. 18.4 The Bode plot

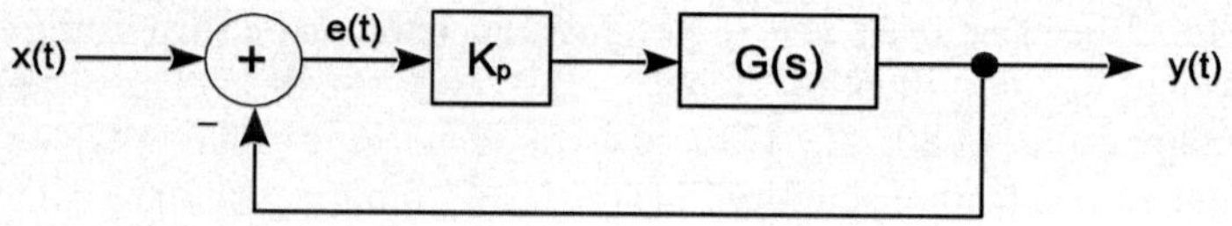

Fig. 18.5 Amplifying

Do you see the problem? On the one hand, we want a large amplification for $y(t)$ to follow $x(t)$, but a large amplification might render the system unstable. That means that we need to be a little shrewder than 'just amplifying'.

The simple K_P amplifier above only affected the gain diagram. In this chapter, we will learn to use other 'amplifiers' that also affect the phase diagram in such a way that we avoid instability when we amplify the signal.

In Fig. 18.5, the controller just amplifies the error signal; the controller produces an output signal *proportional* to the error and is therefore called a *proportional controller*, or just *P controller*. Let's look at a general system.

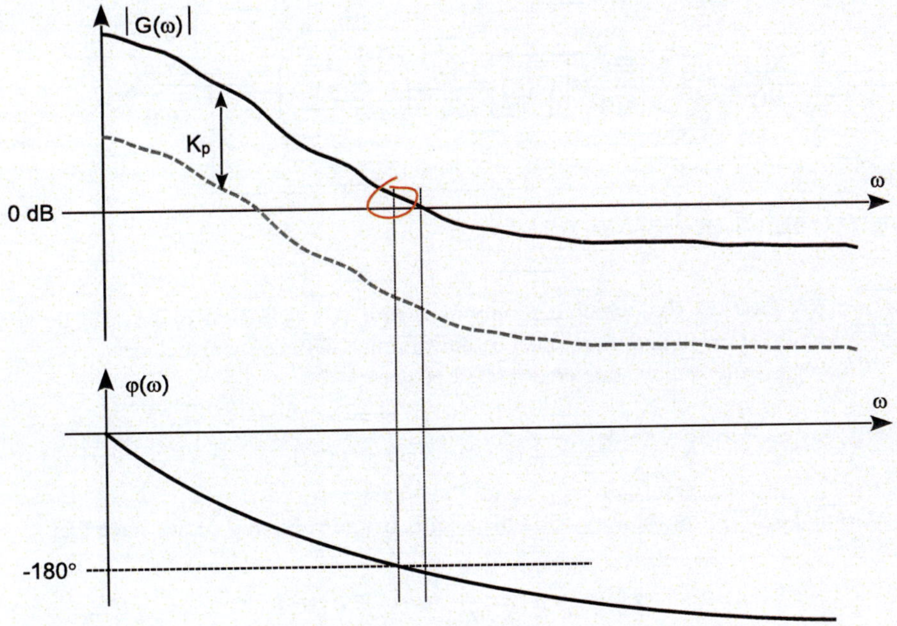

Fig. 18.6 The Bode plot after amplification

18.3 Control Systems

Figure 18.7 illustrates our control system.

In Fig. 18.5, $C(s) = K_P$ and apart from pushing the system closer towards instability, it also has another problem; it has a 'steady state' error. That means that when things have 'settled down', typically after a change in the set value, there would still be an error between the set and process values, i.e., $e(t)$ would not be zero and $x(t) \neq y(t)$ (which after all is the whole point of the system) unless $K_P = \infty$. To see why that is, and to figure out exactly what the remaining steady state error is, we need a system to work with. Let's assume the plant is a simple first-order system, i.e., $G(s) = 1/(s + a)$ and $C(s) = K_P$, see Fig. 18.8.

Let's assume first that the error $e(t)$ is $= 0$ in Fig. 18.8. Since the control system just multiplies it by a constant, the output from the control system will also be $= 0$ and multiplying $G(s)$ with 0 is of course also $= 0$; $y(t) = 0$! But in that case, the error

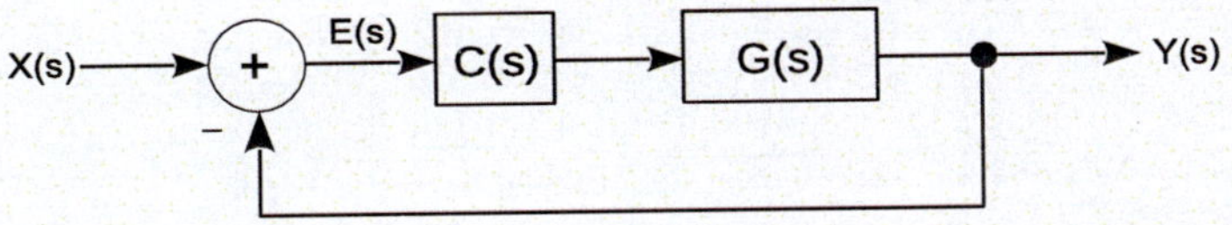

Fig. 18.7 Our control system

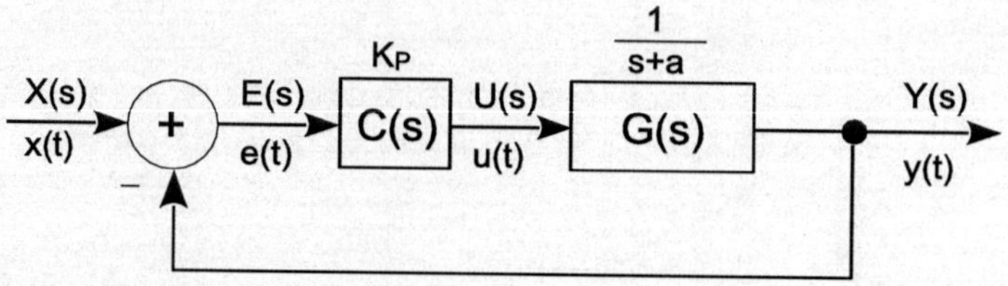

Fig. 18.8 The plant is a first-order system

$e(t) = x(t) - y(t) = x(t)$, which can only be zero if $x(t) = 0$. Hence, if $x(t) \neq 0$, we cannot have $e(t) = 0$, and there must be a difference between $y(t)$ and $x(t)$.

So, what is the steady state error $e(t)$? Let's see:

$$(X(s) - Y(s)) \cdot K_P \cdot \frac{1}{s+a} = Y(s) \Rightarrow (X(s) - Y(s)) \cdot K_P = sY(s) + aY(s)$$

Going back to time–space (using the results from Problem 7.6), we have that

$$x(t) - y(t) = \frac{1}{K_P}(y\prime(t) + ay(t))$$

'Steady state' implies, by definition, that $y'(t) = 0$, so

$$x(t) = y(t)\left(\frac{a}{K_P} + 1\right) \Rightarrow y(t) = \frac{1}{a/K_P + 1}x(t) = \frac{K_P}{a + K_P}x(t) \qquad (18.3)$$

which confirms that K_P needs to be infinite for $y(t) = x(t)$ (but that would make the system unstable). For example, if $a = K_P = 1$, and we set $x(t) = 1$, $y(t)$ would be 0.5. So, the P regulator has some problems and in the next section we will fix that, but first, we will present a theorem that will simplify our analysis a little bit.

In the above analysis, we considered the entire system's transfer function $Y(s)/X(s)$, the *closed loop* system. That is not necessary; we only need to consider the *open loop* system $G_{OL}(s) = C(s)\cdot G(s)$ (if $F(s) = 1$). Nyquist's (simplified) stability criterion states that a feedback system is stable if $|G_{OL}(\omega)| < 1$ at the frequency where $\varphi_{OL}(\omega) = -180°$. (Makes sense; the subtraction in the feedback adds the other 180°.)

18.4 The PI Controller

The trick that eliminates the steady state error is to integrate the error signal:

$$u(t) = K_P \cdot e(t) + K_I \int e(t)dt = K_P\left(e(t) + \frac{1}{T_I}\int e(t)dt\right) \qquad (18.4)$$

where T_I is the *integration time constant*, and this is a PI control function (Proportional and Integrating). Taking the Laplace transform of both sides gives us:

$$U(s) = K_P\left(E(s) + \frac{1}{T_I s}E(s)\right) = K_p\left(1 + \frac{1}{T_I s}\right)E(s) \Rightarrow$$

$$C(s) = \frac{U(s)}{E(s)} = K_P\left(1 + \frac{1}{T_I s}\right) \quad (18.5)$$

(See Problem 7.7 for the Laplace transform of an integral.) Inserting Eq. (18.5) into Fig. 18.8 gives us:

$$(X(s) - Y(s)) \cdot C(s) \cdot G(s) = Y(s) \Rightarrow (X(s) - Y(s))K_P\left(1 + \frac{1}{T_I s}\right)\frac{1}{s+a} = Y(s)$$

$$(X(s) - Y(s))K_P\left(1 + \frac{1}{T_I s}\right) = sY(s) + aY(s)$$

$$(X(s) - Y(s))K_P\left(s + \frac{1}{T_I}\right) = s^2Y(s) + asY(s)$$

$$K_P(sX(s) - sY(s)) + \frac{K_P}{T_I}(X(s) - Y(s)) = s^2Y(s) + asY(s)$$

Going back to time–space:

$$K_P(x\prime(t) - y\prime(t)) + \frac{K_P}{T_I}(x(t) - y(t)) = y''(t) + ay\prime(t)$$

Again, in steady state, all derivatives are zero:

$$\frac{K_P}{T_I}(x(t) - y(t)) = 0 \Rightarrow y(t) = x(t) \quad (18.6)$$

Hence, in steady state $y(t) = x(t)$, there is no steady state error. Unfortunately, while fixing the steady state error, we introduced another problem. In Fig. 18.8, the open-loop transfer function is $C(s)G(s)$, i.e.,

$$G_\mathrm{OL}(s) = |C(s)|\mathrm{e}^{\mathrm{j}\varphi_C} \cdot |G(s)|\mathrm{e}^{\mathrm{j}\varphi_G} = |\ldots| \cdot \mathrm{e}^{\mathrm{j}(\varphi_C + \varphi_G)} \quad (18.7)$$

Hence, the open loop phase diagram is the sum of the phases from $C(s)$ and $G(s)$. Let's take a closer look at the phase function of $C(s)$:

$$C(s) = K_p\left(1 + \frac{1}{T_I S}\right) = K_p\frac{T_I s + 1}{T_I s}$$

$$\Rightarrow C(\omega) = K_P \frac{j\omega T_I + 1}{j\omega T_I}$$
$$= K_P|\ldots| \cdot e^{j(\tan^{-1}\omega T_I - 90°)}$$
$$\Rightarrow \underbrace{\tan^{-1}\omega T_I - 90°}_{\leq 0!} \tag{18.8}$$

From Eq. (18.8), we can see that the PI regulator adds a *negative* contribution to the phase diagram, and it will push the phase diagram in Fig. 18.4 *downwards, decreasing* the phase margin, and hence *increasing* the risk of instability. The integration fixed the steady state error, but the resulting system is more likely to be unstable. We will fix that in a moment, but let's first see what happens if we instead of integrating the error, *differentiate* the error signal.

18.5 The PD Controller

If we, instead of integrating the error signal, differentiate it, we get the control function

$$u(t) = K_P e(t) + K_D e\prime(t) = K_P(e(t) + T_D e\prime(t)) \tag{18.9}$$

where T_D is the *differentiation time constant*. We find the transfer function by taking the Laplace transform of both sides:

$$U(s) = K_P(1 + T_D s)E(s) \Rightarrow C(s) = \frac{U(s)}{E(s)} = K_P(1 + T_D s) \tag{18.10}$$

and the frequency response function is:

$$C(\omega) = K_P(1 + j\omega T_D) = |\ldots| \cdot e^{j\cdot\tan^{-1}\omega T_D}$$
$$\Rightarrow \varphi(\omega) = \underbrace{\tan^{-1}\omega T_D}_{\geq 0} \tag{18.11}$$

When we multiply the plant function *G*(*s*) with *C*(*s*), the phase diagrams will add, and from 18.11, we can see that the control function will now *add* a phase angle >0, which means that the phase diagram in Fig. 18.4 is pushed *upwards*, and hence the phase margin is *increased*; the system is less likely to become unstable.

Next step is of course to combine integration and differentiation.

18.6 The PID Controller

In a PID controller, we both integrate and differentiate the error signal:

$$u(t) = K_P\left(e(t) + \frac{1}{T_I}\int e(t)dt + T_D e\prime(t)\right) \Rightarrow$$

$$C(s) = \frac{U(s)}{E(s)} = K_P\left(1 + \frac{1}{T_I s} + T_D s\right) = K_P\frac{1 + T_I s + T_I T_D s^2}{T_I s} \tag{18.12}$$

It is not obvious what happens to the phase diagram in this transfer function, and before we derive the phase diagram expression, we are going to make a small approximation; the transfer function of most plants is such that the required control function needs to have $T_I >> T_D$. In that case, we can make the following approximation: $T_I s \approx (T_I + T_D)s$, and in that case

$$1 + T_I s + T_I T_D s^2 \approx 1 + T_I s + T_D s + T_I T_D s^2 = (1 + T_I s)(1 + T_D s)$$

Inserting this approximation into Eq. (18.12) gives us the following control function:

$$C(s) \approx K_P\frac{(1 + T_I s)(1 + T_D s)}{T_I s} = K_P\left(1 + \frac{1}{T_I s}\right)(1 + T_D s) \tag{18.13}$$

From Eq. (18.13), we can see that with the assumption that $T_I >> T_D$, the PID controller is approximately the same as cascading a PI and a PD, controller, which means that the total phase diagram is the sum of the PI and PD controller's phase diagrams; the damage the integration does to our phase diagram (in Eq. 18.8) is undone by the differentiation (in Eq. 18.11), and our system is less likely to be unstable.

Let's summarize our conclusions: The P controller only amplifies the error signal by some factor K_P. Increasing K_P makes the system react faster to changes in the set value (rise time and settling time improve) but increasing K_P comes with a prize; the system is pushed closer to instability and the overshoot increases. Most of all though, a P regulator suffers from an inherent incapability of eliminating steady state errors. That's why we almost always need an integrating part; by also integrating the error, we can eliminate the steady state error. However, the integration has a negative influence on the phase margin, it makes the system response slower (risetime and settling time increase) and it also increases the overshoot. (The only good thing about the integration is that it takes care of the steady state error.) The differentiation part is everything the integration is not; it has a positive influence on the phase margin, it improves the response times, and it suppresses the overshoot.

Figure 18.9 summarizes our PID model and Fig. 18.10 is our approximation model when $T_I >> T_D$. Next, we need to figure out how to find the PID parameters (K_P, T_I, and T_D) for a given plant system.

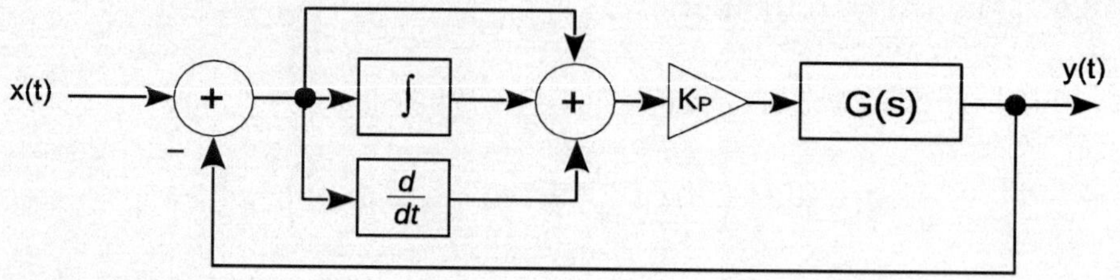

Fig. 18.9 PID controller model

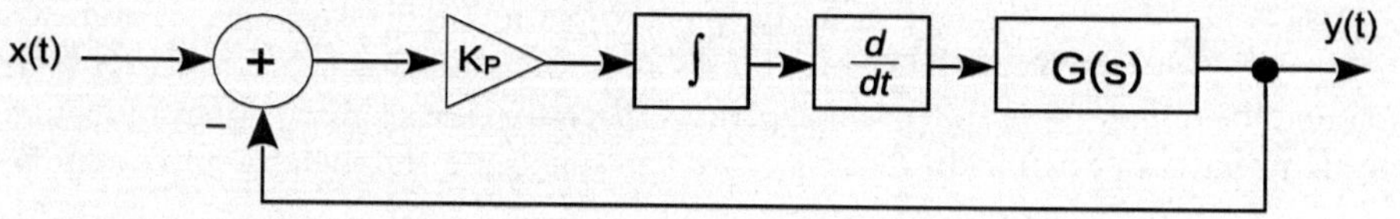

Fig. 18.10 Approximate model when $T_I >> T_D$

18.7 Identifying the System

18.7.1 First-Order Systems

Before we start designing the controller, we must know what kind of system we have, i.e., the plant transfer function $G(s)$. The process of finding the transfer function of an unknown system is called *system identification*. This is a research field of its own, and there is no lack of literature treating this subject in detail. Here, we will keep it short and only illustrate the basic ideas (that are likely to solve the most common problems in a physics lab). Our strategy here will be to use our a priori knowledge of the system (by experience or reasonable assumptions) when we identify our system: 'Because …., it is reasonable to assume that this is a first order system'. We will start gently, by assuming that we have a 'plant' where we have good reasons to assume that it is a first-order system. Hence, the plant function is

$$G(s) = \frac{b}{s+a} = \frac{b/a}{s/a+1} = \frac{K}{Ts+1} \tag{18.14}$$

where K is the amplification and T is the system's time constant. To find K and T, we look at the *step response*; the output when the input is a step signal. The step signal has Laplace transform $1/s$ (see Problem 7.8) and we get the step response by multiplying $G(s)$ with $1/s$:

$$Y_{\text{step}}(s) = \frac{1}{s} \cdot \frac{K}{Ts+1} = \frac{A}{s} + \frac{B}{Ts+1} = \frac{ATs + A + Bs}{s(Ts+1)} \tag{18.15}$$

By comparing the numerators in Eq. (18.15), we can see that $A = K$ and $AT + B = 0 \Rightarrow B = -AT = -KT$. Inserted into Eq. (18.15) gives us

$$Y_{\text{step}}(s) = \frac{K}{s} - \frac{KT}{Ts+1} = K\left(\frac{1}{s} - \frac{1}{s+1/T}\right) \tag{18.16}$$

Inverse Laplace transform of Eq. (18.16) gives us the time function:

$$y_{\text{step}}(t) = K\left(1 - \mathrm{e}^{-t/T}\right) \tag{18.17}$$

This step response is plotted in Fig. 18.11, and in Fig. 18.12 we have plotted the Bode diagram for the case where K and T are both $= 1$.

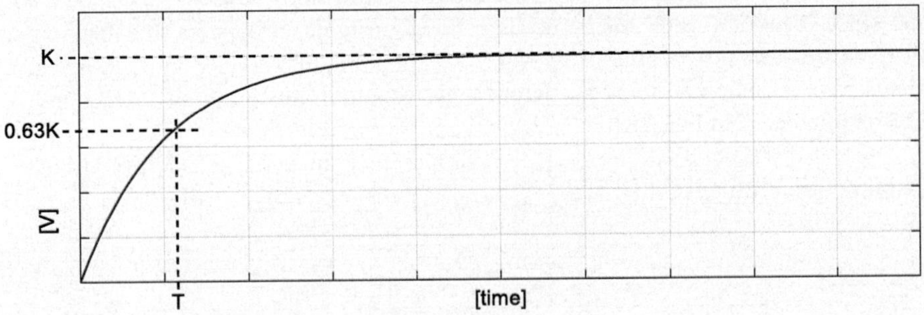

Fig. 18.11 Step response of first-order system

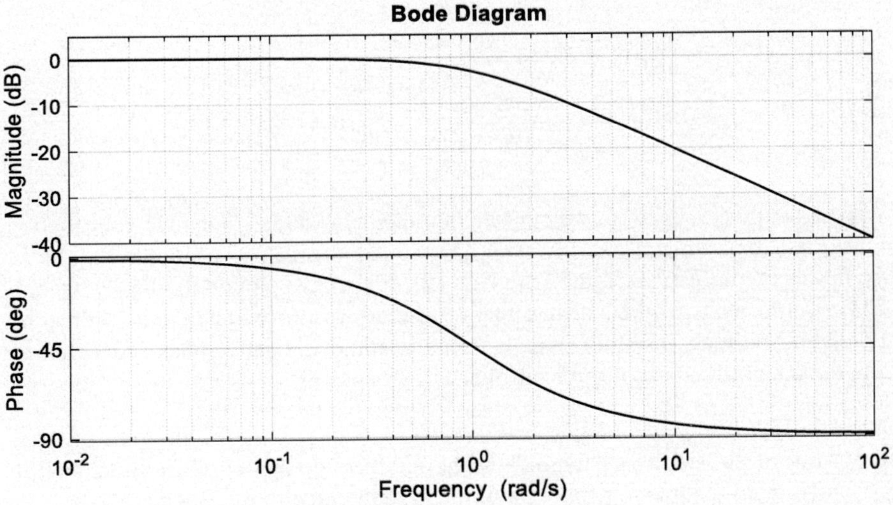

Fig. 18.12 Bode plot of first-order system

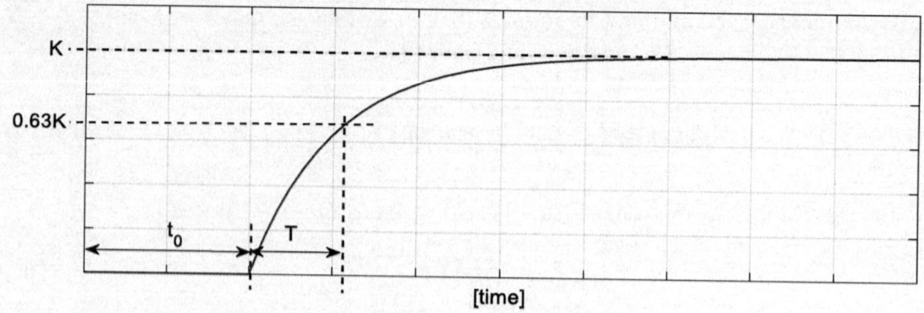

Fig. 18.13 Step response of first-order system with 'dead time'

From the Bode diagram, we can see that the phase shift is never less than $-90°$ and since it takes $-180°$ for the system to be unstable, it appears that first-order systems are inherently stable. However, in a real plant, that is only an illusion; most plants have an inherent delay t_0, before they respond to a step input ('dead time'). This is illustrated in Fig. 18.13.

The transfer function of the delayed system is e^{-st_0} times the non-delayed transfer function (see Problem 7.11):

$$G(s) = \frac{K \cdot e^{-st_0}}{Ts + 1} \tag{18.18}$$

This inherent delay can have a dramatic impact on stability. To see that we find the Bode plot functions:

$$G(\omega) = \frac{K \cdot e^{-j\omega t_0}}{j\omega T + 1}$$

$$\Rightarrow |G(\omega)| = \frac{|K|}{\sqrt{\omega^2 T^2 + 1}} \qquad \varphi(\omega) = \underbrace{-\omega t_0}_{\text{From delay}} - \tan^{-1} \omega T$$

From the phase function, we can see that the delay adds a negative contribution to the phase diagram and stability is no longer guaranteed. The system in Fig. 18.12 also has an amplification of just 1. In Fig. 18.14, we have plotted the Bode diagram of a first-order system with a dead time of two seconds and an inherent amplification of 5 and it is already instable. (We also need to add the negative phase contribution from the integration part of the controller.)

Example 18.1 Figure 18.15 below illustrates a setup to identify the transfer function of a furnace heating system. The step and the step response are illustrated in Fig. 18.16. Find the transfer function of the system and plot the Bode diagram.

Solution We have an amplification of (5–0.5)/(3–1) = 2.25, $t_0 = 3.5$ s and $T = 18$ s:

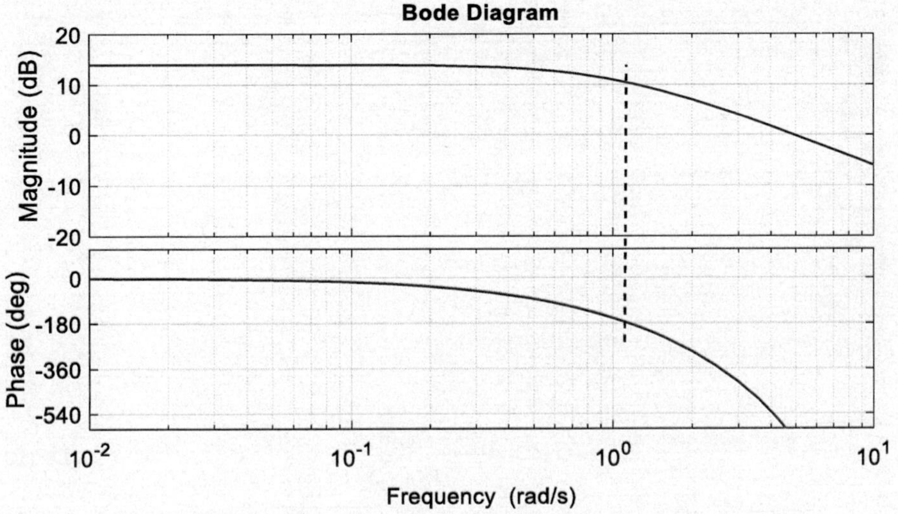

Fig. 18.14 Bode plot of first-order system with 'dead time'

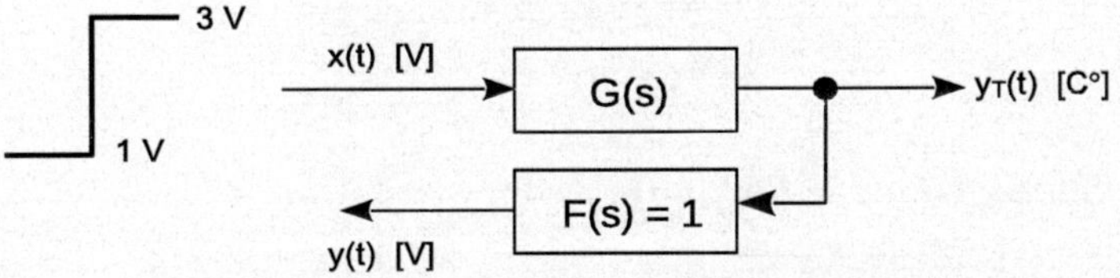

Fig. 18.15 Identifying the system

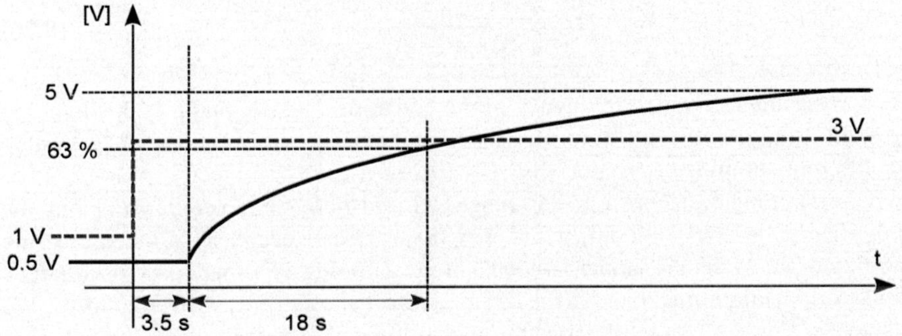

Fig. 18.16 The step and the step response

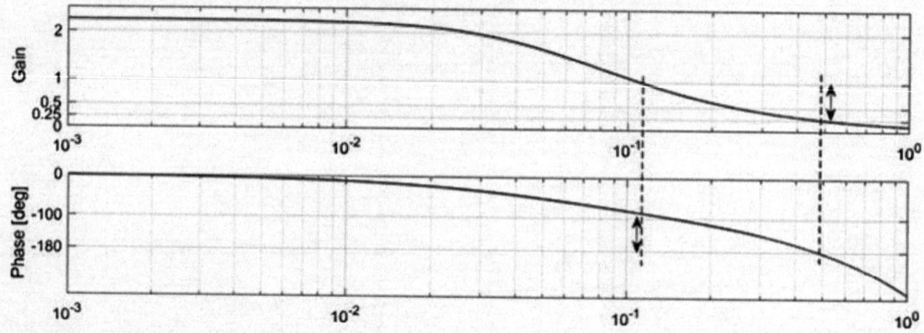

Fig. 18.17 The bode plot

$$G(s) = \frac{2.25 \cdot e^{-3.5s}}{18s+1} \Rightarrow G(\omega) = \frac{2.25}{\sqrt{18^2\omega^2+1}} \cdot e^{-j(3.5\omega+\tan^{-1}18\omega)} \quad (18.19)$$

The Bode diagram is illustrated in Fig. 18.17. We have a phase margin of approximately 90° and a gain margin of approximately 4. The closed-loop system would be stable.

18.7.2 Second-Order Systems

Second-order systems are described by the general transfer function

$$H(s) = \frac{\omega_0^2}{s^2 + 2\zeta\omega_0 + \omega_0^2} \quad (18.20)$$

where ω_0 is the resonance frequency of the undamped system and ζ is the damping constant. In Fig. 18.18, we have plotted the step response of a second-order system for different damping constants ($\zeta = d$) with $\omega_0 = 1$.

If the damping is 0, there is no damping at all, and the systems oscillate indefinitely. If the damping is <1 (but >0), there will be a gradual decrease in the oscillation amplitude (the system is 'underdamped'), if the damping = 1, there is no oscillations at all ('critical damping') and if the damping is >1 the system is 'overdamped'. We can see in Fig. 18.18 that the system response time decreases when the damping increases. The damping also has some impact on the oscillation frequency. For a certain damping, the oscillation frequency is ($\zeta < 1$)

$$\omega_d = \omega_0\sqrt{1-\zeta^2} \quad (18.21)$$

Figure 18.19 illustrates the step response parameters you need to identify a second-order system. First you determine the 'overshoot ratio', see Fig. 18.19. From the

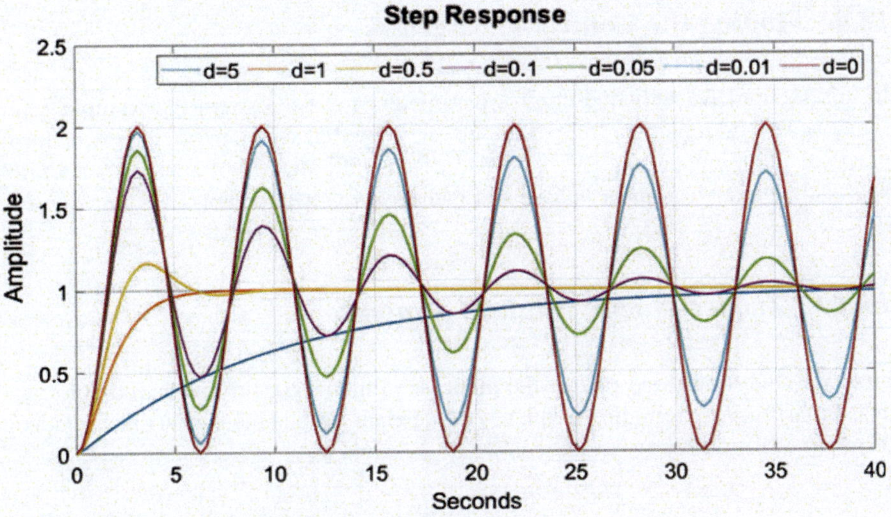

Fig. 18.18 The step response for different damping constants

overshoot ratio (OS), you find the damping constant:

$$\zeta = \sqrt{\frac{(\ln \text{OS})^2}{\pi^2 + (\ln \text{OS})^2}} \tag{18.22}$$

(don't ask), and from the period T you find ω_d which gives you ω_0 (Eq. 18.21).

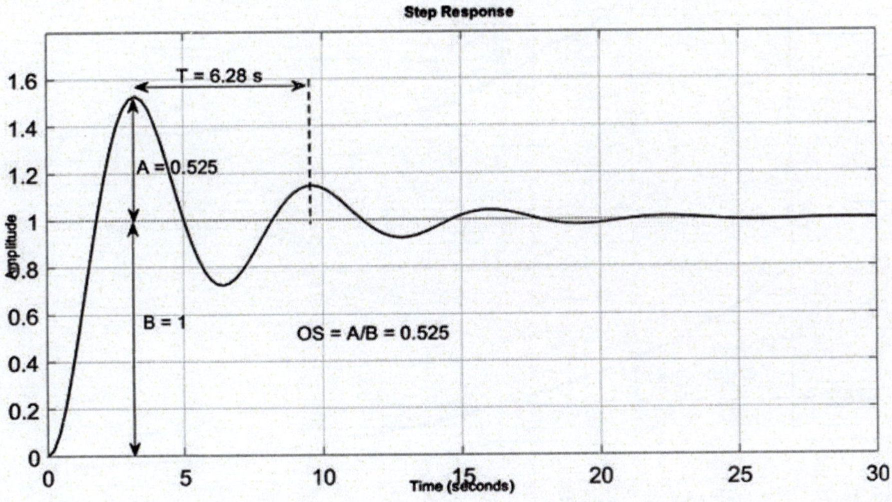

Fig. 18.19 Step response parameters for a second-order system

18.8 Finding the Control Parameters

In this section, we will find control parameters for our system in Example 18.1. First, we will do it using a 'rule of thumb' strategy that seems to be 'plan A' in most practical implementations. We will also discuss the details of a more 'scientific' approach where we consider phase and gain margin requirements.

18.8.1 Ziegler–Nichol's Rule of Thumb

First, we study the Bode plot of the plant, see Fig. 18.20. From the Bode plot, we determine the self-oscillation frequency of the plant, ω_0 and the gain at that frequency $|G_P(\omega_0)|$. From these two numbers, we find T_0 and K_0:

$$T_0 = \frac{2\pi}{\omega_0} \tag{18.23}$$

$$K_0 = \frac{1}{|G_P(\omega_0)|} \tag{18.24}$$

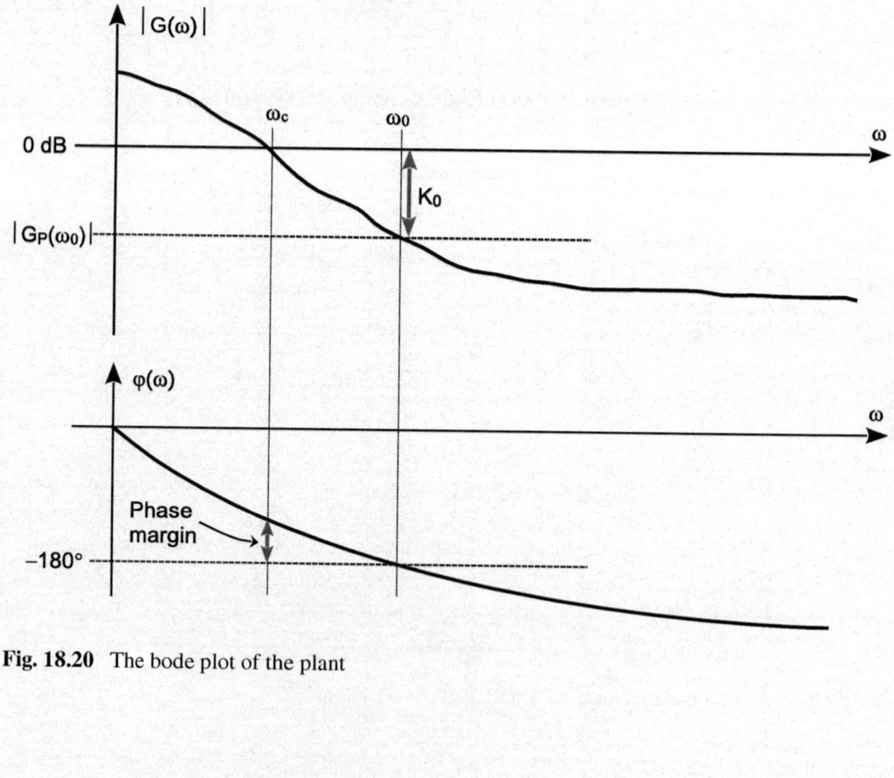

Fig. 18.20 The bode plot of the plant

Table 18.1 Ziegler and Nichol's PID parameter table

	Parameters		
	K	T_I	T_D
P	$0.5K_0$	–	–
PI	$0.45K_0$	$0.85T_0$	–
PID	$0.6K_0$	$0.5T_0$	$0.125T_0$

From T_0 and K_0, we use Ziegler and Nichol's table to find control parameters for different controllers:

Example 18.2 Suggest a PI and a PID controller for the system in Example 18.1 using Ziegler and Nichol's method.

Solution From the Bode plot in Fig. 18.17, we get $\omega_0 = 5 \cdot 10^{-1} = 0.5$ rad/s and $|G_P(\omega_0)| = 0.25$. Hence:

$$T_0 = \frac{2\pi}{0.5} = 13 \text{ seconds and } K_0 = \frac{1}{0.25} = 4$$

Using Table 18.1 gives us the following control functions:

$$\text{PI} : C(s) = 0.45 \cdot 4\left(1 + \frac{1}{0.85 \cdot 13s}\right) = 1.8\left(1 + \frac{1}{11s}\right) \tag{18.25}$$

$$\Rightarrow u_{\text{PI}}(t) = 1.8\left(e(t) + \frac{1}{11}\int e(t)dt\right) \tag{18.26}$$

$$\text{PID} : C(s) = 2.4\left(1 + \frac{1}{6.5s} + 1.6s\right) \tag{18.27}$$

$$\Rightarrow u_{\text{PID}}(t) = 2.4\left(e(t) + \frac{1}{6.5}\int e(t)dt + 1.6e'(t)\right) \tag{18.28}$$

In Figs. 18.21 and 18.22, we have plotted the resulting Bode plot for the plant and the PI and PID controller functions, respectively. First, compare Fig. 18.21 with Fig. 18.17; we can see the effect of the integration part. In Fig. 18.17, we have a phase margin of approximately 90°. In Fig. 18.21, this phase margin has been reduced to 30°. In Fig. 18.22, some of that phase margin has been restored to about 40°.

18.8.2 Using Phase and Gain Margin Criteria

The Ziegler and Nichol's rules of thumb generate stable systems as is illustrated in Figs. 18.21 and 18.22. Another approach is to start with the Bode plot of the system

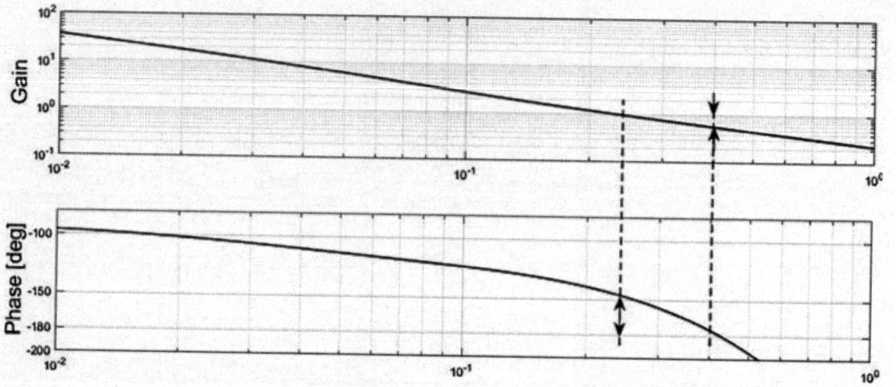

Fig. 18.21 The bode plot of the plant + the PI controller

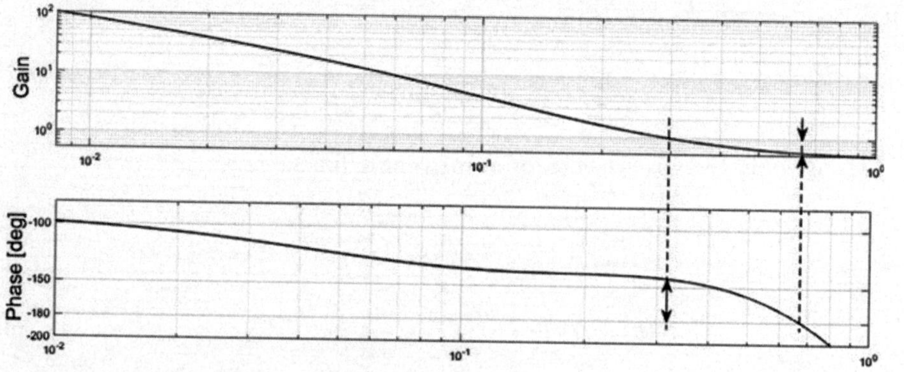

Fig. 18.22 The bode plot of the plant + the PID controller

(Fig. 18.17) and aim for some specific gain and/or phase margins. The general rule in this case is that the integrator part lowers the phase diagram by approximately 11 degrees at the cross-over frequency (where the amplification is 0 dB, see Fig. 18.20). If the system (the plant) itself has a phase margin of 40°, adding an integrator would decrease the phase margin to under 30°. If the system specifications dictate a phase margin of 50°, then you must use the derivative part to increase the phase margin (Eq. (18.11)).

Example 18.3 Given conditions mentioned above, what derivation time T_D would you need for the differentiation part? The cross-over frequency was 0.3 rad/second.

Solution 40–11 = 29°. To meet the phase margin demand, we need to raise the phase diagram by 50–29 = 21°. Equation (18.11) gives us

$$\varphi = \tan^{-1}\omega T_D \Rightarrow T_D = \frac{\tan\varphi}{\omega} = \frac{\tan 21°}{0.3} = 1.3 \text{ sec}$$

18.9 Discretizing

To implement a controller into a computer system (a microcontroller, like an Arduino or a Raspberry Pi), we must 'translate' the control functions to discrete time. There are several ways to do that, and we will present two ways here. We will only consider the PI and PID controller, Eqs. (18.4) and (18.12):

$$u(t) = K_P\left(e(t) + \frac{1}{T_I}\int e(t)dt\right) \tag{18.29}$$

$$u(t) = K_P\left(e(t) + \frac{1}{T_I}\int e(t)dt + T_D e'(t)\right) \tag{18.30}$$

18.9.1 Euler Transformation

A computer system must sample the error signal; in Fig. 18.23, we have sampled the error signal with the sampling rate $f_S = 1/T_S$. From Fig. 18.23, we can see that the integral of $e(t)$ is approximately equal to the sum of the rectangles.

Hence,

$$\int e(t)dt \approx \sum_{i=0}^{n} e(i)\cdot T_S = T_S\sum_{i=0}^{n} e(i) \tag{18.31}$$

Similarly, we can see in Fig. 18.24 that the derivative can be approximated with a straight line, and hence

$$e'(t) \approx \frac{e(n) - e(n-1)}{T_S} \tag{18.32}$$

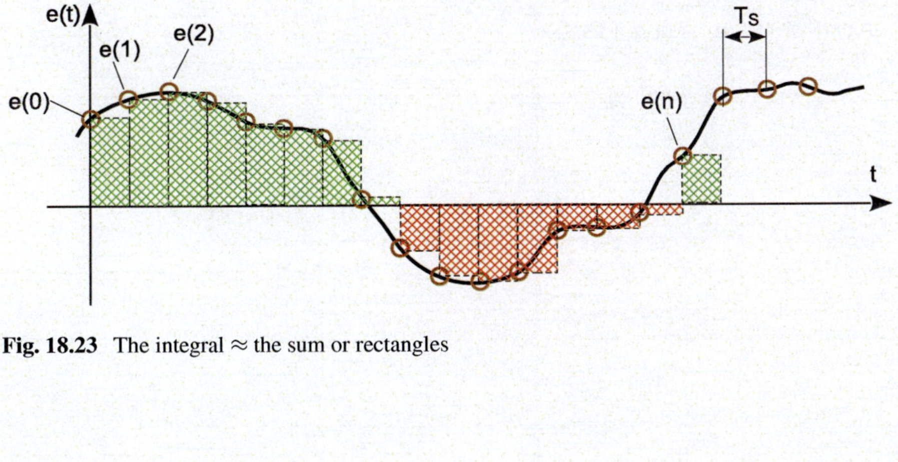

Fig. 18.23 The integral ≈ the sum or rectangles

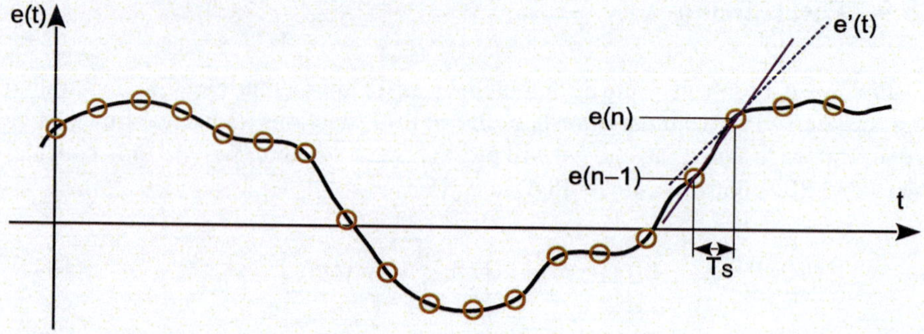

Fig. 18.24 The derivative ≈ the straight line between two samples

That gives us the following 'computer friendly' PI and PID control algorithms (= the 'Euler transformation'):

$$u(n) = K_P\left(e(n) + \frac{T_S}{T_I}\sum\nolimits_{i=0}^{n} e(i)\right) \tag{18.33}$$

$$u(n) = K_P\left(e(n) + \frac{T_S}{T_I}\sum\nolimits_{i=0}^{n} e(i) + \frac{T_D}{T_S}(e(n) - e(n-1))\right) \tag{18.34}$$

Example 18.4 Use the Euler transformation to find computer-friendly algorithms for the PI and PID controllers in Example 18.2.

Solution We obviously must find the sampling rate first. Considering that $\omega_0 =$ 0.5 rad/s, a sampling rate of 2 S/s is enough, i.e., $T_S = 0.5$ s. First, we discretize Eq. (18.26):

$$u_n = 1.8\left(e_n + \frac{0.5}{11}\sum_i e_i\right) = 1.8\left(e_n + 0.045\sum\nolimits_i e_i\right)$$

Next, we discretize Eq. (18.28):

$$\begin{aligned} u_n &= 2.4\left(e_n + \frac{0.5}{6.5}\sum_i e_i + \frac{1.6}{0.5}(e_n - e_{n-1})\right) \\ &= 2.4\left(e_n + 0.077\sum_i e_i + 3.2(e_n - e_{n-1})\right) \end{aligned}$$

18.9.2 Bilinear Transformation

In the bilinear transformation, we start from the Laplace transfer functions (Eqs. 18.5 and 18.12) and first transfer them to the corresponding z transforms using the bilinear transformation (see Chap. 10):

$$s = \frac{2}{T_S}\frac{z-1}{z+1} \tag{18.35}$$

For the PI controller (Eq. 18.5), we get

$$\begin{aligned} C(z) &= K_P\left(1+\frac{1}{T_I\frac{2}{T_S}\frac{z-1}{z+1}}\right) = K_P\left(1+\frac{T_S}{2T_I}\frac{z+1}{z-1}\right) \\ &= K_P\left(\frac{2T_I(z-1)}{2T_I(z-1)}+\frac{T_S(z+1)}{2T_I(z-1)}\right) \\ &= \frac{K_P}{2T_I}\cdot\frac{(2T_I+T_S)+(T_S-2T_I)\cdot z^{-1}}{1-z^{-1}} = \frac{U(z)}{E(z)} \end{aligned} \tag{18.36}$$

An inverse z transformation on expression (18.36) gives us the following difference equation:

$$u_n = u_{n-1} + \frac{K_P}{2T_I}((2T_I+T_S)\cdot e_n + (T_S-2T_I)\cdot e_{n-1}) \tag{18.37}$$

The corresponding expression for the PID controller (Eq. (18.12)) is

$$\begin{aligned} u_n = u_{n-2} + K_P\{&(2T_IT_S + T_S^2 + 4T_IT_D)e_n + (2T_S^2 - 8T_IT_D)e_{n-1} \\ &+(4T_IT_D + T_S^2 - 2T_IT_S)e_{n-2}\} \end{aligned} \tag{18.38}$$

Example 18.5 Use the bilinear transform to find computer-friendly algorithms for the PI and PID controllers in Example 18.2.

Solution Using the same sample rate as in Example 18.4, a bilinear transformation of Eqs. (18.25) and (18.27) give us:

$$\begin{aligned} \text{PI: } u_n &= u_{n-1} + \frac{1.8}{2\cdot 11}\cdot((2\cdot 11+0.5)e_n + (0.5-2\cdot 11)e_{n-1}) \\ &= u_{n-1} + 1.84e_n - 1.76e_{n-1} \end{aligned}$$

$$\text{PID: } u_n = u_{n-2} + 2.4(\ldots) = u_{n-2} + 116e_n - 83e_{n-1} + 35e_{n-2}$$

Comments: In computer algorithms, there is always a breakpoint when you need to use floating-point calculations. Floating-point calculations are time and memory consuming and should be avoided if possible. In the above examples, where we have simulated the heating control of some oven, the use of floating-point calculations would not be a problem since the sampling rate is only 2 S/s.

Also notice the main difference between the Euler and the bilinear transformations above: The bilinear also uses old output samples (and could thereby potentially become instable.)

Appendix
Operational Amplifiers

Abstract A lot of the analog electronics used in electrical measurement systems are based on operational amplifiers (op amps) and a basic understanding of this fundamental device is necessary; op amps will occur repeatedly throughout this book. Because of the extremely high inherent amplification of the differential-ended input signal, an op amp is almost always used with negative feedback. This chapter will demonstrate that two simple rules are all you need to understand and solve any op amp circuit. These simple rules are then used to exemplify the versatility of the op amp.

1. Introduction

The operational amplifier ('op amp') is one of the most versatile analog electronic components and omnipresent in electrical measurement systems. If you are going to work with electrical measurement systems, it is inevitable that you will sooner or later come across op amps and a basic understanding of this multifaceted component is imperative. Figure A.1 illustrates the op amp symbol.

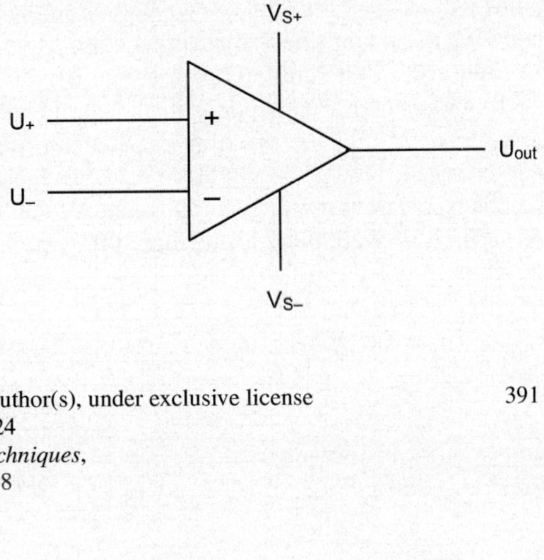

Fig. A.1 Signal model

L. Bengtsson, *Electrical Measurement Techniques*,
https://doi.org/10.1007/978-981-99-8187-8

Fig. A.2 Op amp with negative feedback

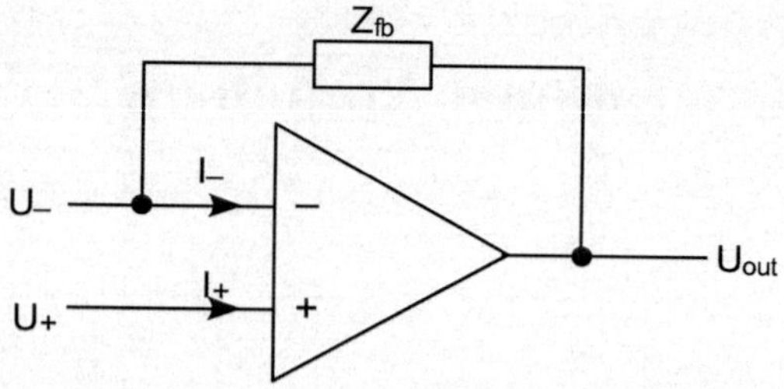

It *is* a *differential* amplifier; it amplifies the potential difference ($U_+ - U_-$), but the differential amplification is so large (>10^6), so for any real-world signal, the output will be driven 'high' ($= V_{S+}$) or 'low' ($= V_{S-}$). So, operated as a differential amplifier, it will act as a 'comparator'. If it is a differential amplifier, you are looking for (with a 'reasonable' amplification) then you are not looking for an op amp, you are looking for an *instrumentation amplifier*, see Chap. 4.

You see, op amps are (almost) always used with *feedback*. That means that (part of) the output signal is fed back to the input. In principle, you can feed the output back in two different ways; either to the '+' input (= 'positive feedback') or to the '−' input ('negative' feedback). With positive feedback, the amplification is *increased* which might seem redundant considering the enormous open-loop gain, but it has some applications (for example in oscillator designs). However, most op amp designs have negative feedback and that's what we will describe here, since that is what you will almost always use in a physics lab. Before we go into that, we will just mention a few characteristic properties of the ideal op amp.

First, the op amp needs dual power supply, ±12 V or ±15 V. For most op amps, the maximum output level is approximately 1–2 V lower than V_{S+} (and the minimum is 1–2 V higher than V_{S-}). If you need the output levels to reach the supply voltage level, look for an op amp that is 'rail-to-rail.'

An ideal op amp has infinite differential gain, infinite bandwidth, infinite input impedance and zero output impedance. A real op amp comes very close to infinite gain and infinite input impedance but does not have zero output impedance (≈ 100 Ω) and is nowhere near infinite bandwidth. However, for our presentation of 'op amps with negative feedback', we only need the 'infinite input impedance' property, and CMOS op amps have that (or as close as you can hope for).

Figure A.2 illustrates an op amp with negative feedback.

To understand what happens in an op amp with negative feedback, it is important to understand that the op amp is not really an *amplifier*. It is a *controller*; it is not designed to amplify anything; it is designed to make $U_+ = U_-$. With negative feedback, the output will generate whatever voltage/current necessary to make $U_+ = U_-$. That's it! Well, there is one more thing; because of the infinite input impedance, we can always assume that the currents I_+ and $I_- = 0$. There is never any current in or out of the inputs. These two rules ($U_+ = U_-$ and $I_+ = I_- = 0$) are sometimes referred to as the 'golden rules' of op amps. These two rules are all you need to design

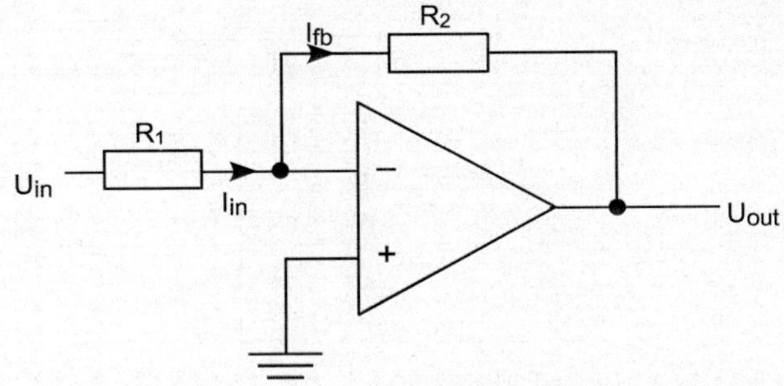

Fig. A.3 The inverting amplifier

'anything'. The op amp has no inherent 'amplification' (except for the infinite open-loop amplification) but you can take advantage of the above-described properties to *design* an amplifier with arbitrary gain. But keep in mind; the op amp (with negative feedback) doesn't care about gain, it only cares about making $U_+ = U_-$, and that's it!

2. Amplifiers

Op amp textbooks always start with the circuit in Fig. A.3.

In Fig. A.3, the '+' input is grounded, so $U_+ = 0$ V. That means that the output will make sure that also $U_- = 0$ (the '−' input is 'virtually grounded'). In that case, the current $I_{in} = (U_{in} - 0)/R_1 = U_{in}/R_1$. When this current reaches the '−' input, it has nowhere else to go but to the R_2 resistor (because the current in/out of the input is zero); $I_{fb} = I_{in}$. Now we can find the output voltage. If we start at the '−' input, where the potential = 0 V, and move to the output, then:

$$0 - I_{fb}R_2 = -\frac{U_{in}}{R_1}R_2 = U_{out} \Rightarrow U_{out} = -\frac{R_2}{R_1}U_{in} \tag{A.1}$$

and we have an amplifier, an *inverting* amplifier because of the minus sign, and we can set the gain arbitrarily with the resistors. Notice how we only used the op amp's 'control' property to design an amplifier.

If you want a *non-inverting* amplifier, you use the circuit in Fig. A.4.

In this case, $U_- = U_+ = U_{in}$. That means that the current I_{in} is U_{in}/R_1, and this current can only come from the output I_{fb} (because the current from the inputs is still = 0 A). So, if we start on the '−' input, where the potential is U_{in}, we can find the output voltage:

$$U_{in} + I_{fb}R_2 = U_{out} \Rightarrow U_{out} = U_{in} + \frac{U_{in}}{R_1}R_2 = U_{in}\left(1 + \frac{R_2}{R_1}\right) \tag{A.2}$$

Again, we can set the gain arbitrarily with the resistors R_1 and R_2.

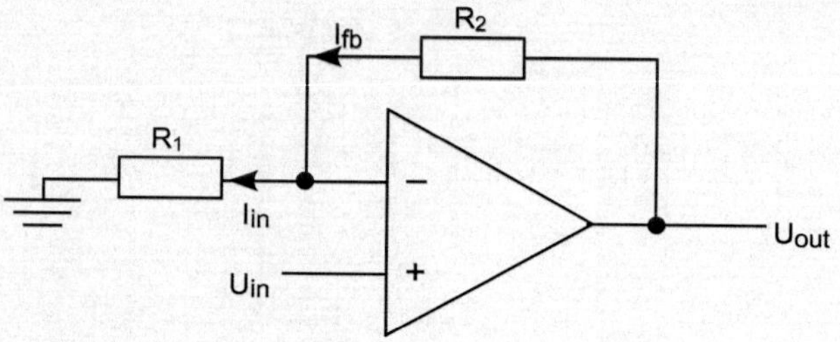

Fig. A.4 A non-inverting amplifier

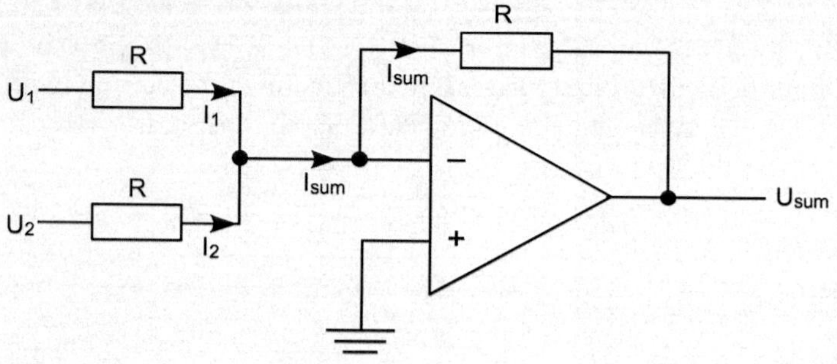

Fig. A.5 Summing circuit

3. Summing

Summing voltages is a common application of op amps. The circuit in Fig. A.5 produces the sum of two voltages.

Since the '−' input is virtually grounded, the currents I_1 and I_2 are U_1/R and U_2/R, respectively. These currents will add at the '−' input and this sum of currents has nowhere else to go but to the output (the output sinks the current):

$$0 - I_{\text{sum}} R = U_{\text{sum}} = -\left(\frac{U_1}{R} + \frac{U_2}{R}\right) R = -(U_1 + U_2) \tag{A.3}$$

4. Integrals and Derivatives

Sometimes we want to find the derivative or the integral of a signal. The circuit in Fig. A.6 will differentiate the input signal. First, the output voltage is $-I_{\text{fb}}R$, and second, the voltage over the capacitor equals the input voltage U_{in}. The current I_{in} is, by definition, the change of charge per time unit:

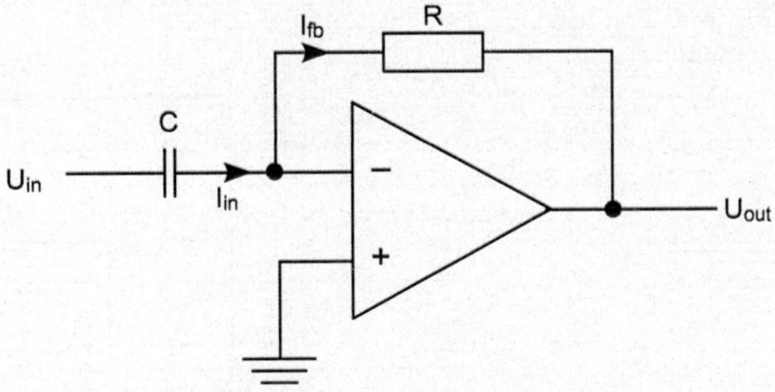

Fig. A.6 Differentiating circuit

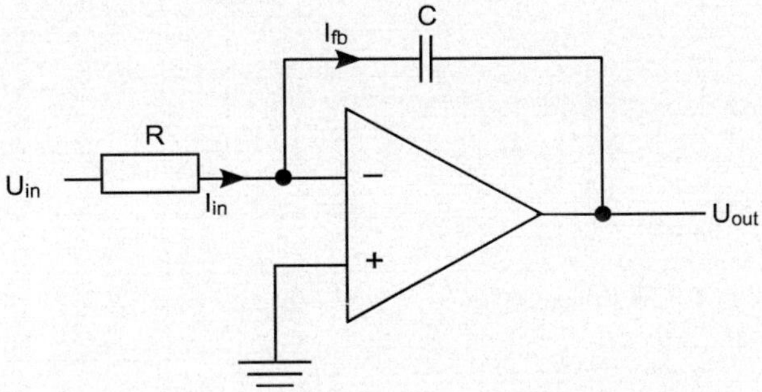

Fig. A.7 Integrating circuit

$$U_{\text{out}} = -I_{\text{fb}}R = -I_{\text{in}}R = -R\frac{dQ}{dt} = -R\frac{d}{dt}(U_c C) = -RC\frac{d}{dt}U_{\text{in}} \quad \text{(A.4)}$$

And hence the input signal is differentiated.

If we change places with the resistor and the capacitor, we get an integrating circuit, see Fig. A.7. In this circuit, the output signal $U_{\text{out}} = -U_{\text{C}}$, and charge Q is, by definition, the integral of current:

$$U_{\text{out}} = -U_{\text{C}} = -\frac{Q}{C} = -\frac{1}{C}\int I_{\text{in}}dt = -\frac{1}{C}\int \frac{U_{\text{in}}}{R}dt = -\frac{1}{RC}\int U_{\text{in}}dt \quad \text{(A.5)}$$

5. Constant Current Generator

Figure A.8 illustrates a constant current generator.

The op amp will keep the current through the R_{sense} resistor constant: $I_{\text{sense}} = (U_{\text{S}} - U_{\text{in}})/R_{\text{sense}}$, and since the collector current is $\approx I_{\text{emitter}} = I_{\text{sense}}$, the current through the load impedance will also be constant, independent of the size of the load.

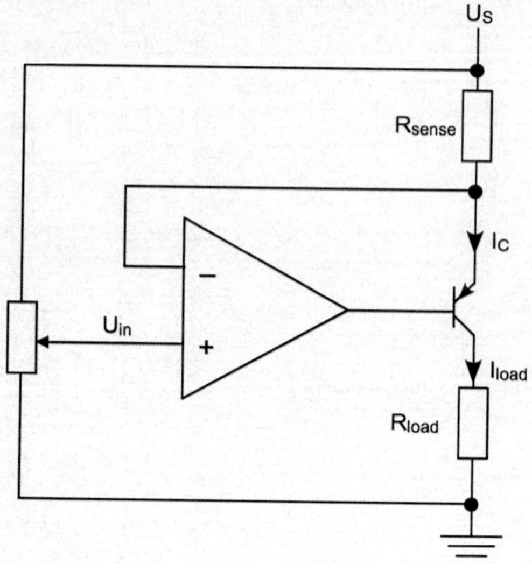

Fig. A.8 Constant current generator

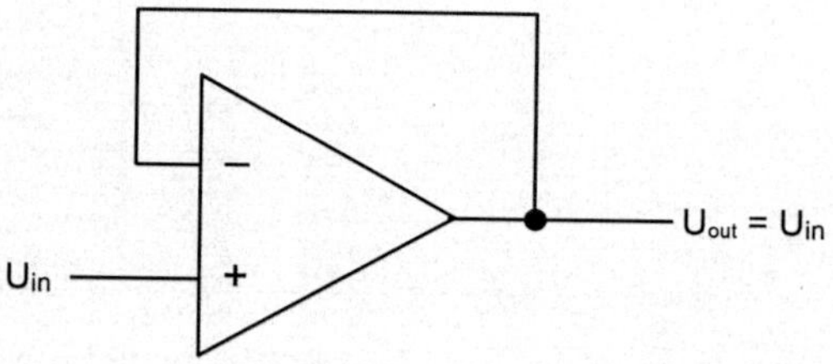

Fig. A.9 A voltage follower

6. Voltage Follower

Figure A.9 illustrates a 'voltage follower'; the output is always equal to the input voltage. The usefulness of that might at first glance be questionable, but this is extremely useful.

The voltage follower is used as an 'impedance converter'. Sometimes we have a signal source with a 'high' output impedance and/or a receiving component with a 'low' input impedance. Using a voltage follower means the source is not loaded (no current is required from the source since the op amp input current is zero) and at the receiving end, a low impedance signal source is connected to the receiver (because the output impedance of the op amp is 'low').

The voltage follower is sometime called a 'buffer'.

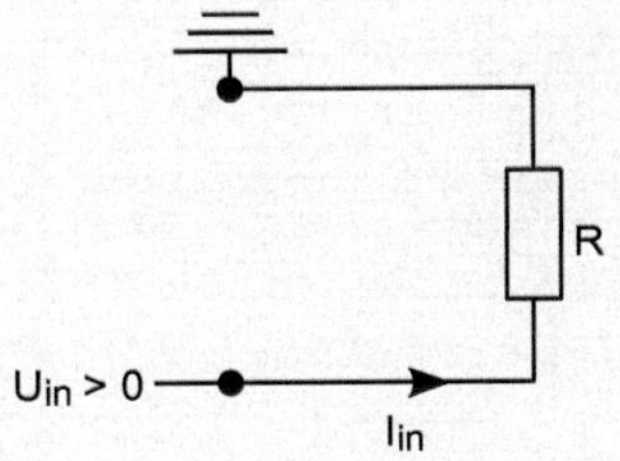

Fig. A.10 Positive resistance

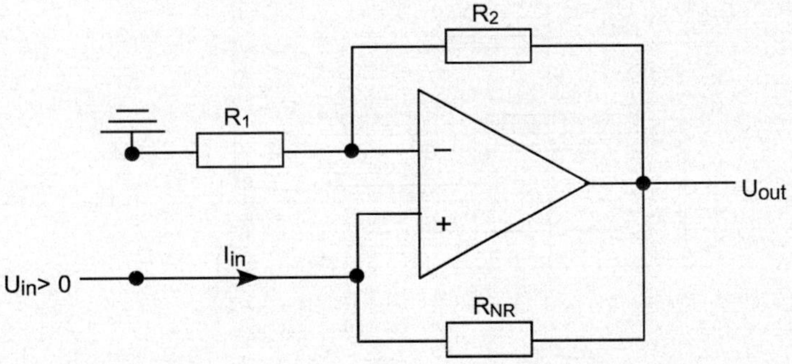

Fig. A.11 Negative resistance

7. Negative Resistance

Consider the circuit in Fig. A.10.

We know that in a 'normal' circuit, the current $I_{\text{in}} = U_{\text{in}}/R$ in Fig. A.10 is >0 if $U_{\text{in}} > 0$; U_{in} *sources* current into the resistor. We would have a *negative* resistance if $I_{\text{in}} < 0$, i.e., if U_{in} would *sink* current. With an op amp, we can design a negative resistance. Consider the circuit in Fig. A.11.

If we can prove that the current I_{in} in Fig. A.11 is <0, then the circuit behaves as a negative resistance. U_{out} is obviously $U_{\text{in}}(1 + R_2/R_1)$. Then

$$I_{\text{in}} = \frac{U_{\text{in}} - U_{\text{out}}}{R_{\text{NR}}} = -\frac{R_2/R_1}{R_{\text{NR}}} U_{\text{in}} < 0 \Rightarrow R_{\text{in}} = \frac{U_{\text{in}}}{I_{\text{in}}} = -\frac{R_1}{R_2} R_{\text{NR}} \tag{A.6}$$

Hence, this circuit acts as a negative resistance, *sinking* current at the input.

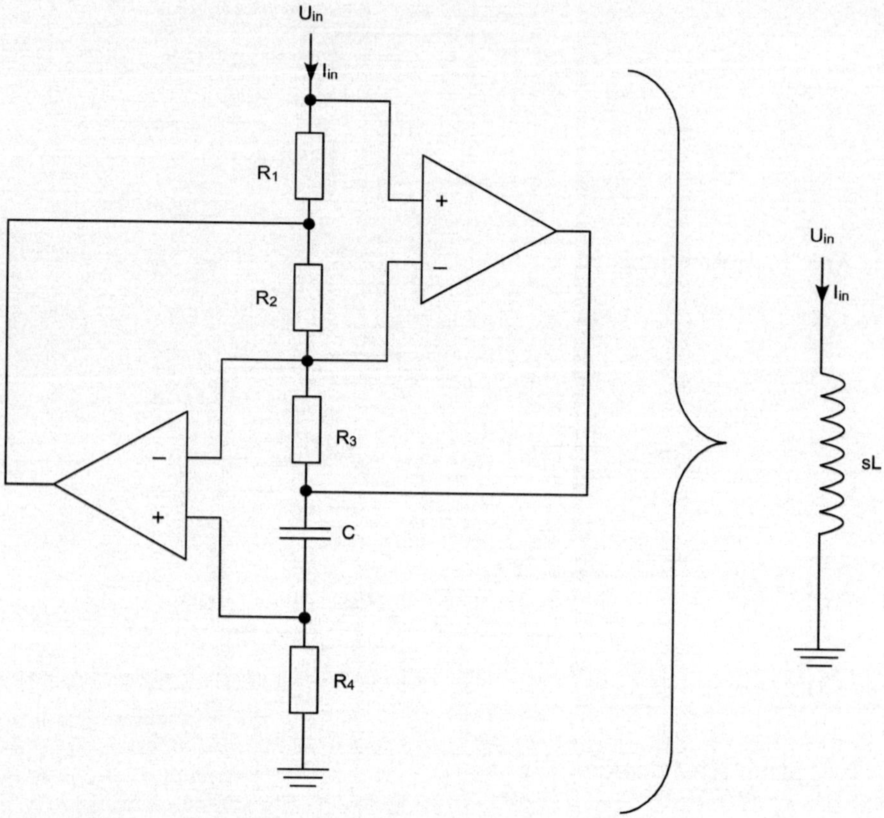

Fig. A.12 Inductor replacement circuit

8. Inductor Replacement

Resistors and capacitors are easily integrated on silicon, but inductors are harder. For that reason, inductor replacement circuits have been developed. Figure A.12 illustrates 'Antoniou's inductor replacement' circuit. This is an inductor replacement circuit if it behaves like an inductor, i.e., if the impedance $U_{in}/I_{in} = sL$. To prove that we need to indicate some potentials and currents, see Fig. A.13.

First, we can see that both op amps' inputs must have the same potential U_{in} (because of the negative feedback). That means that the current I_4 is U_{in}/R_4, and this current must come from the capacitor branch. Then the potential U_C must be:

$$U_C = U_{in} + I_4 \frac{1}{sC} = U_{in}\left(1 + \frac{1}{sR_4C}\right) \tag{A.7}$$

Then the current I_3 must be:

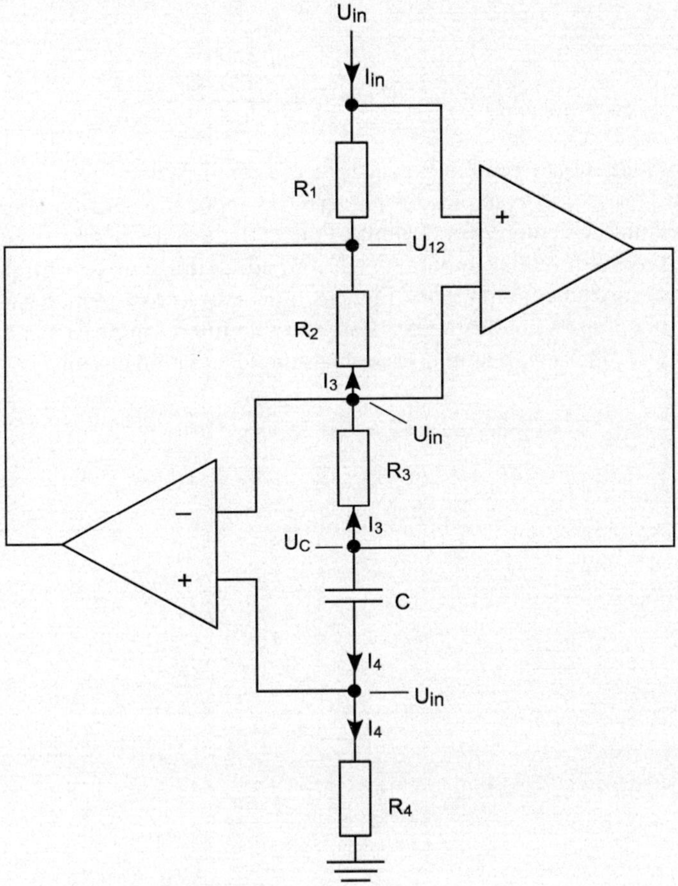

Fig. A.13 Inductor replacement circuit

$$I_3 = \frac{U_C - U_{\text{in}}}{R_3} = \frac{U_{\text{in}}}{sR_3R_4C} \tag{A.8}$$

Then U_{12} is.

$$U_{12} = U_{\text{in}} - R_2I_3 = U_{\text{in}} - \frac{R_2}{sR_3R_4C}U_{\text{in}} = U_{\text{in}}\left(1 - \frac{R_2}{sR_3R_4C}\right) \tag{A.9}$$

Then I_{in} must be:

$$I_{\text{in}} = \frac{U_{\text{in}} - U_{\text{in}}\left(1 - \frac{R_2}{sR_3R_4C}\right)}{R_1} = \frac{R_2}{sR_1R_3R_4C}U_{\text{in}} \tag{A.10}$$

and hence

$$\frac{U_{\text{in}}}{I_{\text{in}}} = sL \text{ where } L = \frac{R_1 R_3 R_4 C}{R_2} \tag{A.11}$$

This proves that the circuit in Fig. A.12 acts as an inductor.

These are just some examples of the versatile applications of op amps and this textbook contains a lot more. For example, Fig. 3.60 illustrates a difference circuit, in Fig. 4.5, we have differential amplifier, Figs. 4.6 and 4.8 illustrate two different implementations of instrument amplifiers, Fig. 6.8 is an active probe, Fig. 6.11 illustrates a current probe, Fig. 9.11 illustrates a state variable filter, Fig. 9.13 is a Sallen-Key filter and in Fig. 11.2 an op amp is used in a sample and hold circuit.

Index

L. Bengtsson, *Electrical Measurement Techniques*,
https://doi.org/10.1007/978-981-99-8187-8

J

K

L

M

N

O

P

T

U